# Recent Advances and Trends in Nonparametric Statistics

Cover figure by B. Schölkopf, generated using a Support Vector Machine as described in "An Introduction to Support Vector Machines" (pp. 3-17) in this book.

# Recent Advances and Trends in Nonparametric Statistics

*Edited by*

**Michael G. Akritas**
*Department of Statistics*
*Penn State University*
*University Park*
*Pennsylvania, USA*

**Dimitris N. Politis**
*Department of Mathematics*
*University of California at San Diego*
*La Jolla, California, USA*

2003

ELSEVIER

Amsterdam • Boston •Heidelberg • London • New York • Oxford • Paris
San Diego • San Francisco • Singapore • Sydney • Tokyo

ELSEVIER B.V.  
Sara Burgerhartstraat 25  
P.O. Box 211, 1000 AE Amsterdam  
The Netherlands  

ELSEVIER Inc.  
525 B Street, Suite 1900  
San Diego, CA 92101-4495  
USA  

ELSEVIER Ltd  
The Boulevard, Langford Lane  
Kidlington, Oxford OX5 1GB  
UK  

ELSEVIER Ltd  
84 Theobalds Road  
London WC1X 8RR  
UK  

© 2003 Elsevier B.V. All rights reserved.

This work is protected under copyright by Elsevier B.V., and the following terms and conditions apply to its use:

Photocopying  
Single photocopies of single chapters may be made for personal use as allowed by national copyright laws. Permission of the Publisher and payment of a fee is required for all other photocopying, including multiple or systematic copying, copying for advertising or promotional purposes, resale, and all forms of document delivery. Special rates are available for educational institutions that wish to make photocopies for non-profit educational classroom use.

Permissions may be sought directly from Elsevier's Rights Department in Oxford, UK: phone (+44) 1865 843830, fax (+44) 1865 853333, e-mail: permissions@elsevier.com. Requests may also be completed on-line via the Elsevier homepage (http://www.elsevier.com/locate/permissions).

In the USA, users may clear permissions and make payments through the Copyright Clearance Center, Inc., 222 Rosewood Drive, Danvers, MA 01923, USA; phone: (+1) (978) 7508400, fax: (+1) (978) 7504744, and in the UK through the Copyright Licensing Agency Rapid Clearance Service (CLARCS), 90 Tottenham Court Road, London W1P 0LP, UK; phone: (+44) 20 7631 5555; fax: (+44) 20 7631 5500. Other countries may have a local reprographic rights agency for payments.

Derivative Works  
Tables of contents may be reproduced for internal circulation, but permission of the Publisher is required for external resale or distribution of such material. Permission of the Publisher is required for all other derivative works, including compilations and translations.

Electronic Storage or Usage  
Permission of the Publisher is required to store or use electronically any material contained in this work, including any chapter or part of a chapter.

Except as outlined above, no part of this work may be reproduced, stored in a retrieval system or transmitted in any form or by any means, electronic, mechanical, photocopying, recording or otherwise, without prior written permission of the Publisher.  
Address permissions requests to: Elsevier's Rights Department, at the fax and e-mail addresses noted above.

Notice  
No responsibility is assumed by the Publisher for any injury and/or damage to persons or property as a matter of products liability, negligence or otherwise, or from any use or operation of any methods, products, instructions or ideas contained in the material herein. Because of rapid advances in the medical sciences, in particular, independent verification of diagnoses and drug dosages should be made.

First edition 2003

Library of Congress Cataloging in Publication Data  
A catalog record is available from the Library of Congress.

British Library Cataloguing in Publication Data  
```
Recent advances and trends in nonparametric statistics
   1.Nonparametric statistics - Congresses
   I.Akritas, Michael G. II.Politis, Dimitris N.
   III.International Conference on Recent Advances and Trends
   in Nonparametric Statistics (2002 : Crete, Greece)
   519.5'4

   ISBN 0444513787
```

ISBN: 0 444 51378 7

∞ The paper used in this publication meets the requirements of ANSI/NISO Z39.48-1992 (Permanence of Paper).  
Printed in The Netherlands.

# Preface

Nonparametric Statistics is certainly not a new subject; pioneering works by M. Bartlett, E. Parzen, M. Rosenblatt and other prominent researchers have opened up exciting lines of research even since the middle of the 20th century. For the most part, however, the theory and practice of 20th century statistics was dominated by the parametric system of ideas put forth by R.A. Fisher, K. Pearson, J. Neyman and other statistical legends. The parametric statistical system is self-contained, very elegant mathematically, and —quite importantly—its practical implementation involves calculations that can be carried out with pen and paper and the occasional slide-rule/calculator. Its limitation, however, lies in forcing the practitioner to make prior assumptions regarding the data, assumptions that are often quite unrealistic.

Not surprisingly, the advent of high-speed, affordable computers in the last two decades has given a new boost to the nonparametric way of thinking. Classical nonparametric procedures such as function smoothing suddenly lost their abstract flavor as they became practically implementable. In addition, many previously unthinkable possibilities became mainstream; prime examples include the bootstrap and resampling methods, wavelets and nonlinear smoothers, graphical methods, data mining, bioinformatics, as well as more recent algorithmic approaches such as bagging, boosting, and support vector machines.

Because of this very recent boom in new technologies/methodologies associated with nonparametric statistics, the undersigned decided to organize an international conference where those new ideas could be exchanged and disseminated. The objective of the meeting was to highlight the current developments and trends in the field and set the stage for collaborations among researchers from the U.S. and Europe. The "International Conference on Recent Advances and Trends in Nonparametric Statistics" took place on the island of Crete, Greece, the week of July 15-19, 2002; a detailed program can be found at: www.stat.psu.edu/~npconf. The conference was co-sponsored by the Institute of Mathematical Statistics, the Nonparametrics Section of the American Statistical Association, the Bernoulli Society, and the Greek Statistical Institute; partial funding was provided by the National Science Foundation and the National Security Agency.

The original plan was to host a small, workshop-like, conference involving about a hundred researchers. Nevertheless, due to the enthusiastic response of keynote speakers, session organizers, invited and contributed speakers, the actual number of speakers was three-fold. Participants represented all five continents, although—as expected—the overwhelming majority were from Europe and North America. Despite its special topic flavor, the "International Conference on Recent Advances and Trends in Nonparametric Statistics" became an event roughly comparable in size to the 2002 European Meeting of Statisticians that took place in Prague a month later involving contributions from the whole, unrestricted realm of statistics, parametric or not.

The present volume is a collection of short articles, mostly review papers and/or more technical papers having a review component. in The volume is not a Proceedings of the Conference; some of the articles are related to conference presentations but many are not. Most of the authors, however, were somehow involved in the Crete Conference be it as speakers, session organizers, or members of the conference scientific advisory committee.

Our hope is that the volume gives an accurate reflection of the state-of-the-art of the field of Nonparametric Statistics at the beginning of the new millennium.

Michael G. Akritas
Penn State University

Dimitris N. Politis
University of California, San Diego

# Table of Contents

PREFACE · v

## 1. ALGORITHMIC APPROACHES TO STATISTICS

**An introduction to support vector machines**
Bernhard Schölkopf · 3

**Bagging, subagging and bragging for improving some prediction algorithms**
Peter Bühlmann · 19

**Data compression by geometric quantization**
Nkem-Amin Khumbah and Edward J. Wegman · 35

## 2. FUNCTIONAL DATA ANALYSIS

**Functional data analysis in evolutionary biology**
Nancy E. Heckman · 49

**Functional nonparametric statistics: a double infinite dimensional framework**
F. Ferraty and Philippe Vieu · 61

## 3. NONPARAMETRIC MODEL BUILDING AND INFERENCE

**Nonparametric models for ANOVA and ANCOVA: a review**
Michael G. Akritas and Edgar Brunner · 79

**Isotonic additive interaction models**
Ilya Gluhovsky · 93

**A nonparametric alternative to analysis of covariance**
Arne Bathke and Edgar Brunner · 109

## 4. GOODNESS OF FIT

Assessing structural relationships between distributions
–a quantile process approach based on Mallows distance
G. Freitag, Axel Munk, and M. Vogt     123

Almost sure representations in survival analysis under
censoring and truncation: applications to goodness-of-fit tests
R. Cao, W. González Manteiga, and C. Iglesias Pérez     139

## 5. HIGH-DIMENSIONAL DATA AND VISUALIZATION

Data depth: center-outward ordering of multivariate
data and nonparametric multivariate statistics
Regina Y. Liu     155

Visual exploration of data through their graph
representations
George Michailidis     169

## 6. NONPARAMETRIC REGRESSION

Inference for nonsmooth regression curves and surfaces
using kernel-based methods
Irene Gijbels     183

Nonparametric smoothing methods for a class of
non-standard curve estimation problems
Oliver Linton and Enno Mammen     203

Weighted local linear approach to censored
nonparametric regression
Zongwu Cai     217

## 7. TOPICS IN NONPARAMETRICS

Adaptive quantile regression
Sara van de Geer     235

Set estimation: an overview and some recent developments
Antonio Cuevas and Alberto Rodríguez-Casal     251

Nonparametric methods for heavy tailed vector data:
a survey with applications from finance and hydrology
Mark M. Meerschaert and Hans-Peter Scheffler   265

## 8. NONPARAMETRICS IN FINANCE AND RISK MANAGEMENT

Nonparametric methods in continuous-time finance:
a selective review
Zongwu Cai and Yongmiao Hong   283

Nonparametric estimation in a stochastic volatility model
Jürgen Franke, Wolfgang Härdle, and Jens-Peter Kreiss   303

Dynamic nonparametric filtering with application to
volatility estimation
Ming-Yen Cheng, Jianqing Fan, and Vladimir Spokoiny   315

A normalizing and variance-stabilizing transformation for
financial time series
Dimitris N. Politis   335

## 9. BIOINFORMATICS AND BIOSTATISTICS

Biostochastics and nonparametrics: oranges and apples?
Pranab Kumar Sen   351

Some issues concerning length-biased sampling in survival
analysis
Masoud Asgharian and David B. Wolfson   367

Covariate centering and scaling in varying-coefficient
regression with application to longitudinal growth studies
Colin O. Wu, Kai F. Yu, and Vivian W.S. Yuan   377

Directed peeling and covering of patient rules
Michael LeBlanc, James Moon, and John Crowley   393

## 10. RESAMPLING AND SUBSAMPLING

**Statistical analysis of survival models with Bayesian bootstrap**
Jaeyong Lee and Yongdai Kim  411

**On optimal variance estimation under different spatial subsampling schemes**
Daniel J. Nordman and Soumendra N. Lahiri  421

**Locally stationary processes and the local block bootstrap**
Arif Dowla, Efstathios Paparoditis, and Dimitris N. Politis  437

## 11. TIME SERIES AND STOCHASTIC PROCESSES

**Spectral analysis and a class of nonstationary processes**
Murray Rosenblatt  447

**Curve estimation for locally stationary time series models**
Rainer Dahlhaus  451

**Assessing spatial isotropy**
Michael Sherman, Yongtao Guan, and James A. Calvin  467

## 12. WAVELET AND MULTIRESOLUTION METHODS

**Automatic landmark registration of 1D curves**
Jérémie Bigot  479

**Stochastic multiresolution models for turbulence**
B. Whitcher, J.B. Weiss, D.W. Nychka, and T.J. Hoar  497

## AUTHOR INDEX  511

# 1. Algorithmic Approaches to Statistics

# An Introduction to Support Vector Machines

Bernhard Schölkopf[a]

[a]Max-Planck-Institut für biologische Kybernetik,
Spemannstr. 38, Tübingen, Germany
bernhard.schoelkopf@tuebingen.mpg.de

This article gives a short introduction to the main ideas of statistical learning theory, support vector machines, and kernel feature spaces.[1]

## 1. An Introductory Example

Suppose we are given empirical data

$$(x_1, y_1), \ldots, (x_m, y_m) \in \mathcal{X} \times \{\pm 1\}. \tag{1}$$

Here, the *domain* $\mathcal{X}$ is some nonempty set that the *patterns* $x_i$ are taken from; the $y_i$ are called *labels* or *targets*. Unless stated otherwise, indices $i$ and $j$ will always be understood to run over the training set, i.e., $i, j = 1, \ldots, m$.

Note that we have not made any assumptions on the domain $\mathcal{X}$ other than it being a set. In order to study the problem of learning, we need additional structure. In learning, we want to be able to *generalize* to unseen data points. In the case of pattern recognition, given some new pattern $x \in \mathcal{X}$, we want to predict the corresponding $y \in \{\pm 1\}$. By this we mean, loosely speaking, that we choose $y$ such that $(x, y)$ is in some sense *similar* to the training examples. To this end, we need similarity measures in $\mathcal{X}$ and in $\{\pm 1\}$. The latter is easier, as two target values can only be identical or different.[2] For the former, we require a similarity measure

$$\begin{aligned} k : \mathcal{X} \times \mathcal{X} &\to \mathbb{R}, \\ (x, x') &\mapsto k(x, x'), \end{aligned} \tag{2}$$

i.e., a function that, given two examples $x$ and $x'$, returns a real number characterizing their similarity. For reasons that will become clear later, the function $k$ is called a *kernel* [12,1,6].

A type of similarity measure that is of particular mathematical appeal are dot products. For instance, given two vectors $\mathbf{x}, \mathbf{x}' \in \mathbb{R}^N$, the canonical dot product is defined as

$$(\mathbf{x} \cdot \mathbf{x}') := \sum_{n=1}^{N} [\mathbf{x}]_n [\mathbf{x}']_n. \tag{3}$$

---
[1]The present article is partly based on Microsoft TR-2000-23, Redmond, WA.
[2]In the case where the outputs are taken from a general set $\mathcal{Y}$, the situation is more complex. For algorithms to deal with this case, cf. [22].

Here, $[\mathbf{x}]_n$ denotes the $n$-th entry of $\mathbf{x}$.

The geometrical interpretation of this dot product is that it computes the cosine of the angle between the vectors $\mathbf{x}$ and $\mathbf{x}'$, provided they are normalized to length 1. Moreover, it allows computation of the length of a vector $\mathbf{x}$ as $\sqrt{(\mathbf{x} \cdot \mathbf{x})}$, and of the distance between two vectors as the length of the difference vector. Therefore, being able to compute dot products amounts to being able to carry out all geometrical constructions that can be formulated in terms of angles, lengths and distances.

Note, however, that we have not made the assumption that the patterns live in a dot product space. In order to be able to use a dot product as a similarity measure, we therefore first need to embed them into some dot product space $\mathcal{H}$, which need not be identical to $\mathbb{R}^N$. To this end, we use a map

$$\Phi : \mathcal{X} \to \mathcal{H}$$
$$x \mapsto \mathbf{x} := \Phi(x). \tag{4}$$

The space $\mathcal{H}$ is called a *feature space*. To summarize, embedding the data into $\mathcal{H}$ has three benefits.

1. It lets us define a similarity measure from the dot product in $\mathcal{H}$,

$$k(x, x') := (\mathbf{x} \cdot \mathbf{x}') = (\Phi(x) \cdot \Phi(x')). \tag{5}$$

2. It allows us to deal with the patterns geometrically, and thus lets us study learning algorithms using linear algebra and analytic geometry.

3. The freedom to choose the mapping $\Phi$ will enable us to design a large variety of learning algorithms. For instance, consider a situation where the inputs already live in a dot product space. In that case, we could directly define a similarity measure as the dot product. However, we might still choose to first apply a nonlinear map $\Phi$ to change the representation into one that is more suitable for a given problem and learning algorithm.

We are now in the position to describe a simple pattern recognition algorithm. The idea is to compute the means of the two classes in feature space,

$$\mathbf{c}_1 = \frac{1}{m_1} \sum_{\{i : y_i = +1\}} \mathbf{x}_i, \tag{6}$$

$$\mathbf{c}_2 = \frac{1}{m_2} \sum_{\{i : y_i = -1\}} \mathbf{x}_i, \tag{7}$$

where $m_1$ and $m_2$ are the number of examples with positive and negative labels, respectively. We then assign a new point $\mathbf{x}$ to the class whose mean is closer to it. This geometrical construction can be formulated in terms of dot products. Half-way in between $\mathbf{c}_1$ and $\mathbf{c}_2$ lies the point $\mathbf{c} := (\mathbf{c}_1 + \mathbf{c}_2)/2$. We compute the class of $\mathbf{x}$ by checking whether the vector connecting $\mathbf{c}$ and $\mathbf{x}$ encloses an angle smaller than $\pi/2$ with the vector $\mathbf{w} := \mathbf{c}_1 - \mathbf{c}_2$ connecting the class means, in other words

$$\begin{aligned} y &= \operatorname{sgn}\left((\mathbf{x} - \mathbf{c}) \cdot \mathbf{w}\right) \\ y &= \operatorname{sgn}\left((\mathbf{x} - (\mathbf{c}_1 + \mathbf{c}_2)/2) \cdot (\mathbf{c}_1 - \mathbf{c}_2)\right) \\ &= \operatorname{sgn}\left((\mathbf{x} \cdot \mathbf{c}_1) - (\mathbf{x} \cdot \mathbf{c}_2) + b\right). \end{aligned} \quad (8)$$

Here, we have defined the offset

$$b := \frac{1}{2}\left(\|\mathbf{c}_2\|^2 - \|\mathbf{c}_1\|^2\right). \quad (9)$$

It will prove instructive to rewrite this expression in terms of the patterns $x_i$ in the input domain $\mathcal{X}$. Note that we do not have a dot product in $\mathcal{X}$, all we have is the similarity measure $k$ (cf. (5)). Therefore, we need to rewrite everything in terms of the kernel $k$ evaluated on input patterns. To this end, substitute (6) and (7) into (8) to get the *decision function*

$$\begin{aligned} y &= \operatorname{sgn}\left(\frac{1}{m_1}\sum_{\{i:y_i=+1\}} (\mathbf{x} \cdot \mathbf{x}_i) - \frac{1}{m_2}\sum_{\{i:y_i=-1\}} (\mathbf{x} \cdot \mathbf{x}_i) + b\right) \\ &= \operatorname{sgn}\left(\frac{1}{m_1}\sum_{\{i:y_i=+1\}} k(x, x_i) - \frac{1}{m_2}\sum_{\{i:y_i=-1\}} k(x, x_i) + b\right). \end{aligned} \quad (10)$$

Similarly, the offset becomes

$$b := \frac{1}{2}\left(\frac{1}{m_2^2}\sum_{\{(i,j):y_i=y_j=-1\}} k(x_i, x_j) - \frac{1}{m_1^2}\sum_{\{(i,j):y_i=y_j=+1\}} k(x_i, x_j)\right). \quad (11)$$

Let us consider one well-known special case of this type of classifier. Assume that the class means have the same distance to the origin (hence $b = 0$), and that $k$ can be viewed as a density, i.e., it is positive and has integral 1,

$$\int_{\mathcal{X}} k(x, x')dx = 1 \text{ for all } x' \in \mathcal{X}. \quad (12)$$

In order to state this assumption, we have to require that we can define an integral on $\mathcal{X}$.

If the above holds true, then (10) corresponds to the so-called Bayes decision boundary separating the two classes, subject to the assumption that the two classes were generated from two probability distributions that are correctly estimated by the *Parzen windows* estimators of the two classes,

$$p_1(x) := \frac{1}{m_1}\sum_{\{i:y_i=+1\}} k(x, x_i) \quad (13)$$

$$p_2(x) := \frac{1}{m_2}\sum_{\{i:y_i=-1\}} k(x, x_i). \quad (14)$$

Given some point $x$, the label is then simply computed by checking which of the two, $p_1(x)$ or $p_2(x)$, is larger, leading to (10). Note that this decision is the best we can do if we have no prior information about the probabilities of the two classes, or a uniform prior distribution. For further details, see [15].

The classifier (10) is quite close to the types of learning machines that we will be interested in. It is linear in the feature space (Equation (8)), while in the input domain, it is represented by a kernel expansion (Equation (10)). It is example-based in the sense that the kernels are centered on the training examples, i.e., one of the two arguments of the kernels is always a training example. This is a general property of kernel methods, due to the Representer Theorem [11,15]. The main point where the more sophisticated techniques to be discussed later will deviate from (10) is in the selection of the examples that the kernels are centered on, and in the weight that is put on the individual kernels in the decision function. Namely, it will no longer be the case that *all* training examples appear in the kernel expansion, and the weights of the kernels in the expansion will no longer be uniform. In the feature space representation, this statement corresponds to saying that we will study all normal vectors **w** of decision hyperplanes that can be represented as linear combinations of the training examples. For instance, we might want to remove the influence of patterns that are very far away from the decision boundary, either since we expect that they will not improve the generalization error of the decision function, or since we would like to reduce the computational cost of evaluating the decision function (cf. (10)). The hyperplane will then only depend on a subset of training examples, called *support vectors*.

## 2. Learning Pattern Recognition from Examples

With the above example in mind, let us now consider the problem of pattern recognition in a more formal setting, highlighting some ideas developed in statistical learning theory [18]. In two-class pattern recognition, we seek to estimate a function

$$f : \mathcal{X} \to \{\pm 1\} \tag{15}$$

based on input-output training data (1). We assume that the data were generated independently from some unknown (but fixed) probability distribution $P(x, y)$. Our goal is to learn a function that will correctly classify unseen examples $(x, y)$, i.e., we want $f(x) = y$ for examples $(x, y)$ that were also generated from $P(x, y)$.

If we put no restriction on the class of functions that we choose our estimate $f$ from, however, even a function which does well on the training data, e.g. by satisfying $f(x_i) = y_i$ for all $i = 1, \ldots, m$, need not generalize well to unseen examples. To see this, note that for each function $f$ and any test set $(\bar{x}_1, \bar{y}_1), \ldots, (\bar{x}_{\bar{m}}, \bar{y}_{\bar{m}}) \in \mathbb{R}^N \times \{\pm 1\}$, satisfying $\{\bar{x}_1, \ldots, \bar{x}_{\bar{m}}\} \cap \{x_1, \ldots, x_m\} = \{\}$, there exists another function $f^*$ such that $f^*(x_i) = f(x_i)$ for all $i = 1, \ldots, m$, yet $f^*(\bar{x}_i) \neq f(\bar{x}_i)$ for all $i = 1, \ldots, \bar{m}$. As we are only given the training data, we have no means of selecting which of the two functions (and hence which of the completely different sets of test label predictions) is preferable. Hence, only minimizing the training error (or *empirical risk*),

$$R_{emp}[f] = \frac{1}{m} \sum_{i=1}^{m} \frac{1}{2} |f(x_i) - y_i|, \tag{16}$$

does not imply a small test error (called *risk*), averaged over test examples drawn from the underlying distribution $P(x,y)$,

$$R[f] = \int \frac{1}{2}|f(x) - y| \, dP(x,y). \tag{17}$$

Statistical learning theory ([16], [18], [19]), or VC (Vapnik-Chervonenkis) theory, shows that it is imperative to restrict the class of functions that $f$ is chosen from to one which has a *capacity* that is suitable for the amount of available training data. VC theory provides *bounds* on the test error. The minimization of these bounds, which depend on both the empirical risk and the capacity of the function class, leads to the principle of *structural risk minimization*. The best-known capacity concept of VC theory is the *VC dimension*, defined as the largest number $h$ of points that can be separated in all possible ways using functions of the given class. An example of a VC bound is the following: if $h < m$ is the VC dimension of the class of functions that the learning machine can implement, then for all functions of that class, with a probability of at least $1 - \eta$, the bound

$$R(f) \leq R_{emp}(f) + \psi(h, m, \eta) \tag{18}$$

holds, where $m$ is the number of training examples and the *confidence term* $\psi$ is defined as

$$\psi(h, m, \eta) = \sqrt{\frac{h\left(\log\frac{2m}{h} + 1\right) - \log(\eta/4)}{m}}. \tag{19}$$

Tighter bounds can be formulated in terms of other concepts, such as the *annealed VC entropy* or the *Growth function*. These are usually considered to be harder to evaluate, but they play a fundamental role in the conceptual part of VC theory [18]. Alternative capacity concepts that can be used to formulate bounds include the *fat shattering dimension* [2].

The bound (18) deserves some further explanatory remarks. Suppose we wanted to learn a "dependency" where $P(x,y) = P(x) \cdot P(y)$, i.e., where the pattern $x$ contains no information about the label $y$, with uniform $P(y)$. Given a training sample of fixed size, we can then surely come up with a learning machine which achieves zero training error (provided we have no examples contradicting each other). However, in order to reproduce the random labellings, this machine will necessarily require a large VC dimension $h$. Thus, the confidence term (19), increasing monotonically with $h$, will be large, and the bound (18) will *not* support possible hopes that due to the small training error, we should expect a small test error. This makes it understandable how (18) can hold independently of assumptions about the underlying distribution $P(x,y)$: it always holds (provided that $h < m$), but it does not always make a nontrivial prediction — a bound on an error rate becomes void if it is larger than the maximum error rate. In order to get nontrivial predictions from (18), the function space must be restricted such that the capacity (e.g. VC dimension) is small enough (in relation to the available amount of data).

## 3. Optimal Margin Hyperplane Classifiers

In the present section, we shall describe a hyperplane learning algorithm that can be performed in a dot product space (such as the feature space that we introduced previously). As described in the previous section, to design learning algorithms, one needs to come up with a class of functions whose capacity can be computed.

Vapnik and Lerner [17] considered the class of hyperplanes

$$(\mathbf{w} \cdot \mathbf{x}) + b = 0 \quad \mathbf{w} \in \mathbb{R}^N, b \in \mathbb{R}, \tag{20}$$

corresponding to decision functions

$$f(\mathbf{x}) = \operatorname{sgn}\left((\mathbf{w} \cdot \mathbf{x}) + b\right), \tag{21}$$

and proposed a learning algorithm for separable problems, termed the *Generalized Portrait*, for constructing $f$ from empirical data. It is based on two facts. First, among all hyperplanes separating the data (assuming that the data is separable), there exists a unique one yielding the maximum margin of separation between the classes,

$$\max_{\mathbf{w},b} \min\{\|\mathbf{x} - \mathbf{x}_i\| : \mathbf{x} \in \mathbb{R}^N, (\mathbf{w} \cdot \mathbf{x}) + b = 0, i = 1, \ldots, m\}. \tag{22}$$

Second, the capacity can be shown to decrease with increasing margin.

To construct this *Optimum Margin Hyperplane* (cf. Figure 1), one solves the following optimization problem:

$$\text{minimize} \quad \tau(\mathbf{w}) = \frac{1}{2}\|\mathbf{w}\|^2 \tag{23}$$

$$\text{subject to} \quad y_i \cdot ((\mathbf{w} \cdot \mathbf{x}_i) + b) \geq 1, \ i = 1, \ldots, m. \tag{24}$$

This constrained optimization problem is dealt with by introducing Lagrange multipliers $\alpha_i \geq 0$ and a Lagrangian

$$L(\mathbf{w}, b, \boldsymbol{\alpha}) = \frac{1}{2}\|\mathbf{w}\|^2 - \sum_{i=1}^{m} \alpha_i \left(y_i \cdot ((\mathbf{x}_i \cdot \mathbf{w}) + b) - 1\right). \tag{25}$$

The Lagrangian $L$ has to be minimized with respect to the *primal variables* $\mathbf{w}$ and $b$ and maximized with respect to the *dual variables* $\alpha_i$ (i.e., a saddle point has to be found). Let us try to get some intuition for this. If a constraint (24) is violated, then $y_i \cdot ((\mathbf{w} \cdot \mathbf{x}_i) + b) - 1 < 0$, in which case $L$ can be increased by increasing the corresponding $\alpha_i$. At the same time, $\mathbf{w}$ and $b$ will have to change such that $L$ decreases. To prevent $-\alpha_i (y_i \cdot ((\mathbf{w} \cdot \mathbf{x}_i) + b) - 1)$ from becoming arbitrarily large, the change in $\mathbf{w}$ and $b$ will ensure that, provided the problem is separable, the constraint will eventually be satisfied. Similarly, one can understand that for all constraints which are not precisely met as equalities, i.e., for which $y_i \cdot ((\mathbf{w} \cdot \mathbf{x}_i) + b) - 1 > 0$, the corresponding $\alpha_i$ must be 0, for this is the value of $\alpha_i$ that maximizes $L$. This is the statement of the Karush-Kuhn-Tucker conditions of optimization theory [5].

The condition that at the saddle point, the derivatives of $L$ with respect to the primal variables must vanish,

$$\frac{\partial}{\partial b}L(\mathbf{w}, b, \boldsymbol{\alpha}) = 0, \ \frac{\partial}{\partial \mathbf{w}}L(\mathbf{w}, b, \boldsymbol{\alpha}) = 0, \tag{26}$$

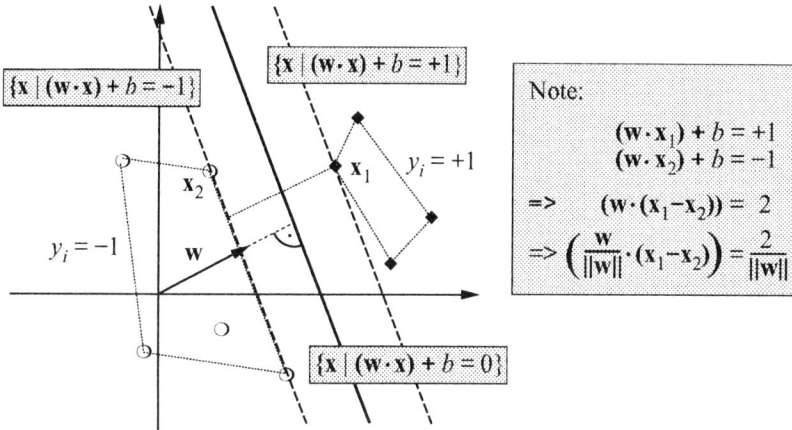

Figure 1. A binary classification toy problem: separate balls from diamonds. The *Optimum Margin Hyperplane* is orthogonal to the shortest line connecting the convex hulls of the two classes (dotted), and intersects it half-way between the two classes. The problem being separable, there exists a weight vector $\mathbf{w}$ and a threshold $b$ such that $y_i \cdot ((\mathbf{w} \cdot \mathbf{x}_i) + b) > 0$ $(i = 1, \ldots, m)$. Rescaling $\mathbf{w}$ and $b$ such that the point(s) closest to the hyperplane satisfy $|(\mathbf{w} \cdot \mathbf{x}_i) + b| = 1$, we obtain a *canonical* form $(\mathbf{w}, b)$ of the hyperplane, satisfying $y_i \cdot ((\mathbf{w} \cdot \mathbf{x}_i) + b) \geq 1$. Note that in this case, the *margin*, measured perpendicularly to the hyperplane, equals $2/\|\mathbf{w}\|$. This can be seen by considering two points $\mathbf{x}_1, \mathbf{x}_2$ on opposite sides of the margin, i.e., $(\mathbf{w} \cdot \mathbf{x}_1) + b = 1, (\mathbf{w} \cdot \mathbf{x}_2) + b = -1$, and projecting them onto the hyperplane normal vector $\mathbf{w}/\|\mathbf{w}\|$ (from [13]).

leads to

$$\sum_{i=1}^{m} \alpha_i y_i = 0 \qquad (27)$$

and

$$\mathbf{w} = \sum_{i=1}^{m} \alpha_i y_i \mathbf{x}_i. \qquad (28)$$

The solution vector thus has an expansion in terms of a subset of the training patterns, namely those patterns whose $\alpha_i$ is non-zero, called *Support Vectors*. By the Karush-Kuhn-Tucker conditions

$$\alpha_i \cdot [y_i((\mathbf{x}_i \cdot \mathbf{w}) + b) - 1] = 0, \ i = 1, \ldots, m, \qquad (29)$$

the Support Vectors satisfy $y_i((\mathbf{x}_i \cdot \mathbf{w}) + b) - 1 = 0$, i.e., they lie on the margin (cf. Figure 1). All remaining examples of the training set are irrelevant: their constraint (24) does not play a role in the optimization, and they do not appear in the expansion (28). This nicely captures our intuition of the problem: as the hyperplane (cf. Figure 1) is geometrically completely determined by the patterns closest to it, the solution should not depend on the other examples.

By substituting (27) and (28) into $L$, one eliminates the primal variables and arrives at the Wolfe dual of the optimization problem (e.g., [5]): find multipliers $\alpha_i$ which

$$\text{maximize} \quad W(\alpha) = \sum_{i=1}^{m} \alpha_i - \frac{1}{2} \sum_{i,j=1}^{m} \alpha_i \alpha_j y_i y_j (\mathbf{x}_i \cdot \mathbf{x}_j) \tag{30}$$

$$\text{subject to} \quad \alpha_i \geq 0, \ i = 1, \ldots, m, \ \text{and} \ \sum_{i=1}^{m} \alpha_i y_i = 0. \tag{31}$$

By substituting (28) into (21), the hyperplane decision function can thus be written as

$$f(\mathbf{x}) = \operatorname{sgn}\left(\sum_{i=1}^{m} y_i \alpha_i \cdot (\mathbf{x} \cdot \mathbf{x}_i) + b\right), \tag{32}$$

where $b$ is computed using (29).

The structure of the optimization problem closely resembles those that typically arise in Lagrange's formulation of mechanics. There, often only a subset of the constraints become active. For instance, if we keep a ball in a box, then it will typically roll into one of the corners. The constraints corresponding to the walls which are not touched by the ball are irrelevant, the walls could just as well be removed.

Seen in this light, it is not too surprising that it is possible to give a mechanical interpretation of optimal margin hyperplanes [7]: If we assume that each support vector $\mathbf{x}_i$ exerts a perpendicular force of size $\alpha_i$ and sign $y_i$ on a solid plane sheet lying along the hyperplane, then the solution satisfies the requirements of mechanical stability. The constraint (27) states that the forces on the sheet sum to zero; and (28) implies that the torques also sum to zero, via $\sum_i \mathbf{x}_i \times y_i \alpha_i \cdot \mathbf{w}/\|\mathbf{w}\| = \mathbf{w} \times \mathbf{w}/\|\mathbf{w}\| = 0$.

There are theoretical arguments supporting the good generalization performance of the optimal hyperplane ([16], [23], [3], [20]). In addition, it is computationally attractive, since it can be constructed by solving a quadratic programming problem.

## 4. Support Vector Classifiers

We now have all the tools to describe support vector machines [6,18,14,9,15]. Everything in the last section was formulated in a dot product space. We think of this space as the feature space $\mathcal{H}$ described in Section 1. To express the formulas in terms of the input patterns living in $\mathcal{X}$, we thus need to employ (5), which expresses the dot product of bold face feature vectors $\mathbf{x}, \mathbf{x}'$ in terms of the kernel $k$ evaluated on input patterns $x, x'$,

$$k(x, x') = (\mathbf{x} \cdot \mathbf{x}'). \tag{33}$$

This can be done since all feature vectors only occured in dot products. The weight vector (cf. (28)) then becomes an expansion in feature space, and will thus typically no longer correspond to the image of a single vector from input space. We thus obtain decision

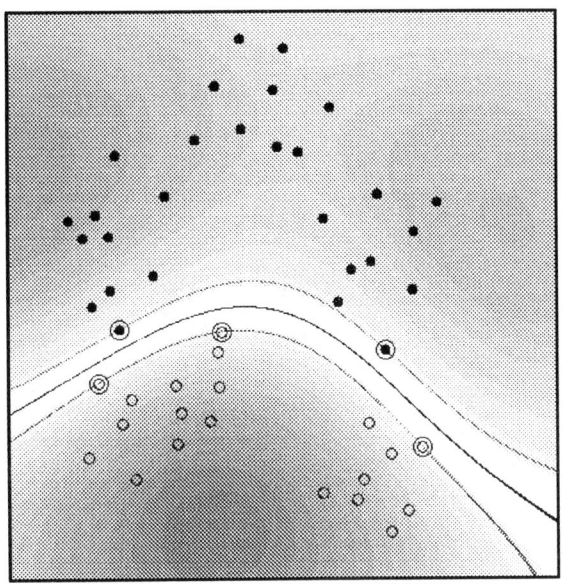

Figure 2. Example of a Support Vector classifier found by using a radial basis function kernel $k(x,x') = \exp(-\|x-x'\|^2)$. Both coordinate axes range from -1 to +1. Circles and disks are two classes of training examples; the middle line is the decision surface; the outer lines precisely meet the constraint (24). Note that the Support Vectors found by the algorithm (marked by extra circles) are not centers of clusters, but examples which are critical for the given classification task. Grey values code the modulus of the argument $\sum_{i=1}^{m} y_i \alpha_i \cdot k(x,x_i) + b$ of the decision function (34) (from [13]).)

functions of the more general form (cf. (32))

$$f(x) = \text{sgn}\left(\sum_{i=1}^{m} y_i \alpha_i \cdot (\Phi(x) \cdot \Phi(x_i)) + b\right)$$
$$= \text{sgn}\left(\sum_{i=1}^{m} y_i \alpha_i \cdot k(x,x_i) + b\right), \tag{34}$$

and the following quadratic program (cf. (30)):

$$\text{maximize} \quad W(\alpha) = \sum_{i=1}^{m} \alpha_i - \frac{1}{2} \sum_{i,j=1}^{m} \alpha_i \alpha_j y_i y_j k(x_i, x_j) \tag{35}$$

$$\text{subject to} \quad \alpha_i \geq 0, \; i=1,\ldots,m, \text{ and } \sum_{i=1}^{m} \alpha_i y_i = 0. \tag{36}$$

In practice, a separating hyperplane may not exist, e.g. if a high noise level causes a large overlap of the classes. To allow for the possibility of examples violating (24), one introduces slack variables [8]

$$\xi_i \geq 0, \; i=1,\ldots,m \tag{37}$$

in order to relax the constraints to

$$y_i \cdot ((\mathbf{w} \cdot \mathbf{x}_i) + b) \geq 1 - \xi_i, \; i=1,\ldots,m. \tag{38}$$

A classifier which generalizes well is then found by controlling both the classifier capacity (via $\|\mathbf{w}\|$) and the sum of the slacks $\sum_i \xi_i$. The latter is done as it can be shown to provide an upper bound on the number of training errors which leads to a convex optimization problem.

One possible realization of a *soft margin* classifier is minimizing the objective function

$$\tau(\mathbf{w}, \boldsymbol{\xi}) = \frac{1}{2}\|\mathbf{w}\|^2 + C \sum_{i=1}^{m} \xi_i \qquad (39)$$

subject to the constraints (37) and (38), for some value of the constant $C > 0$ determining the trade-off. Here, we use the shorthand $\boldsymbol{\xi} = (\xi_1, \ldots, \xi_m)$. Incorporating kernels, and rewriting it in terms of Lagrange multipliers, this again leads to the problem of maximizing (35), subject to the constraints

$$0 \leq \alpha_i \leq C, \ i = 1, \ldots, m, \text{ and } \sum_{i=1}^{m} \alpha_i y_i = 0. \qquad (40)$$

The only difference from the separable case is the upper bound $C$ on the Lagrange multipliers $\alpha_i$. This way, the influence of the individual patterns (which could be outliers) gets limited. As above, the solution takes the form (34). The threshold $b$ can be computed by exploiting the fact that for all SVs $x_i$ with $\alpha_i < C$, the slack variable $\xi_i$ is zero (this again follows from the Karush-Kuhn-Tucker complementarity conditions), and hence

$$\sum_{j=1}^{m} y_j \alpha_j \cdot k(x_i, x_j) + b = y_i. \qquad (41)$$

Another possible realization of a soft margin variant of the optimal hyperplane uses the $\nu$-parametrization [15]. In it, the parameter $C$ is replaced by a parameter $\nu \in [0, 1]$ which can be shown to lower and upper bound the number of examples that will be SVs and that will come to lie on the wrong side of the hyperplane, respectively. It uses a primal objective function with the error term $\frac{1}{\nu m}\sum_i \xi_i - \rho$, and separation constraints

$$y_i \cdot ((\mathbf{w} \cdot \mathbf{x}_i) + b) \geq \rho - \xi_i, \ i = 1, \ldots, m. \qquad (42)$$

The margin parameter $\rho$ is a variable of the optimization problem. The dual can be shown to consist of maximizing the quadratic part of (35), subject to $0 \leq \alpha_i \leq 1/(\nu m)$, $\sum_i \alpha_i y_i = 0$ and the additional constraint $\sum_i \alpha_i = 1$. The advantage of the $\nu$-SVM is its more intuitive parametrization.

We conclude this section by noting that the SV algorithm has been generalized to problems such as regression estimation [18] as well as one-class problems and novelty detection [15]. The algorithms involved are similar to the case of pattern recognition described above. Moreover, the kernel method for computing dot products in feature spaces is not restricted to SV machines. Indeed, it has been pointed out that it can be used to develop nonlinear generalizations of any algorithm that can be cast in terms of dot products, such as principal component analysis [15], and a number of developments have followed this example.

## 5. Polynomial Kernels

We now take a closer look at the issue of the similarity measure, or kernel, $k$.

In this section, we think of $\mathcal{X}$ as a subset of the vector space $\mathbb{R}^N$, ($N \in \mathbb{N}$), endowed with the canonical dot product (3).

## 5.1. Product Features

Suppose we are given patterns $x \in \mathbb{R}^N$ where most information is contained in the $d$-th order products (monomials) of entries $[x]_j$ of $x$,

$$[x]_{j_1} \cdot \ldots \cdot [x]_{j_d}, \tag{43}$$

where $j_1, \ldots, j_d \in \{1, \ldots, N\}$. In that case, we might prefer to *extract* these product features, and work in the feature space $F$ of all products of $d$ entries. In visual recognition problems, where images are often represented as vectors, this would amount to extracting features which are products of individual pixels.

For instance, in $\mathbb{R}^2$, we can collect all monomial feature extractors of degree 2 in the nonlinear map

$$\Phi : \mathbb{R}^2 \to F = \mathbb{R}^3 \tag{44}$$
$$([x]_1, [x]_2) \mapsto ([x]_1^2, [x]_2^2, [x]_1 [x]_2). \tag{45}$$

This approach works fine for small toy examples, but it fails for realistically sized problems: for $N$-dimensional input patterns, there exist

$$N_F = \frac{(N + d - 1)!}{d!(N - 1)!} \tag{46}$$

different monomials (43), comprising a feature space $F$ of dimensionality $N_F$. For instance, already $16 \times 16$ pixel input images and a monomial degree $d = 5$ yield a dimensionality of $10^{10}$.

In certain cases described below, there exists, however, a way of *computing dot products* in these high-dimensional feature spaces without explicitly mapping into them: by means of kernels nonlinear in the input space $\mathbb{R}^N$. Thus, if the subsequent processing can be carried out using dot products exclusively, we are able to deal with the high dimensionality.

The following section describes how dot products in polynomial feature spaces can be computed efficiently.

## 5.2. Polynomial Feature Spaces Induced by Kernels

In order to compute dot products of the form $(\Phi(x) \cdot \Phi(x'))$, we employ kernel representations of the form

$$k(x, x') = (\Phi(x) \cdot \Phi(x')), \tag{47}$$

which allow us to compute the value of the dot product in $F$ without having to carry out the map $\Phi$. This method was used by Boser, Guyon and Vapnik [6] to extend the *Generalized Portrait* hyperplane classifier of Vapnik and Chervonenkis [16] to nonlinear Support Vector machines. Aizerman et al. [1] call $F$ the *linearization space*, and used in the context of the potential function classification method to express the dot product between elements of $F$ in terms of elements of the input space.

What does $k$ look like for the case of polynomial features? We start by giving an example [18] for $N = d = 2$. For the map

$$C_2 : ([x]_1, [x]_2) \mapsto ([x]_1^2, [x]_2^2, [x]_1 [x]_2, [x]_2 [x]_1), \tag{48}$$

dot products in $F$ take the form

$$(C_2(x) \cdot C_2(x')) = [x]_1^2[x']_1^2 + [x]_2^2[x']_2^2 + 2[x]_1[x]_2[x']_1[x']_2 = (x \cdot x')^2, \tag{49}$$

i.e., the desired kernel $k$ is simply the square of the dot product in input space. The same works for arbitrary $N, d \in \mathbb{N}$ [6]:

**Proposition 1** *Define $C_d$ to map $x \in \mathbb{R}^N$ to the vector $C_d(x)$ whose entries are all possible d-th degree ordered products of the entries of $x$. Then the corresponding kernel computing the dot product of vectors mapped by $C_d$ is*

$$k(x, x') = (C_d(x) \cdot C_d(x')) = (x \cdot x')^d. \tag{50}$$

**Proof.** We directly compute

$$(C_d(x) \cdot C_d(x')) = \sum_{j_1,\ldots,j_d=1}^{N} [x]_{j_1} \cdot \ldots \cdot [x]_{j_d} \cdot [x']_{j_1} \cdot \ldots \cdot [x']_{j_d} \tag{51}$$

$$= \left( \sum_{j=1}^{N} [x]_j \cdot [x']_j \right)^d = (x \cdot x')^d. \tag{52}$$

∎

Instead of ordered products, we can use unordered ones to obtain a map $\Phi_d$ which yields the same value of the dot product. To this end, we have to compensate for the multiple occurence of certain monomials in $C_d$ by scaling the respective entries of $\Phi_d$ with the square roots of their numbers of occurence. Then, by this definition of $\Phi_d$, and (50),

$$(\Phi_d(x) \cdot \Phi_d(x')) = (C_d(x) \cdot C_d(x')) = (x \cdot x')^d. \tag{53}$$

For instance, if $n$ of the $j_i$ in (43) are equal, and the remaining ones are different, then the coefficient in the corresponding component of $\Phi_d$ is $\sqrt{(d-n+1)!}$ (for the general case, cf. [15]). For $\Phi_2$, this simply means that [18]

$$\Phi_2(x) = ([x]_1^2, [x]_2^2, \sqrt{2}\,[x]_1[x]_2). \tag{54}$$

If $x$ represents an image with the entries being pixel values, we can use the kernel $(x \cdot x')^d$ to work in the space spanned by products of any $d$ pixels — provided that we are able to do our work solely in terms of dot products, without any explicit usage of a mapped pattern $\Phi_d(x)$. Using kernels of the form (50), we take into account higher-order statistics without the combinatorial explosion (cf. (46)) of time and memory complexity which goes along already with moderately high $N$ and $d$.

To conclude this section, note that it is possible to modify (50) such that it maps into the space of all monomials *up to* degree $d$, defining

$$k(x, x') = ((x \cdot x') + 1)^d. \tag{55}$$

## 6. Examples of Kernels

When considering feature maps, it is also possible to look at things the other way around, and start with the kernel. Given a kernel function satisfying a condition termed *positive definiteness*, it is possible to construct a feature space such that the kernel computes the dot product in that feature space. This has been brought to the attention of the machine learning community by [1], [6], and [18]. In functional analysis, the issue has been studied under the heading of *Hilbert space representations* of kernels. A good monograph on the theory of kernels is [4]. A treatment which focuses on the aspects relevant to machine learning can be found in [15].

The condition of positive definiteness is equivalent to saying that no matter what are the training patterns $x_i$, all eigenvalues of the Gram matrix $(k(x_i, x_j))_{ij}$ are nonnegative. Besides (50), a popular kernel satisfying this condition is the Gaussion radial basis function (RBF) [1]

$$k(x, x') = \exp\left(-\frac{\|x - x'\|^2}{2\,\sigma^2}\right), \tag{56}$$

where $\sigma > 0$. Note that this kernel is translation invariant. It has some additional properties, which may serve as an example of how the choice of kernel can have rather strong implications on the geometry of the domain occupied by the data mapped into the feature space.

As the Gaussian RBF kernel satisfies $k(x, x) = 1$ for all $x \in \mathcal{X}$, each mapped example has unit length, $\|\Phi(x)\| = 1$. In addition, as $k(x, x') > 0$ for all $x, x' \in \mathcal{X}$, all points lie inside the same orthant in feature space. To see this, recall that for unit length vectors, the dot product (3) equals the cosine of the enclosed angle. Hence

$$\cos(\angle(\Phi(x), \Phi(x'))) = (\Phi(x) \cdot \Phi(x')) = k(x, x') > 0, \tag{57}$$

which amounts to saying that the enclosed angle between any two mapped examples is smaller than $\pi/2$.

The examples given so far apply to the case of vectorial data. In fact it is possible to construct kernels that are used to compute similarity scores for data drawn from rather different domains. This generalizes kernel learning algorithms to a large number of situations where a vectorial representation is not readily available ([13], [10], [21]). Let us next give an example where $\mathcal{X}$ is not a vector space.

**Example 1 (Similarity of probabilistic events)** *If $\mathcal{A}$ is a $\sigma$-algebra, and $P$ a probability measure on $\mathcal{A}$, and $A$ and $B$ two events in $\mathcal{A}$, then*

$$k(A, B) = P(A \cap B) - P(A)P(B) \tag{58}$$

*is a positive definite kernel.*

Further examples include kernels for string matching, as proposed by [21] and [10].

There is an analogue of the kernel trick for distances rather than dot products, i.e., dissimilarities rather than similarities. This leads to the class of *conditionally positive definite kernels.* In a nutshell, these kernels can be represented by distances in Hilbert spaces, and they contain positive definite kernels as special cases. Interestingly, it turns out that SVMs and kernel PCA can be applied also with this larger class of kernels, due to their being translation invariant in feature space [15]. An even larger class of dissimilarity functions, namely all semi-metrics, is obtained by not requiring the embedding to be in a Hilbert space, but in the more general class of Banach spaces.

## 7. Conclusion

One of the most appealing features of kernel algorithms is the solid foundation provided by both statistical learning theory and functional analysis. Kernel methods let us interpret (and design) learning algorithms geometrically in feature spaces nonlinearly related to the input space, and combine statistics and geometry in a promising way. Kernels provide an elegant framework for studying three fundamental issues of machine learning:

- *Similarity measures* — the kernel can be viewed as a (nonlinear) similarity measure, and should ideally incorporate prior knowledge about the problem at hand

- *Data representation* — as described above, kernels induce representations of the data in a linear space

- *Function class* — due to the representer theorem, the kernel implicitly also determines the function class which is used for learning.

**Acknowledgements.** Thanks to Assaf Naor for useful remarks about the representation of dissimilarities, and to Isabelle Guyon, Alex Smola, and Jason Weston for help with earlier versions of the manuscript.

## REFERENCES

1. M. A. Aizerman, É. M. Braverman, and L. I. Rozonoér. Theoretical foundations of the potential function method in pattern recognition learning. *Automation and Remote Control*, 25:821–837, 1964.
2. N. Alon, S. Ben-David, N. Cesa-Bianchi, and D. Haussler. Scale-sensitive dimensions, uniform convergence, and learnability. *Journal of the ACM*, 44(4):615–631, 1997.
3. P. L. Bartlett and J. Shawe-Taylor. Generalization performance of support vector machines and other pattern classifiers. In B. Schölkopf, C. J. C. Burges, and A. J. Smola, editors, *Advances in Kernel Methods — Support Vector Learning*, pages 43–54, Cambridge, MA, 1999. MIT Press.
4. C. Berg, J. P. R. Christensen, and P. Ressel. *Harmonic Analysis on Semigroups*. Springer-Verlag, New York, 1984.
5. D. P. Bertsekas. *Nonlinear Programming*. Athena Scientific, Belmont, MA, 1995.
6. B. E. Boser, I. M. Guyon, and V. Vapnik. A training algorithm for optimal margin classifiers. In D. Haussler, editor, *Proceedings of the 5th Annual ACM Workshop on Computational Learning Theory*, pages 144–152, Pittsburgh, PA, July 1992. ACM Press.

7. C. J. C. Burges and B. Schölkopf. Improving the accuracy and speed of support vector learning machines. In M. Mozer, M. Jordan, and T. Petsche, editors, *Advances in Neural Information Processing Systems 9*, pages 375–381, Cambridge, MA, 1997. MIT Press.
8. C. Cortes and V. Vapnik. Support vector networks. *Machine Learning*, 20:273–297, 1995.
9. N. Cristianini and J. Shawe-Taylor. *An Introduction to Support Vector Machines and other kernel-based learning methods*. Cambridge University Press, Cambridge, UK, 2000.
10. D. Haussler. Convolutional kernels on discrete structures. Technical Report UCSC-CRL-99-10, Computer Science Department, University of California at Santa Cruz, 1999.
11. G. S. Kimeldorf and G. Wahba. Some results on Tchebycheffian spline functions. *Journal of Mathematical Analysis and Applications*, 33:82–95, 1971.
12. J. Mercer. Functions of positive and negative type and their connection with the theory of integral equations. *Philosophical Transactions of the Royal Society, London*, A 209:415–446, 1909.
13. B. Schölkopf. *Support Vector Learning*. R. Oldenbourg Verlag, München, 1997. Doktorarbeit, Technische Universität Berlin. Available from http://www.kyb.tuebingen.mpg.de/~bs.
14. B. Schölkopf, C. J. C. Burges, and A. J. Smola. *Advances in Kernel Methods — Support Vector Learning*. MIT Press, Cambridge, MA, 1999.
15. B. Schölkopf and A. J. Smola. *Learning with Kernels*. MIT Press, Cambridge, MA, 2002.
16. V. Vapnik and A. Chervonenkis. *Theory of Pattern Recognition [in Russian]*. Nauka, Moscow, 1974. (German Translation: W. Wapnik & A. Tscherwonenkis, *Theorie der Zeichenerkennung*, Akademie-Verlag, Berlin, 1979).
17. V. Vapnik and A. Lerner. Pattern recognition using generalized portrait method. *Automation and Remote Control*, 24:774–780, 1963.
18. V. N. Vapnik. *The Nature of Statistical Learning Theory*. Springer Verlag, New York, 1995.
19. V. N. Vapnik. *Statistical Learning Theory*. Wiley, New York, 1998.
20. U. von Luxburg, O. Bousquet, and B. Schölkopf. A compression approach to support vector model selection. Technical Report 101, Max Planck Institute for Biological Cybernetics, 2002.
21. C. Watkins. Dynamic alignment kernels. In A. J. Smola, P. L. Bartlett, B. Schölkopf, and D. Schuurmans, editors, *Advances in Large Margin Classifiers*, pages 39–50, Cambridge, MA, 2000. MIT Press.
22. J. Weston, O. Chapelle, A. Elisseeff, B. Schölkopf, and V. Vapnik. Kernel dependency estimation. In S. Becker, S. Thrun, and K. Obermayer, editors, *Advances in Neural Information Processing Systems*, volume 15, Cambridge, MA, USA, 2003. MIT Press.
23. R. C. Williamson, A. J. Smola, and B. Schölkopf. Generalization performance of regularization networks and support vector machines via entropy numbers of compact operators. *IEEE Transactions on Information Theory*, 47(6):2516–2532, 2001.

*Recent Advances and Trends in Nonparametric Statistics*
Michael G. Akritas and Dimitris N. Politis (Editors)
© 2003 Elsevier Science B.V. All rights reserved.

# Bagging, Subagging and Bragging for Improving some Prediction Algorithms

Peter Bühlmann

Seminar für Statistik, ETH Zürich, CH-8092 Zürich, Switzerland

Bagging (**b**ootstrap **agg**regat**ing**), proposed by Breiman [1], is a method to improve the predictive power of some special estimators or algorithms such as regression or classification trees. First, we review a recently developed theory explaining why bagging decision trees, or also the subagging (**sub**sample **agg**regat**ing**) variant, yields smooth decisions, reducing the variance and mean squared error. We then propose bragging (**b**ootstrap **r**obust **agg**regat**ing**) as a new version of bagging which, in contrast to bagging, is empirically demonstrated to improve also the MARS algorithm which itself already yields continuous function estimates. Finally, bagging is demonstrated as a "module" in conjunction with boosting for an example about tumor classification using microarray gene expressions.

## 1. Introduction

Bagging [1], a sobriquet for **b**ootstrap **agg**regat**ing**, is a method for improving unstable estimation or classification schemes. It is very useful for large, high dimensional data set problems where finding a good model or classifier is difficult.

Breiman [1] motivated bagging as a variance reduction technique for a given basis algorithm (i.e. an estimator) such as decision trees or a method that does variable selection and fitting in a linear model. It has attracted much attention and is quite frequently applied, although theoretical insights have been lacking until very recently [4-6]. We present here parts of a theory from Bühlmann and Yu [4], indicating that bagging is a smoothing operation which turns out to be advantageous when aiming to improve the predictive performance of regression or classification trees. In case of regression trees, this theory confirms Breiman's intuition that bagging is a variance reduction technique, reducing also the mean squared error (MSE). The same also holds for subagging (**sub**sample **agg**regat**ing**) which is a computationally cheaper version than bagging. However, for other "complex" basis algorithms, the variance and MSE reduction effect of bagging is not necessarily true; this has also been shown by Buja and Stuetzle [5] in the simpler case where the estimator is a $U$-statistics.

Moreover, we propose bragging (**b**ootstrap **r**obust **agg**regat**ing**) as a simple, yet new modification which improves an estimation procedure not necessarily because of its smoothing effects but also due to averaging over unstable selection of variables or terms in complex models or algorithms; empirical examples are given for bragging MARS, where bagging is often not a useful device, whereas bragging turns out to be effective.

Bagging can also be useful as a "module" in other algorithms: BagBoosting [3] is a boosting algorithm [8] with a bagged learner, often a bagged regression tree. From our theory it will become evident that BagBoosting using bagged regression trees, which have smaller MSEs than trees, is better than boosting with regression trees. We demonstrate improvements of BagBoosting over boosting for a problem about tumor classification using microarray gene expression predictors.

## 2. Bagging and Subagging

Consider the regression or classification setting. The data is given as i.i.d. pairs $(X_i, Y_i)$ $(i = 1, \ldots, n)$, where $X_i \in \mathbb{R}^d$ denotes the $d$-dimensional predictor variable and $Y_i \in \mathbb{R}$ (regression) or $Y_i \in \{0, 1, \ldots, J-1\}$ (classification with $J$ classes). The goal is function estimation and the target function is usually $\mathbb{E}[Y|X = x]$ for regression or the multivariate function $\mathbb{P}[Y = j|X = x]$ $(j = 1, \ldots, J-1)$ for classification. We denote in the sequel a function estimator, which is the outcome from a given basis algorithm, by

$$\hat{\theta}_n(\cdot) = h_n((X_1, Y_1), \ldots, (X_n, Y_n))(\cdot) : \mathbb{R}^d \to \mathbb{R},$$

defined by the function $h_n(\cdot)$. Examples of such estimators include linear regression with variable selection, regression trees such as CART [2] or MARS [9].

**Definition 1** *(Bagging). Theoretically, bagging is defined as follows.*

*(I) Construct a bootstrap sample $(X_1^*, Y_1^*), \ldots, (X_n^*, Y_n^*)$ by random drawing $n$ times with replacement from the data $(X_1, Y_1), \ldots, (X_n, Y_n)$.*

*(II) Compute the bootstrapped estimator $\hat{\theta}_n^*(\cdot)$ by the plug-in principle:*
$\hat{\theta}_n^*(\cdot) = h_n((X_1^*, Y_1^*), \ldots, (X_n^*, Y_n^*))(\cdot).$

*(III) The bagged estimator is $\hat{\theta}_{n;Bag}(\cdot) = \mathbb{E}^*[\hat{\theta}_n^*(\cdot)].$*

In practice, the bootstrap expectation in (III) is implemented by Monte Carlo: for every bootstrap simulation $b \in \{1, \ldots, B\}$ from (I), we compute $\hat{\theta}_n^{*b}(\cdot)$ $(b = 1, \ldots, B)$ as in (II) to approximate

$$\hat{\theta}_{n;Bag}(\cdot) \approx B^{-1} \sum_{b=1}^{B} \hat{\theta}_n^{*b}(\cdot). \tag{1}$$

The number $B$ is often chosen as 50 or 100, depending on sample size and on the computational cost to evaluate the estimator $\hat{\theta}_n(\cdot)$. The theoretical quantity in (III) corresponds to $B = \infty$: the finite number $B$ in practice governs the accuracy of the Monte Carlo approximation but otherwise, it shouldn't be viewed as a tuning parameter for bagging.

This is exactly Breiman's [1] definition for bagging regression estimators $\hat{\theta}_n(\cdot)$. For classification, we propose to average the bootstrapped probabilities $\hat{\theta}_{n,j}^{*b}(\cdot) = \hat{\mathbb{P}}^*[Y^{*b} = j|X = \cdot]$ $(j = 0, \ldots, J-1)$ yielding an estimator for $\mathbb{P}[Y = j|X = \cdot]$, whereas Breiman [1] proposed to vote among classifiers for constructing the bagged classifier.

A trivial equality indicates the somewhat unusual approach of using the bootstrap methodology:

$$\hat{\theta}_{n;Bag}(\cdot) = \hat{\theta}_n(\cdot) + (\mathbb{E}^*[\hat{\theta}_n^*(\cdot)] - \hat{\theta}_n(\cdot)) = \hat{\theta}_n(\cdot) + \text{Bias}_n^*(\cdot),$$

where $\text{Bias}_n^*(\cdot)$ is the usual bootstrap bias estimate of $\hat{\theta}_n(\cdot)$. Instead of the usual bias correction with a negative sign, bagging comes along with the wrong sign and *adds* the bootstrap bias estimate. Thus, we would expect that bagging has a higher bias than $\hat{\theta}_n(\cdot)$, which we will argue to be true in some sense. But according to the usual interplay between bias and variance in nonparametric statistics, the hope is to gain more by reducing the variance than increasing the bias so that overall, bagging would pay-off in terms of the MSE. Again, this hope turns out to be true for some basis algorithms (or estimation methods). In fact, Breiman [1] describes heuristically the performance of bagging as follows. The variance of the bagged estimator $\hat{\theta}_{n;Bag}(\cdot)$ is equal or smaller than that for the original estimator $\hat{\theta}_n(\cdot)$; and there can be a drastic variance reduction if the original estimator is "unstable".

Breiman [1] only gives a heuristic definition of instability. Bühlmann and Yu [4] define the following asymptotic notion of instability.

**Definition 2** *(Stability of an estimator). An estimator $\hat{\theta}_n(x)$ is called stable at $x$ if $\hat{\theta}_n(x) = \theta(x) + o_P(1)$ $(n \to \infty)$ for some fixed value $\theta(x)$.*

Although this definition resembles very much the one for consistency, it is different in spirit since the value $\theta(x)$ here is only a stable limit and not necessarily the parameter of interest. Instability thus takes place whenever the procedure $\hat{\theta}_n(\cdot)$ is not converging to a fixed value: other realizations from the data generating distribution (even for infinite sample size) would produce a different value of the procedure, with positive probability.

## 2.1. Unstable estimators with hard decision indicator

Instability often occurs when hard decisions with indicator functions are involved as in regression or classification trees. One of the main underlying ideas why bagging works can be demonstrated with a simple example.

### 2.1.1. Toy example: a simple, instructive analysis

Consider the estimator

$$\hat{\theta}_n(x) = \mathbf{1}_{[\overline{Y}_n \leq x]}, \quad x \in \mathbb{R}, \tag{2}$$

where $\overline{Y}_n = n^{-1} \sum_{i=1}^n Y_i$ (no predictor variables $X_i$ are used for this example). The target we have in mind is $\theta(x) = \lim_{n \to \infty} \mathbb{E}[\hat{\theta}_n(x)]$. If we take the view of fixed $x$, after a proper scaling, a simple yet precise analysis below shows that bagging is a smoothing operation. Due to the central limit theorem we have

$$n^{1/2}(\overline{Y}_n - \mu) \to_D \mathcal{N}(0, \sigma^2) \; (n \to \infty) \tag{3}$$

with $\mu = \mathbb{E}[Y_1]$ and $\sigma^2 = \text{Var}(Y_1)$. Then, for $x$ in a $n^{-1/2}$-neighborhood of $\mu$,

$$x = x_n(c) = \mu + c\sigma n^{-1/2}, \tag{4}$$

we have the distributional approximation

$$\hat{\theta}_n(x_n(c)) \to_D g(Z) = \mathbf{1}_{[Z \leq c]} \ (n \to \infty), \ Z \sim \mathcal{N}(0,1). \tag{5}$$

Obviously, for a fixed $c$, this is a hard decision function of $Z$. Denoting by $\Phi(\cdot)$ the c.d.f. of a standard normal distribution, it follows that

$$\begin{aligned} \mathbb{E}[\hat{\theta}_n(x_n(c))] &\to \mathbb{P}[Z \leq c] = \Phi(c) \ (n \to \infty), \\ \mathrm{Var}(\hat{\theta}_n(x_n(c))) &\to \Phi(c)(1-\Phi(c)) \ (n \to \infty). \end{aligned} \tag{6}$$

Since the variance does not converge to zero, $\hat{\theta}_n(x_n(c))$ is unstable in the sense of Definition 2: the predictor takes the values 0 and 1 with a positive probability, even as $n$ tends to infinity. On the other hand, averaging for the bagged estimator looks as follows:

$$\begin{aligned} \hat{\theta}_{n;Bag}(x_n(c)) &= \mathbb{E}^*[\mathbf{1}_{[\overline{Y}_n^* \leq x_n(c)]}] = \mathbb{E}^*[\mathbf{1}_{[n^{1/2}(\overline{Y}_n^* - \overline{Y}_n)/\sigma \leq n^{1/2}(x_n(c) - \overline{Y}_n)/\sigma]}] \\ &= \Phi(n^{1/2}(x_n(c) - \overline{Y}_n)) + o_P(1) \\ &\to_D g_{Bag}(Z) = \Phi(c - Z) \ (n \to \infty), \ Z \sim \mathcal{N}(0,1), \end{aligned} \tag{7}$$

where the first approximation (second line) follows because the bootstrap works for the arithmetic mean $\overline{Y}_n$, i.e.,

$$\sup_{x \in \mathbb{R}} |\mathbb{P}^*[n^{1/2}(\overline{Y}_n^* - \overline{Y}_n) \leq x] - \mathbb{P}[n^{1/2}(\overline{Y}_n - \mu) \leq x]| = o_P(1) \ (n \to \infty), \tag{8}$$

and the second approximation (third line in (7)) holds, because of (3) and the definition of $x_n(c)$ in (4). Comparing with (5), bagging produces a soft decision function of $Z$: it is a shifted inverse probit, similar to a sigmoid-type function. Figure 1 illustrates the two functions $g(\cdot)$ and $g_{Bag}(\cdot)$. We see that bagging is a smoothing operation. The amount of smoothing is determined "automatically" and turns out to be very reasonable (we are not claiming any optimality here). The effect of smoothing is that bagging reduces

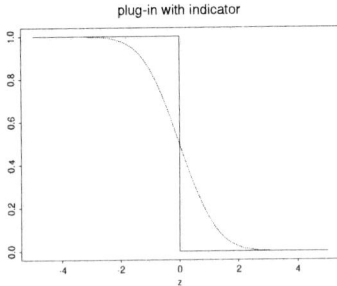

Figure 1. Indicator estimator from (2) at $x = x_n(0)$ as in (4). Function $g(z) = \mathbf{1}_{[z \leq 0]}$ (solid line) and $g_{Bag}(z)$ (dotted line) defining the asymptotics of the estimator in (5) and its bagged version in (7).

variance due to a soft- instead of a hard-thresholding operation. An instructive case is with $x = x_n(0) = \mu$, i.e., $x$ is exactly at the most unstable location, where $\mathrm{Var}(\hat{\theta}_n(x))$ is maximal. Formula (7) gives

$$\hat{\theta}_{n;B}(x_n(0)) \to_D \Phi(-Z) = U, \ U \sim \mathrm{Uniform}([0,1]).$$

Thus,

$$\mathbf{E}[\hat{\theta}_{n;B}(x_n(0))] \to \mathbf{E}[U] = 1/2 \ (n \to \infty)$$
$$\mathrm{Var}(\hat{\theta}_{n;B}(x_n(0))) \to \mathrm{Var}(U) = 1/12 \ (n \to \infty).$$

Comparing with (6), bagging is asymptotically unbiased (for the asymptotic parameter $\lim_{n\to\infty} \mathbf{E}[\hat{\theta}_n(x_n(0))] = \Phi(0) = 1/2$), but the asymptotic variance is reduced by a factor 3!

More generally, we can compute the first two asymptotic moments in the unstable region with $x = x_n(c)$. Denote the convolution of $g_1$ and $g_2$ by $g_1 * g_2(\cdot) = \int_\mathbb{R} g_1(\cdot - y)g_2(y)dy$, and the standard normal density by $\varphi(\cdot)$.

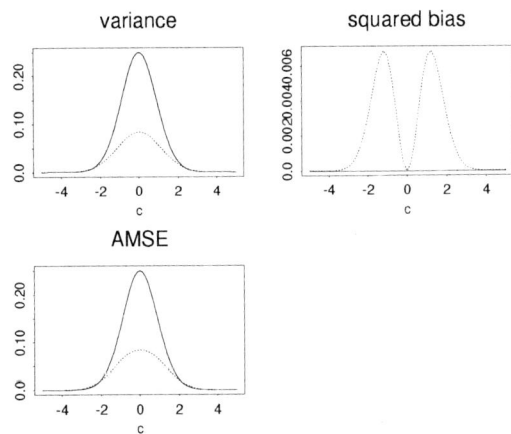

Figure 2. Indicator estimator from (2) at $x = x_n(c)$ as in (4). Asymptotic variance, squared bias and mean squared error (AMSE) (the target is $\lim_{n\to\infty} \mathbf{E}[\hat{\theta}_n(x)]$) for the estimator $\hat{\theta}_n(x_n(c))$ from (2) (solid line) and for the bagged estimator $\hat{\theta}_{n;B}(x_n(c))$ (dotted line) as a function of $c$.

**Corollary 1** *For the estimator in (2) with $x = x_n(c)$ as in formula (4),*

(i) $\lim_{n\to\infty} \mathbf{E}[\hat{\theta}_n(x_n(c))] = \Phi(c)$,
$\lim_{n\to\infty} \mathrm{Var}(\hat{\theta}_n(x_n(c))) = \Phi(c)(1 - \Phi(c))$.

*(ii)* $\lim_{n\to\infty} \mathbb{E}[\hat{\theta}_{n;B}(x_n(c))] = \Phi * \varphi(c)$,
$\lim_{n\to\infty} \text{Var}(\hat{\theta}_{n;B}(x_n(c))) = \Phi^2 * \varphi(c) - (\Phi * \varphi(c))^2$.

Proof: Assertion (i) is restating (6). Assertion (ii) follows by (7) together with the boundedness of the function $g_{Bag}(\cdot)$ therein. □

Numerical evaluations of these first two asymptotic moments and the mean squared error (MSE) are given in Figure 2. We see that in the approximate range where $|c| \leq 2.3$, bagging improves the asymptotic MSE. The biggest gain is at the most unstable point $x = \mu = \mathbb{E}[Y_1]$, corresponding to $c = 0$. The squared bias with bagging has only a negligible effect on the MSE (note the different scales in Figure 2). Note that we always give an a-priori advantage to the original estimator which is asymptotically unbiased for the target as defined.

In [4], this kind of analysis has been given for more general estimators than $\overline{Y}_n$ in (2) and also for estimation in linear models after testing. Hard decision indicator functions are involved there as well and bagging reduces variance due to its smoothing effect. The key to derive this property is always the fact that the bootstrap is asymptotically consistent as in (8).

### 2.1.2. Regression trees

We address here the effect of bagging in the case of decision trees which are most commonly used in practice in conjunction with bagging. Decision trees consist of piecewise constant fitted functions whose supports (for the piecewise constants) are given by indicator functions similar to (2). Hence we expect bagging to bring a significant variance reduction as in section 2.1.1.

For simplicity of exposition, we consider first a one-dimensional predictor space and a so-called regression stump which is a regression tree with one split and two terminal nodes. The stump estimator (or algorithm) is then defined as the decision tree,

$$\hat{\theta}_n(x) = \hat{\beta}_\ell \mathbf{1}_{[x < \hat{d}_n]} + \hat{\beta}_u \mathbf{1}_{[x \geq \hat{d}_n]} = \hat{\beta}_\ell + (\hat{\beta}_u - \hat{\beta}_\ell)\mathbf{1}_{[\hat{d}_n \leq x]}, \tag{9}$$

where the estimates are obtained by least squares as

$$(\hat{\beta}_\ell, \hat{\beta}_u, \hat{d}_n) = \text{argmin}_{\beta_\ell, \beta_u, d} \sum_{i=1}^n (Y_i - \beta_\ell \mathbf{1}_{[X_i < d]} - \beta_u \mathbf{1}_{[X_i \geq d]})^2.$$

These values are estimates for the best projected parameters defined by

$$(\beta_\ell^0, \beta_u^0, d^0) = \text{argmin}_{\beta_\ell, \beta_u, d} \mathbb{E}[(Y - \beta_\ell \mathbf{1}_{[X < d]} - \beta_u \mathbf{1}_{[X \geq d]})^2]. \tag{10}$$

The main mathematical difference of the stump in (9) to the toy estimator in (2) is the behavior of $\hat{d}_n$ in comparison to the behavior of $\overline{Y}_n$ (and not the constants $\hat{\beta}_\ell$ and $\hat{\beta}_u$ involved in the stump). It is shown in [4] that $\hat{d}_n$ has convergence rate $n^{-1/3}$ (in case of a smooth regression function) and a limiting distribution which is non-Gaussian. This also explains that the bootstrap is not consistent, but consistency as in (8) turned out to be crucial in our analysis in section 2.1.1. Summarizing, the asymptotic analysis of bagging a stump is difficult and still unsolved. However, a computationally attractive version of bagging, which has been found as good as bagging, turns out to be more tractable from a theoretical point of view.

## 2.2. Subagging

Subagging is a sobriquet for **sub**sample **agg**regat**ing** where subsampling is used instead of the bootstrap for the aggregation. An estimator $\hat{\theta}_n(\cdot) = h_n((X_1, Y_1), \ldots, (X_n, Y_n))(\cdot)$ is aggregated as follows:

$$\hat{\theta}_{n;SB(m)}(\cdot) = \binom{n}{m}^{-1} \sum_{(i_1,\ldots,i_m)\in\mathcal{I}} h_m((X_{i_1}, Y_{i_1}), \ldots, (X_{i_m}, Y_{i_m}))(\cdot),$$

where $\mathcal{I}$ is the set of $m$-tuples whose elements in $\{1, \ldots, n\}$ are all distinct. This aggregation can be approximated by a stochastic computation. The subagging algorithm is as follows.

(I) For $b = 1, \ldots, B$ ($B = 50$ or $100$) do:

  (i) Generate a random subsample $(X_1^{*b}, Y_1^{*b}), \ldots, (X_m^{*b}, Y_m^{*b})$ by random drawing $m$ times without replacement from the data $(X_1, Y_1), \ldots, (X_n, Y_n)$ (instead of resampling with replacement in bagging).
  
  (ii) Compute the subsampled estimator $\hat{\theta}_m^{*b}(\cdot) = h_m((X_1^{*b}, Y_1^{*b}), \ldots, (X_m^{*b}, Y_m^{*b}))(\cdot)$.

(II) Average the subsampled estimators to approximate $\hat{\theta}_{n;SB(m)}(\cdot) \approx B^{-1} \sum_{b=1}^{B} \hat{\theta}_n^{*b}(\cdot)$.

As indicated in the notation, subagging depends on the subsample size $m$ which is a tuning parameter (in contrast to $B$).

An interesting case is *half subagging* with $m = [n/2]$. More generally, we could also use $m = [an]$ with $0 < a < 1$ (i.e. $m$ a fraction of $n$) and we will argue why the usual choice $m = o(n)$ in subsampling for distribution estimation [12] is a bad choice. Half subagging with $m = [n/2]$ has been studied also by Buja and Stuetzle [5]: they showed that in case where $\hat{\theta}_n$ is a $U$-statistic, half subagging is exactly equivalent to bagging. Moreover, they observe, consistently with our experience, that half subagging yields very similar empirical results to bagging when the estimator $\hat{\theta}_n(\cdot)$ is a decision tree. Thus, if we don't want to optimize over the tuning parameter $m$, very often a good choice in practice is $m = [n/2]$. Consequently, half subagging typically saves more than half of the computing time because the computational order of an estimator $\hat{\theta}_n$ is usually at least linear in $n$.

### 2.2.1. Subagging regression trees

We describe here in a non-technical way the main mathematical result from [4] about subagging regression trees.

The underlying assumptions for our mathematical theory are as follows. The data generating regression model is

$$Y_i = f(X_i) + \varepsilon_i,$$

where $X_1, \ldots, X_n$ and $\varepsilon_1, \ldots, \varepsilon_n$ are i.i.d. variables, independent from each other, and $\mathbf{E}[\varepsilon_1] = 0$, $\mathbf{E}|\varepsilon_1|^2 < \infty$. The regression function $f(\cdot)$ is assumed to be smooth and the distribution of $X_i$ and $\varepsilon_i$ are assumed to have suitably regular densities.

It is then shown in [4] that for $m = [an]$ ($0 < a < 1$),

$$\limsup_{n \to \infty} \frac{\mathbf{E}[(\hat{\theta}_{n;SB(m)}(x) - \theta(x))^2]}{\mathbf{E}[(\hat{\theta}_n(x) - \theta(x))^2]} < 1,$$

for $x$ in suitable neighborhoods (depending on the fraction $a$) around the best projected split points of a regression tree (e.g. the parameter $d^0$ in (10) for a stump), and where $\theta(x) = \lim_{n\to\infty} \mathbb{E}[\hat{\theta}_n(x)]$. That is, subagging asymptotically reduces the MSE for $x$ in neighborhoods around the unstable split points, a fact which we may also compare with Figure 2. Moreover, one can argue that globally,

$$\mathbb{E}[(\hat{\theta}_{n;SB(m)}(X) - \theta(X))^2] \stackrel{\text{approx.}}{<} \mathbb{E}[(\hat{\theta}_n(X) - \theta(X))^2]$$

for $n$ large and where the expectations are taken also over (new) predictors $X$.

For subagging with small order $m = o(n)$, such a result is no longer true: the reason is that small order subagging will then be dominated by a large bias (while variance reduction is even better than for fraction subagging with $m = [an]$, $0 < a < 1$).

Similarly as for the toy example in section 2.1.1, subagging smoothes the hard decisions in a regression tree resulting in reduced variance and MSE.

### 2.3. Bagging basis functions in MARS

We discuss here the effect of bagging on the basic ingredient in MARS [9]. For a one-dimensional predictor variable, the basis function in MARS is a piecewise linear spline function $[x - d]_+ = (x - d)\mathbf{1}_{[d \leq x]}$. Its estimated version takes the form

$$\hat{\theta}_n(x) = \hat{\beta}_n[x - \hat{d}_n]_+, \ x \in \mathbb{R}, \tag{11}$$

with the least squares estimates

$$(\hat{\beta}_n, \hat{d}_n) = \text{argmin}_{\beta, d} \sum_{i=1}^n (Y_i - \beta[X_i - d]_+)^2$$

for the best projected values $(\beta^0, d^0) = \text{argmin}_{\beta, d} \mathbb{E}[(Y - \beta[X - d]_+)^2]$. It is possible to

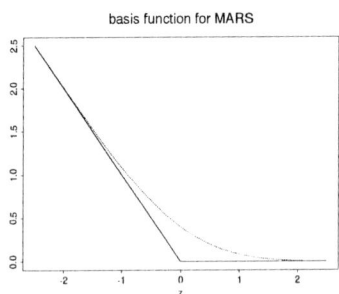

Figure 3. MARS basis function from (11) at $x = x_n(0) = d^0$ as in (12). Function $g(z)$ and $g_{Bag}(z)$ (dotted line) from (13), defining the asymptotics of $\hat{\theta}_n(x_n(0))$ and its bagged version, respectively.

derive the asymptotic behavior of this MARS basis function for $x$ in a neighborhood of the best projected knot $d^0$, i.e. for

$$x = x_n(c) = d^0 + c\sigma_d n^{-1/2}, \tag{12}$$

where $\sigma_d^2$ is the asymptotic variance of $\hat{d}_n$. The smoothing effect of bagging with the MARS basis function can be described as follows, assuming some regularity conditions:

$$n^{1/2}\sigma_d^{-1}\hat{\theta}_n(x_n(c)) \to_D g(Z) = \beta^0(c-Z)\mathbf{1}_{[Z \le c]},$$
$$n^{1/2}\sigma_d^{-1}\hat{\theta}_{n;Bag}(x_n(c)) \to_D g_{Bag}(Z) = \beta^0\{(c-Z)\Phi(c-Z) + \varphi(c-Z)\}, \tag{13}$$

where $Z \sim \mathcal{N}(0,1)$, see [4]. The functions $g(\cdot)$ and $g_{Bag}(\cdot)$ are displayed in Figure 3. We see that already the original MARS basis function estimator corresponds to a continuous limiting function $g(\cdot)$, in contrast to regression stumps in (9). Also, the smoothing effect of bagging, described by the asymptotic function $g_{Bag}(\cdot)$, turns out to have a small effect only. The MSEs for the MARS basis function estimator and its bagged version, which can be computed from (13), are displayed in Figure 4. We clearly see that the bagging improvement is at most only very marginal.

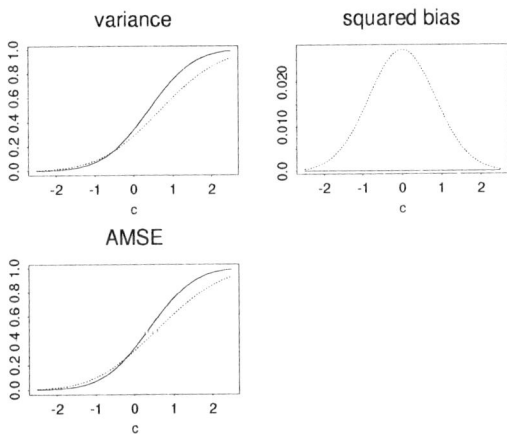

Figure 4. MARS estimator $\hat{\theta}_n(x_n(c))$ from (11) with $x_n(c)$ from (12). Asymptotic variance, squared bias and mean squared error (AMSE) (the target is $\theta(x) = \lim_{n\to\infty}\mathbf{E}[\hat{\theta}_n(x)]$), standardized by the factor $n\sigma_d^{-2}$, for the estimator $\hat{\theta}_n(x_n(c))$ in (11) (solid line) and for the bagged estimator $\hat{\theta}_{n;B}(x_n(c))$ (dotted line), as a function of $c$.

### 2.3.1. Effect of unstable variable selection

The analysis in section 2.3 only covers the case with one-dimensional predictor variables and one linear piecewise function. It can be viewed as an analogue for regression stumps with one predictor. While for the latter, (su-)bagging does improve the MSE substantially (in theory and also in simulated finite sample cases [4]), we see no relevant asymptotic gain for bagging with one MARS basis function. With higher dimensional predictors or more basis functions per predictor, there could be *another* effect of bagging, besides the smoothing effect which is very dominant in regression trees, but negligible for spline functions in MARS. We will demonstrate empirically in section 3, that bagging can be very unstable for MARS (in terms of choosing basis functions) with more than one predictor variable and has a chance to perform very poorly; this is in sharp contrast to bagging decision trees which we have found to be *always* better in terms of MSE compared to original trees. But a version of bagging based on robust aggregation, see section 3.1, is found to be effective for MARS, yielding better MSE performance than the original MARS. This better performance is not due to bagging-type smoothing (at least not asymptotically as argued above); it rather seems to help reducing the negative effect of unstable variable selection in MARS.

## 3. Numerical results and Bragging

We show here the bagging procedure in action for synthetic data from three different models:

(M1) $X = (X^{(1)}, \ldots, X^{(10)}) \sim \text{Unif}([0,1]^{10})$, $\varepsilon \sim \mathcal{N}(0,6)$,
$Y = 10\sin(\pi X^{(1)} X^{(2)}) + 20(X^{(3)} - .5)^2 + 10 X^{(4)} + 5 X^{(5)} + \varepsilon$.

(M2) $X = (X^{(1)}, \ldots, X^{(5)}) \sim \mathcal{N}_5(0, I)$, $\varepsilon \sim \mathcal{N}(0,4)$,
$Y = X^{(1)} + .5 X^{(2)} + .8 X^{(3)} + .5 X^{(4)} + .3 X^{(5)} + 2 X^{(1)} X^{(2)} + 3 X^{(2)} X^{(3)} + \varepsilon$.

(M3) $X^{(1)}, X^{(2)}$ i.i.d. $\sim \text{Unif}(\{0,1\})$, $X^{(3)}, X^{(4)}$ i.i.d. $\sim \text{Unif}(\{0,1,2,3\})$,
$X^{(5)} \sim \text{Unif}(\{0,1,2,3,4,5,6,7\})$,
$X^{(1)}, X^{(2)}, \ldots, X^{(5)}$ independent, $\varepsilon \sim \mathcal{N}(0,4)$,
$Y = 1_{[X^{(1)}=0]} - 3 1_{[X^{(1)}=1]} + .5 1_{[X^{(2)}=0]} + 2 1_{[X^{(2)}=1]} + .8 1_{[X^{(3)}=0]} - 2 1_{[X^{(3)}=1]}$
$+ 2 1_{X^{(3)}=2]} - 1 1_{[X^{(3)}=3]} + .5 1_{[X^{(4)}=0]} + 1.2 1_{[X^{(4)}=1]} - .9 1_{[X^{(4)}=2]} + 1.8 1_{[X^{(4)}=3]}$
$+ .3 1_{[X^{(5)}=0]} - .6 1_{[X^{(5)}=1]} + .9 1_{[X^{(5)}=2]} - 1.2 1_{[X^{(5)}=3]} + 1.5 1_{[X^{(5)}=4]} - 1.8 1_{[X^{(5)}=5]}$
$+ 2.1 1_{[X^{(5)}=6]} - 2.4 1_{[X^{(5)}=7]} + 2 1_{[X^{(1)}=0, X^{(2)}=1]} + 3 1_{[X^{(2)}=0, X^{(3)}=1]} + \varepsilon$.

Sample size is always chosen as $n = 300$. Model (M1) also known as Friedman #1, has 5 ineffective predictor variables is; in (M2), all predictor variables are effective and some of them occur also as strong interactions; in (M3), all predictor variables are factors and effective, some of them also occurring as strong interactions. The signal to noise ratios $\text{Var}(f(X))/\text{Var}(\varepsilon)$, where $f(\cdot) = \mathbf{E}[Y|X = \cdot]$, are 3.97 for (M1), 3.77 for (M2) and 3.08 for (M3) : thus, all models have similar orders of signal to noise ratio. All our numerical results are based on 100 independent model simulations and we always use 100 bootstrap replications for bagging.

We report the MSE $\mathbf{E}[(\hat{\theta}(X) - f(X))^2]$, where the expectation is also over new predictors $X$; sometimes, we also consider the random variable of the (test set) squared error $(\hat{\theta}(X) - f(X))^2$.

In case of trees, we never lose with bagging in terms of MSE (and also in terms of the test set squared error; this cannot be seen from Table 1). The model (M3) with discrete predictors (factors) is beyond what has been treated in theory. Nevertheless, we also see here that bagging "works well". Bagging MARS is highly unstable: the MSE is worse with bagging while the median squared error performance is better.

As pointed out in section 2.3, bagging a piecewise linear spline does not reduce the MSE asymptotically in case of a one-dimensional predictor variable. Also, it is expected that from an asymptotic point of view with fixed number of predictors $d < \infty$, the same predictor variables will be selected for the original and the bootstrap samples. However, the asymptotic stability of variable selection may be misleading for practical applications: when running MARS on a bootstrap sample, we will often see that the terms selected by MARS are different from the ones selected from the original sample. In particular, the chance that this happens increases when more terms in the MARS algorithm are allowed. Bagging MARS has a potential to reduce the test set squared error, due to averaging over unstable variable selection, for some individual datasets. But there is also a chance for extremely bad performance on some data: overall (on average), the MSE of MARS becomes large. But there is a simple trick to improve the highly unstable behavior of bagging MARS, as described next.

Table 1
Mean squared error (MSE) and quantiles of squared error in models (M1)-(M3) for various methods. "tree" indicates the default regression tree in R using the function `rpart`; "MARS (deg=2)" indicates MARS constraining the interaction terms to be at most of order 2 (function `mars` in R with "degree" parameter set to 2). Number of bootstrap replications in bagging is 100. Number of simulations is 100.

| model & method | MSE | (0, 0.25, 0.5, 0.75, 1)-quantiles of squared error |
|---|---|---|
| (M1), tree | 10.56 | (8.66, 9.84, 10.52, 11.12, 14.22) |
| (M1), bagged tree | 5.81 | (4.55, 5.42, 5.68, 6.09, 7.22) |
| (M1), bragged tree | 6.13 | (4.86, 5.78, 6.07, 6.47, 7.77) |
| (M1), MARS (deg=2) | 1.60 | (0.77, 1.25, 1.47, 1.82, 5.71) |
| (M1), bagged MARS (deg=2) | 1.83 | (0.45, 0.68, 0.85, 1.05, 71.09) |
| (M1), bragged MARS (deg=2) | 0.81 | (0.41, 0.64, 0.79, 0.92, 1.55) |
| (M2), tree | 11.30 | (7.07, 8.67, 10.91, 13.14, 19.84) |
| (M2), bagged tree | 7.61 | (4.52, 6.48, 7.50, 8.56, 11.52) |
| (M2), bragged tree | 7.32 | (4.29, 5.95, 7.09, 8.26, 11.13) |
| (M2), MARS (deg=2) | 4.40 | (0.18, 0.69, 1.01, 1.30, 217.79) |
| (M2), bagged MARS (deg=2) | 22.33 | (0.27, 0.50, 0.67, 1.43, 1193.73) |
| (M2), bragged MARS (deg=2) | 0.43 | (0.18, 0.32, 0.39, 0.54, 1.35) |
| (M3), tree | 3.04 | (2.23, 2.75, 3.00, 3.33, 4.03) |
| (M3), bagged tree | 1.70 | (1.17, 1.53, 1.68, 1.82, 2.29) |
| (M3), bragged tree | 1.90 | (1.40, 1.69, 1.88, 2.05, 2.61) |

### 3.1. Bragging

Bragging is a sobriquet for bootstrap robust aggregating. Instead of the sample mean in (1) in the bagging algorithm (see Definition 1), we use a robust location estimator for the realized bootstrapped estimators $\hat{\theta}_n^{*b}(\cdot)$, $b = 1, \ldots B$. We propose to use

$$\hat{\theta}_{n;Brag} = \text{median}(\{\hat{\theta}_n^{*b}(\cdot); b = 1, \ldots, B\}).$$

We also looked at other robust location estimators for aggregating the bootstrapped estimates $\hat{\theta}_n^{*b}(\cdot)$'s such as Huber's estimator and Hampel's redescending M-estimator, but aggregation with the sample median was found to be slightly better.

We show the results for bragging in Table 1 as well. We see that bragging MARS brings a substantial improvement over bagging MARS and over original MARS. The bootstrapped MARS estimates can be very poor resulting in a bad bagged estimator; since aggregation with the sample median is highly robust, bragging MARS does not suffer from such instabilities. In case of trees, bagging is stable ("always" reducing MSE), and bragging and bagging have about the same performance.

#### 3.1.1. Some further experiments with the Friedman #1 model

We consider a version of model (M1), including also dimensionality $d = 20$, and with $\text{Var}(\varepsilon) = 1$:

(M4,d)   $X = (X^{(1)}, \ldots, X^{(d)}) \sim \text{Unif}([0,1]^d)$, $d \in \{10, 20\}$, $\varepsilon \sim \mathcal{N}(0,1)$,
$Y = 10\sin(\pi X^{(1)} X^{(2)}) + 20(X^{(3)} - .5)^2 + 10 X^{(4)} + 5 X^{(5)} + \varepsilon.$

We consider three different sample sizes $n \in \{50, 300, 1000\}$. Particularly for $d = 20$ and $n = 50$, the data is very high-dimensional relative to sample size.

As basis algorithms, we consider regression trees, MARS (deg=1) which has no interaction terms (i.e. an additive model with forward variable selection) and MARS (deg=2) which allows up to second order interaction terms.

Figure 5 displays the performance of the MSEs for the basis algorithms and their bagged and bragged versions. We see that bagging or bragging trees pays-off even asymptotically: the explanation for this is given in section 2. For MARS, bagging or bragging becomes more ineffective for larger sample size: this is again consistent with theory (for $d < \infty$), because the selection of predictor variables is asymptotically stable and bagging is not inducing any smoothing effect (see section 2.3) which would help to reduce the MSE. For MARS allowing second order effects (deg=2), we need a larger sample size than $n = 1000$ to make the selection among the $d(d+1)/2$ terms stable (instead of the $d$ terms when no interaction terms are allowed).

Our little demonstration should help to point out that from an asymptotic point of view, bagging or bragging MARS is not yielding much gain in terms of MSE. The reason is that bagging MARS is not inducing any relevant smoothing effect as for trees, which would be beneficial in terms of MSE. With trees, we exploit again empirically, that bagging is paying off even when $n$ is large.

From a more practical point of view, when sample size is large, we would typically allow interaction terms of order 3 or 4 or higher (or more terms per predictor); this would then put us back to the case where predictor terms selection would be unstable. Thus, for many applications in practice, bragging MARS will be very useful; and bagging MARS

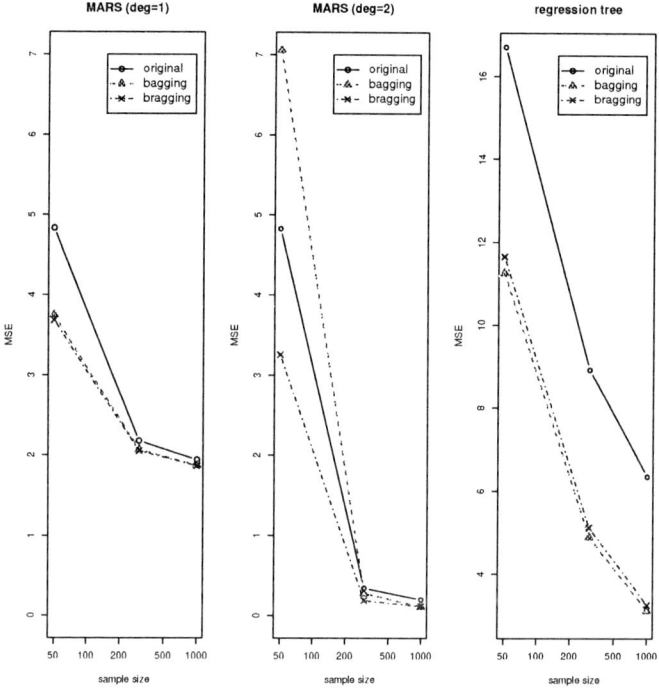

Figure 5. Mean squared error (MSE) in model (M4,d=10) with sample sizes $n \in \{50, 300, 1000\}$. MARS (deg=1) is MARS without interaction terms, MARS (deg=2) is MARS with interactions up to degree 2 (function mars in R), and regression tree is with at least 5 observations in every terminal node (function rpart in R with "minbucket" parameter set to 5).

can be dangerous as indicated in Table 1. To support the last point, we also looked at the MSE in model (M4,d=20) for sample size $n = 50$, i.e. for a very high-dimensional problem relative to sample size:

| (M4,d=20) | MARS (deg=2) | bagged MARS (deg=2) | bragged MARS (deg=2) |
|---|---|---|---|
| MSE | 7.10 | 57.95 | 4.64 |

## 4. BagBoosting and an application for tumor classification using microarray gene expression data with very large $d$

BagBoosting is using bagging as a "module" in a boosting algorithm. Boosting, originally proposed by Freund and Schapire [8] is an algorithm which builds a linear combination of estimators,

$$F_m(\cdot) = \sum_{j=1}^{m} \alpha_j \hat{\theta}_{j;X,U}(\cdot), m = 1, 2, \ldots,$$

where $\alpha_j > 0$ are adaptively chosen weights and $\hat{\theta}_{j;X,U}(\cdot)$ are function estimates based on current generalized residuals $U_1, \ldots, U_n$ and the available predictors $(X_1, \ldots, X_n)$. In boosting terminology, the function estimator is called a "learner". For example, the learner could be a regression tree, yielding function estimates $\hat{\theta}_{j;X,U}(\cdot)$ based on the predictor variables and current generalized residuals $(X_i, U_i)$, $i = 1, \ldots, n$. The linear combination weights $\alpha_j$ are from a one-dimensional line search in a gradient descent algorithm, and the number of iterations $m$ is a tuning parameter of boosting, describing when to stop the algorithm. For more details, see for example [11].

**BagBoosting.** BagBoosting, proposed in Bühlmann and Yu [3], is a boosting algorithm where the learner is a bagged function estimator. For example, the learner could be a bagged stump, i.e. a bagged two-node regression tree.

Our BagBoosting algorithm is different from stochastic gradient descent (Friedman [10]) who uses a regression tree learner based on *one* subsample, randomly chosen in every boosting iteration, whereas BagBoosting uses an aggregated (su-)bagged tree in every boosting iteration.

We consider here a dataset consisting of microarray gene expression levels of 40 tumor and 22 normal colon tissues: for each tissue 6'500 human genes are measured using the Affymetrix technology. A selection of 2'000 genes with highest minimal intensity across the samples has been made, and these data are publicly available at http://microarray.princeton.edu/oncology. We pre-processed the data by carrying out a base 10 logarithmic transformation and standardizing each tissue sample to zero mean and unit variance across the gene expressions. Summarizing, we have data $(X_i, Y_i)$ with $Y_i \in \{0, 1\}$ (cancerous and normal type) and $X_i \in \mathbb{R}^{2000}$ (2'000 gene expression levels) with $\sum_{j=1}^{2000} X_i^{(j)} = 0$, $\sum_{j=1}^{2000} (X_i^{(j)})^2 / 1999 = 1$ $(i = 1, \ldots, n = 62)$.

The entire classification method is then a two-step scheme where first the 200 most significant genes are selected according to a two-sample Wilcoxon test, followed by Logit-Boost (Friedman et al. [11]) with stumps, or BagBoosting using LogitBoost with bagged stumps, or a (single) classification tree (these three latter methods are run on the 200 selected genes only). We emphasize that the selection of the 200 genes is part of the classification method and based on training data only (in cross-validation experiments). The theory in section 2 tells us that a bagged stump is better (in terms of MSE) than a stump, and it is then heuristically clear that BagBoosting with a bagged stump is better than boosting with original stumps. In the context of microarray gene expression data, this has been generally confirmed in [7]. Figure 6 exhibits an advantage of BagBoosting over LogitBoost, and both are much better than a single classification tree. Our cross-validated misclassification rates become even smaller when using leave-one-out cross-validation and are then very competitive with other published results [7]; but we think that the 2/3 training 1/3 test set cross-validation yields more reliable estimates for the unknown misclassification error.

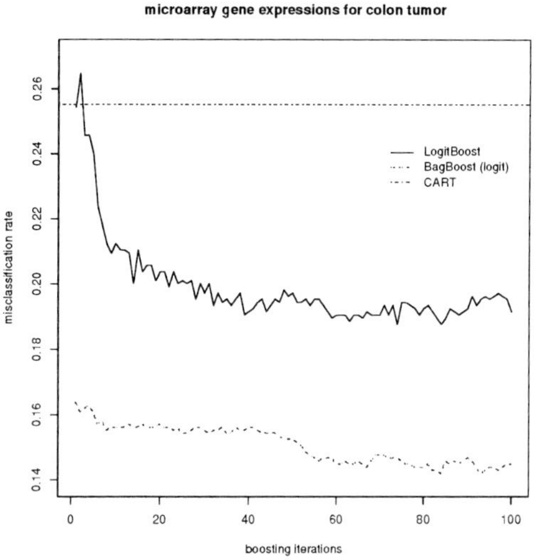

Figure 6. Misclassification errors for classifying cancerous versus normal colon tissue from 200 pre-selected microarray gene expressions, based on cross-validation with 50 random divisions into 2/3 training and 1/3 test sets. LogitBoost with stumps, BagBoosting is LogitBoost with bagged stumps, and classification tree with default settings for function `rpart` from R.

## 5. Conclusions

The theory and heuristics in section 2 describes that bagging is a smoothing operation in cases where the original estimator involves hard-decision indicator functions. The smoothing implies a variance and MSE reduction. The same applies also to subagging which is based on subsampling $m$ data-points instead of resampling $n$ points: the choice $m = [n/2]$ is in many cases almost equivalent to bagging, see also [5].

It has been empirically demonstrated in [4] that smoothing due to bagging and its corresponding variance and MSE reduction can be well seen in finite sample problems, even in low-dimensional settings with stumps for one predictor variable; see also section 3. It has been long "believed" that bagging would only help for high-dimensional, complex algorithms.

For smoother estimators than decision trees, bagging doesn't yield a first order smoothing effect: Buja and Stuetzle [5] have made this rigorous by showing that bagging a U-statistic has only second-order asymptotic effects on the MSE and bagging a U-statistics sometimes even increases MSE compared to the original U-statistics. A similar behavior also applies for linear spline functions which are used in MARS, as described in section 2.3: bagging MARS is likely not to be as effective as bagging decision trees and can sometimes even result in very poor performance, as empirically shown in section 3.

Bragging based on robust aggregation, which is a new version of bagging proposed here, seems to have the ability to achieve another effect than smoothing. We showed empirically that bragging MARS often improves upon the original MARS algorithm. The probable reason which renders bragging MARS successful is that it averages (in a robust way) the instabilities of selected terms in MARS which then helps for variance reduction.

## REFERENCES

1. Breiman, L. (1996). Bagging predictors. Machine Learning **24**, 123–140.
2. Breiman, L., Friedman, J.H., Olshen, R.A. & Stone, C.J. (1984). *Classification and Regression Trees*. Wadsworth, Belmont (CA).
3. Bühlmann, P. and Yu, B. (2000). Discussion on Additive logistic regression: a statistical view of boosting, auths. J. Friedman, T. Hastie and R. Tibshirani. Annals of Statistics **28**, 377–386.
4. Bühlmann, P. and Yu, B. (2002). Analyzing bagging. Annals of Statistics **30**, 927–961.
5. Buja, A. and Stuetzle, W. (2002). Observations on bagging. Preprint. Available from http://ljsavage.wharton.upenn.edu/~buja/
6. Chen, S.X. and Hall, P. (2003). Effects of bagging and bias correction on estimators defined by estimating equations. To appear in Statistica Sinica.
7. Dettling, M. (2003). BagBoosting for tumor classification with gene expression data. In preparation.
8. Freund, Y. and Schapire, R.E. (1996). Experiments with a new boosting algorithm. In Machine Learning: Proc. Thirteenth International Conference, pp. 148–156. Morgan Kauffman, San Francisco.
9. Friedman, J.H. (1991). Multivariate adaptive regression splines (with Discussion). Annals of Statistics **19**, 1–141 (with discussion).
10. Friedman, J.H. (2002). Stochastic gradient boosting. Computational Statistics & Data Analysis **38**, 367–378.
11. Friedman, J.H., Hastie, T. and Tibshirani, R. (2000). Additive logistic regression: a statistical view of boosting. Annals of Statistics **28**, 337–407 (with discussion).
12. Politis, D.N., Romano, J.P. and Wolf, M. (1999). *Subsampling*. Springer, New York.

## Data Compression by Geometric Quantization

Nkem-Amin Khumbah[a] and Edward J. Wegman[b]

[a]Department of Mathematics and Computer Science, North Georgia College and State University, 212 Newton Oakes Center, Dahlonega, Georgia 30597 USA

[b]Center for Computational Statistics, George Mason University, Fairfax, VA 22030 USA

In this paper, we propose a nonparametric method for data quantization so as to reduce massive data sets to more manageable sizes. We investigate the probabilistic foundation and demonstrate statistical results for the quantization process. We discuss optimal geometric quantization procedures and discuss the computational and storage complexity of these procedures.

## 1. INTRODUCTION

The conjunction of large data set size, high dimensional data sets, and high-order algorithmic complexity makes exploratory data analysis and data mining infeasible for a number of data sets. Wegman (1995) argues that there are two significant thresholds when exploring data sets. One, he argues occurs around $10^6$ to $10^7$ bytes when issues of computational feasibility, data transfer, and human visual resolution become major issues. Of course, since 1995, computer chip speed has improved by roughly two orders of magnitude, and network communication speed and data transfer speed from hard drives has improved by at least one order of magnitude, but regrettably the human visual system has probably even degraded somewhat as populations age. Even so, computer screen resolution has actually only improved marginally. The rough figure of $10^6$ to $10^7$ bytes is still a reasonable threshold for interactive, visual-exploratory analysis including visual data mining. The second threshold Wegman (1995) argues occurs when data is sufficiently massive that it can no longer be stored on hard drives and must be stored in magnetic tape silos. In 1995, this threshold occurred at about one terabyte. This threshold too has moved upwards since 1995 and it is now feasible for a PC computer to have a terabyte of on-line storage installed in a single machine. However, while this threshold described in Wegman (1995) has moved up by perhaps two orders of magnitude, it is clear that data acquisition capability has also moved by similar orders of magnitude, so that the overall two-threshold premise is basically valid for the foreseeable future with only slight adjustments in the exact placement of the thresholds.

It is therefore of interest to be able to compress data in such a way that inference made about the data remain essentially unchanged, but that the data are amenable to issues of computational speed, transfer speed, storage capability, and capability for visualization. In short we would like to have the ability to compress data to around $10^6$ to $10^7$

observations without losing substantial inference capabilities. This is essentially a process of binning data. The phrase *quantization* is used in electrical engineering literature and carries much more of a sense of what we mean than the word binning typically used within the statistics community. In essence we are thinking of binning at a very fine scale consistent with sample size, say a million bins for $10^{12}$ observations. For comparison, digital images are usually quantized at something on the order of $2^{24}$ color levels, which is well beyond the capability of the human visual system to distinguish distinct colors. Similarly, CD quality audio is typically quantized at $2^{16}$ bits, which for all intents is beyond the capability of the human auditory system to distinguish.

The electrical engineering literature has discussed vector quantization, which is a process of forming clusters and then associating with each cluster a representor of that cluster. We note that Braverman (2001, 2002) has made very effective use of vector quantization for certain classes of satellite data for statistical purposes. However, vector quantization has the drawback that most clustering algorithms are $O(n^2)$ complexity so that a data set with $10^{12}$ observations would require $O(10^{24})$ computations, which even with teraflop computers is not feasible. We are proposing a nonparametric, geometry-based quantization with $O(n)$ computational complexity. In what follows, we adopt the convention that $n$ refers to data set size, $d$ refers to dimension of the data, and $k$ will refer to the number of bins or tiles into which the data is quantized. We begin by laying out the probabilistic framework.

## 2. QUANTIZATION AND PROBABILITY SPACES

The basic idea of quantization is to divide the underlying space into a finite number of representors. This can be done either in a Euclidean $d$-space or in the underlying probability space depending which is more convenient for developing theory or practice. In either case we want the theory to be consistent. We describe here the basic structure. First we consider an abstract probability space described by $(\Omega, \mathfrak{A}, P)$, where $\Omega$ is an abstract set, $\mathfrak{A}$ is a $\sigma$-field, and $P$ is a probability measure. Consider a mapping $Q_k : \Omega \longrightarrow \Omega$, which is a many-to-one mapping and maps a measurable subset $A \in \Omega$ into a single point, $Q(A)$. $Q_k$ will be our quantizer map. Later, we will suppress the subscript $k$. We will eventually choose only a finite number, $k$, of $A \in \mathfrak{A}$ according to a geometric criterion. Let $\Omega_k = \{b_j\}$ be the finite collection of images of $Q_k$. The family $\mathfrak{A}_k$ is given by

$$\mathfrak{A}_k = \{Q_k(A) : A \in \mathfrak{A}\}$$

so that for each $B \in \mathfrak{A}_k$, $Q_k^{-1}(B) \subset \Omega$ is measurable. For each $B \subset \mathfrak{A}_k$, $B = \bigcup_{j \in I_B} \{b_j\}$, with $b_j$ the singleton images of some $A_j \in \mathfrak{A}$ and $I_B$ and index set for $B$. Thus

$$Q_k^{-1}(B) = Q_k^{-1}(\bigcup_{j \in I_B} \{b_j\}) = \bigcup_{j \in I_B} A_j \in \mathfrak{A}.$$

Thus $\mathfrak{A}_k$ contains countable (actually finite) unions. Similarly, for each $b_j \in \Omega_k$, we define $p_j = P(Q_k^{-1}(b_j))$ and let $P_k$ be the finite probability measure associated with measurable sets in $\mathfrak{A}_k$. The points $b_j$ are the representors of the set $A_j$ in the original probability space. For $B \in \mathfrak{A}_k$, $B = \underset{j \in I_B}{\cup} \{b_j\}$, we define

$$P_k(B) = P(Q_k(B)) = \sum_{j \in I_B} P(Q_k(b_j)) = \sum_{j \in I_B} p_j.$$

**Lemma 1**: $\mathfrak{A}_k$ is a field of sets.

**Proof**: Suppose first that $\{A_j\}$ is a partition of $\Omega$ where each $A_j \in \mathfrak{A}$. Then $\Omega_k = Q_k(\Omega)$. Then $\Omega_k \in \mathfrak{A}_k$ since by definition $\Omega \in \mathfrak{A}$. Next suppose $B_1, B_2 \in \mathfrak{A}_k$. Then $\exists\, A_1, A_2 \in \mathfrak{A}$ such that $B_1 = Q_k(A_1)$ and $B_2 = Q_k(A_2)$ and $B_1 \cup B_2 = Q_k(A_1) \cup Q_k(A_2)$. Let $b \in B_1 \cup B_2$. Then either $b \in B_1$ or $b \in B_2$ or both. If $b \in B_1 \Rightarrow b \in Q_k(A_1) \exists\, \omega \in A_1$ such that $b = Q_k(\omega) \Rightarrow \omega \in A_1 \cup A_2 \Rightarrow b \in Q_k(A_1 \cup A_2)$. Similarly if $b \in B_2$. Thus

$$Q_k(A_1) \cup Q_k(A_2) \subset Q_k(A_1 \cup A_2).$$

Conversely, suppose $b \in Q_k(A_1 \cup A_2)$. Then $\exists\, \omega \in A_1 \cup A_2$ such that $b = Q_k(\omega)$. If $\omega \in A_1$, then $b \in Q_k(A_1)$. If $\omega \in Q_k(A_2)$, then $b \in Q_k(A_2)$. In either case, $b \in Q_k(A_1) \cup Q_k(A_2)$. Thus

$$B_1 \cup B_2 = Q_k(A_1) \cup Q_k(A_2) = Q_k(A_1 \cup A_2) \subset \mathfrak{A}_k.$$

Thus, by induction, $\mathfrak{A}_k$ is closed under countable (finite) unions. Now consider $B_1$ and $B_2$ and suppose $b \in B_1 - B_2$. Then there exists $A_1 \in \mathfrak{A}$ such that $B_1 = Q_k(A_1)$ and $A_2 \in \mathfrak{A}$ such that $B_2 = Q_k(A_2)$. Since $b \in B_1$, for some $\omega \in A_1$, $b = Q_k(\omega)$. Suppose for that $\omega$, $\omega \in A_2$. Then $b \in Q_k(A_2) \Rightarrow$ by above $b \in Q_k(A_1 \cup A_2) = B_1 \cup B_2$. This is contradiction $b \in B_1 - B_2$. Thus $Q_k(A_1) - Q_k(A_2) = Q_k(A_1 - A_2)$. This $\mathfrak{A}_k$ is closed under complementation. Finally, $\phi = \Omega_k - \Omega_k = Q(\Omega - \Omega) = Q_k(\phi)$.

It is straightforward to show that $P_k$ is a finite probability measure. Now suppose $Y$ is a random vector mapping $(\Omega, \mathfrak{A}, P)$ into $\mathcal{R}^d$, $d$-dimensional Euclidean space. Let $F$ be the induced probability distribution. We will have in mind partitioning $\mathcal{R}^d$ into a finite number of sets, which we will call $S_j$, $j = 1, \ldots, k$. The $S_j$ will be a tessellation of $\mathcal{R}^d$ and for each $S_j$ we will choose a representor. In particular, $q_k(\boldsymbol{x}) = y_j$ if $\boldsymbol{x} \in S_j$. Here we denote the vector $\boldsymbol{x}$ by a bold $\boldsymbol{x}$. Let $W = \{y_1, \ldots, y_k\}$ the quantized space. Equivalently, $q_k^{-1}(y_j) = \{\boldsymbol{x} \in \mathcal{R}^d : q_k(\boldsymbol{x}) = y_j\}$, $j = 1, \ldots, k$. As above there is an induced probability $P(y_j) = \int_{S_j} dF(\boldsymbol{x})$. The geometric tessellation on $\mathcal{R}^d$ induces a partition on $\Omega$ as follows: $A_j = Y^{-1} \circ q_k^{-1}(y_j)$ which in turn creates the finite probability space $(\Omega_k, \mathfrak{A}_k, P_k)$ described above. This in turn induces a finite random variable $Y_k : \Omega_k \longrightarrow W$. In short the following diagram commutes.

$$\begin{array}{ccc} (\Omega, \mathfrak{A}, P) & \xrightarrow{Q_k} & (\Omega_K, \mathfrak{A}_k, P_k) \\ \downarrow Y & & \downarrow Y_k \\ \mathcal{R}^d & \xrightarrow{q_k} & W. \end{array}$$

The choice of representor $y_j$ could reasonably be any element of $S_j$. However, an excellent choice is $y_j = \int_{S_j} \boldsymbol{x}\, dF(\boldsymbol{x}) = E(Y | Q_k = y_j)$. We are now able to suppress the subscript $k$ and rewrite this equation as $E(Y|Q) = Q$. That is the expected value of a quantized random vector is just the quantized vector itself. Another way of thinking is that within a tile of our tessellation, we choose the representor to be the mean value of the conditional distribution in the tile. This feature is called self-consistency, which has some important implications for quantization. For more details on self consistency, see Tarpey and Flury (1996). This idea was exploited in Braverman (2000, 2002) for vector quantization.

## 3. SELF CONSISTENCY AND ITS IMPLICATIONS

Clearly, we have as an easy result $E(Y) = E_Q E(Y|Q) = E(Q)$. Thus if we have a sample $Y_1, Y_2, \ldots, Y_n$ and $\hat{\theta}(Y_1, \ldots, Y_n)$ is a linear unbiased estimator of a parameter $\theta$, the same estimator based on the quantized version, say $E[\hat{\theta}|Q]$ will also be a linear unbiased estimator.

**Theorem 1:**

1. $E(Y) = E(Q)$
2. If $\hat{\theta}$ is a linear unbiased estimator of $\theta$, then so is $E(\hat{\theta}|Q)$
3. If $h$ is a convex function, then $E(h(Q)) \leq E(h(Y))$. In particular $E(Q^2) \leq E(Y^2)$ so that $var(Q) \leq var(Y)$
4. $E[Q(Q-Y)] = 0$
5. $cov(Y - Q) = cov(Y) - cov(Q)$
6. $E(Y-P)^2 \geq E(Y-Q)^2$ where $P$ is any other quantizer.

**Proof:**

We have just shown 1 and 2 above. To see 3, recall from Jensen's Inequality $E(h(Y)) \geq hE(Y)$. Thus $E[h(Q)] = E[h(E(Y|Q))] \leq E[E(h(Y)|Q] = E(h(Y))$. Because $h(y) = y^2$ is a convex function, it follows that $E(Q^2) \leq E(Y^2)$. Moreover, because $E(Q) = E(Y)$, it follows that $var(Q) \leq var(Y)$.

To see 4, consider

$$E(Q(Q-Y)) = E_Q E(Q(Q-Y)|Q)$$
$$= E_Q E(Q^2|Q) - E_Q E(QY|Q)$$
$$= E_Q(Q^2) - E_Q[QE(Y|Q)]$$
$$= E_Q(Q^2) - E_Q(Q^2)$$
$$= 0.$$

To see 5,

$$cov(Y-Q) = E[(Y-Q)'(Y-Q)]$$
$$= E(Y-Q)'Y - E[(Y-Q)'Q]$$

$$\begin{aligned}
&= E(Y'Y) - E(Y'Q) \\
&= E(Y'Y) - E_Q E(Y'Q|Q) \\
&= E(Y'Y) - E_Q[E(Y'|Q)Q] \\
&= E(Y'Y) - E(Q^2).
\end{aligned}$$

Because $[E(Y)]^2 = [E(Q)]^2$, it follows that $cov(Y - Q) = cov(Y) - cov(Q)$.

Finally to see 6, consider

$$\begin{aligned}
E[Y - P]^2 &= E[Y - Q + Q - P]^2 \\
&= E[Y - Q]^2 + E[Q - P]^2 + 2E(Y - Q)(Q - P) \\
&= E[Y - Q]^2 + E[Q - P]^2 + 2\{E[(Y - Q)Q] - E[(Y - Q)P]\} \\
&= E[Y - Q]^2 + E[Q - P]^2 - 2E[(Y - Q)P] \\
&= E[Y - Q]^2 + E[Q - P]^2 - 2E_P(P[E(Y - Q)]|P) \\
&= E[Y - Q]^2 + E[Q - P]^2 \\
&\geq E[Y - Q]^2.
\end{aligned}$$

Result 6 indicates that the optimal strategy for quantizing in terms of reducing mean square error is to use the conditional expectation of the random vector for a given tile as the representor of the tile.

One slightly troubling result of this theorem is that quantization is variance reducing. Ideally, we would prefer that the variance structure remained identical between the quantized data and the original data. Experimental results shown in Figure 1 suggest that indeed for data set sizes we are considering, there need be little concern.

## 4. SOME NOTES ON CONVERGENCE

An important fact to recognize is that quantization turns continuous spaces on $\Omega$ into a discrete space $\Omega_k$. This makes convergence substantially easier to deal with. Consider a sequence of random variables $Y_i : \Omega_k \longrightarrow W$. Then we have the following result.

**Theorem 2**: Suppose $Y_i$ converges to $Y$ in any of the following modes of convergence. Then it converges in all the other modes as well. That is the following modes of convergence are equivalent.

(a) Convergence in probability
(b) Convergence almost surely
(c) Convergence in $p$-norm
(d) Almost uniform convergence
(e) Uniform convergence

**Proof**:

Standard results show that (e) $\Rightarrow$ (d) $\Rightarrow$ (b) $\Rightarrow$ (a).

Show that (a) $\Rightarrow$ (b). Suppose $Y_i \longrightarrow Y$ as $i \longrightarrow \infty$ in probability. Then for every $\epsilon > 0$,

$$\lim_{i \to \infty} P_k(|Y_i - Y| \leq \epsilon) = 1.$$

Let $\Omega_k = \{\omega_j\}_{j=1}^k$. We may assume the $\omega_j$ are all distinct. Let us consider the corresponding $y_j \in W$. Because $y_j$ is the representor of tile $S_j$, there is an induced metric on $W$. Suppose $d(y_j, y_m)$ is that metric. We may assume that the $y_j$ like the $\omega_j$ are distinct. Choose $\epsilon > 0$ such that $\epsilon < \min d(y_j, y_m)$, $j, m = 1, \ldots, k$. Then

$$\lim_{i \to \infty} P_k[|Y_i - Y| \leq \min d(y_j, y_m)] = 1.$$

This implies there exists an $N_\epsilon$ such that for every $i > N_\epsilon$, $P_k(Y_i = Y) = 1$. Thus (a)$\Rightarrow$(b).

Show that (b)$\Rightarrow$(e). Suppose $Y_i \longrightarrow Y$ pointwise almost surely. Then given $\epsilon > 0$ and for every $j = 1, \ldots, k$, $\exists N_j$ such that whenever $i > N_j$, $|Y_i(\omega_j) - Y(\omega_j)| < \epsilon$. Take $N = \max_{1 \leq j \leq k} N_j$. Then whenever $i \geq N$

$$|Y_i(\omega_j) - Y(\omega_j)| \leq \max_{1 \leq j \leq k} |Y_i(\omega_j) - Y(\omega_j)| \leq \epsilon.$$

Thus (b)$\Rightarrow$(e)

Show that (c)$\Rightarrow$(a). Suppose $Y_i \longrightarrow Y$ in $p$-norm. Thus $\lim_{i \to \infty} \sum_{j=1}^k |Y_i(\omega_j) - Y(\omega_j)|^p p_j = 0$. Then for any $\epsilon > 0$, let $A_i = \{\omega_j : |Y_i(\omega_j) - Y(\omega_j)|^p > \epsilon\}$. Then

$$\sum_{j \in A_i} |Y_i(\omega_j) - Y(\omega_j)|^p p_j \geq \epsilon P_k(A_i).$$

Hence

$$\sum_{j=1}^k |Y_i(\omega_j) - Y(\omega_j)|^p p_j =$$
$$= \sum_{j \in A_i} |Y_i(\omega_j) - Y(\omega_j)|^p p_j + \sum_{j \in \Omega_k - A_i} |Y_i(\omega_j) - Y(\omega_j)|^p p_j$$
$$\geq \epsilon P_k(A_i) + \sum_{j \in \Omega_k - A_i} |Y_i(\omega_j) - Y(\omega_j)|^p p_j$$
$$\geq \epsilon P_k(A_i).$$

Thus $0 \leq \lim_{i \to \infty} \epsilon P_k(A_i) \leq \lim_{i \to \infty} \sum_{j=1}^k |Y_i(\omega_j) - Y(\omega_j)|^p p_j = 0$. Thus we have (c)$\Rightarrow$(a).

Finally show (b)$\Rightarrow$(c). Suppose $Y_i \longrightarrow Y$ almost surely. Then given $\epsilon > 0$, $\exists N_\epsilon$ such that whenever $i > N_\epsilon$,

$$|Y_i(\omega_j) - Y(\omega_j)| < \epsilon \text{ for every } \omega_j \in \Omega_k.$$

Then

$$\sum_{j=1}^k |Y_i(\omega_j) - Y(\omega_j)|^p p_j \leq |\epsilon|^p \sum_{j=1}^k p_j = |\epsilon|^p.$$

Thus $Y_i \longrightarrow Y$ in $p$-norm.

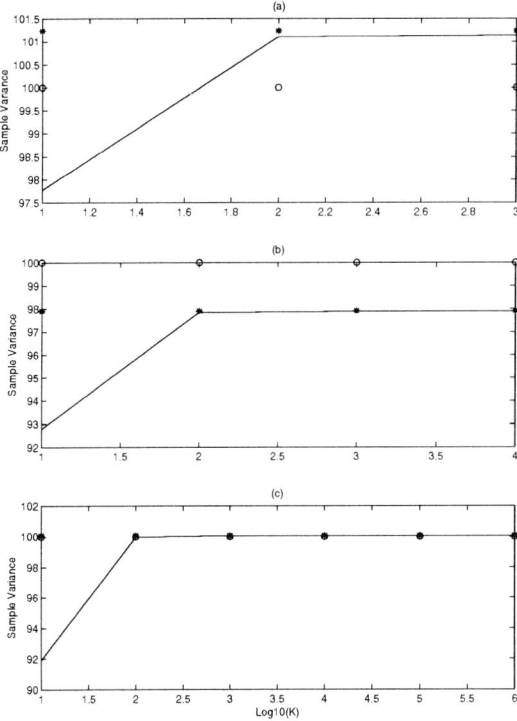

Figure 1. Comparison of true variance (given by circle), estimated variance (given by asterisks) and quantized estimated variance (given by solid line). The three panels starting at top have $n = 10^3$, $n = 10^4$, and $n = 10^6$

## 5. MINIMIZING DISTORTION AND GEOMETRY

We will focus on distortion as the mean square error difference between the unquantized data and the quantized data. We have previously shown that by choosing $E(Y|Q) = Q$, we minimize $E(Y-Q)^2$, i.e. the distortion. That is, we choose the representor of a tile to be the average values within the tile. However, the shape of the tile impacts the distortion as well. Clearly if the tile is a sphere and the distribution is symmetric within the sphere, then the geometric centroid and the conditional expectation will coincide, and the ideal representor would simply be the geometric centroid. In general this is not the case. However it is shown by Gersho (1979) and Gish and Pierce (1968), that in $\mathcal{L}_2$-vector spaces, the geometric shape that minimizes distortion is a sphere. Unfortunately, the sphere does not tessellate any Euclidean space except $\mathcal{R}^1$, in which case it is a straight line segment. Ideally, geometric quantization has the following constraints:

1. We must tessellate the space with tiles that are a spherical as possible,

2. The tiles should preferably be congruent,

3. The tessellation must be space filling so as not to miss observations.

As a measure of sphericity, we can take the dimensionless second moment. Consider a regular polytope in $d$-dimensions. The (hyper)-volume and second moment can be expressed in terms of the volume and second moments of its faces (the part of the polytope that lies in one of its enclosing hyperplanes), then in terms of its $(d-2)$-dimensional faces and so on. See Sloane and Conway (1999) for details.

Suppose $S$ is a $d$-dimensional polytope with $N_1$ congruent faces, $F_1^1, \ldots, F_1^{N_1}$; $N_2$ congruent faces, $F_2^1, \ldots, F_2^{N_2}$ and so on. Suppose that $S$ contains a point, $A$, such that all the generalized pyramids $AF_1^j$, $j = 1, \ldots, N_1$ are congruent, and $AF_2^j$, $j = 1, \ldots, N_2$ are congruent, and so on. Let $a_i \in F_i$ be the foot of the perpendicular from $A$ to $F_i$, $V_{d-1}(i)$ be the volume of the face $F_i$, $U_{d-1}(i)$ be the second moment of $F_i$ about $a_i$ and define $h_i = \|A - a_i\|$. Then the volume and second moment of $S$ about $A$ are given in Sloane and Conway (1999) by

$$vol(S) = \sum_i \frac{N_i h_i}{d} V_{d-1}(i)$$

and

$$U(S) = \sum_i \frac{N_i h_i}{d+2} \left[ h_i^2 V_{d-1}(i) + U_{d-1}(i) \right].$$

As mentioned above, in 1-dimensional space the only polytope of interest is a straight line, which also happens to be a one-dimensional sphere. In two dimensions, the only regular polytopes that tessellate 2-space are equilateral triangles, squares, and hexagons. Of these, hexagons have the smallest dimensionless second moment, and from distortion considerations, this would be the tessellation of choice. In three dimensions, perhaps the most interesting, the regular polytopes that can tessellate three space are tetrahedrons, cubes, hexagonal prisms, rhombic dodecahedrons, and truncated octahedrons.

Table 1
Dimensionless Second Moments for Several Regular 3-D Polytopes. Those marked with an * tessellate 3-space.

| | |
|---|---|
| Tetrahedron* | .1040042... |
| Cube* | .0833333... |
| Octahedron | .0825482... |
| Hexagonal Prism* | .0812227... |
| Rhombic Dodecahedron* | .0787451... |
| Truncated Octahedron* | .0785433... |
| Dodecahedron | .0781285... |
| Icosahedron | .0778185... |
| Sphere | .0769670... |

As can be see from Table 1, the truncated octahedron would be the polytope of choice for tessellating 3-space from the perspective of minimizing distortion. In four dimensions, 4-simplices, hypercubes and 24-cells tessellate 4-dimensional space, with the 24-cell having the minimal dimensionless second moment. Many of these polytopes are illustrated in Wegman (2001) with animations on a CD.

## 6. COMPUTATIONAL AND STORAGE COMPLEXITY OF GEOMETRIC QUANTIZATION

Of course, the main point of the geometric quantization is to reduce the data set size effectively so that computationally complex algorithms may be used, so that data transfer in reasonable time frames is feasible, and so that visualization methods may be used. All of this depends on the quantizing algorithms to be computationally feasible. As mentioned above the engineering concept of vector quantization relies on clustering algorithms, which, generally, are $O(n^2)$. Fortunately, the geometric quantizing algorithms are $O(n)$.

Consider first the one-dimensional case. Suppose the data are located within the set $[a, b]$ and that we desire to quantize the data into $k$ tiles. Then the algorithm

$$j = fixed[k * (x_i - a)/(b - a)]$$

calculates $j$ the index of the tile into which $x_i$ falls. The $fixed$ is an operation which changes a floating point number into a fixed point number. The term $(b - a)$ need only be calculated once. Thus there is one subtraction, one division, one multiplication and one $fixed$ operation for each observation. In total for a data set of size $n$ there are $4n + 1 = O(n)$ operations required. Notice also that the storage requirement drops from $n$ to $3k$. For each of the $k$ quantized representors, we must store the location of the tile, the number of observations in the tile, and the value of the representor. Generally we want to choose $k \ll n$. Our operational target is $k = 10^6$.

In the two-dimensional case, if we tessellate with hexagons, each hexagon has three pairs of parallel sides. Thus the tile must be triply indexed and the above algorithm applied in each of three directions. Thus the computational complexity for hexagons is $12n + 3 = O(n)$ and the storage complexity remains $3k$. Of course, if we tessellate with squares or rectangles, there are only two pairs of parallel sides, so that the tiles need be only doubly indexed. Thus the computational complexity for squares or rectangles is $8n + 2 = O(n)$ with the storage requirement remaining at $3k$.

In the three-dimensional case, the most spherical polytope is the truncated octahedron. This polytope has 3 pairs of square sides and 4 pairs of hexagonal sides. Thus tiles when tessellated using truncated octahedrons must be indexed by 7 indices so that the one-dimensional algorithm must be applied 7 times. This yields a computational complexity of $28n + 7 = O(n)$ and storage requirements of $3k$. Of course if cubes are used to tessellate 3-space, they need only be triply indexed so the computational complexity is $12n + 3 = O(n)$.

Of course in general if $d$-dimensional hypercubes are used to tessellate $d$-space, there are $d$ orthogonal directions so the general form for the computational complexity will be $(4n+1)d = O(n)$ with storage requirements always remaining at $3k$. Unfortunately, while the computational complexity is always $O(n)$, the number of tiles, assuming a constant tile size per orthogonal direction, grows exponentially with dimension. Thus for high dimensions the number of tiles can easily be greater than the number of observations making quantization a useless exercise. Geometric quantization is probably not useful beyond about 5 dimensions unless the partition in each direction is severely limited. Indeed, within the computer science literature, the so-called datacube is essentially a hyper-rectangular tiling with each variable limited to a small number of categories, usually 3 or 4.

One extremely positive aspect of this geometric quantization is that the tiles are determined independent of data. Thus geometric quantization can be used for streaming data, i.e. as a data point comes in, it can be placed into a tile and the data for that tile recursively updated. The mean value for a tile can be recursively computed.

## 7. DISCUSSION AND CONCLUSIONS

In this paper we have introduced the idea of quantizing massive data sets in order to compress data sets to a feasible size from several points of view. This quantization is essentially a nonparametric procedure. We have laid out the probabilistic foundations and exploited the self-consistency property to show that most statistical properties for quantized data remain unchanged from what they would have been with unquantized data. Quantization of course reduces the underlying probability space to a finite space. This has great advantages from a convergence perspective. These results hold no matter the mode of quantization. However, in order to minimize distortion, i.e. mean square error, the optimal geometry involves tessellating space with the most spherical possible tessellating polytopes. We have identified these in dimensions 1 through 4 and shown that computational complexity in general is $O(n)$. We also commented that geometric quantization is probably not practical beyond 5 dimensions unless extremely coarse tiles are used.

Practically speaking, it is probably better to use hypercubes rather than the distortion optimal polytope because of reduced complexity and greater interpretability. Alternate geometric tessellations are possible. One procedure might be to select some data points at random and create a Delauney tessellation. This procedure was used for multivariate density estimation in Hearne and Wegman (1991, 1992, 1994). The rate of growth of the number of tiles with random tessellations has been shown experimentally to be slower than with regular tessellations. However, there are much more severe departures from spherical shapes. An alternate procedure being currently explored is to use density estimates to determine local modes and then use the local modes for Voronoi tessellations. Because kernel density estimates are also $O(n)$ computationally, this approach to forming clusters may lead to an alternate to distance-based clustering that would make vector quantization feasible for large data sets.

## ACKNOWLEDGEMENTS

This paper is based on an invited talk prepared by Dr. Wegman for the International Conference on Current Advances and Trends in Nonparametric Statistics held July 15-19, 2002 in Crete, Greece. Due to an illness in the family, Dr. Wegman was unable to attend and gratefully acknowledges Dr. Don Faxon's assistance by presenting the paper on Dr. Wegman's behalf. We also gratefully acknowledge the kind invitation and the flexibility of the organizers in adapting to the substitute speaker. This paper is based in part on the Ph.D. dissertation of Nkem-Amin Khumbah (Khumbah, 2000) under the direction of Dr. Wegman. Dr. Wegman's work was funded by the Office of Naval Research under contract DAAD19-99-1-0314 administered by the Army Research Office, by the Air Force Office of Scientific Research under contract F49620-01-1-0274 and contract DAAD19-01-1-0464, the latter also administered by the Army Research Office and finally by the Defense Advanced Research Projects Agency through cooperative agreement 8105-48267 with the Johns Hopkins University.

## REFERENCES

Braverman, Amy (2000), "Compressing massive geophysical data using quantization," *Computing Science and Statistics (CD)*, 32, 28-36

Braverman, Amy (2002), "Compressing massive geophysical datasets using vector quantization," *Journal of Computational and Graphical Statistics*, 11(1), 44-62

Gersho, Allen (1979), "Asymptotically optimal block quantization," *IEEE Transactions on Information Theory*, IT-25(4), 373-380

Gish, H. and Pierce, John N. (1968), "Asymptotically efficient quantizing," *IEEE Transactions on Information Theory*, IT-12(5), 678-681, 683

Hearne, Leonard B. and Wegman, Edward J. (1991), "Adaptive probability density estimation in lower dimensions using random tessellations," *Computing Science and Statistics: Proceedings of the 23rd Symposium on the Interface*, 241-245

Hearne, Leonard B. and Wegman, Edward J. (1992), "Maximum entropy density estimation using random tessellations," *Computing Science and Statistics*, 24, 483-487

Hearne, Leonard B. and Wegman, Edward J. (1994), "Fast multidimensional density estimation based on random-width bins," *Computing Science and Statistics*, 26, 150-155

Khumbah, Nkem-Amin N. (2000), *Mathematical Quantization for Massive Data Sets*, Ph.D. dissertation, School of Information Technology and Engineering, George Mason University

Sloane, N. J. A. and Conway, J. H. (1999), *Sphere Packings, Lattices and Groups*, (Third Edition) Springer-Verlag: New York

Tarpey, T. and Flury, B. (1996), "Self-consistency: A fundamental concept in statistics," *Statistical Science*, 11, 229-243

Wegman, Edward J. (1995), "Huge data sets and the frontiers of computational feasibility," *Journal of Computational and Graphical Statistics*, 4(4), 281-295

Wegman, Edward J. (2001), "Data compression by quantization," *Computing Science and Statistics(CD)*, 33, \Interface2001\I2001HTML\EWegman.htm

# 2. Functional Data Analysis

*Recent Advances and Trends in Nonparametric Statistics*
Michael G. Akritas and Dimitris N. Politis (Editors)
© 2003 Elsevier Science B.V. All rights reserved.

# Functional data analysis in evolutionary biology

Nancy E. Heckman[a]*

[a]Statistics Department, University of British Columbia,
Vancouver, B.C. Canada

The evolution of physical traits from one generation to the next depends on a selection mechanism and the existence of genetic variability in the parent population. The modelling and statistical analysis of this evolution is well-developed for qualitative traits and for real or vector valued traits. The methodology can be found in quantitative genetics textbooks, as well as in evolutionary biology and animal breeding literature. The development of methodology for function-valued traits is just beginning. I will explain, in statistician's terms, some basic concepts of quantitative genetics, and discuss connections to functional data analysis.

## 1. INTRODUCTION

Animal breeders and evolutionary biologists study how physical traits change from generation to generation. Which dairy cattle should we breed to maximize milk production in the next generation? Caterpillars grow at a rate that is temperature dependent. Does the pattern of this dependence have a genetic component? Is the pattern related to survival?

A physical trait is called a phenotype. It can be qualitative such as hair colour, quantitative such as height, a vector such as (height,weight), or a function such as growth as a function of temperature. Understanding how phenotypes evolve is a basic goal of quantitative genetics.

Two components of evolution of phenotypes are a selection mechanism and genetic variability. If there is no selection, the distribution of the phenotype remains the same from generation to generation. If there is selection, the selection mechanism can be chosen, as in animal breeding, or it can be unknown, to be estimated, as in evolutionary biology. However, the presence of selection doesn't guarantee evolution of the phenotype. If there is no genetic variability, the phenotype can not evolve save by mutation. Thus, knowledge about the genetic variability of the phenotype is important in determining how phenotypes evolve. Here, evolution refers to adaptive evolution, that is, to evolution governed by selection and genetic variability as opposed to, e.g., mutation.

Figure 1 shows two sets of curves. Each curve shows the average distance (kilometers per week) run on a wheel as a function of age. The average is over mice in a particular breeding line. The curves are constructed from data first discussed in [1] and analyzed

---
*This research was supported by Natural Sciences and Engineering Research Council of Canada. Thanks to Richard Gomulkiewicz, Wei Liu, Joel Kingsolver, and Patrick Carter for their helpful suggestions.

in [2]. The top four curves come from four lines of mice bred for fast wheel-running at age eight weeks. The bottom four curves come from a control group of four lines of mice. The researchers are interested in studying the over-all effect of the selective breeding. From the plots, it seems that the biggest effect is an over-all increase in wheel-running at all ages. Therefore, any sensible statistical analysis will probably conclude that wheel-running activity at age eight weeks is genetically correlated with wheel-running activity at all other ages. Can we make further statements about the distribution of the genetic component of wheel-running?

Figure 2 shows the short term growth rates of two types of caterpillars as a function of temperature. These curves are called thermal performance curves. They were calculated from data collected by Joel Kingsolver [3] under laboratory conditions. Caterpillars were then placed in a natural setting, with temperature varying, and their survival noted. The researchers would like to relate an individual's chance of survival in the natural setting to the individual's thermal performance curve. In addition, the researchers would like to predict the thermal performance curves of future generations under current naturally occuring temperature patterns and under other patterns, such as those caused by global warming.

Methodology for the analysis of the evolution of qualitative, real-valued and vector-valued traits is fairly well-established. Their analysis is discussed in many quantitative genetics book (see, for instance, [4]). In a sequence of papers ([5],[6]), Lande and Arnold showed the importance of considering vector-valued traits: due to genetic correlation, how one trait evolves can effect how another trait evolves. Therefore much information is lost by studying traits marginally rather than jointly. The study of function-valued traits takes this logic one-step further. The continuity of the trait as a function of, e.g., age, provides valuable information that shouldn't be ignored. Methodological research for function-valued traits is still in its infancy. The purpose of the current paper is to make accessible to statisticians the relevant parts of quantitative genetics for vector-valued traits, and to point a path for work to be done on function-valued traits.

Section 2 contains notation, standard terminology and some important results in quantitative genetics, namely the Breeder's Equation, the Robertson-Price Identity, an equation of Lande's ([5] and [7]) relating fitness to the selection gradient and the calculation of a relationship coefficient in a simple setting. Estimation procedures for vector-valued traits are discussed in section 3. Section 4 contains a summary of some of the existing work on function-valued traits.

## 2. SOME RESULTS OF QUANTITATIVE GENETICS

Let $z$ denote the observable physical trait, that is, the phenotype, and write $z = g + e$ where $g$ is called the additive genetic component of $z$ and $e$ is the environmental component. We think of $z$, $g$, and $e$ as being random (variables, vectors or functions). Assume that $g$ and $e$ are uncorrelated and that $e$ is mean 0. For each individual we are only able to observe $z$. Throughout this section and section 3, suppose that $z, g$ and $e \in \Re^k$. Extensions to $z, g$ and $e$ functions are given in section 4.

A fundamental equation in quantitative genetics is the Breeder's Equation. It relates the means of the phenotypes in three populations: parents before selection (with mean phenotype $\mu_p$), parents after selection (with mean phenotype $\mu_{sp}$), and offspring of these selected parents (with mean phenotype $\mu_{so}$). Specifically the Breeder's Equation relates $\mu_{sp} - \mu_p$ to $\mu_{so} - \mu_p$. The usual notation for $\mu_{sp} - \mu_p$ is $S$, and it is called the selection differential. The usual notation for $\mu_{so} - \mu_p$ is $R$, and it is called the response to selection. The relationship between $R$ and $S$ involves $P$, the $k \times k$ covariance matrix of the $z$'s in the parent population, and $G$, the $k \times k$ covariance matrix of the $g$'s in the parent population. The Breeder's Equation is:

$$\mu_{so} - \mu_p = GP^{-1}(\mu_{sp} - \mu_p) \quad \text{that is} \quad R = GP^{-1}S \equiv G\beta \tag{1}$$

where $\beta$ is called the selection gradient. The Breeder's Equation is derived in section 2.1.

Animal breeders use the Breeder's Equation to choose the value of $\mu_{sp}$ that yields the desired $\mu_{so}$. For instance, suppose that $z \in \Re^5$ has components equal to annual milk production of an individual cow in years 1, 2, 3, 4 and 5. We determine the components of $\mu_{sp} - \mu_p$ that maximize the sum of the components of $\mu_{so}$. We then choose cows for breeding according to our calculations.

For evolutionary biologists, the Breeder's Equation has importance beyond predicting phenotypes one generation ahead. The equation, along with knowledge of the genetic covariance $G$, brings an understanding of the possibilities for evolution of $z$. For instance, if $\beta$ is an eigenvector of $G$ corresponding to a zero eigenvalue, then there is no genetic variability in the direction of $\beta$ and so no evolution will occur. In this case, $\beta$ is called an evolutionary constraint ([8]). If $\beta$ is an eigenvector of $G$ corresponding to a large eigenvalue, then there is a lot of genetic variability in the direction of $\beta$. In this case, $\beta$ is called a genetic line of least resistance ([9]). As another example of the insight provided by the Breeder's Equation and knowledge of $G$, suppose that two species are subject to the same selection gradient $\beta$. If the $G$'s for the two species are different, then the species will experience different responses to selection. This difference in response can lead to the formation of subpopulations with very different phenotypes.

The Breeder's Equation involves unknown parameters that must be estimated from data. For animal breeders, data consist of individuals' $z_i$'s, along with the relationships between individuals. From this information, they can estimate $\mu_p$, $P$ and $G$. The estimation of $G$ relies on the calculation of a relationship coefficient $\theta$ in equations of the form $\text{cov}(g_1, g_2) = \theta \text{cov}(g_1)$ where $g_1$ and $g_2$ are genotypic values of two related individuals. Calculation of $\theta$ in a simple setting is discussed in section 2.3. Evolutionary biologists require more than the $z_i$'s and the relationship information. These biologists must also infer the selection mechanism, so they must also measure each individual's fitness. For instance, through the banding of birds, a biologist can label every bird in a closed community, wait for a season, and then determine which birds survived through the season. Estimation of $\mu_p$, $P$, $G$, $\beta$ and fitness is discussed in section 3.

## 2.1. Derivation of the Breeder's Equation

The derivation of the Breeder's Equation uses the relationship between $z_o$, the offspring phenotype, and $z_{mid} \equiv (z_m + z_f)/2$, the midparent phenotype, which is simply the average of the phenotypes of the mother and father. Assume that the offspring phenotype and the midparent phenotype are linearly related:

$$E(z_o | z_{mid}) = \alpha + H z_{mid} \tag{2}$$

where $H$ is a $k$ by $k$ matrix. Also, assume that the males and females in the parent population have the same distribution of phenotypes with mean $\mu_p$ and covariance matrix $P$, and the same distribution of genotypic values with genetic covariance matrix $G$. Assume that selection acts on males and females in the same way. In addition, suppose that mating is at random, so that the covariance between $z_m$ and $z_f$ is zero.

Consider a hypothetical population, the offspring of parents in the absence of selection, with population mean phenotype $\mu_o$. I will show that

$$\mu_{so} - \mu_o = H(\mu_{sp} - \mu_p) \quad \text{and} \quad H = GP^{-1}. \tag{3}$$

Any biologically acceptable model has $\mu_p = \mu_o$ if there is no selection. This will put restrictions on $\alpha$ and $H$, namely $(I - H)\mu_p = \alpha$. But we do not need this restriction to prove (3). However, with this restriction, (3) is equivalent to (1).

To show the first part of (3), let $f_p$ and $f_{sp}$ denote the densities of $z_m$ in the parent population and the selected parent population respectively. By assumption, these are also the densities of $z_f$ in the two populations. Then

$$\begin{aligned}
\mu_{so} - \mu_o &= \int\int E(z_o|z_m, z_f)\, f_{sp}(z_m)\, f_{sp}(z_f)\, dz_m\, dz_f - \int\int E(z_o|z_m, z_f)\, f_p(z_m)\, f_p(z_f)\, dz_m\, dz_f \\
&= H\left[\frac{1}{2}\int\int(z_m + z_f)\, f_{sp}(z_m)\, f_{sp}(z_f)\, dz_m\, dz_f - \frac{1}{2}\int\int(z_m + z_f)\, f_p(z_m)\, f_p(z_f)\, dz_m\, dz_f\right] \\
&= H\left[\int z\, f_{sp}(z)\, dz - \int z\, f_p(z)\, dz\right] \\
&= H[\mu_{sp} - \mu_p].
\end{aligned}$$

To prove the second part of (3) I first show that $H = \text{cov}(z_o, z_{mid})[\text{cov}(z_{mid})]^{-1}$:

$$\begin{aligned}
\text{cov}(z_o, z_{mid}) &= E(z_o z'_{mid}) - E(z_o)E(z'_{mid}) \\
&= E[(\alpha + Hz_{mid})z'_{mid}] - [\alpha + HE(z_{mid})]E(z'_{mid}) \\
&= H[E(z_{mid}z'_{mid}) - E(z_{mid})E(z'_{mid})] \\
&= H\,\text{cov}(z_{mid}).
\end{aligned}$$

But

$$\text{cov}(z_{mid}) = \frac{1}{4}\text{cov}(z_f + z_m) = \frac{1}{2}\text{cov}(z_f) = \frac{1}{2}P$$

and

$$\text{cov}(z_o, z_{mid}) = \text{cov}(z_o, z_f)$$

which is equal to $G/2$ (shown in section 2.3 in a simple case). Thus $H = \text{cov}(z_o, z_{mid})[\text{cov}(z_{mid})]^{-1} = (G/2)(P/2)^{-1} = GP^{-1}$.

The linearity assumption in (2) is often replaced with the more stringent assumption that $(z'_o, z'_{mid})'$ has a multivariate normal distribution. To see that this implies the linearity assumption, note that

$$\begin{aligned}
E(z_o|z_{mid}) &= \mu_o + \text{cov}(z_o, z_{mid})[\text{cov}(z_{mid})]^{-1}(z_{mid} - \mu_p) \\
&= \mu_o - \text{cov}(z_o, z_{mid})[\text{cov}(z_{mid})]^{-1}\mu_p + \text{cov}(z_o, z_{mid})[\text{cov}(z_{mid})]^{-1}z_{mid}.
\end{aligned}$$

## 2.2. The selection gradient, $\beta$

Now consider how selection determines $S = \mu_{sp} - \mu_p$ and $\beta = P^{-1}S$. The selection mechanism is quantified in terms of a fitness function $W$. For the purposes of evolution, an individual is deemed fit if he can pass along his genes to offspring that are able to reproduce. In practice, however, $W$ is often defined as something more easily measurable, such as body mass or age of pupation. Here, take the direct definition and suppose that an individual that successfully reproduces is given a value of $\delta = 1$, while an unsuccessful individual is given a value of $\delta = 0$. Then

$P\{\delta = 1|z\} = W(z)$.

Other important quantities are the mean fitness

$$\bar{W} \equiv E(W(z_p)) = \int W(z)\, f_p(z)\, dz$$

and the relative fitness $W(z)/\bar{W}$.

Below I derive two relationships between the selection differential $S$ and the fitness $W$:

$$\operatorname{cov}(z_p, W(z_p)/\bar{W}) = \mu_{so} - \mu_p \equiv S \tag{4}$$

and

$$\nabla \log \bar{W} = P^{-1}S = \beta = \text{the selection gradient} \tag{5}$$

where $\nabla$ denotes the gradient with respect to the mean vector $\mu_p$. Equation (4) was first discussed in [10], [11] and [12]. Equation (5) was first derived and discussed in [5], [6] and [7].

The derivation of (4) involves no assumptions, save that the covariances are finite. This relationship implies that, if phenotype and relative fitness are positively correlated, then the mean of the parent population will increase with selection.

The derivation of (5) assumes that $z_p$ has a multivariate normal distribution. Equation (5) provides more specific information than (4). To have some insight into this equation suppose that $P$ is the identity matrix and note that the gradient of a function always points in the direction of maximal instantaneous change. From (5), we see that, under selection as defined by $W$, $S = \mu_{sp} - \mu_p = \nabla \log \bar{W}$. Therefore, under the selection associated with $W$, the population mean changes in an "optimal" direction - the direction that maximizes the instantaneous change in the logarithm of mean fitness. If $P$ is not equal to the identity, then a covariance between traits might cause the population mean to move in a direction that is not exactly optimal.

To carry out the calculations to prove (4) and (5), first write $f_{sp}$, the density of the phenotypes of the selected parents, in terms of $f_p$, the density of the phenotypes of the original population of parents.

$$f_{sp}(z) = f_{z_p|\delta=1}(z) \propto P\{\delta = 1|z_p = z\}f_p(z) = W(z)f_p(z)$$

so $f_{sp}(z) = W(z)f_p(z)/\bar{W}$.

To prove (4), write

$$\begin{aligned}\operatorname{cov}(z_p, W(z_p)/\bar{W}) &= E(z_p W(z_p)/\bar{W}) - E(z_p)E[W(z_p)/\bar{W}] \\ &= \int z \frac{W(z)}{\bar{W}} f_p(z)\, dz - E(z_p) \times 1 = E(z_{sp}) - E(z_p) = S.\end{aligned}$$

To prove (5), use the normality assumption to write

$$f_p(z) \equiv f(z; \mu_p) = \frac{1}{\sqrt{2\pi|P|}} \exp\left[-\frac{1}{2}(z-\mu_p)'P^{-1}(z-\mu_p)\right].$$

Then

$$\begin{aligned}\nabla \log \bar{W} &= \frac{1}{\bar{W}} \nabla \int W(z) f(z; \mu_p) \, dz = \frac{1}{\bar{W}} \int W(z) \nabla f(z; \mu_p) \, dz \\ &= \frac{1}{\bar{W}} \int W(z) P^{-1}(z - \mu_p) f(z; \mu_p) \, dz \\ &= P^{-1}[\mu_{sp} - \mu_p] = P^{-1} S.\end{aligned}$$

### 2.3. The relationship coefficient $\theta$

Consider two related individuals with genotypic values $g_1$ and $g_2$ respectively. Models used in quantitative genetics imply that $\text{cov}(g_1) = \theta_1 \text{cov}(g_2) = \theta_2 \text{cov}(g_1, g_2)$ where the $\theta_i$'s, known, depend on the relationship. These covariance equations are crucial in the estimation of $G$. I will carry out a calculation for a very simple case.

Consider $g_m$ the genotypic value of the mother and $g_o$ the genotypic value of her offspring when the father is chosen at random from the parent population. Suppose that the $g$'s are real-valued and correspond to a locus with two alleles $A$ and $a$, with genotype $AA$ yielding $g = x$, $Aa$ yielding $g = 0$ and $aa$ yielding $g = -x$. Suppose that, in the parent population, there is a proportion $p$ of $A$ alleles and $q = 1 - p$ $a$ alleles and that the population is in Hardy-Weinberg equilibrium, that is the proportion of $AA$'s is $p^2$, the proportion of $Aa$'s is $2pq$ and the proportion of $aa$'s is $q^2$. I will show that $E(g_m) = E(g_o)$, $\text{var}(g_m) = \text{var}(g_o)$ and $\text{cov}(g_m, g_o) = \text{var}(g_m)/2$.

The joint distribution of $g_m$ and $g_o$ is given in Table 1. For instance, the probability that $g_m = 0$ and $g_o = 0$ is given as $2pq * 1/2$. The table calculations proceed via conditioning. For instance

$$\begin{aligned} P\{g_o = 0 \text{ and } g_m = 0\} &= P\{g_m = 0\} P\{g_o = 0 | g_m = 0\} \\ &= 2pq * P\{\text{offspring is } Aa | \text{mother is } Aa\} \\ &= 2pq * \Big[P\{\text{mother contributes } A \text{ and father contributes } a | \text{mother is } Aa\} \\ &\qquad + P\{\text{mother contributes } a \text{ and father contributes } A | \text{mother is } Aa\}\Big] \\ &= 2pq * \Big[P\{\text{mother contributes } A | \text{mother is } Aa\} P\{\text{father contributes } a\} \\ &\qquad + P\{\text{mother contributes } a | \text{mother is } Aa\} P\{\text{father contributes } A\}\Big] \\ &= 2pq * \left[\frac{1}{2}q + \frac{1}{2}p\right] = 2pq \frac{1}{2}.\end{aligned}$$

Using Table 1 we see that, in the offspring, the proportion of $AA$'s is $p^2$, of $Aa$'s is $2pq$, and of $aa$'s is $q^2$, just as in the parent population. So $E(g_m) = E(g_o)$ and $\text{var}(g_m) = \text{var}(g_o)$. We easily calculate that $\text{var}(g_m) = 2pqx^2$ and $\text{cov}(g_m, g_o) = pqx^2$.

Table 1
Joint distribution of mother/offspring genotypic values

| Parental genotype | Offspring genotype | | |
|---|---|---|---|
| | $AA$ ($g_o = x$) | $Aa$ ($g_o = 0$) | $aa$ ($g_o = -x$) |
| $AA$ ($g_m = x$) | $p^2 * p$ | $p^2 * q$ | 0 |
| $Aa$ ($g_m = 0$) | $2pq * p/2$ | $2pq * 1/2$ | $2pq * q/2$ |
| $aa$ ($g_m = -x$) | 0 | $q^2 * p$ | $q^2 * q$ |

## 3. ESTIMATION

The estimation of $\mu_p$ and $P$ is straightforward - they are the population mean and covariance matrix of observable phenotype vectors. If we have a random sample of individuals' phenotypes from the parent population, we simply use the sample mean and sample covariance matrix as estimates.

### 3.1. Estimation of $G$

If we had a random sample of genotypic values, we could simply estimate $G$ by the sample covariance. Unfortunately, this is impossible as genotypic values are unobservable. However, we can use phenotypes of related individuals to "see" the individuals' genotypes and so make inferences about $G$. The estimation relies heavily on knowing the relationship coefficients (see section 2.3). For a detailed discussion of the results in this section see [4].

Consider a group of related individuals with the phenotype of the $j$th individual equal to $z_j = g_j + e_j$. As usual, we assume that $e_j$ is mean zero and that $e_j$ and $g_j$ are uncorrelated. Assume the individuals are raised in independent but similar environments, that is, assume that the $e_j$'s are uncorrelated and identically distributed. Since the individuals are related, the $g_j$'s will be correlated. The form of the correlation is determined by the relationship between the individuals.

Consider a simple example where $z_m$ and $z_f$ are the mother's and father's phenotypes and $z_o$ is the offspring's phenotype. Under the assumption that the distribution of genotypes of males is the same as that of females, that the population is in Hardy-Weinberg equilibrium and that mating is at random, letting $G$ be the genetic covariance in the parent population,

$$\text{cov}(g_m) = \text{cov}(g_f) = \text{cov}(g_o) \equiv G \text{ and } \text{cov}(g_m, g_f) = 0.$$

Standard genetic arguments similar to those in section 2.3 show that

$$\text{cov}(g_m, g_o) = \text{cov}(g_f, g_o) = \frac{1}{2}G.$$

We can use these covariance relationships to find the covariance between $z_o$ and the midparent phenotype $z_{mid} \equiv (z_m + z_f)/2$:

$$\text{cov}(z_{mid}, z_o) = \text{cov}(z_m, z_o) = \text{cov}(g_m, g_o) = \frac{1}{2}G.$$

Thus we can easily estimate $G$ based on data from a random sample of triples of parents and offspring: $(z_{mi}, z_{fi}, z_{oi})$, $i = 1, \ldots, n$. Our estimate is two times the sample covariance between the $z_{mid,i}$'s and $z_{oi}$'s. Note that this method of moments estimate need not be symmetric, nor need it be positive definite.

As another example, consider $z_1, \ldots, z_L$ equal to the phenotypes of half-siblings, that is, of siblings with the same father but different (randomly selected) mothers. Using standard genetic arguments, one can show that $\text{cov}(g_j) = G$, $\text{cov}(g_j, g_{j^*}) = G/4$ and thus

$$\text{cov}(z_j, z_{j^*}) = G/4 \text{ for } j \neq j^*,$$

where $G$ is the genetic covariance in the parent population. We can split $g_j$ into two parts, one from the father ($s$ for "sire") and one from the mother ($g_{mj}$): $g_j = s + g_{mj}$. Note that $s$ doesn't depend on $j$, as all half-siblings have the same father. Thus we can write $z_j = s + \tilde{e}_j$ where the $\tilde{e}_j$'s include both the environmental effect and the genetic contribution of the randomly chosen mother. The $\tilde{e}_j$'s are uncorrelated and uncorrelated with $s$. Using this, we see that

$$\text{cov}(z_j, z_{j^*}) = \text{cov}(s) \quad j \neq j^* \quad \text{and so } \text{cov}(s) = G/4.$$

Therefore we can estimate $G$ by 4 times any estimate of $\text{cov}(s)$.

We can estimate $\text{cov}(s)$ via MANOVA. Suppose that we have $n$ families of half-siblings. Denote the phenotypes of the offspring of the $i$th father as $z_{ij} = s_i + \tilde{e}_{ij}$, $j = 1, \ldots, L$, $i = 1, \ldots, n$. These data have the structure of a oneway random effects MANOVA, where the treatment/group is "father". Then usual calculations of MANOVA yield an unbiased estimate of $\text{cov}(s)$:

$$\widehat{\text{cov}}(s) = \frac{SS_g}{L(n-1)} - \frac{SS_e}{nL(L-1)}$$

with

$$SS_g = L \sum_{i=1}^{n} (\bar{z}_i - \bar{z})(\bar{z}_i - \bar{z})' \quad SS_e = \sum_{i=1}^{n} \sum_{j=1}^{L} (z_{ij} - \bar{z}_i)(z_{ij} - \bar{z}_i)'$$

where $\bar{z}_i = \sum_j z_{ij}/L$ and $\bar{z} = \sum_{i,j} z_{ij}/(nL)$. So $\hat{G} = 4\widehat{\text{cov}}(s)$ is an unbiased estimate of $G$. Unfortunately, we cannot guarantee that $\hat{G}$ is non-negative definite.

The above analyses don't use all of the genetic information in the relationships of the individuals. For instance, in the half-sibling analysis, we do not make use of the fact that $\text{cov}(z_j) = G + \text{cov}(e_j)$. A likelihood-based approach would make complete use of this information. As an example, suppose that there are $m = 3$ half-siblings per family, $z \in \Re$ and $\text{cov}(e_j) = \sigma^2$. Then the phenotypes of the three half-siblings all have mean $\mu$ and the covariance matrix is of the form

$$\text{cov} \begin{pmatrix} z_1 \\ z_2 \\ z_3 \end{pmatrix} = \begin{bmatrix} G+\sigma^2 & G/4 & G/4 \\ G/4 & G+\sigma^2 & G/4 \\ G/4 & G/4 & G+\sigma^2 \end{bmatrix}$$

$$= G \begin{bmatrix} 1 & 1/4 & 1/4 \\ 1/4 & 1 & 1/4 \\ 1/4 & 1/4 & 1 \end{bmatrix} + \sigma^2 I \equiv AG + \sigma^2 I.$$

The matrix $A$ is called the relationship matrix and depends only on the fact that we are considering half-siblings. If, in addition, we assume that the phenotypes are normally distributed, we can write down the likelihood and maximize it to estimate $G$.

Similar likelihood calculations hold for data from a wide variety of relationships. All of the likelihoods will contain the unknown $G$ and the known relationship matrix $A$. A similar approach to maximum likelihood is restricted maximum likelihood (REML). For a discussion of these techniques, see Chapter 27 of [4].

### 3.2. Inference about the selection gradient $\beta$ and the fitness $W$

Estimation of $\beta$, the selection gradient, is straightforward using the Robertson-Price Identity (4). This identity, along with (1), tells us that $\beta = P^{-1}\text{cov}(z, W(z)/\bar{W})$. Suppose we have a random sample of $n$ individuals, with their $z_i$'s and fitness measures $W(z_i)$. We estimate mean fitness by $\widehat{\bar{W}} = \sum W(z_i)/n$ and estimate an individual's relative fitness by $w_i = W(z_i)/\widehat{\bar{W}}$. We now perform a least-squares regression of $w_i$ on $z_i$. The slope coefficients of the regression are the estimates of the components of $\beta$.

The fitness function $W$ is of interest in its own right. If $W$ has a unique maximum at $z = \mu$, then one can expect that population phenotypes will evolve towards $\mu$. If $W$ has two local maxima, we may eventually end up with two distinct populations. $W$ is easy to estimate from survival data, with standard parametric logistic regression. Survival data are readily available from field and laboratory studies. For univariate and vector-valued traits [13] and [14] demonstrated the usefulness of smoothing methods for flexibly estimating the fitness function.

## 4. FUNCTION-VALUED TRAITS

A function-valued trait $z(\cdot)$ can be thought of as a stochastic process. It has a mean function $\mu(\cdot)$ and a covariance function $P(s,t) = \text{cov}(z(s), z(t))$. Similarly, the genotypic value $g$ and environmental effect $e$ have mean and covariance functions. As usual, we assume that $g$ and $e$ are uncorrelated and the mean function of $e$ is zero.

The importance of modelling function-valued traits in evolutionary biology was first put forward in [15]. That paper contains function-valued versions of equations (1) and (4). Equation (5) was generalized to function-valued traits in [16] and computed for certain classes of fitness functions in [17]. The latter paper also extended (1) to larger classes of mean and covariance functions than discussed in [15].

Three main approaches exist for analyzing data consisting of function-valued traits: the random regression approach ([18] and [19]), the parametric character process approach ([20] and [21]) and a third, somewhat ad hoc approach ([8], [15] and [22]).

In the ad hoc approach, we require observations on individual $i$'s phenotype at a grid of common values, that is, we observe $z_i(t_1), \ldots, z_i(t_K)$. We use standard techniques for vector-valued traits to estimate $\tilde{G}$, the $K$ by $K$ genetic covariance matrix $\tilde{G}_{ij} = \text{cov}(g(t_i), g(t_j))$ and $\tilde{\beta}$, the $K$-dimensional selection gradient vector. We then smooth or interpolate these estimates to estimate the bivariate genetic covariance function and the selection gradient function. This simple methodology seems to offer improvement over a completely vector-based analysis for, e.g., finding eigenfunctions of $G$. This approach was used in [3] for the analysis of thermal performance curves in caterpillars.

In the random regression approach we model individual $i$'s genotypic value as $g_i(t) = \sum_{j=1}^{m} \alpha_{ij} \phi_j(t)$ for some pre-specified basis functions $\phi_j$. Here, $\alpha_i \equiv (\alpha_{i1}, \ldots, \alpha_{im})'$ is random and $\Sigma^{ii}$, the covariance matrix of $\alpha_i$, determines the covariance function of $g_i$:

$$\text{cov}(g_i(s), g_i(t)) = \text{cov}\left(\sum_j \alpha_{ij} \phi_j(s), \sum_j \alpha_{ij} \phi_j(t)\right) = (\phi_1(s), \ldots, \phi_m(s)) \Sigma^{ii} (\phi_1(t), \ldots, \phi_m(t))'.$$

Similarly, the covariance between $\alpha_i$ and $\alpha_{i^*}$, $\Sigma^{ii^*}$, determines the covariance between $g_i$ and $g_{i^*}$:

$$\text{cov}(g_i(s), g_{i^*}(t)) = (\phi_1(s), \ldots, \phi_m(s)) \Sigma^{ii^*} (\phi_1(t), \ldots, \phi_m(t))'.$$

Just as for vector-valued traits, we can use the relationships between individuals to find equations relating different covariances. For instance, suppose that individuals $i$ and $i^*$ are half-siblings, that $G$ is the covariance function in the parent population, and that $\Sigma$ is the covariance matrix of $\alpha$ in the parent population. Then, for $i \neq i^*$,

$$\text{cov}(g_i(s), g_{i^*}(t)) = \frac{1}{4} G(s, t) = \frac{1}{4} \text{cov}(g_i(s), g_i(t))$$

and so

$$\text{cov}(\alpha_i, \alpha_{i^*}) = \frac{1}{4} \Sigma = \frac{1}{4} \text{cov}(\alpha_i).$$

Using these relationships and assuming normality, we can write down a likelihood for our phenotype data and maximize it to find an estimate of $\Sigma$ and thus an estimate of the underlying $G$. This approach is well-served in the animal breeding literature.

In the character process approach, all covariance functions are assumed to have a specific parametric form, such as $G(s, t) = \theta_0 \exp(-\theta_1 (s-t)^2)$, with unknown parameters. Again, the knowledge of the relationships among individuals gives us further information about $G$. This information can be used to write down a parametric likelihood, which is maximized to estimate the unknown parameters.

**REFERENCES**

1. J.G. Swallow, P.A. Carter and T. Garland, Behavior Genetics No. 28 (1998) 227.
2. T. Morgan, T. Garland Jr. and P. Carter, Evolution No. 57 (2003) 646.
3. J.G. Kingsolver, R. Gomulkiewicz and P.A. Carter, Genetica No. 112 (2001) 87.
4. M. Lynch and B. Walsh, Genetics and Analysis of Quantitative Traits, Sinauer Associates, Sunderland Massachusetts, 1998.
5. R. Lande, Evolution No. 33 (1979) 402.
6. R. Lande and S. Arnold, Evolution No. 37 (1983) 1210.
7. R. Lande, Evolution No. 30 (1976) 314.
8. M. Kirkpatrick and D. Lofsvold, Evolution No. 46 (1992) 954.
9. D. Schluter, Evolution No. 50 (1996) 1766.
10. A.Robertson, Anim. Prod. (1966) No. 8 95.
11. G.R. Price, Nature No. 227 (1970) 520.
12. G.R. Price, Ann. Hum Genet. No. 35 (1972) 485.

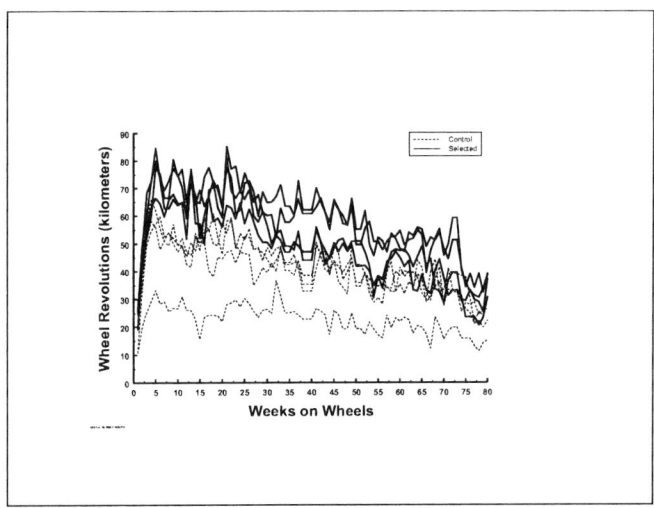

Figure 1: Mean wheel-running (km/week) as a function of age for four selected lines (solid lines) and four control lines (dashed lines) of mice.

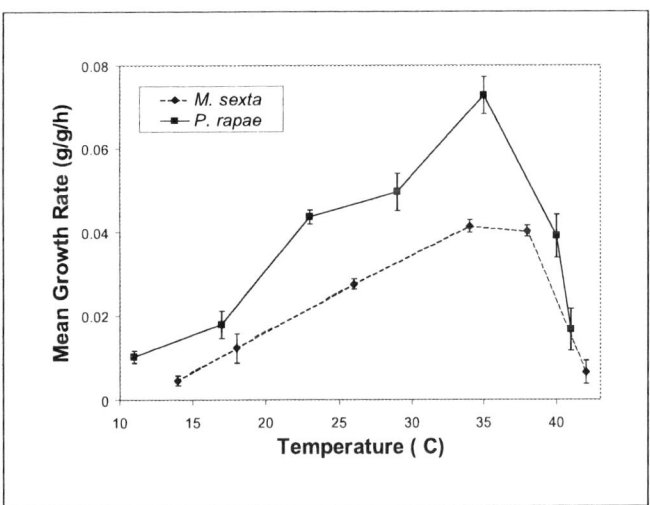

Figure 2: Mean($\pm$ 1 se) short-term relative growty rates as a function of temperature for *Manduca sexta* (diamonds) and *Pieris rapae* (squares) caterpillars. Figures 1 and 2 are reproduced with kind permission of Kluwer Academic publishers, from Variation, selection and evolution of function-valued traits, by Kingsolver, Gomulkiewicz and Carter, *Genetica* 112-113 (2001) 87-104.

13. D. Schluter, Evolution No. 42 (1988) 849.
14. D. Schluter and D. Nychka, American Naturalist No. 143 (1994) 597.
15. M. Kirkpatrick and N. Heckman, J. Math. Bio. No. 27 (1989) 429.
16. R. Gomulkiewicz and J.H. Beder, SIAM Journal of Applied Math No. 56 (1996) 509.
17. J.H. Beder and R. Gomulkiewicz, J. Math. Biol. No. 36 (1998) 299.
18. K. Meyer and W. Hill, Livest. Prof No. 47 (1997) 185.
19. K. Meyer, Genet. Select. Evol No. 30 (1998) 221.
20. F. Jaffrézic and S.D. Pletcher, Genetics No. 156 (2000) 913.
21. S.D. Pletcher and C.J. Geyer, Genetics No. 153 (1999) 825.
22. M. Kirkpatrick, D. Lofsvold and M. Bulmer, Genetics No. 124 (1990) 979.

*Recent Advances and Trends in Nonparametric Statistics*
Michael G. Akritas and Dimitris N. Politis (Editors)
© 2003 Elsevier Science B.V. All rights reserved.

# Functional nonparametric statistics: a double infinite dimensional framework

F. Ferraty [a,b] and Ph. Vieu [b] *

[a]Équipe GRIMM, Université Toulouse Le Mirail, 31058 Toulouse Cedex, France

[b]Laboratoire de Statistique et Probabilités, UMR CNRS C55830, Université Paul Sabatier, 118 route de Narbonne, 31062 Toulouse Cedex, France

Functional aspects are more and more frequent and varied in modern Statistics so much so that the designation of *Functional Statistics* had emerged recently (see [46]). A symbolical example of this new field of Statistics concerns the problems of nonparametric estimation in presence of functional data which are doubly infinite dimensional problems since the functional aspects appears twice: in the nature of the observed data and in the object to be estimated which derives from the statistical model.

We will give a comprehensive introduction to nonparametric modelling for functional variables, together with the presentation of the state of art in this recent field of Statistics. We will present several different situations (such as regression, conditional cumulative distribution, conditional density, conditional mode and quantiles, density estimation, time series prediction, supervised curves classification, ...), but we will pay special attention to the regression problem. We present one selected recent theoretical result in regression in order to explain the double dimensional effects. The description of the state of art, as well in regression as for other mentionned functional problems, will be articulated through this selected asymptotic result. At the end, several open problems and suggestions for future researches are discussed.

## 1. Introduction to Functional Statistics

In the few past years, functional aspects had taken an important place in the Statistical Science. In a first side, since the beginning of the sixties, a lot of attention has been paid to free-distribution statitiscal models and/or methods. The functional feature of these methods comes from the nature of the object to be estimated (such as for instance a density function, a regression function, ...) which is not assumed to be parametrizable by a finite number of real quantities. In this setting, one is usually speaking of *Nonparametric Statistics* (a recent description of the abondant literature existing in this field can be found for instance in [45] and in other chapters of this book). From an other side, there is actually an increasing number of situations coming from different fields of applied

---

*The authors would like to express their gratitude to all the participants of the working group STAPH on Functional Statistics in the Laboratory of Statistics and Probability of Toulouse (http://www.lsp.ups-tlse.fr/Fp/Ferraty/staph.html).

statistical (environmetrics, chemometrics, biometrics, medicin, econometrics, ...) in which the collected data are curves. In this situation, the functional feature is linked with the observations. Since the middle of the nineties, this has motivated different statistical developments, that we could quickly name as *Statistics for Functional Variables* (or *Data*). Most often at this time, these functional models were combined with parametric modelling for the object to be estimated (see [6,39] for general monographies in this setting), and a commonly shared point of view was that it was too difficult to attack both together the functional feature of the data and a nonparametric modelisation for the object to be estimated. This was extremely well summarized in the following words given in the earlier paper [5] "These being nonparametric by themselves, it seems rather heavy to introduce a nonparametric model for observations lying in functional space ..."

The aim of this chapter is to discuss recent advances in *Functional Nonparametric Statistics*, that is on models and/or methods which can capture both the functional (*i.e.* the nonparametric) feature of the statistical target and the functional nature of statistical samples in which curves and more generally functional objects are directly observed. The spirit of this paper is to give a comprehensive introduction to nonparametric modelling for functional variables, together with the presentation of the state of art in this recent field of modern Statistics. Note that this purpose of Functional Nonparametric Statistics is just covering one part of the recent enfatuation existing actually on *Functional Statistics* which is reachable for the authors through the activities of the STAPH working group (see [46] for the birth of the *Functional Statistics* appelation, and see [47] for access to recent developments in this field and for its close connexions with *Operatorial Statistics*).

In the following we will discuss several statistical problems that we will attack from a nonparametric point of view (regression estimation, density estimation, supervised classification, time series prediction, ...) and which are concerned with a sample of functional data. We have choosen to use a presentation that highlightes particularly the double dimension effects. This choice will allow to see how the infinite dimensions, both of the nonparametric model and of the nature of the statistical sample, play a prominent role on the behaviour of the methods.

Even in the usual finite dimensional situation, it could be the case that the word "nonparametric" has different meanings. Because here the double infinite dimension may increase such ambiguity, we will propose now a formal definition. Let $\Phi$ be a map defined on some space $\mathcal{E}$ (either real, vectorial or functional). A statistical model for the estimation of such an object $\Phi$ is usually given by some condition of the form

$$\Phi \in \mathcal{C}, \tag{1}$$

where $\mathcal{C}$ is a subset of maps. The model will be said either nonparametric or parametric according to the structure of $\mathcal{C}$, while it will be said functional, multivariate or real according to the nature of the space $\mathcal{E}$. Formally, we propose the following definition, which will be discussed through some specific examples in Section 2.1 below (see also the coursebooks [22,19,17] for previous introduction and comments on this definition).

**Definition 1.1** *The model defined by (1) is said to be a parametric model for the estimation of $\Phi$, if the class $\mathcal{C}$ can be indexed by a finite number of parameters in $\mathcal{E}$, and it is called nonparametric in the opposite case. The model (1) is said to be functional if $\mathcal{E}$*

*is an infinite dimensional space.*

There are many reasons, basically linked with its large scope of possible applications and with the existence of several appealing theoretical asymptotic results, for which we decided to emphasize on the regression estimation problem (see Section 2). After some introductory section, discussing the state of art in Functional Regression (see Section 2.1) we will present a Nonparametric Functional Regression Model together with some kernel type estimator (see Section 2.2). We will present one asymptotic result (see Section 2.3) which has been selected in the recent existing literature not to be the most general but to be the most adapted for providing easy understanding of the basic effects of such a doubly functional problem. Sections 2.4 and 2.5 will be concerned with some comments about this previous result together with an exhaustive survey both about other existing theoretical works and about practical applications of functional nonparametric methods. For the other problems discussed in Section 3, the existing literature is quite more reduced, and we will just discuss for each situation the model and the adapted estimator before presenting a survey of the literature. Finally, Section 4 discusses general comments and open problems.

## 2. Functional Nonparametric Regression

### 2.1. Introduction to functional regression

Let us consider the problem of estimating the regression function of some real random variable $Y$ given some (eventually infinite dimensional) variable $X$. Given a sample $(X_i, Y_i)$, $i = 1, \ldots n$ of independent couples of random variables each having the same distribution as $(X, Y)$, we are interested in the estimation of the function $r$ (assumed to exist) defined by

$$\forall i, Y_i = r(X_i) + \epsilon_i, \tag{2}$$

where the $\epsilon_i$ are zero mean variables satisfying for all $i$, $E(\epsilon_i|X_i) = 0$. There is an extensive literature about this problem when the variable $X$ is valued in the real line or finite dimensional space. Most of the nonparametric models are based on smoothness restrictions on the function $r$ to be estimated. A wide scope of references in this field can be found in [45] as well as in several other chapters of this book itself. Starting with the monography by [39] several authors had looked at this problem when the variable $X$ is functional (*i.e.* when the variable $X$ takes its values in an infinite dimensional space) but when the model assumes that $r$ is linear and continuous. More precisely, such a *Functional Linear Regression Model* consists in considering a variable $X$ which takes its values in some Hilbertian space $\mathcal{H}$ with scalar product $<.,.>$ and in assuming that

r is a continuous and linear operator. (3)

Note that, even if the object to be estimated (*i.e.* $r$) is functional, this is really a parametric model in the sense of the Definition 1.1, since the condition 3 insures (via the usual Riesz's theorem) that the operator $r$ is uniquely determined by one single element $r^* \in \mathcal{H}$ from the formula:

$$\forall x \in \mathcal{H}, \ r(x) = <x, r^*>. \tag{4}$$

Under this model several parametric estimates of $r$ have been introduced. They are mainly based on preliminary estimates of $r^*$. The first functional writting of such a model seems to goes back to the discussion in [33]. The most recent theoretical advances in this respect can be found in [8] and [9] (see also [11] but in the different setting of fixed functional design) while a large scope of possible real data applications is given in [40].

Such a modelisation presents two main drawbacks. The first one relies with the linearity assumption made in (3) which can be hard to verify in practice, even harder than in usual finite dimensional case since no visualisation support can be available here. The second one relies directly to the Hilbertian context which avoids for any semi-metric modelisation which can be of first importance for applied studies (see discussion in Section 2.5). The recent advances that we will discuss below have as objective to provide solutions to both of these problems by mean of the introduction of the so-called *Functional Nonparametric Regression Model* defined in (5) below.

## 2.2. Functional nonparametric regression: modelling and estimating

Because it does not complicate in any mean the presentation of our study, and because it will allow for additional interesting possibilities of applications (see for instance the discussion in Section 2.5), we will attack the nonparametric modelisation in the general semi-metric context. Assume that $X$ takes values in some semi-metric space $\mathcal{K}$, with semi-metric $d$. The model is defined by some *Lipschitz* type hypothesis:

$$\exists C > 0, \exists \beta > 0, \forall (x,y) \in \mathcal{K} \times \mathcal{K}, \ |r(x) - r(y)| \leq C d(x,y)^\beta. \tag{5}$$

This is clearly a nonparametric functional model in the sense of the Definition 1.1.

Both to save space and not to shade our main purpose, we will now look only at some fixed point $x \in \mathcal{K}$ (more complete situations will discussed later in 2.4). The sparsness of the data will of course play an important role in the behaviour of any estimate in such an infinite dimensional nonparametric model. To see this clearly, we will introduce the following condition about the concentration of the marginal distribution of the infinite dimensional process $X$, which is a quite unrestrictive assumption and which is the key for proving asymptotic results (see the discussion in Section 4.1). Assume that there exists some real function $\phi_x$ such that:

$$\exists (\beta_1, \beta_2) \in R^{+^2}, \ 0 < \beta_1 \phi_x(h) \leq P\left[d(x,X) < h\right] \leq \beta_2 \phi_x(h). \tag{6}$$

Following the ideas introduced in [21], a natural way for constructing nonparametric estimates of the operator $r$ is to use convolution kernel ideas. More precisely, we introduce the estimate:

$$\hat{r}(x) = \frac{\sum_{i=1}^n Y_i K\left(d(x,X_i)/h\right)}{\sum_{i=1}^n K\left(d(x,X_i)/h\right)}, \tag{7}$$

where $K$ and $h$ are such that:

$K$ has support $[0,1]$, and is strictly decreasing, \hfill (8)

$$\exists (\theta_1, \theta_2), \ \forall t \in [0,1], \ 0 < \theta_1 < K(t) < \theta_2 < \infty, \tag{9}$$

$$\lim_{n \to \infty} h = 0 \ \text{and} \ \lim_{n \to \infty} \frac{\log n}{n \phi_x(h)} = 0. \tag{10}$$

We will stay here with scalar response variable $Y$ (see however the discussion in Section 2.4), and we assume that

$$\exists p \geq 2,\ E|Y|^p < \infty. \tag{11}$$

### 2.3. One selected asymptotic result

We will now present some selected asymptotic result (with rates). We have selected this result, even if it is stated in a slightly more restrictive form than what could be possible, because it is well adapted to our purpose. In one hand, it highlightes the double infinite dimension effects. In the other hand, it allows us to detail the proof in a short way. The earliest result of this type was given in [21] (see also [23] for a recent more general formulation).

**Theorem 2.1** *Under the conditions (5), (6), (8), (9), (10) and (11) we have*

$$\hat{r}(x) - r(x) = O\left(h^\beta\right) + O\left(\sqrt{\frac{\log n}{n\phi_x(h)}}\right),\ a.s. \tag{12}$$

**Proof of Theorem 2.1.** We will use the following notations:

$$\hat{r}_1(x) = \frac{1}{nE\Delta_1(x)} \sum_{i=1}^{n} \Delta_i(x),\ \text{with}\ \Delta_i(x) = K\left(d(x, X_i)/h\right)$$

and

$$\hat{r}_2(x) = \frac{1}{nE\Delta_1(x)} \sum_{i=1}^{n} \Gamma_i(x),\ \text{with}\ \Gamma_i(x) = Y_i K\left(d(x, X_i)/h\right).$$

Because of the decompozition

$$\hat{r}(x) - r(x) = \frac{1}{\hat{r}_1(x)} \left[(\hat{r}_2(x) - E\hat{r}_2(x)) - (r(x) - E\hat{r}_2(x))\right]$$

$$+ \frac{r(x)}{\hat{r}_1(x)} \left[(\hat{r}_1(x) - E\hat{r}_1(x)) - (E\hat{r}_1(x) - 1)\right],$$

and by noting that $E\hat{r}_1(x) = 1$, the claimed result will follow directly from:

$$E\hat{r}_2(x) - r(x) = O\left(h^\beta\right); \tag{13}$$

$$|\hat{r}_2(x) - E\hat{r}_2(x)| = O\left(\sqrt{\frac{\log n}{n\phi_x(h)}}\right),\ a.s.; \tag{14}$$

$$|\hat{r}_1(x) - E\hat{r}_1(x)| = O\left(\sqrt{\frac{\log n}{n\phi_x(h)}}\right),\ a.s. \tag{15}$$

- <u>Proof of (13)</u>. By using the equidistribution of $(X_i, Y_i)$, $i = 1, \ldots n$ we have:

$$E\hat{r}_2(x) = \frac{1}{nE\Delta_1(x)} \sum_{i=1}^{n} E\Gamma_i(x) = \frac{1}{E\Delta_1(x)} E\Gamma_1(x), \tag{16}$$

while in the other hand we have:

$$\begin{aligned} E\Gamma_1(x) &= Er(X_1)K\left(d(x,X_1)/h\right) \\ &= E(r(X_1) - r(x))K\left(d(x,X_1)/h\right) + r(x)E\Delta_1(x). \end{aligned}$$

Using now the Lipschitz condition (5) and the conditions on $K$, we arrive at

$$\begin{aligned} |E\Gamma_1(x) - r(x)E\Delta_1(x)| &\leq E\left(sup_{t\in B(x,h)} |r(x) - r(t)| K\left(d(x,X_1)/h\right)\right) \\ &\leq Ch^\beta E\Delta_1(x). \end{aligned} \quad (17)$$

- Proof of (14). The idea of this proof is based on the application of a Bernstein'type exponential inequality. There are many inequalities of this type available in the literature. One of the most recent is given in formula (6.19a) in [42]. This formula gives directly, for any $\lambda > 0$ and for any $r > 1$:

$$\begin{aligned} P[|\hat{r}_2(x) - E\hat{r}_2(x)| > \lambda] &= P\left[|\sum_{i=1}^n \Gamma_i(x) - E\Gamma_i(x)| > \lambda n\, E\Delta_1(x)\right] \\ &\leq C\left(1 + \frac{C(n\,E\Delta_1(x)\,\lambda)^2}{n\,r\,\mathrm{Var}(\Gamma_1(x))}\right)^{-\frac{r}{2}}. \end{aligned}$$

Note that the conditions on $K$ allow to write that

$$E\Delta_1(x) \geq C\phi_x(h) > 0 \text{ and } E\Gamma_1(x)^2 \geq C\phi_x(h) > 0. \quad (18)$$

So, we can easily find some $\lambda_0 > 0$ and some $\epsilon > 0$ such that:

$$P\left[|\hat{r}_2(x) - E\hat{r}_2(x)| > \lambda_0 \sqrt{\frac{\log n}{n\,\phi_x(h)}}\right] \leq Cn^{-1-\epsilon}. \quad (19)$$

Complete calculus are given in a more general framework in [25].

- Proof of (15). This proof is a direct application of the previous result (14) in the special case when $Y \equiv 1$. •

## 2.4. Discussion on theoretical aspects of functional regression estimation

Let us recall that there exist several other results of the kind of Theorem 2.1. They include for instance uniform version of (12) (see [23,16]), extension of (12) to non necessarily independant pairs $(Y_i, X_i)$ (see for instance Section 3.3 below) or generalization of the proposed model/estimate to the situation when $Y$ is also functional (see [41]). Note also [28] who discusses and compares both linear and nonparametric approaches.

Our purpose here is to centrate our comments on the effects of the double infinite dimensional structure. First of all note that, as usually in nonparametric estimation, the rates of convergence are splitted into two parts. There is a bias component in $O\left(h^\beta\right)$ which depends mainly on the smoothness degree assumed for $r$ and is linked with the infinite dimension structure of the model (5). There is a stochastic component in $O\left(\sqrt{\log n/(n\phi_x(h))}\right)$

which is linked with the concentration of the probability distribution of the variable $X$ on small balls. This link appears clearly by mean of the function $\phi_x(.)$. This is really through this component that the functional nature of the covariate $X$ has an influence, and so the condition (6) plays a crucial role.

This kind of condition will be the subject of the Section 4.1 but let us note here that it is very unrestrictive since even in the special (simplest) situation when $\mathcal{K} = R$, it is less restrictive that the usual assumption of existency of density function (see [45] for relevant literature on this real case). Let us also note that in the classical multivariate (but finite) situation when $\mathcal{K} = R^p$, the condition (6) can be shown (see [23]) to hold under standard hypotheses with $\phi(h) \equiv Ch^d$, and therefore one finds the well-known curse of dimensionality (see [45] for relevant literature on this multivariate real case).

## 2.5. Discussion on applied aspects

For space reasons, it is out of purpose to present here a serious applied study. However, it is worth being noted that the method proposed here has been successfully used in several different kinds of applied problems, including spectrometric data analysis ([23]), economic prediction data ([16]), climatic data ([3]), air quality data ([12,1]) or power ground pollution data ([15]).

It turns out from these applications, that the good behaviour on finite samples of such functional nonparametric method relies on two parameters: the bandwidth and the semimetric. The effects of the bandwidth which are well-known in any nonparametric setting are mainly linked with the functional nature of the model (5), and one has to choose bandwidth balancing trade-off between bias and dispersion. Proper to the functional nature of the variable $X$, and as discussed theoretically in Section 2.4 before, the semimetric $d$ appears in all these examples to be of great importance. Note that, because applications are mainly concerned with variables $X$ which are curves, it is natural to base a measure of proximity between curves on their derivatives and this is hardly possible with usual hilbertian or even metric spaces formalisation (see [23] for a deep discussion and for the construction of some class of semi-metrics). So finally, let us emphasize on the fact that these examples show that there is a real applied support for having developed the theory in abstract semi-metric space rather than only in Hilbert or Banach spaces.

## 3. Other statistical problems of functional nature

### 3.1. Functional nonparametric conditional cumulative distribution estimation

With the same notations as defined in Section 2.1, our functional nonparametric ideas can be used for estimating several others features of the conditional probability measure of $Y$ given $X$. First of all, it is natural to try to estimate the cumulative conditional distribution of $Y$ given $X$, defined as

$$F^x(y) = P[Y \leq y | X = x]. \tag{20}$$

To this end, by combining the functional ideas described before in Section 2.1 for regression setting together with ideas discussed by many authors in finite dimensional cumulative distribution estimation problem (see for instance [43,44]), a kernel type estimate can be defined in the following way:

$$\hat{F}^x(y) = \frac{\sum_{i=1}^{n} K\left(d(x, X_i)/h\right) H\left((y - Y_i)/h_H\right)}{\sum_{i=1}^{n} K\left(d(x, X_i)/h\right)}, \qquad (21)$$

where $K$ and $h$ are defined as before, $h_H$ is a new positive smoothing factor and $H$ is some known cumulative distribution function. In [18] results of the same type as Theorem 2.1 are proved. More precisely, these authors showed that for the functional nonparametric model defined by

$$\forall (y_1, y_2), \forall (x_1, x_2), |F^{x_1}(y_1) - F^{x_2}(y_2)| \leq C_x \left( d(x_1, x_2)^b + |y_1 - y_2|^b \right) \qquad (22)$$

and under the key distribution concentration condition (6) together with suitable restrictions on the kernels and the bandwidths, the following result holds:

$$\hat{F}^x(y) - F^x(y) = O(h^b) + O(h_H^b) + O\left(\sqrt{\frac{\log n}{n\phi_x(h)}}\right), \quad a.s. \qquad (23)$$

This proof performs roughly over the same step as the proof of Theorem 2.1 before. It is given with all details (even under more general formulation) in [18]. These authors had also shown an uniform version of this result from which they get a direct application to conditional functional quantile estimation.

### 3.2. Functional nonparametric conditional density estimation

Another field of research concerns the estimation of the conditional density $f^x(y)$ of this distribution, which can be estimated by mean of the derivative of (21):

$$\hat{f}^x(y) = \frac{h^{-1} \sum_{i=1}^{n} K\left(d(x, X_i)/h\right) H'\left((y - Y_i)/h_H\right)}{\sum_{i=1}^{n} K\left(d(x, X_i)/h\right)}. \qquad (24)$$

Under suitable conditions on the kernels and the bandwidths, and for the functional nonparametric model defined by the following smoothness condition

$$\forall (y_1, y_2), \forall (x_1, x_2), |f^{x_1}(y_1) - f^{x_2}(y_2)| \leq C_x \left( d(x_1, x_2)^b + |y_1 - y_2|^b \right) \qquad (25)$$

and by the key concentration condition (6), the following result holds:

$$|\hat{f}^x(y) - f^x(y)| = O(h^{b_1}) + O(h_H^{b_2}) + O\left(\sqrt{\frac{\log n}{n\, h_H\, \phi_x(h)}}\right), \quad a.s. \qquad (26)$$

This proof performs roughly over the same step as the proof of Theorem 2.1 but by incorporating specific arguments inspired by ideas used in [48] in the finite dimensional situation. The complete proof is in [18]. Note also that these authors had also shown an uniform version of this result from which they get a direct application to conditional functional mode estimation.

## 3.3. Time series prediction

i) Introduction. The problem of predicting future values of a real valued time series can be attacked by nonparametric methods. The usual way to provide theoretical support for that is to extend works existing in regression estimation to dependent data. By such a procedure, and starting with [10] there is a quite large literature for doing that when one takes just a finite number of past values of the process for the prediction of the future (see [4] for recent monography and [17,19,22] for coursebooks in this field). There is of course a natural interest for attacking this problem in a functional way, that is by taking a continuous set of past values as a predictor. This leads to the study of Functional Regression Model but with dependent samples. First works in this direction are well summarized in the recent monography by [6] (see also [2]), and they were only concerning linear processes (see however [7] for first nonparametric ideas in this field). In the following, we want to consider the problem both by considering continuous set of predictive past values and without using any restrictive parametrization for the relationship between past and future. More precisely, we will show how the functional nonparametric regression model studied in Section 2 before can be extended to the dependent case in order to attain this objective.

ii) The model. Let $Z$ be a continuous-time real valued random stationary process observed on a compact set $\mathcal{T}_N$ which may be divided into $N$ intervals with constant width $\tau > 0$, so that $\mathcal{T}_N = [0; N\tau)$. From $Z$, $N$ functional identically distributed random variables $(W_i)_{i=1,\ldots,N}$ can be constructed as follows:

$$\forall t \in [0; \tau), \qquad W_i(t) = Z((i-1)\tau + t), \qquad i = 1, \ldots, N. \tag{27}$$

Given some characteristic of the future of $W_i$ denoted by $G(W_{i+1})$, one way to model the time series $Z$ is given by the following *Functional Nonparametric Time Series Model*:

$$G(W_{i+1}) = R(W_i) + \varepsilon_i, \qquad i = 1, \ldots, N, \tag{28}$$

where $G$ is known, $R$ is unknown (to be estimated) and the $\varepsilon_i$'s are centered dependent errors. In practice, we have to precise $G$, afterthen, many applications can be derived from this model. Indeed, if we take $\forall i$, $G(W_{i+1}) = W_{i+1}(T)$ where $T \in [0; \tau)$, the aim becomes the prediction of future values (see for instance the electricity consumption data application discussed in [16]). If we consider $\forall i$, $G(W_{i+1}) = \sup_{[0;\tau)} W_{i+1}(t)$ we model the case of the forecast of the supremum of some processes (see for instance the problem of predicting peaks of ozone concentration treated in [1]). Another usefull shape for $G$ is given by $G(W_{i+1}) = (W_{i+1}(\tau) - W_{i+1}(0))/W_{i+1}(0)$, which can be interpreted as the problem of predicting the variation rate on each interval, which can be for instance helpful to know the temporary trend of an economical index.

iii) A kernel type estimate. Now, taking $X_i = W_i$ and $Y_i = G(W_{i+1})$, it is obvious that (28) is a regression problem with functional regressor $X_i$, scalar response $Y_i$ and with dependent sample. To study the functional nonparametric time series model, it suffices to characterize the model of dependence between the pairs $\{(X_i, Y_i)\}_i$. One way to do this, is to assume that the sequence $\{(X_i, Y_i)\}_i$ is $\alpha - mixing$. Finally, model (28) is included in the following general *Functional Nonparametric Regression Model for Dependent data*:

$$Y_i = R(X_i) + \epsilon_i, \qquad i = 1, \ldots, N, \tag{29}$$

where the sequence $(X_i, Y_i)_i$ is assumed to be $\alpha - mixing$. Thus, keeping in mind the estimator introduced in Section 2.2, we can deduce easily the estimator $\hat{R}$ of $R$ as follows:

$$\hat{R}(x) = \frac{\sum_{i=1}^{n} G(W_{i+1}) K(d(x, W_i)/h)}{\sum_{i=1}^{n} K(d(x, W_i)/h)}. \tag{30}$$

The recent work [25] is stating asymptotic results for such an estimator. To do this, we need the key concentration condition (6) together with the following similar one:

$$\exists (\eta_1, \eta_2) \in R^{+^2}, \ 0 < \eta_1 \psi_x(h) \leq \sup_{i \neq j} P[\{d(x, W_i) < h\} \cap \{d(x, W_j) < h\}] \leq \eta_2 \psi_x(h). \tag{31}$$

The goal of this assumption is to control the local concentration of the joint distribution. Thus, under (6), (31) and some other hypotheses, we obtain a new version of Theorem 2.1 which can be expressed as follows

$$\hat{R}(x) - R(x) = O\left(h^\beta\right) + O\left(\sqrt{\frac{\log n}{n\phi_x(h)}}\right) + O(V_n), \ a.s., \tag{32}$$

where $\beta$ comes from the Lipschitz type hypothesis (5), $R$ being replaced $r$ and $\phi_x(h)$ comes from (6). Moreover, the term $V_n$ is a sequence which depends on $n$, $\phi_x(h)$, $\psi_x(h)$ and on the mixing structure. The assumption (31) allows us to control the covariance effects.

### 3.4. Curves discrimination

i) Practical motivations. Given a learning sample containing observed functional random variables with known class memberships, our problem is to predict the group membership of a new incoming functional observation. This setting covers many applications. For instance, [32] and more recently [24] proposed a method to solve a phoneme classification problem. Other examples (radar range profiles, spectrometric curves) can be found in [30], [24] and references therein. Because the objects to be classified are evidently of functional nature, it is natural to attack this problem by using a functional framework. Our aim in this section is to present recent advances in this direction.

ii) The nonparametric functional model. The aim is to propose a kernel estimator of the posterior probabilities that an incoming functional observation belongs to a given class. To do this, we consider that the sample $(X_1, Y_1), \ldots, (X_n, Y_n)$ is drawn from the pair $(X, Y)$ where $X$ takes its values in the semi-metric space $(\mathcal{K}, d)$ and $Y$ is a categorical response valued in $\overline{G} = \{1, \ldots, G\}$ (and thus, defining $G$ groups). Given an observed $x$, the purpose is to estimate nonparametrically the following posterior probabilities

$$p_g(x) = P(Y = g | X = x), \quad g \in \overline{G}, \tag{33}$$

and then to assign the new observation $x$ to the class with highest estimated probability.

iii) A kernel functional discrimination approach. Roughly speaking, two approaches can be found in the literature. The first one (see [30]) is an heuristic method. It consists in projecting the functional random variable into a finite dimensional subspace and in using afterthen a multivariate kernel estimator. The second one (see [24]) takes into account the full functional nature of the data and it can be expressed as follows. It is clear that

$$P(Y = g | X = x) = E\left(1_{[Y = g]} | X = x\right), \quad g \in \overline{G}, \tag{34}$$

with $1_{[Y=g]}$ equals to 1 if $Y = g$ and 0 elsewhere. The fact that the posterior probabilities can be formulated in terms of conditional expectation allows us to introduce a kernel type estimator derived from the one proposed in Section 2.2 and using previous work in real case (see e.g. [13]). For all $g \in \overline{G}$ and $x \in \mathcal{K}$, the estimator $\hat{p}_{g,h}(x)$ of $p_g(x)$ is given by

$$\hat{p}_{g,h}(x) = \frac{\sum_{i=1}^{n} 1_{[Y_i=g]} K\left(d(X_i, x)/h\right)}{\sum_{i=1}^{n} K\left(d(X_i, x)/h\right)}. \tag{35}$$

Thus, it is easy to derive from Section 2.3 asymptotic properties for the kernel estimator $\hat{p}_{g,h}$. That means that, under a nonparametric functional model assuming that each function $p_g(.)$ satisfies the following smoothness condition

$$\exists C_g > 0, \exists \beta > 0, \forall (x,y) \in \mathcal{K} \times \mathcal{K}, \ |p_g(x) - p_g(y)| \leq C_g d(x,y)^\beta, \tag{36}$$

under the key concentration condition (6), and under suitable restrictions on $h$ and $K$, the following result holds:

$$\hat{p}_{g,h}(x) - p_g(x) = O\left(h^\beta\right) + O\left(\sqrt{\frac{\log n}{n\phi_x(h)}}\right), \ a.s. \tag{37}$$

This result has not to be proved since it is a direct corollary of Theorem (2.1). The reader will find in [25] a complete study (including real data applications) for this functional nonparametric supervised classification method.

### 3.5. Density estimation

i) *Some applied motivations.* Nonparametric density estimation covers an important field in the nonparametric statistical literature. For instance, given observations spanned by a real unknown distribution, it is of a great interest to detect extremal points, modes, quantiles,..., and same motivations can be invoked in the functional framework. Indeed, given a sample of random curves it could be interesting to estimate "modal curves" or "extremal curves". For instance, let us discuss briefly two recent applied studies in this field. [26] proposed an heuristic nonparametric method to estimate the mode of a distribution of random curves. Using a kernel type estimator for the density function, their procedure is applied to "Female shoulder growth velocities" data in relation with the age. In particular, they detected two modes, which allowed to emphasize two clusters (see [29] for other application concerning rainfall data). These two papers show that modal curves (and hence the density function of the curves distribution) are successful tools for classify a curves sample.

ii) *The theoretical framework.* Almost simultaneously, theoretical aspects emerged (note however the previous works [27,34]). Before to go on, we introduce some notations. So, let $\mu$ be a $\sigma$-finite measure defined on $\mathcal{K}$ such that $0 < \mu(B) < \infty$ for all opened ball $B \subset \mathcal{K}$. Let $X_1, \ldots, X_n$ be $n$ functional random variables identically and independently distributed as a variable $X$ which takes its values in $\mathcal{K}$. From now on, we assume that $X$ has a density $f$ relatively to $\mu$ and the main goal is to estimate $f$. The crucial point here concerns the existence of such a density $f$, and more precisely the choice of the dominating measure $\mu$ defined on an infinite dimensional space (see Chapter 1 in [37] for discussion and references about this problem). This point is the reason why advances in density estimation are much less developped than in other fields. However if we have to deal with a diffusion process, and as pointed out in [36] recall that, it is known that under some general assumptions (see [35]) such a process has a density with respect to the Wiener measure.

iii) A kernel type estimate. After this essential theoretical consideration, we can focus on a kernel type estimator $\hat{f}$ of the density $f$ which can be defined as

$$\hat{f}(x) = \frac{1}{n\,\gamma_n(x)} \sum_{i=1}^{n} K_n\left(d(x, X_i)\right), \tag{38}$$

where $\{\gamma_n(x)\}_n$ is a sequence of positive real numbers depending on $x$ and $K_n$ is a positive function defined on $R^+$.

In the case of $\gamma_n(x) = h^{\delta(x)}$, $K_n(u) = K(u/h)$ with $\delta(x) > 0$, a first study of the asymptotic behaviour of the quantity (38) has been obtained by [21] as by-products of similar study but in a regression framework. Note that in [26] an analogous kernel type estimator has been introduced with $\gamma_n(x) = h^{\alpha+1} \int u^\alpha K(u) du$ for some $\alpha > 0$ and $d(x, X_i) = \|x - X_i\|$, where $\|.\|$ is a norm but without any asymptotic result with respect to $\hat{f}$.

The first asymptotic results in the general density context are due to [36] without any specification for $\gamma_n(x)$. It is out of purpose here to present in details these results and the technical assumptions accompanying it, but it should be noted that the general expression of the rates of convergence is of the same kind as those presented in Theorem 2.1 for regression, with a bias component depending on additional smoothness restrictions on $f$ and a stochastic component which is linked with the concentration of the distribution of $X$ around small balls (and this, because existency of density, can be quantified in terms of the concentration properties of the dominant measure $\mu$).

## 4. Concluding section

We conclude this overview about functional nonparametric statistics in two parts. The first one focuses on the crucial assumptions needed for the distribution defined on an infinite dimensional space. Indeed, these conditions allow us to override the so-called "curse of dimensionality". The second part proposes many open problems which promise a very long life to functional nonparametric statistics.

### 4.1. Distribution conditions in infinite dimensional space

In all the theoritical parts, it appears always a quantity which plays a major role, namely $P(d(x, X) < \rho)$. More precisely, when we make a pointwise estimation at $x$, we explicit the local behaviour of the distribution through assumptions on $P(d(x, X) < \rho)$. Different kinds of hypotheses concerning this probability can be found in the recent literature. For instance, as explained in Section 2.2, condition (6) can be used and in the context of time series, the similar hypothesis (31) for the joint distribution has been introduced. However, in order to improve the rate of convergence, a useful tool is to consider fractal dimension type assumptions for $X$, which concerns, once again, the probability $P(d(x, X) < \rho)$. In [26], the authors assume that $P(d(x, X) < \rho) \sim C_x \rho^\delta$ for some $\delta > 0$ and where $d(x, X) = \|x - X\|$ for some norm $\|.\|$. Independently and for different purpose, [21] consider the following local version of this assumption: $P(d(x, X) < \rho) \sim C_x \rho^{\delta(x)}$. According to [38], this last hypothesis coincides with the pointwise fractal dimension of the distribution of $X$. Moreover, [21] gives an other fractal type condition $P(d(x,X) < \rho) = C_x \rho^{\delta(x)} + O\left(\rho^{\delta(x)+\zeta(x)}\right)$ in order to improve the accuracy of the rate of convergence. In addition, similar assumptions can be formulated for the joint distribution in the time series framework (see [16]). All these fractal type conditions are useful ways to explain the fact that any estimation at $x$ needs a sufficiently large number of observations around $x$. An underlying but crucial problem in all these assumptions concerns the choice of the semi-metric $d$ which always appears in the quantity $P(d(x,X) < \rho)$. Indeed, a good choice of $d$ could change the concentration property of $X$, and then improve the rate of convergence, overriding by this way the curse of dimensionality. Of course it is very difficult to give a general solution for choosing $d$. Nevertheless, a first partial solution can be found in [23] and [24]. It consists in parametrizing by an integer some family of semi-metrics. Another way to build in a systematic way some adapted semi-metric is to consider in the regression context the functional single index model (see [20]). The main difference with the above mentionned methods is that here, the family of semi-metric is indexed by a functional parameter. These are promising startpoints to developp more general methods.

## 4.2. Open problems

As pointed out just before, excepted the concentration property of the probability, most other problems are the same as in well-known nonparametric finite dimensional setting. Because of these similarities it is reasonable to hope that in the next future these problems can find reasonable solutions inspired from what nonparametric statisticians know perfectly now in finite dimension. The aim of this short concluding section is to clear the way for future researchers by listing some key open problems and by proposing some ideas of solution.

The first remark concerns the kind of smoothness condition imposed in the above presented results on the object to be estimated, and which is always some Lipschitz type condition. There is obviously necessity to develop similar theory under other smoothness models (such for instance existency of derivatives, in a sense to be precised, for the functional operator to be estimated). It could be hoped that under such type of models, all the results presented before could be extended with precise evaluation of the constants involved in the rates of convergence. Such a precision will be of particular importance for many applied purpose such as, for instance, the construction of confidence areas (by mean of some asymptotic normality result) or the selection of the parameters of the estimates (bandwidth, kernel and semi-metric). Of course other possibilities could be developped for selecting these parameters, such as the usual cross-validation (see for instance [31]) that should be easily adpated to our functional context.

At the actual state of art in this field, most of existing results are concerning kernel type estimates, and this is the reason we just discussed such estimates here. The only exception we know is the density estimation context, for which [36] gives some asymptotics results for some orthogonal series type density estimate. There are really challenging open problems in the next future concerning the construction and the study of alternative estimates in any of the above discussed functional estimation problems. For instance, by taking inspiration on what exists in the finite dimensional case (see [14]), it should certainly be possible to construct some kind of functional local polynomial theory.

Finally, it may be the case that in practical situations one has to deal with more than one functional variable. Of course, the theory presented before in general abstract semi-metric spaces would apply in such a multi-functional problem, but the rates of convergence could be spoiled by high number of variables. This is what is known in multi (but finite) dimensional problems as the curse of dimensionality. Here also is a challenging open problem, whose resolution would open the way for many interesting practical applications. It seems to us that, according to the general discussion in Section 4.1, this *Curse of Multi-Functionality* will appear through some concentration hypothesis and a possible way to attack it would be to developp some functional version of the existing finite dimensional reduction dimension models. Previous works in this direction are involving [20] about some single functional index model, and [1] about practical application of some functional additive regression model.

## REFERENCES

1. G. Aneiros Pérez, H. Cardot, G. Estévez Pérez and Ph. Vieu, Maximum ozone concentration forecasting by functional nonparametric approaches, Proceedings of TIES 2002 meeting, Genova Italia, 2002.
2. A. Antoniadis, Autoregressive processes in Hilbert spaces, Contribution to the International Conference on Current Advances and Trends in Nonparametric Statistics, Crete July 15-19, 2002.
3. Ph. Besse, H. Cardot and D. Stephenson, Autoregressive forecasting of some functional climatic variations, Scandinavian Journal of Statistics, No. 27 673 687 (2000).
4. D. Bosq, Nonparametric Statistics for Stochastic Processes: Estimation and Prediction (2nd edition), Lecture Notes in Statistics, 110, 1998.
5. D. Bosq, Modelization, nonparametric estimation and prediction for continuous time processes, In Nonparametric functional estimation and related topics (Ed. G.G. Roussas), NATO ASI Series Kluwer Academic Publishers, 1990.
6. D. Bosq, Linear Processes in Function Spaces: Theory and Applications, Lecture Notes in Statistics 149, Springer-Verlag, New-York, 2000.
7. D. Bosq and M. Delecroix, Nonparametric prediction of a Hilbert space valued random variable, Stochatic Process Appl. No. 19 271 280 (1985).
8. H., Cardot, F. Ferraty, A. Mas and P. Sarda, Testing Hypotheses in the Functional Linear Model, Scandinavian Journal of Statistics, in print, 2002.
9. H., Cardot, F. Ferraty and P. Sarda, Spline Estimators for the Functional Linear Model, Preprint, 2003.
10. G. Collomb, Propriétés de convergence presque complète du prédicteur à noyau, Z. Wahrsch. verw. Geb. No. 66, 441 460 (1984).
11. A. Cuevas, M. Febrero and R. Fraiman, Linear function regression: the case of fixed design and functional response, Canad. J. of Statistics, No. 30 285 300 (2002).
12. J. Damon and S. Guillas, The inclusion of exogenous variables in functional autoregressive ozone forecasting, Environmetrics, No. 13 759 774 (2002).
13. L. Devroye and T.J. Wagner, Distribution-free consistency results in noparametric discrimination and regression function estimation, Ann. of Statist, 8 231 239 (1980).
14. J. Fan and I. Gijbels, Local polynomial modeling and its applications, Chapman and

Hall, London, 1996.
15. B. Fernández de Castro, S. Guillas and W. González Manteiga, Functional samples and bootstrap for the prediction of SO2 levels, Preprint (2003).
16. F. Ferraty, A. Goia and Ph. Vieu, Functional model for time series: a fractal approach to dimension reduction. Test, No. 11 317 344 (2002).
17. F. Ferraty, A. Goia and Ph. Vieu, Statistica Funzionale: modelli di regressione non-parametrici (in italian), Collana Statistica, Franco-Angeli, Milano, 2002.
18. F. Ferraty, A. Laksaci and Ph. Vieu, Estimating some characteristics of the conditional distribution in nonparametric functional models, Prepint (2003).
19. F. Ferraty, V. Núñez Antón and Ph. Vieu, Regresión no paramétrica: desde la dimensión uno hasta la dimensión infinita (in spanish), Serv. Edit. Univers. País Vasco, Bilbao, 2001.
20. F. Ferraty, A. Peuch and Ph. Vieu, Modèle à indice fonctionnel simple (in french), Preprint (2003).
21. F. Ferraty and Ph. Vieu, Dimension fractale et estimation de la régression dans des espaces vectoriels semi-normés (in french), C.R. Acad. Sci. Paris, No. 330 403 406 (2000).
22. F. Ferraty and Ph. Vieu, Statistique Fonctionnelle: Modèles de régression pour variables aléatoires uni, multi et infiniment dimensionées (in french), Coursebook at Univ. P. Sabatier, Toulouse, France, Free access on line at http://www.lsp.ups-tlse.fr/Fp/Ferraty/staph.html, 2001.
23. F. Ferraty and Ph. Vieu, The functional nonparametric model and application to spectrometric data, Computational Statistics, No. 17 545 564 (2002).
24. F. Ferraty and Ph. Vieu, Curves discrimination: a nonparametric functional approach, Computational Statistics & Data Analysis, In print, (2003).
25. F. Ferraty and Ph. Vieu, Nonparametric models for functional data, with applications in regression, time series prediction and curves discrimination. J. Nonparametric Statistics, In print (2003).
26. T. Gasser, P. Hall and B. Presnell, Nonparametric estimation of the mode of a distribution of random curves. J. R. Statist. Soc. B, No. 60 681 691 (1998).
27. J. Geffroy, Estimation de la densité dans une espace métrique, C.R. Acad. Sci. Paris, No. 278 1149 1452 (1974).
28. A. Goia, Contribution à l'étude des modèles de régression pour variables aléatoires fonctionnelles (in french), PhD Toulouse III, 2003.
29. P. Hall and N. Heckman, Estimating and depicting the structure of a distribution of random functions, Biometrika, No. 89 145 158 (2002).
30. P. Hall, P. Poskitt and D. Presnell, A functional data-analytic approach to signal discrimination, Technometrics, No. 43 1 9 (2001).
31. W. Härdle and J. Marron, Optimal bandwidth selection, Ann. of Statist., No. 13(4) 1465 1487 (1985).
32. T. Hastie, A. Buja and R. Tibshirani, Penalized discriminant analysis. Ann. Statist., 23 73 102 (1995).
33. T. Hastie and C. Mallows, Discuss. Technometrics No. 35 140 143 (1993).
34. P. Jacob and P. Oliveira, General approach to nonparametric histogram estimation, Statistics, No. 27 73 92 (1995).

35. R. Lipster and A. Shiryayev, On the absolute continuity of measures corresponding to processes of diffusion type relative to a Wiener measure. Izv Akad Nauk SSSR Ser Mat, No. 36 4 (1972).
36. S. Niang, Estimation de la densité en dimension infinie: Application aux processus de type diffusion (in french), C.R. Acad. Sci. Paris, No. 334 213 216 (2002).
37. S. Niang, Sur l'estimation de la densité en dimension infinie: application aux diffusions (in french), PhD Paris VI, 2002.
38. Ya.B. Pesin, On rigourous mathematical definitions of correlation dimension and generalized spectrum for dimensions, J. Statis. Phys., 71 529 547 (1993)
39. J. Ramsay and B. Silverman, Functional Data Analysis, Springer-Verlag, New-York, 1997.
40. J. Ramsay and B. Silverman, Applied functional data analysis: Methods and case studies, Springer-Verlag, New York, 2002.
41. N. Rhomari, Kernel regression in Banach space, Contribution to the International Conference on Current Advances and Trends in Nonparametric Statistics, Crete July 15-19, 2002.
42. E. Rio, Théorie asymptotique des processus faiblement dépendants (in french), SMAI Math & Applications 31 Springer, 1999.
43. G. Roussas, Nonparametric estimation of the transition distribution function of a Markov process, Ann. Math. Statist. No. 40 1386 1400 (1969).
44. M. Samanta, Non-parametric estimation of conditional quantiles, Statist. & Proba. Lett. No. 7 407 412 (1989).
45. M. Schimek (ed.), Smoothing and regression: Approaches, Computation, and Application, Wiley Series in Probability and Statistics, Wiley, New-York, 2000.
46. Staph, Birth and firts activities of the working group in Functional Statistics, Tech. Report (see http://www.lsp.ups-tlse.fr/Fp/Ferraty/staph.html), Univ. P. Sabatier, Toulouse, France, No. 05 (2001).
47. Staph, Proceedings of the working group in Functional and Operatorial Statistics, Tech. Report (see http://www.lsp.ups-tlse.fr/Fp/Ferraty/staph.html), Univ. P. Sabatier, Toulouse, France, No. 12 (2002).
48. E. Youndjé, Propriétés de convergence de l'estimateur à noyau de la densité conditionnelle (in french), Rev. Roumaine Math. Pures Appl. No. 41 535 566 (1996).

# 3. Nonparametric Model Building

Recent Advances and Trends in Nonparametric Statistics
Michael G. Akritas and Dimitris N. Politis (Editors)
© 2003 Elsevier Science B.V. All rights reserved.

# Nonparametric Models for ANOVA and ANCOVA: A Review

Michael G. Akritas[a] and Edgar Brunner[b]*

[a]Department of Statistics, Penn State University, 326 Thomas Bldg.,
University Park, PA 16802, USA, email: mga@stat.psu.edu

[b]Department of Medical Statistics, University of Göttingen, Humboldtallee 32,
D-37073 Göttingen, Germany, email: brunner@ams.med.uni-goettingen.de

Recently developed nonparametric models, hypotheses and test statistics for ANOVA and ANCOVA designs with independent and dependent ordinal data (continuous or not) are reviewed and discussed.

## 1. INTRODUCTION

Univariate and multivariate analysis of variance (ANOVA and MANOVA), as well as analysis of covariance (ANCOVA) form cornerstones of applied statistics. They are used in many disciplines (biological, medical, social or psychological studies) for addressing complex questions about the effects of factors and covariates and about the interaction between them. The data from such studies can be either independent or dependent (longitudinal studies). In two critical ways, the logic of the classical approach requires that the response variable has units of measurement that are equal (or have comparable meaning) across its entire range. First, the parameters of the model that constitute the effects being estimated or tested are defined in terms of mean values on that metric. Second, the desirable statistical properties of least squares estimators depend on assumptions about the distribution of the residuals (normality, homogeneity, and independence), and those residuals are defined by differences in this metric.

Unfortunately, measurements in many applied sciences rarely justify the assumption of an equal interval scale. For example, measures in the social sciences generally have arbitrary metrics because the theoretical processes being studied are specified without regard to any particular scale of measurement. As a consequence, applied scientists increasingly turn to parametric and semi-parametric alternatives to classical least squares statistics, such as generalized linear models, generalized additive models, and frailty models. These techniques avoid those assumptions of the classical model that are overly inconsistent with the nature of the dependent variable. For instance, ordinal regression with a logit link function (Winship & Mare [44]) provides a model that is plausible for Likert-type rating scales, improving on the classical model by assuming a discrete distribution for the response variable in place of a continuous error distribution and by using the logit link to avoid the possibility of fitted values falling outside the range of the response scale.

---
*Supported in part by DFG/Br-655/12

It must not be forgotten, however, that the parametric and semi-parametric alternatives also depend on assumptions, and those assumptions may or may not be correct for any given application. In the case of the ordinal model with a logit link, the logistic function is assumed to characterize the relationship between the linear form of the model and the response probabilities, and relationships to the independent variables are assumed to be homogeneous across response categories and consistent with the response thresholds corresponding to the logit link. There are several examples in the published literature where different models have produced conflicting results.

In this article we review recent developments in statistical modelling that are completely nonparametric (also nonlinear and nonadditive). Because this approach defines the effects of interest nonparametrically, in a manner general across possible transformation of the response scale, it is well suited to the ambiguity of scaling typical in many applied sciences. It further avoids assumptions about the form and consistency of distributions of the response variable, which applied scientists are rarely in a position to specify in advance.

The basic nonparametric models for ANOVA and ANCOVA with independent data are reviewed in Section 2. Section 3 presents the extension of the ANOVA model to repeated measures designs. Related work is discussed in Section 5, while other extensions are briefly outlined in Section 4.

## 2. THE BASIC NONPARAMETRIC MODELS

In this section we present the nonparametric models for two-way ANOVA and one-way ANCOVA. These are the simplest designs where all features of the nonparametric modelling can be appreciated. Following the idea of Ruymgaart [37], all distribution functions (so also conditional ones) in this paper are taken as the average of their left- and right-continuous version, $F(x) = \frac{1}{2}[F^+(x)+F^-(x)]$. This permits a unified presentation of the model and test procedures for all types of data. For the two-way ANOVA design, the observations will be denoted by $Y_{ijk}$, $i = 1, \ldots, I, j = 1, \ldots, J, k = 1, \ldots, n_{ij}$. For the one-way ANCOVA design, the observations will be denoted by $(Y_{ij}, X_{ij})$, $i = 1, \ldots, I, j = 1, \ldots, n_i$, where $X_{ij}$ is the covariate, $Y_{ij}$ is the response, $i$ enumerates the factor levels and $j$ the observations within each factor level.

The nonparametric model for the two-way ANOVA design specifies only that

$$Y_{ijk} \sim F_{ij}, \qquad (2.1)$$

for some distribution function $F_{ij}$ (Akritas & Arnold [4], Akritas, Arnold & Brunner [6]).

The nonparametric model for the one-way ANCOVA design specifies only

$$Y_{ij}|X_{ij} = x \sim F_{ix}, \qquad (2.2)$$

i.e., that conditionally on $X_{ij} = x$, $Y_{ij}$ has distribution function that depends on $i$ and $x$ (Akritas, Arnold & Du [7]). Note that models (2.1) and (2.2) do not specify how the response distribution changes when the levels, or covariate value changes. Thus they are completely nonparametric (also nonlinear and nonadditive).

For the two-way ANOVA design set $\overline{F}_{i\cdot}(y) = J^{-1}\sum_j F_{ij}(y)$, and $\overline{F}_{\cdot j}(y) = I^{-1}\sum_i F_{ij}(y)$. For the one-way ANCOVA design, choose a distribution function $G(x)$ and let

$$\overline{F}_{i\cdot}^G(y) = \int F_{ix}(y)\, dG(x), \text{ and } \overline{F}_{\cdot x}(y) = \frac{1}{k}\sum_{i=1}^{k} F_{ix}(y). \qquad (2.3)$$

If the $X_{ij}$ are a random sample, we can choose $G(x)$ to be the overall distribution function of the covariate. Thus, if the covariate has the same distribution in all groups, $\overline{F}_{i.}(y)$ is the marginal distribution function of $Y_{ij}$. The hypotheses of interest in model (2.1) or (2.2) are:

1. The $\overline{F}_{i.}(y)$ do not depend on $i$, or $\overline{F}_{i.}^G$ do not depend on $i$ (no main effect);

2. The $\overline{F}_{.j}(y)$ do not depend on $j$, or $\overline{F}_{.x}(y)$ do not depend on $x$ (no main effect);

3. The $F_{ij}(y) = \overline{F}_{i.}(y) + K_j(y)$, or $F_{ix}(y) = \overline{F}_{i.}^G(y) + K_x(y)$ (no interaction);

4. $F_{ij}(y)$ is independent of $i$, or $F_{ix}(y)$ is independent of $i$ (no simple effect);

5. $F_{ij}(y)$ is independent of $j$, or $F_{ix}(y)$ is independent of $x$ (no simple effect).

Note that for the ANCOVA setting, the first hypothesis is sensible even when the model is not additive (i.e., even when the slopes are not equal in the classical case), while the third hypothesis is the generalization of testing the equality of slopes in the classical model above. An important advantage of the non-parametric models is that these hypotheses and the procedures we suggest for analyzing them are unchanged by monotone transformations in the response. In the classical model, it is often necessary to find an appropriate transformation which simultaneously linearizes the expectation and equalizes the variances. In the non-parametric model such a transformation is not necessary.

The above hypotheses can also be described in terms of corresponding nonparametric effects being zero. These nonparametric effects are defined from decompositions of $F_{ij}$ and $F_{ix}$. The decomposition of $F_{ij}$ (Akritas & Arnold [4]) is

$$F_{ij}(y) = M(y) + A_i(y) + B_j(y) + C_{ij}(y). \tag{2.4}$$

where $M = \overline{F}_{..}$, $A_i = \overline{F}_{i.} - M$, $B_j = \overline{F}_{.j} - M$, and $C_{ij} = F_{ij} - \overline{F}_{i.} - \overline{F}_{.j} + M$. $A_i$, $B_j$, and $C_{ij}$ are, respectively, the nonparametric main row, main column, and interaction effects. The decomposition of $F_{ix}$ (Akritas, Arnold & Du [7]) is

$$F_{ix}(y) = M^G(y) + A_i^G(y) + B_x^G(y) + C_{ix}^G(y). \tag{2.5}$$

where $M^G(y) = I^{-1}\sum_{i=1}^{I} \int F_{ix}(y)dG(x)$, $A_i^G(y) = \overline{F}_{i.}^G(y) - M(y)$, $B_x^G(y) = \overline{F}_{.x}(y) - M^G(y)$, and $C_{ix}^G(y) = F_{ix}(y) - M^G(y) - A_i^G(y) - B_x^G(y)$ are, respectively, the nonparametric main factor, main covariate, and interaction effects.

Because decomposition (2.4) bears close resemblance to the decomposition of a rectangular array of means, $\mu_{ij}$, it is straight forward to see how nonparametric effects are defined in higher-way layouts. A decomposition for two-way ANCOVA that displays the covariate adjusted main effects and interactions of the two factors is easily obtained from (2.5) by replacing the single index $i$ by $(i,j)$ and further decomposing $A_{(ij)}^G(y)$. This yields

$$F_{ijx}(y) = M^G(y) + A_i^G(y) + B_j^G(y) + (AB)_{ij}^G(y) + D_x^G(y) + C_{ijx}^G(y), \tag{2.6}$$

where $M^G(y) = I^{-1}J^{-1}\sum_{i=1}^{I}\sum_{j=1}^{J} \int F_{ijx}(y)dG(x)$, $A_i^G(y) = J^{-1}\sum_{j=1}^{J} \int F_{ijx}(y)dG(x) - M^G(y)$, $B_j^G(y) = I^{-1}\sum_{i=1}^{I} \int F_{ijx}(y)dG(x) - M^G(y)$, $(AB)_{ij}^G = \int F_{ijx}(y)dG(x) - A_i^G(y) - $

$B_j^G(y) + M^G(y)$, $D_x^G(y) = I^{-1}J^{-1}\sum_{i=1}^{I}\sum_{j=1}^{J} F_{ijx}(y) - M^G(y)$, and $C_{ijx}^G(y) = F_{ijx}(y) - M^G(y) - A_i^G(y) - B_j^G(y) - (AB)_{ij}^G(y) - D_x^G(y)$.

Estimates of the nonparametric effects provide useful graphical quantification of the effects; see Du, Akritas, Arnold & Osgood [27] for some such plots. Akritas & Arnold [4], Akritas, Arnold & Brunner [6], and Akritas, Arnold & Du [7] establish relations between the nonparametric hypotheses and the corresponding parametric ones. The basic result is that the nonparametric hypotheses are stronger than their parametric counterparts in the sense that they imply but are not implied by them. For example, Akritas, Arnold & Brunner [6] show the following

PROPOSITION 2.1 *The nonparametric hypothesis of no interaction is equivalent to the statement that, for any monotone transformation t, $Et(Y_{ijk}) = \mu_t + \alpha_{t,i} + \beta_{t,j}$.*

Thus the nonparametric hypothesis of no interaction is equivalent to the statement that the mean of any transformation of the response can be decomposed in an additive fashion. This strong form of additivity captures the *substantive* meaning of no interaction between factors as a scientist might think of it.

Because the nonparametric hypotheses are invariant under monotone transformations of the response, it is natural to use (mid-)rank test statistics. In addition, rank statistics are robust against outliers and have good power properties. A general theory of testing hypotheses in two- and higher-way ANOVA designs is possible from the observation that all hypotheses can be expressed as $\boldsymbol{CF} = 0$, where $\boldsymbol{C}$ is a contrast matrix and $\boldsymbol{F}$ the column vector of cell distribution functions. For example in two-way ANOVA, $\boldsymbol{F} = (F_{11}, \ldots, F_{1J}, \ldots, F_{I1}, \ldots, F_{IJ})'$. An easy way to generate the appropriate contrast matrix for each hypothesis is given in Akritas, Arnold & Brunner [6]. The statistic for testing nonparametric hypothesis $\boldsymbol{CF} = 0$ is based on the asymptotic distribution of

$$\widehat{\boldsymbol{T}}_C = \boldsymbol{C}\int \widehat{H}\,d\widehat{\boldsymbol{F}}, \tag{2.7}$$

where $\widehat{\boldsymbol{F}}$ is the empirical estimator $\boldsymbol{F}$, i.e. the vector made up of the empirical distribution functions, and $\widehat{H}$ is the combined empirical distribution function. Note that the empirical distribution functions are also defined here as the average of their left- and right-continuous versions. Because the overall rank of $Y_{ijk}$ among all $N$ observations is given by $R_{ijk} = \frac{1}{2} + N\widehat{H}(Y_{ijk})$,

$$\boldsymbol{C}\int \widehat{H}\,d\widehat{\boldsymbol{F}} = N^{-1}\boldsymbol{C}\left(R_{11\cdot} - \frac{1}{2}, \ldots, R_{IJ\cdot} - \frac{1}{2}\right)' = N^{-1}\boldsymbol{C}(R_{11\cdot}, \ldots, R_{IJ\cdot})'.$$

Thus (2.7) is a vector of linear (mid-)rank statistics. Let $\boldsymbol{V} = \mathrm{Diag}(\lambda_{11}^{-1}\sigma_{11}^2, \ldots, \lambda_{ab}^{-1}\sigma_{ab}^2)$, with $\lambda_{ij} = n_{ij}/N$, $\sigma_{ij}^2 = \mathrm{Var}[H(Y_{ijk})]$ for all $k = 1, \ldots, n_{ij}$, and $\widehat{\boldsymbol{V}}$ denote the matrix $\boldsymbol{V}$ with $\sigma_{ij}^2$ replaced by

$$\widehat{\sigma}_{ij}^2 = N^{-2}(n_{ij} - 1)^{-1}\sum_{k=1}^{n_{ij}}(R_{ijk} - R_{ij\cdot})^2. \tag{2.8}$$

Then, Akritas, Arnold & Brunner [6] show

THEOREM 2.2 *Under the hypothesis $\boldsymbol{CF} = 0$, where $\boldsymbol{C}$ has full row rank, the test statistic*

$$Q(\boldsymbol{C}) = N\widehat{\boldsymbol{T}}'_C(\boldsymbol{C}\widehat{\boldsymbol{V}}\boldsymbol{C}')^{-1}\widehat{\boldsymbol{T}}_C \tag{2.9}$$

has, as $N \to \infty$, a $\chi^2_r$ distribution, where $\chi^2_r$ denotes the chi-square distribution with $r$ degrees of freedom, and $r$ is the number of rows of $\boldsymbol{C}$.

It should be noted that the above quadratic form can be computed with SAS, S-plus, Minitab, or any other higher level software system with matrix operation capabilities. Finally, Akritas, Arnold & Brunner [6] describe a finite-sample correction procedure that works remarkably well with small sample sizes. Such a correction is decribed in the more general context of Section 3.

For ANCOVA, we will only discuss testing the first type of hypotheses given below (2.3). This includes also testing for covariate adjusted main effects and interactions between factors in higher-way ANCOVA designs with one covariate. Testing procedures for the other type of hypotheses have been developed much more recently and will be briefly described in Section 4.

To describe the test statistics set $\overline{\boldsymbol{F}}^G_{\cdot} = (\overline{F}^G_{1\cdot}, \ldots, \overline{F}^G_{I\cdot})'$. Then any of the nonparametric hypotheses in one- two- and higher-way ANCOVA designs is of the form

$$H_0 : \boldsymbol{C}\overline{\boldsymbol{F}}^G_{\cdot} = \boldsymbol{0}. \tag{2.10}$$

for some full-rank contrast matrix $\boldsymbol{C}$. The statistic for testing such a hypothesis is based on

$$\widehat{T}^G_C = \boldsymbol{C} \int \widehat{H} d\widehat{\overline{\boldsymbol{F}}}^G_{\cdot}, \tag{2.11}$$

where $\widehat{H}$ is the empirical distribution function of all $Y_{ij}$, $\widehat{\overline{\boldsymbol{F}}}^G_{\cdot} = (\widehat{\overline{F}}^G_{1\cdot}(y), \ldots, \widehat{\overline{F}}^G_{I\cdot}(y))'$, and

$$\widehat{\overline{F}}^G_{i\cdot}(y) = \int \widehat{F}_{ix}(y) d\widehat{G}(x), \text{ with } \widehat{F}_{ix}(y) = \sum_{j=1}^{n_i} \frac{K\left(\frac{x-X_{ij}}{a_{n_i}}\right)}{\sum_{j'} K\left(\frac{x-X_{ij'}}{a_{n_i}}\right)} I(Y_{ij} \leq y), \tag{2.12}$$

where $\widehat{G}$ is the empirical distribution function of all $X_{ij}$, and $K$ is a known probability density function (kernel). Note that (2.11) is a vector of weighted mid-rank statistics. Let $\widehat{\boldsymbol{V}}^G$ denote the estimate of the asymptotic covariance matrix of $\int \widehat{H} d\widehat{\overline{\boldsymbol{F}}}^G_{\cdot}$ given in Akritas, Arnold & Du [7]. Then in the aforementioned paper it is shown that under suitable smoothness assumptions

THEOREM 2.3 *Let $r$ denote the rank of $\boldsymbol{C}$. Under the null hypothesis (2.10),*

$$N\left(\widehat{T}^G_C\right)'\left(\boldsymbol{C}\widehat{\boldsymbol{V}}^G\boldsymbol{C}'\right)^{-1}\widehat{T}^G_C \to \chi^2_r, \text{ in distribution.}$$

Note that even though the kernel estimator of the conditional distribution that enters (2.11) has a slower rate of convergence, the test statistic does have the usual square root of $N$ rate of convergence.

We finish this subsection with a biological illustration (taken from Akritas, Arnold and Brunner (1996)) of the (mid-)rank procedures for testing the nonparametric hypotheses with independent data.

**Example 1.**

Two inhalable test substances, drug 1 and drug 2 (factor $A$), are to be compared with regard to their irritative activity in the respiratory tract of the rat after subchronic inhalative exposure. Reserve cell hyperplasia in the respiratory epithelium of the nose after exposure to 2, 5, and

Table 1
Data and rank means for the reserve cell hyperplasia trial.

| | Test substance | | | | | | | | Results | | |
|---|---|---|---|---|---|---|---|---|---|---|---|
| | drug 1 | | | | drug 2 | | | | | | |
| | no. of rats with scale | | | | no. of rats with scale | | | | Rank Means $R_{ij.}$ | | $\widetilde{R}_{.j.}$ |
| Concentration | 0 | 1 | 2 | 3 | 0 | 1 | 2 | 3 | drug 1 | drug 2 | |
| 2 | 18 | 2 | 0 | 0 | 16 | 3 | 1 | 0 | 33.95 | 39.68 | 36.81 |
| 5 | 12 | 6 | 2 | 0 | 8 | 8 | 3 | 1 | 49.85 | 62.08 | 55.96 |
| 10 | 3 | 7 | 6 | 4 | 1 | 5 | 8 | 6 | 83.18 | 94.28 | 88.73 |
| | | | | | | | | | 55.66 | 65.34 | $\widetilde{R}_{i..}$ |

$10[ppm]$ of the test substances served as a criterion for irritation. The result was histopathologically evaluated by the grading scales: 0 = 'no changes', 1 = 'slight changes', 2 = 'distinct changes', 3 = 'severe changes'. The results for the two drugs and three concentration groups (factor B) with 20 rats each, are given in Table 1.

Mid-rank statistics of the form described in (2.9) below where used to analyze this data set. The test for interaction gave $p$-value 0.77 indicating that the results are quite homogeneous within the three concentrations (no interaction). A significant treatment effect for the drug is proved at the 5% level ($p = 0.028$). For the concentration a highly significant effect is proved ($p < 10^{-7}$). We remark that in this example, both the asymptotic and small sample approximations give quite similar $p$-values.

## 3. REPEATED MEASURES

As mentioned in the Section 1, in many biological experiments and medical or psychological studies, randomly chosen subjects are observed repeatedly under the same or under different treatments. Such designs include growth curves, longitudinal data or repeated measures designs. For the analysis of such designs there exists a variety of classical linear models assuming the multivariate normality of the observed random vectors; a special structure for the dependencies of the multivariate observations, e.g. compound symmetry, may or may not be assumed. In this section we describe nonparametric generalizations of the classical ANOVA and MANOVA procedures where not only the assumption of normality of the error terms is relaxed but also treatment effects and hypotheses are defined in a nonparametric framework. Based on the ideas explained in Section 2, we formulate nonparametric hypotheses in various designs and derive (mid-)rank statistics for testing these hypotheses. We re-emphasize that the procedures to be described are applicable to all ordinal data including discrete such as scores in psychological tests, and quality scales in order to describe the degree of the damage of plants or trees in ecological or environmental studies.

Akritas and Brunner [8] consider a general class of repeated measures designs with some factors having subjects nested within their levels (whole-plot factors) and other factors being crossed with the subjects (sub-plot factors). They also allow cluster sampling, i.e. each subject may receive more than one measurement at each occasion (sub-plot factor level combination). For simplicity we consider here only designs where each subject is observed at all occasions but receives only one observation per occasion. As before, we will use a single index to enumerate the factor-level combinations of possibly several whole-plot factors and one index for the level combinations of sub-plot factors. In such designs we observe random vectors $\boldsymbol{X}_{ik} = (X_{ik1}, \ldots, X_{ikd})'$, where $i$ enumerates the cells of the whole-plot factors, $k$ enumerates the subjects within cell $i$, and the third index, which we will call $s$, enumerates the $d$ cells of the sub-plot factors. Thus $\boldsymbol{X}_{ik}$ is the vector of repeated observations or measurements taken on subject $k$ within cell $i$. The nonparametric mixed model assumes only that the vectors $\boldsymbol{X}_{ik}$, $i = 1, \ldots, r, k = 1, \ldots, n_i$, are independent and

$$\boldsymbol{X}_{ik} = (X_{ik1}, \ldots, X_{ikd})' \sim F_i(\boldsymbol{x}), \; i = 1, \ldots, r, \; k = 1, \ldots, n_i. \tag{3.13}$$

so that the marginal distribution functions $F_{is}(x) = \frac{1}{2}\left[F_{is}^+(x) + F_{is}^-(x)\right]$, $i = 1, \ldots, r$, $s = 1, \ldots, d$, of $X_{iks}$ do not depend on $k$ (Akritas and Brunner [8]). In particular, no special structure is assumed for the dependencies between the components of the vectors $\boldsymbol{X}_{ik}$.

A difference between the marginal distributions $F_{is}$ is quantified by the so-called relative treatment effects $p_{is} = \int H(x) dF_{is}(x)$, where $H(x) = \frac{1}{N}\sum_{i=1}^{r}\sum_{s=1}^{d} n_i F_{is}(x)$ is the weighted average of the $N = d \cdot \sum_{i=1}^{r} n_i$ distribution functions.

Let $\boldsymbol{F} = (F_{11}, \ldots, F_{1d}, \ldots, F_{r1}, \ldots, F_{rd})'$ denote the vector of the marginal distributions and let $\boldsymbol{p} = \int H d\boldsymbol{F} = (p_{11}, \ldots, p_{1d}, \ldots, p_{r1}, \ldots, p_{rd})'$, denote the vector of the relative treatment effects. Then nonparametric hypotheses are formulated by means of the marginal distributions $F_{is}(x)$ in the same way as for independent observations. Let $\boldsymbol{C}$ denote a suitable contrast matrix to formulate a hypothesis. Then, $H_0^F : \boldsymbol{CF} = \boldsymbol{0}$ is the most general form of a nonparametric hypothesis in the mixed model (3.13).

The relative treatment effects $p_{is} = \int H(x) dF_{is}(x)$ are estimated by replacing $F_{is}(x)$ and $H(x)$ by the obvious empirical estimators $\widehat{F}_{is}(x)$, $\widehat{H}(x) = \frac{1}{N}\sum_{i=1}^{r}\sum_{s=1}^{d} n_i \widehat{F}_{is}(x)$. An asymptotically unbiased and $L_2$-consistent estimator of $p_{is}$ is then given by

$$\widehat{p}_{is} = \int \widehat{H} d\widehat{F}_{is} = \frac{1}{n_i}\sum_{k=1}^{n_i} \widehat{H}(X_{iks}) = \frac{1}{n_i}\sum_{k=1}^{n_i} \frac{1}{N}\left(R_{iks} - \frac{1}{2}\right) = \frac{1}{N}\left(\overline{R}_{i \cdot s} - \frac{1}{2}\right). \tag{3.14}$$

where $R_{iks}$ denotes the mid-rank of $X_{iks}$ among all $N$ observations. Thus, $\boldsymbol{p} = \int H d\boldsymbol{F}$ is estimated by

$$\begin{aligned}\widehat{\boldsymbol{p}} &= \int \widehat{H} d\widehat{\boldsymbol{F}} = (\widehat{p}_{11}, \ldots, \widehat{p}_{rd})' \\ &= \frac{1}{N}\left(\overline{R}_{1 \cdot 1} - \tfrac{1}{2}, \ldots, \overline{R}_{1 \cdot d} - \tfrac{1}{2}, \ldots, \overline{R}_{r \cdot 1} - \tfrac{1}{2}, \ldots, \overline{R}_{r \cdot d} - \tfrac{1}{2}\right).\end{aligned} \tag{3.15}$$

Let now $Y_{iks} = H(X_{iks})$, $s = 1, \ldots, d$, be the so-called asymptotic rank transform of $X_{iks}$ (Akritas [1]), and set $\boldsymbol{Y}_{ik} = (Y_{ik1}, \ldots, Y_{ikd})'$, and $\overline{\boldsymbol{Y}}_{\cdot \cdot} = \left(\overline{\boldsymbol{Y}}_{1 \cdot}', \ldots, \overline{\boldsymbol{Y}}_{r \cdot}'\right)'$, where $\overline{\boldsymbol{Y}}_{i \cdot} = \left(\overline{Y}_{i \cdot 1}, \ldots, \overline{Y}_{i \cdot d}\right)' = \frac{1}{n_i}\sum_{k=1}^{n_i} \boldsymbol{Y}_{ik}$. Finally, let $\boldsymbol{V}_i = \mathrm{Cov}(\boldsymbol{Y}_{i1})$, $k = 1, \ldots, n_i$. By the independence of the $\overline{\boldsymbol{Y}}_{i \cdot}$, the covariance matrix $\boldsymbol{V}_n = \mathrm{Cov}(\sqrt{n}\,\overline{\boldsymbol{Y}}_{\cdot \cdot})$ is block diagonal, i.e.

$$\boldsymbol{V}_n = \bigoplus_{i=1}^{r} \frac{n}{n_i} \boldsymbol{V}_i. \tag{3.16}$$

Akritas and Brunner [8] show

THEOREM 3.1 *Let $\boldsymbol{X}_{ik} = (X_{ik1}, \ldots, X_{ikd})'$ be independent and identically distributed random vectors and let $\boldsymbol{V}_n$ be the covariance matrix given in (3.16). Then, under suitable assumptions and under $H_0^F : \boldsymbol{CF} = \boldsymbol{0}$, the statistic $\sqrt{n}\boldsymbol{C}\widehat{\boldsymbol{p}} = \sqrt{n}\boldsymbol{C}\int \widehat{H}d\widehat{\boldsymbol{F}}$ has, asymptotically, a multivariate normal distribution with mean $\boldsymbol{0}$ and covariance matrix $\boldsymbol{CV}_n\boldsymbol{C}'$.*

Since the covariance matrix $\boldsymbol{V}_n$ is unknown, it must be estimated from the data. A consistent estimator is given in the next theorem (Akritas and Brunner [8]).

THEOREM 3.2 *For $i = 1, \ldots, r$, let $\overline{\boldsymbol{R}}_{i\cdot} = n_i^{-1}\sum_{k=1}^{n_i} \boldsymbol{R}_{ik}$ denote the mean of the vectors $\boldsymbol{R}_{ik} = (R_{ik1}, \ldots, R_{ikd})'$, and set*

$$\widehat{\boldsymbol{V}}_i = \frac{1}{N^2(n_i - 1)} \sum_{k=1}^{n_i} (\boldsymbol{R}_{ik} - \overline{\boldsymbol{R}}_{i\cdot})(\boldsymbol{R}_{ik} - \overline{\boldsymbol{R}}_{i\cdot})', \quad \widehat{\boldsymbol{V}}_n = \bigoplus_{i=1}^{r} \frac{n}{n_i}\widehat{\boldsymbol{V}}_i. \qquad (3.17)$$

*Then, under the same assumptions as in Theorem 3.1, $\|\widehat{\boldsymbol{V}}_i - \boldsymbol{V}_i\| \xrightarrow{P} 0$, $i = 1, \ldots, r$, and $\|\widehat{\boldsymbol{V}}_n - \boldsymbol{V}_n\| \xrightarrow{P} 0$, where $\boldsymbol{V}_i = \mathrm{Cov}(\boldsymbol{Y}_{i1})$, $\boldsymbol{V}_n$ is defined in (3.16), and $\|\cdot\|$ denotes the Euclidean norm of a matrix.*

To test the nonparametric hypothesis $H_0^F : \boldsymbol{CF} = \boldsymbol{0}$ explained in Section 2, we consider quadratic forms of the random vectors $\sqrt{n}\boldsymbol{C}\widehat{\boldsymbol{p}}$. Other statistics used in multivariate analysis, like Wilk's $\Lambda$, for example, are not discussed here since they require the equality of the covariance matrices. In nonparametric models, however, this assumption is not reasonable since, in general, the assumption of homoscedastic distribution functions is not transferred to the asymptotic rank transform vectors $\boldsymbol{Y}_{ik} = (Y_{ik1}, \ldots, Y_{ikd})'$, $i = 1, \ldots, r$, $k = 1, \ldots, n_i$, where $Y_{iks} = H(X_{iks})$, $s = 1, \ldots, d$. This is easily seen by the fact that $H(\cdot)$ is a non-linear transformation in general as pointed out already by Akritas [1]. We consider two different quadratic forms.

**Wald-Type Statistics (WTS)**

The simplest way to derive a quadratic form for testing the hypothesis $H_0^F : \boldsymbol{CF} = \boldsymbol{0}$ is to use a generalized inverse of the covariance matrix $\boldsymbol{CV}_n\boldsymbol{C}'$ under $H_0^F$ as the generating matrix of the quadratic form where the unknown covariance matrix $\boldsymbol{V}_n$ is replaced by the consistent estimator $\widehat{\boldsymbol{V}}_n$ given in (3.17). Let $[\boldsymbol{C}\widehat{\boldsymbol{V}}_n\boldsymbol{C}']^+$ denote the Moore-Penrose inverse of $\boldsymbol{C}\widehat{\boldsymbol{V}}_n\boldsymbol{C}'$. Assume that $\boldsymbol{V}_n \to \boldsymbol{V} \neq \boldsymbol{0}$ such that $r(\boldsymbol{CV}_n) = r(\boldsymbol{CV})$, where $r(\cdot)$ denotes the rank of a matrix. Then under $H_0^F : \boldsymbol{CF} = \boldsymbol{0}$, it follows from Theorem 3.1 and Theorem 3.2 that

$$Q_n^W(\boldsymbol{C}) = n\widehat{\boldsymbol{p}}'\boldsymbol{C}' \left[\boldsymbol{C}\widehat{\boldsymbol{V}}_n\boldsymbol{C}'\right]^+ \boldsymbol{C}\widehat{\boldsymbol{p}} \qquad (3.18)$$

has, asymptotically, a central $\chi^2$-distribution with $f = r(\boldsymbol{CV})$ degrees of freedom. This statistic is called the rank version of the WTS. However, extremely large sample sizes are needed for a satisfactory approximation.

**ANOVA-Type Statistics (ATS)**

The estimation of the covariance matrix $\boldsymbol{V}_n$ in (3.18) requires large sample sizes. Therefore, Brunner, Munzel and Puri [19] suggested to leave out the estimator $\widehat{\boldsymbol{V}}_n$ in the generating matrix of the quadratic form and to consider the asymptotic distribution of

$$Q_n^A(\boldsymbol{C}) = n\widehat{\boldsymbol{p}}'\boldsymbol{C}'[\boldsymbol{CC}']^-\boldsymbol{C}\widehat{\boldsymbol{p}} = n\widehat{\boldsymbol{p}}'\boldsymbol{T}\widehat{\boldsymbol{p}}. \qquad (3.19)$$

We note that $T = C'[CC']^-C$ is a projection matrix where $[CC']^-$ denotes some $g$-inverse of $CC'$. The matrix $T$ provides a standard formulation of the hypothesis $CF = 0$ since two hypotheses $C_1F = 0$ and $C_2F = 0$ are equivalent if and only if $T_1 = C'_1[C_1C'_1]^-C_1 = T_2 = C'_2[C_2C'_2]^-C_2$. To see this, note that $TF = 0 \iff CF = 0$ because $C'[CC']^-$ is a generalized inverse of $C$. The asymptotic distribution of $Q_n^A(C)$ is given in the next Theorem.

THEOREM 3.3 *Let $T = C'[CC']^-C$ and let $V_n$ be as in (3.16). Then, under the the same assumptions as in Theorem 3.1, and under the hypothesis $H_0^F : CF = 0$, it follows that*

$$Q_n^A(C) = n\widehat{p}'T\widehat{p} \sim \sum_{i=1}^{r}\sum_{s=1}^{d} \lambda_{is}Z_{is}, \quad as\ n \to \infty. \tag{3.20}$$

*where the $\lambda_{is}$ are the eigenvalues of $TV_nT$ and the $Z_{is} \sim \chi_1^2$ are independent random variables which are $\chi_1^2$ distributed.*

Brunner, Munzel and Puri [19] suggested to approximate the distribution of the random variable $\sum_{i=1}^{r}\sum_{s=1}^{d}\lambda_{is}Z_{is}$ by a scaled $\chi^2$-distribution such that the first two moments coincide. This approximation dates back to Box [18] and is quite accurate. The result is given in the following approximation procedure.

APPROXIMATION PROCEDURE 3.4 *If $tr(TV_n) \geq t_0 > 0$ then, under $H_0^F : CF = 0$, the first two moments of the asymptotic distribution of $tr(TV_n) \cdot Q_n^A(C)/tr(TV_nTV_n)$ and of the $\chi_f^2$-distribution coincide for $f = [tr(TV_n)]^2/tr(TV_nTV_n)$, where $tr(\cdot)$ denotes the trace of a square matrix.*

*The unknown traces $tr(TV_n)$ and $tr(TV_nTV_n)$ can be estimated consistently by replacing $V_n$ with $\widehat{V}_n$ given in (3.17) and it follows under $H_0^F : CF = 0$ that the statistic*

$$F_n(C) = \frac{n \cdot tr(T\widehat{V}_n)}{tr(T\widehat{V}_nT\widehat{V}_n)}\widehat{p}'T\widehat{p} \stackrel{\cdot}{\sim} \chi_f^2 \tag{3.21}$$

*has approximately a central $\chi_f^2$-distribution where $f$ is estimated by*

$$\widehat{f} = \frac{[tr(T\widehat{V}_n)]^2}{tr(T\widehat{V}_nT\widehat{V}_n)}. \tag{3.22}$$

We note that for very small sample sizes the estimator $\widehat{f}$ in (3.22) may be slightly biased.

We note that $Q_n^W(C) = F_n(C)/f$ if $r(C) = 1$ which follows from simple algebraic arguments. See Brunner, Munzel and Puri [19] for details regarding the consistency of the tests based on $Q_n^W(C)$ or $F_n(C)/f$.

In some special cases the so-called compound symmetry of the covariance matrix can be assumed under the hypothesis. In particular, in repeated measures designs with one homogeneous group of subjects and $d$ repeated measures, compound symmetry can be assumed under the hypothesis $H_0^F : F_1 = \cdots = F_d$ if the subjects are blocks which can be split into homogeneous parts and each part is treated separately. In this case, only two quantities have to be estimated: the common variance and the common covariance. For more details, we refer to Brunner, Munzel and Puri [19].

The recent book Brunner, Domhof and Langer [20] presents many examples and discusses software for the computation of the statistics $Q_n^W(C)$ and $F_n(C)/f$.

## 4. EXTENSIONS AND FURTHER DEVELOPMENTS

The nonparametric ANOVA methodology has been extended to designs with independent and dependent censored data (Akritas & Brunner [9], O'Gorman & Akritas [35]), to data missing completely at random (Brunner, Munzel and Puri [19]), and to data missing at random (Akritas, Kuha & Osgood [10], Antoniou & Akritas [14]). An interesting generalization to multivariate ANOVA using spatial ranks is developed in Choi & Marden [22]. The paper by Brunner, Munzel and Puri [19] deals also with general rank-scores.

The nonparametric ANCOVA methodology described in Section 2 has been extended to up to three covariates also for repeated measures designs (Tsangari & Akritas [39], Tsangari & Akritas [40]). For four or more covariates the "curse of dimensionality" effects take over and this methodology, which requires consistent estimation of the conditional distributions, does not work.

As mentioned in Section 2 testing hypotheses involving the covariate in the nonparametric ANCOVA model is much more recent. The theory for testing such hypotheses is closely connected with the theory of ANOVA when the number of factor levels is large. See Boos and Brownie (1995), Akritas & Arnold [5], Bathke [15], Akritas & Papadatos [11], Wang & Akritas [42], Wang & Akritas [41] are representative publications in this new area; the last three deal also with heteroscedastic designs, while the last considers rank statistics for this problem. To see the connection with the nonparametric ANCOVA problem, think of the covariate as a factor with many levels. Then, hypotheses regarding the factor "covariate" in this hypothetical ANOVA design approximate the corresponding hypotheses in the ANCOVA design, and coincide with them asymptotically. The theory of ANOVA when the number of factor levels is large is not directly applicable because it requires at least two observations per factor level combination, whereas in typical ANCOVA designs there is only one observation per covariate value. To remedy this we use smoothness assumptions to augment the hypothetical ANOVA design by considering windows around each covariate value. This induces dependence (since different cells may share observations) and thus the aforementioned theory for ANOVA when the number of factor levels is large is still not directly applicable. However a new theory can be developed; see Wang & Akritas [43]. This approach also yields an alternative approach (and test statistic) to the procedure of Akritas, Arnold & Du [7] for the covariate adjusted main effects and interactions of factors. The interesting aspect of this theory is that it does not require consistent estimation of the conditional distribution functions, as that of Akritas, Arnold and Du [7] does, since the window sizes need not tend to infinity. Thus, there is hope that this new methodology will allow the extension of the nonparametric methodology to ANCOVA designs with more than three covariates.

## 5. RELATED WORK

There is a plethora of related work for ANOVA designs, with both independent and dependent observations, which emphasize invariance under monotone transformations. See Patil and Hoel (1973), Govindarajulu [28], Conover and Iman [24], Brunner & Neumann [21], Kepner & Robinson [31], Akritas [1], Akritas [2], Alvo & Cabilio [12], Thompson [38], Akritas [3], Marden and Muyot (1995), Cliff (1996), Alvo & Cabilio [13], Handcock & Janssen [30]. These, however, do not make reference to the nonparametric model and hypotheses (2.1), (2.4) and thus their applicability is limited. Some of these approaches are discussed in the recent book

Brunner, Domhof and Langer [20] (e.g. Section 5.8) which, however, emphasizes testing the nonparametric hypotheses.

There is a much richer bibliography dealing with rank-related methods for testing the usual parametric hypotheses, such as aligned rank tests. Such statistics cannot be invariant under monotone transformations since they test hypotheses that are not. Thus they belong in the area of robust statistics.

There is also a lot of work in nonparametric ANCOVA. See Hall & Hart [29], King, Hart & Wehrly [32], Young & Bowman [45], Kulasekera [33], Bowman & Azzalini [17], p.80, Dette & Munk [25], Dette & Neumeyer [26], to mention a few. Again, this literature does not make use of the nonparametric model (2.2), (2.5) and, as a consequence, its scope is limited. In particular, these papers deal only with the so-called problem of curve comparison, which corresponds to the fourth of the hypotheses listed under (2.3), i.e. no simple factor effect, and their methods are only applicable to one-way ANCOVA with continuous response. Different from the above, Bathke & Brunner [16] contains an interesting alternative approach to testing for covariate-adjusted factor effects.

**REFERENCES**

1. Akritas, M. G. (1990). The Rank Transform Method in Some Two-Factor Designs. *Journal of the American Statistical Association* 85, 73–78.
2. Akritas, M. G. (1991). Limitations of the Rank Transform Drocedure: A Study of Repeated Measures Designs, Part I. *Journal of the American Statistical Association* 85, 73–78.
3. Akritas, M. G. (1993). Limitations of the Rank Transform Drocedure: A Study of Repeated Measures Designs, Part II, *Statistics & Probability Letters* 17, 149–156.
4. Akritas, M.G. and S.F. Arnold (1994). Fully Nonparametric Hypotheses for Factorial Designs I: Multivariate Repeated Measures Designs. *Journal of the American Statistical Association*, 89, 336–343.
5. Akritas, M. G. and Arnold, S. F. (2000). Asymptotics for ANOVA when the number of levels is large. *Journal of the American Statistical Association* 95, 212-226.
6. Akritas, M.G., Arnold, S.F. and Brunner, E. (1997). Nonparametric Hypotheses and Rank Statistics for Unbalanced Factorial Designs. *Journal of the American Statistical Association* 92, 258–265.
7. Akritas, M. G., Arnold, S. F. and Du, Y. (2000). Nonparametric models and methods for nonlinear analysis of covariance. *Biometrika* 87, 507-526.
8. Akritas, M.G. and Brunner , E. (1997a). A Unified Approach to Rank Tests in Mixed Models, *Journal of Statistical Planning and Inference* 61, 249–277.
9. Akritas, M.G. and Brunner, E. (1997b). Nonparametric methods for designs with censored data. *Journal of the American Statistical Association* 92, 568-576.
10. Akritas, M.G, Kuha, J. and Osgood (2002). A nonparametric approach to matched pairs with censored data. *Sociological Methods & Research* 30, 425-462. (With Discussion)
11. Akritas, M. G. and Papadatos, N. (2002). Heteroskedastic One-Way ANOVA and Lack-of-Fit Tests. *Journal of the American Statistical Association*, tentatively accepted.
12. Alvo, M. and Cabilio, P. (1991). On the balanced incomplete block design for rankings. *Annals of Statistics* 19,1597-1613.
13. Alvo, M. and Cabilio, P. (1999). A general rank based approach to the analysis of block

data. *Commun. Stat.-Theor. M.* 28, 197-215.
14. Antoniou, E. and Akritas, M.G. (2003). Nonparametric methods for designs with data missing at random. Submitted.
15. Bathke, A. (2002). ANOVA for a large number of treatments. *Mathematical Methods of Statistics* 11, 118-132.
16. Bathke, A. and Brunner, E. (2002). A nonparametric alternative to analysis of covariance. Preprint.
17. Bowman, A. W. and Azzalini, A. (1997). *Applied Smoothing Techniques for Data Analysis*. Oxford: Oxford University Press.
18. Box, G. E. P. (1954). Some theorems on quadratic forms applied in the study of analysis of variance problems, I. Effect of inequality of variance in the one-way classification. *Annals of Mathematical Statistics* 25, 290–302.
19. Brunner, E., Munzel, U. and Puri, M.L. (1999). Rank-Score Tests in Factorial Designs With Repeated Measures. *Journal of Multivariate Analysis* 70, 286–317.
20. Brunner, E., Domhof, S. and Langer, F. (2002). *Nonparametric Analysis of Longitudinal Data in Factorial Designs*, Wiley, New York.
21. Brunner, E. and Neumann, N. (1982). Rank Tests for Correlated Random Variables. *Biometrical Journal*, 24, 373–389.
22. Choi, K. and Marden, J. (2002). Multivariate analysis of variance using spatial ranks. *Sociological Methods & Research* 30, 341-366.
23. Cliff, N. (1996). Answering ordingal questions with ordinal data using ordinal statistics. *Multivariate Behavioral Research* 31, 331-350.
24. Conover, W. J. and Iman, R. L. (1981), Rank transformations as a bridge between parametric and nonparametric statistics (with discussion). *American Statistician* 35, 124–133.
25. Dette, D. and Munk, A. (1998). Nonparametric comparison of several regression functions: exact and asymptotic theory. *Annals of Statistics* 26, 2339-2368.
26. Dette, D. and Neumeyer, N. (2001). Nonparametric analysis of covariance. *Annals of Statistics* 29, 1361-1400.
27. Du, Y., Akritas, M. G., Arnold, S. F. and Osgood, D. W. (2002). Analysis of teenage deviant behavior data. *Sociological Methods & Research* 30, 309-340.
28. Govindarajulu, Z. (1975), Robustness of Mann-Whitney-Wilcoxon test to dependence in the variables, *Studia Scientiarum Mathematicarum Hungarica* 10, 39–45.
29. Hall, P. and Hart, J.D. (1990). Bootstrap test for defference between means in nonparametric regression. *Journal of the American Statistical Association* 85, 1039-1049.
30. Handcock, M. S., and Janssen, P. (2002). Statistical inference for the relative density. *Sociological Methods & Research* **30**, 394-424.
31. Kepner, J. L. and Robinson, D. H. (1988). Nonparametric Methods for Detecting Treatment Effects in Repeated Measures Designs. *Journal of the American Statistical Association* 83, 456–461.
32. King, E.C., Hart, J.D. and Wehrly, T.E. (1991). Testing the equality of regression curves using linear smoothers. *Statistics & Probability Letters* 12, 239-247.
33. Kulasekera, K.B. (1995). Comparison of regression curves using quadi residuals. *Journal of the American Statistical Association* 90, 1085-1093.
34. Marden, J. I. and Muyot, M. E. T. (1995). Rank tests for main and interaction effects in Analysis of Variance. *Journal of the American Statistical Association* 90, 1388-1398.

35. O'Gorman, J.T. and Akritas, M.G. (2001). Nonparametric models and methods for designs with dependent censored data. *Biometrics* 57, 88-95
36. Patil, K. M. and Hoel, D. G. (1973). A nonparametric test for interaction in factorial experiments. *Journal of the American Statistical Association* 68, 615-620.
37. Ruymgaart, F.H. (1980). A Unified Approach to the Asymptotic Distribution Theorey of Certain Midrank Statistics, in *Statistique non Parametrique Asymptotique*, 1–18, J.P. Raoult (Ed.), Lecture Notes on Mathematics, No. 821, Springer, Berlin.
38. Thompson, G. L. (1991). A unified approach to rank tests for multivariate and repeated measures designs. *Journal of the American Statistical Association* 86, 410–419.
39. Tsangari, H. and Akritas, M.G. (2003). Nonparametric ANCOVA with two and three covariates. *Journal of Multivariate Analysis*. In press.
40. Tsangari, H. and Akritas, M.G. (2003). Nonparametric models and methods for ANCOVA with dependent data. *Journal of Nonparametric Statistics*. In press.
41. Wang, H. and Akritas, M.G. (2003). Rank tests for ANOVA with large number of factor levels. *Journal of Nonparametric Statistics*. In press.
42. Wang, L. and Akritas, M.G. (2003). Two-way Heteroscedastic ANOVA when the Number of Levels is Large. Submitted.
43. Wang, L. and Akritas, M.G. (2003). Testing for the covariate effect in nonparametric ANCOVA. Submitted.
44. Winship, C. and Mare, R. D. (1984). Regression models with ordinal variables. *American Sociological Review* 49, 512-525.
45. Young, S. G. and Bowman, A. W. (1995). Non-parametric analysis of covariance. *Biometrics* 51, 920-31.

*Recent Advances and Trends in Nonparametric Statistics*
Michael G. Akritas and Dimitris N. Politis (Editors)
© 2003 Elsevier Science B.V. All rights reserved.

## Isotonic Additive Interaction Models

Ilya Gluhovsky [a]

[a]Sun Microsystems Laboratories
2600 Casey Ave MTV29-120, Mountain View, CA 94040

In this chapter we consider parametric and nonparametric approaches to fitting isotonic additive models. The modeled relationship between a predictor vector and a response is assumed to be smooth, additive with low-order interactions, and monotone in some of the predictors. A parametric solution is to fit a monotone regression spline model. A nonparametric solution is to fit an unrestricted additive model first and then isotonize the fit.

### 1. Introduction

In this chapter we describe a regression problem where the regression function $f$ is assumed to possess two properties. First, $f$ is monotone in some, not necessarily all, of its inputs. That is, a function $f(\underline{x})$ of a $P$-dimensional argument $\underline{x}$ is said to be nonincreasing in its first $R \leq P$ arguments if $a < b$ implies $f(x_r = a, x_{-r}) \geq f(x_r = b, x_{-r})$ for any $r \leq R$ and $x_{-r}$, where $x_{-r} = (x_1, \cdots, x_{r-1}, x_{r+1}, \cdots, x_P)$. This assumption is natural in a variety of cases. For example, one may be interested in describing the relationship between a child's height or weight and his age, or that between high school achievements and those in college. Many physical laws also prescribe monotone relationships.

There are circumstances where the relationship is expected to be monotone only in some of the inputs. For example, if the temperature field is studied as a function of the location, we could assume that the temperature increases as one moves towards the equator, but changes in an unrestricted way as we move along a parallel. A more practical example will be given in Section 4.

Second, we assume that the form of $f$ is additive. We also generalize to include typically low-order interaction terms. Thus, $f$ is assumed to be additive up to some order of interactions. As an example, we can think of a bivariate interaction model

$$f(\underline{x}) = \alpha_0 + \sum_{p=1}^{P} f_p(x_p) + \sum_{p<r} f_{pr}(x_p, x_r). \tag{1}$$

Here $P$ is the number of predictors or the dimensionality of the inputs, the $f_p$ are the main effects and the $f_{pr}$ are the bivariate interactions. They will be referred as the additive components of the model.

Additivity is often found to be a useful generalization of many simpler models. A linear model is the simplest additive model whose main effects are constrained to be $f_p(x_p) = a_p x_p$. A quadratic polynomial model follows (1). The central idea of additive modeling is to treat the additive components as unspecified smooth functions of the predictors ([10]). We avoid imposing a particular parametric form on the functions, but rather control the degree of smoothness in an estimating procedure.

Additivity is frequently thought to be a reasonable proposition to control complexity of the model and estimation variance. For example, a linear model is easily fitted, but is often not flexible enough. Then even the best linear model may miss many features of the regression relationship. On the other end of the spectrum, estimating a $P$-dimensional function for, say, $P > 3$ is often imprecise using data sets of moderate sizes. While the fitted function is very flexible, the estimation variance may shed doubt on the usefulness of the results.

An additional advantage of using additive models is their interpretability that they largely share with the linear models. For example, in the main effects only model, the variation of the response surface holding all but one predictor fixed does not depend on the values of the other predictors ([10], p. 88). Thus, one may easily examine the roles of the predictors in modeling the response. A popular way to fit unconstrained additive models is the backfitting algorithm ([10]).

When fitting an isotonic additive (IA) function that involves interactions, it is important to realize that no additive component by itself needs to be isotonic. The restriction is only on their sum. For example, if $f(x_1, x_2, x_3) = f_1(x_1) + f_{12}(x_1, x_2) + f_3(x_3)$, the restriction that $f$ be isotonic in all the inputs translates into that for $f_3$ in $x_3$, for $f_{12}(x_1, x_2)$ in $x_2$, but not necessarily in $x_1$, as $f_1$ may compensate for some of its non-isotonic behavior. Therefore, when fitting these models, it would be unduly restricting the fit to require that all the additive components be isotonic. However, it will be shown that one can always represent an IA function such that all additive components are isotonic.

To build the regression function, suppose that we are given data pairs $(\underline{x}_i, y_i)$, $1 \leq i \leq N$, where $\underline{x}_i$ is a $P$-dimensional vector. In this chapter, we will consider both parametric an nonparametric approaches to fitting IA models. To motivate the discussion, we begin with the goal of [1] of fitting an IA function that is as close as possible to the data points, namely, the one that minimizes the criterion

$$C_0 = \sum_{i=1}^{N} c(h(x_i) - y_i) \qquad (2)$$

over all IA functions $h$, where $c$ is a strictly convex penalty, such as a square. The criterion does not yield the definition outside the data set, but if we "connect the dots", we will obtain a function that is defined (at least) over the convex hull of the data. Such a function will not be smooth and will tend to overfit the data. In Figure 1, a rugged monotone function passing through the data points is contrasted with a smooth fit of these data points. Therefore, one idea is to endow the fitting criterion with a smoothing penalty. One way to do so in the context of additive models may seem to use the backfitting algorithm with a smoothing penalty in each additive component update. Unfortunately, optimizing such a penalty subject to the monotonicity constraint is rather cumbersome. Also, as we mentioned above, restricting the response surface to be isotonic is not the same as restricting the additive components. Therefore, enforcing the constraint within each backfitting iteration is likely to degrade the fit. This is because the "redistribution" among additive components that we can carry out to make them all isotonic (described in Section 3.1) may disturb the additive smoothing penalty used in backfitting. To put it in different wording, requiring the regression function to be both isotonic and smooth is not the same as to require each additive component to be both isotonic and smooth. Therefore, the optimization needs to take place in a single step over all the additive components simultaneously. That only adds to the complexity. We will not pursue this method here.

Alternatively, a parametric method proposed in a univariate setting in [17] can be used. The basic idea is to fit a regression spline model with monotone basis splines and nonnegative coefficients. That will necessarily produce an isotonic function. The single optimization step is easier to carry out when it does not involve a smoothing penalty. The constraint that all additive components are isotonic is no longer an obstacle in principle because as we mentioned and as we will show, optimizing with respect to that constraint is equivalent to the original problem. The reason for saying "in principle" is while *functionally* equivalent, the problem domains may be different when using a particular method for fitting the functions.

A nonparametric approach is proposed in [8]. There we will begin by fitting an unconstrained additive model $f$ and then find the closest isotonic additive function $h$ to $f$ that is of the same functional form as $f$. That allows one to capture the regression function in a smooth additive way and in the process to select the additive model, that is, to pick those main effects and interactions that should be included in the model and prescribe the degree of smoothness for each additive component. Where the unrestricted additive model violates the isotonic requirement, it is corrected in the least intrusive way.

We will describe the pros and cons of the two approaches just outlined in detail below.

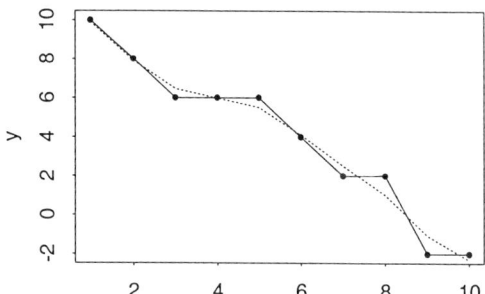

Figure 1. Rugged monotone function (solid); smooth fit (dotted).

## 2. Modeling Based on Monotone Regression Splines

In this section we describe how to build an IA model when isotonic spline terms are used for additive components. Without loss of generality, we will consider nonincreasing models. To simplify the notation, suppose that our goal is to fit a bivariate interactions model (1). Each univariate component $p$ is represented as a non-positive linear combination of $I$-splines. To this end, one chooses an increasing sequence of knots $t_{p,1} < \cdots < t_{p,m_p}$ such that all data points are between $t_{p,1}$ and $t_{p,m_p}$. A spline of degree $k$ is defined to be a piecewise polynomial of degree $k$ that has a continuous $k-1$th derivative. This is not the most general definition, but it is a common practical choice. Thus, the spline is a polynomial between the adjacent knots and the polynomials are joined at a knot in a smooth way. The most important practical choice is $k = 3$ in which case the *cubic* spline is fitted. It is twice-differentiable everywhere and has the continuous second derivative.

The $k$-degree splines of the $p$th predictor form a linear space. A computationally convenient basis for that linear space is called $B$-splines. Define $B_i(x_p, k)$ using the following recursive formula:

$$B_i(x_p, 1) = \begin{cases} (t_{p,i+1} - t_{p,i})^{-1}, & t_{p,i+1} \leq x_p \leq t_{p,i} \\ 0, & \text{otherwise} \end{cases}$$

and

$$B_i(x_p, k) = \frac{k[(x_p - t_{p,i})B_i(x_p, k-1) + (t_{p,i+k} - x_p)B_{i+1}(x_p, k-1)]}{(k-1)(t_{p,i+k} - t_{p,i})}$$

The basis functions $B_i(x_p, k)$ are nonnegative and integrate to 1. Also, they are nonzero only over $(t_{p,i}, t_{p,i+k})$. Therefore, substantial computational savings ensue when handling the corresponding model and smoothing penalty matrices. Our interest, however, is in their integrated versions. Define $I$-spline basis functions

$$I_i(x_p, k) = \int_{t_{p,1}}^{x_p} B(u, k) du$$

It turns out that for the knot sequence considered above, we have the representation in terms of B-splines: For $t_{p,j} \leq x_p \leq t_{p,j+1}$,

$$I_i(x_p, k) = \begin{cases} 0, & j < i \\ \sum_{m=i}^{j}(t_{p,m+k+1} - t_{p,m})B_m(x_p, k+1)/(k+1), & j-k+1 \leq i \leq j \\ 1, & i < j-k+1 \end{cases}$$

The $I$-splines are linearly independent and hence form a basis. Also, they are nondecreasing, so a non-positive linear combination is nonincreasing. Unfortunately, not every nonincreasing spline can be represented in that form, but that loss is usually not significant.

In order to model interaction terms, one may resort to a tensor product basis for low-dimensional terms. To build a basis for a $(p, r)$-interaction of predictors $p$ and $r$, define $I_{ij}(x_p, x_r, k) = I_i(x_p, k) \times I_j(x_r, k)$ assuming that we are using the same degree $k$ for all dimensions.

When an interaction model is fitted, some main effects and lower-order interactions may become *aliased* by the higher-order interactions. For example, consider again the model $f(x_1, x_2, x_3) = f_1(x_1) + f_{12}(x_1, x_2) + f_3(x_3)$. $f_1(x_1)$ is aliased by $f_{12}(x_1, x_2)$. The reason we care about extracting $f_1(x_1)$ out of $f_{12}(x_1, x_2)$ is interpretation. A natural way of sensibly separating the main effects and lower-order interactions from the higher-order interactions is to project the higher-order interactions onto the lower order terms. In the example we are considering, the joint effect $\hat{f}_{12}(x_1, x_2)$ would be projected onto the space of basis functions $I_i(x_1)$. Because, the main effect $f_1(x_1)$ must be nonincreasing, optimizing with respect to a non-positive weights rather than projecting is necessary. In the data case, we will find the closest non-positive linear combination to the fitted vector $\mathbf{f_{12}}$ of evaluations of the joint effect $\hat{f}_{12}(x_1, x_2)$ at the data points.

Model selection analysis is likely to be necessary especially when considering tensor product bases, since the full model will have large number of terms and therefore is likely to overfit the data. An additional difficulty with fitting regression spline models in general is selection of the number and placement of the knots. A common practice is to select the number of knots based on some global criterion such as AIC, cross-validation, or suchlike. Then for predictor $p$ to place the $j$th of $m$ knots at the $j/(m+1)$th quantile of the data vector $x_p$. Ideally, however, one should group knots closer together in places where the target function is more volatile to allow capturing of these local features. However, it is a hard problem to anticipate these from the data. In this respect, nonparametric methods have an edge.

It must be mentioned that the problem of potentially misplacing the knots is propagated to questionable model selection. For example, the functional form of the additive model may not be chosen correctly because the model failed to find a significant interaction obliterated, for example, by the knot placement.

## 3. Nonparametric 2-Step Fitting

In this section we describe a nonparametric approach to fitting isotonic additive models [8]. The method consists of two steps. First, we fit an unconstrained additive model (UA) to the data ([10]). As part of fitting a UA model, we carry out model selection thereby choosing the appropriate main effects and interactions to be included as additive components and also choosing the amount of smoothness for each additive component. This is traditionally done via the stepwise and other techniques described in [10], Section 9.4. To be specific, let us again consider the binomial interaction model (1) and assume that we have chosen main effects $U$ (for univariate) and interactions $I$. Then the chosen model has the form

$$f(\underline{x}) = \alpha_0 + \sum_{p \in U} f_p(x_p) + \sum_{(p,r) \in I} f_{pr}(x_p, x_r). \tag{3}$$

It is fitted with the prescribed smoothness for each additive component to produce

$$\hat{f}(\underline{x}) = \hat{\alpha}_0 + \sum_{p \in U} \hat{f}_p(x_p) + \sum_{(p,r) \in I} \hat{f}_{pr}(x_p, x_r). \tag{4}$$

The second step of the technique is to find the isotonic function $\hat{h}$ of the same functional form (3) as $\hat{f}$, such that the distance $\rho(\hat{h}, \hat{f})$ between $\hat{h}$ and $\hat{f}$ is minimal. The distance $\rho$ is defined as

$$\rho(h, f) = \sum_i c(h(\underline{z}_i) - f(\underline{z}_i)), \tag{5}$$

where $c$ is given in definition (2), a square being the conventional choice, and $\underline{z}_i$ are a set of points where the function needs to be evaluated. We first describe the algorithm that minimizes $\rho$ and then provide justification to the procedure and discuss the choice of $\underline{z}_i$.

### 3.1. Isotonic Algorithm

We now describe the details of optimizing for the isotonic additive (IA) $\hat{h}$ given an additive function $\hat{f}$. $\hat{h}$ is further required to be of the same functional form as $\hat{f}$. For notational convenience, we will again assume a bivariate interactions model. To simplify the presentation, we will mainly consider the square penalty in (5). The goal is to minimize the following least squares criterion:

$$\rho(h, \hat{f}) = \sum_i \left( \sum_{p \in U} h_p(z_p^{(i)}) + \sum_{p,r \in I} h_{pr}(z_p^{(i)}, z_r^{(i)}) - \left( \sum_{p \in U} \hat{f}_p(z_p^{(i)}) + \sum_{p,r \in I} \hat{f}_{pr}(z_p^{(i)}, z_r^{(i)}) \right) \right)^2 \quad (6)$$

where $\underline{z}^{(i)} = \underline{z}_i$ are the given points over which $\hat{f}$ is approximated as discussed above and $U$ ($I$) is the set of univariate effects (interactions) included in the model. As was noted above, functions $h_p$ and $h_{pr}$ will not be uniquely determined if the same coordinate enters several components. For example, for functional form $h = h_{12} + h_{13}$, there are many pairs of $h_{12}$ and $h_{13}$, different from each other by univariate functions of predictor 1, yielding the same $\hat{h}$. The first step of the algorithm is to extract function $\hat{h}$ in some representation. In particular, it makes computational sense to omit any univariate terms that are aliased by the interaction terms. Let us denote the remaining univariate terms by $U' = \{p \in U : \not\exists r \text{ such that } (p, r) \in I\}$ (we do not distinguish between pairs $(p, r)$ and $(r, p)$). More generally, one can consider other criteria than least squares, i.e.

$$\rho(h, \hat{f}) = \sum_i c_i \left( \sum_{p \in U} h_p(z_p^{(i)}) + \sum_{p,r \in I} h_{pr}(z_p^{(i)}, z_r^{(i)}) - \left( \sum_{p \in U} \hat{f}_p(z_p^{(i)}) + \sum_{p,r \in I} \hat{f}_{pr}(z_p^{(i)}, z_r^{(i)}) \right) \right) \quad (7)$$

for strictly convex differentiable functions $c_i$.

The algorithm follows a cyclic optimization technique described in [1]. As we mentioned in the introduction, the technique was applied there to the raw data and thus, typically resulted in rough and overfitting estimates. Here it is applied to the smooth unconstrained additive fit $\hat{f}$.

We outline a cycle of this technique. Functions $h_p$, $p \in U'$ and $h_{rs}$, $r, s \in I$ are updated one at a time by minimizing the criterion $\rho$ with respect to $h_p(\cdot)$ or $h_{rs}(\cdot, \cdot)$ respectively. The technique is cyclic because it "cycles" through all the additive components. During each step of the cycle, only one of the additive components is updated while the others are held at their previous values. The order within a cycle is established and we denote by $h_{[j]}$ the $j$th visited additive component. Thus, the $j$th visit during the $m$th cycle optimizes

$$h_{[j]}^{(m)} = \operatorname{argmin}_{h_{[j]}} \sum_i \left( \hat{f}(\underline{z}_i) - \sum_{k<j} h_{[k]}^{(m)}(\underline{z}_i) - h_{[j]}(\underline{z}_i) - \sum_{k>j} h_{[k]}^{(m-1)}(\underline{z}_i) \right)^2 \quad (8)$$

since all $h_{[k]}, k < j$ have been updated during the $m$th cycle and $h_{[k]}, k > j$ have not. If a set of criteria $c_i$ rather than least squares is used,

$$h_{[j]}^{(m)} = \operatorname{argmin}_{h_{[j]}} \sum_i c_i \left( \hat{f}(\underline{z}_i) - \sum_{k<j} h_{[k]}^{(m)}(\underline{z}_i) - h_{[j]}(\underline{z}_i) - \sum_{k>j} h_{[k]}^{(m-1)}(\underline{z}_i) \right) \quad (9)$$

Let $h^{(m)} = \sum_j h_{[j]}^{(m)}$, the estimate after the $m$th cycle. The properties of this algorithm are now outlined.

**Theorem 1.** If each minimizer $h_{[j]}^{(m)}$ in (9) is unique, $h^{(m)} \to \hat{h}$, a minimizer of (7).

To prove the theorem, we first show that the restriction that $h$ be isotonic is equivalent to the restriction that all $h_{[j]}$ be isotonic, that is, for any isotonic $h$ there exist isotonic components $h_{[j]}$ such that $h = \sum_j h_{[j]}$. This is the fact that was referred in the introduction.

**Lemma 1.** The restriction that $h$ be isotonic is equivalent to the restriction that all $h_{[j]}$ be isotonic.

Proof. For each predictor $p$ consider all $h_{pr_1}, \cdots, h_{pr_k}$ that depend on $x_p$. $f_p(x_p) = \sum_{i=1}^{k} h_{pr_i}(x_p, x_{r_i})$ is monotone for any $x_{r_1}, \cdots, x_{r_k}$, i.e.

$$\max_{x_{r_1},\cdots,x_{r_k}} (\sum_i h_{pr_i}(x_p = a_l, x_{r_i}) - \sum h_{pr_i}(x_p = a_{l+1}, x_{r_i})) =$$

$$\sum_i \max_{x_{r_i}} (h_{pr_i}(x_p = a_l, x_{r_i}) - h_{pr_i}(x_p = a_{l+1}, x_{r_i})) \le 0,$$

where $a_l$ and $a_{l+1}$ are neighboring values of $x_p$.
Whenever $m_i = \max_{x_{r_i}} (h_{pr_i}(x_p = a_l, x_{r_i}) - h_{pr_i}(x_p = a_{l+1}, x_{r_i})) > 0$, subtract that $m_i$ from $h_{pr_i}(x_p = a_{l+1}, x_{r_i})$:

$$h_{pr_i}(x_p = a_{l+1}, x_{r_i}) \leftarrow h_{pr_i}(x_p = a_{l+1}, x_{r_i}) - m_i$$

and add it to any other $h_{pr_j}(x_p = a_{l+1}, x_{r_j})$'s so that all $m_i \le 0$. Since we are adding/subtracting univariate functions, the functional form remains the same and so does the sum $h$.

Proof of Theorem 1. The algorithm has the same structure as in [1] and the proof carries over. It is an instance of global convergence of descent algorithms [15].

**Corollary.** (a) $h^{(m)} \to \hat{h}$, the unique minimizer of (7).
(b) Each $h_{[j]}^{(i)} \to h_{[j]}$ for some $h_{[j]}$.

Proof. See [8].

Observe that optimization steps (9) are an application of pool adjacent violators (PAV) algorithm ([5]). To apply the algorithm, function values $\hat{f}(z_i) - \sum_{k<j} h_{[k]}^{(m)}(z_i) - \sum_{k>j} h_{[k]}^{(m-1)}(z_i)$ are projected onto the coordinates $x_{i_j}$ of which $h_{[j]}$ is a function. In the case of ties in these coordinates, the weights are incremented appropriately. Now the problem is treated as isotonic regression in the $x_{i_j}$. The dimensionality of this step is that of the corresponding interaction. If $h$ is not required to be monotone along all of the $x_{i_j}$, PAV is carried out only in the direction of those along which it is.

A special case of the PAV algorithm can be briefly described as follows ([7]). Suppose we are given finite sequence $a_1, \cdots, a_m$, and our goal it to find nonincreasing $\hat{a}_1 \geq \cdots \geq \hat{a}_m$, such that $\sum_i (a_i - \hat{a}_i)^2$ is minimized. Starting with $a_1$, we move to the right and stop at the first "violation" $a_i < a_{i+1}$. We pool $a_i$ and $a_{i+1}$ together and replace them with their average: $a_i, a_{i+1} \leftarrow (a_i + a_{i+1})/2$. We then move to the left to make sure that $a_{i-1} \geq a_i$ - if not, we pool $a_{i-1}$ together with the $a_i$ and $a_{i+1}$ and replace the three with their average. The backtracking continues until the violation is removed. We then continue to the next sequence entry $i + 2$. In Figure 2, PAV is applied to the dotted line (at the integer arguments) to produce the dashed line. In the case of the weighted problem that arises above when there are ties in the data set, the target sum becomes $\sum_i w_i(a_i - \hat{a}_i)^2$, where $w_i$ are the weights. The PAV pooling together takes form $a_i, a_{i+1} \leftarrow (w_i a_i + w_{i+1} a_{i+1})/(w_i + w_{i+1})$, etc. The corresponding problem in more than one dimension is solved in [5] by applying PAV along one dimension at a time and redefining the target sequences appropriately during each iteration.

Part (a) of the corollary above implies that this algorithm converges to the isotonic additive function $\hat{h}$ that best approximates the original additive function $\hat{f}$. Part (b) says that in addition, there is a limit for each additive component. However, since there are many ways to represent $\hat{h}$ and also because this limit would generally depend on the ordering of the additive components $h_{[j]}$, these limits may not in general be interpreted as the corresponding main or interaction effects. The interaction and univariate effects are now mixed in together, and our next goal is to redefine them, so that they can be given the same interpretation as the corresponding additive components of $\hat{f}$. Note that function $\hat{h}$ is not changed during this process, so the solution to the estimation (as opposed to interpretation) problem is not affected by this step.

We first consider those predictors $p$ for which the univariate effects are included into the model. In unconstrained additive modeling, [10], Section 9.5.3, suggest imposing the following identifiability constraints on, say, term $h_{12}(x_1, x_2)$: $E(h_{12}(X_1, X_2)|X_i) = 0, i = 1, 2$. To this end, the main effect additive model is fitted to $h_{12}$ via backfitting, since this constraint can be satisfied by first choosing $h_i(X_i)$ such that $E(h_{12}(X_1, X_2) - h_i(X_i)|X_{3-i}) = h_{3-i}(X_{3-i}), i = 1, 2$ and then redefining $h_{12}(x_1, x_2) \leftarrow h_{12}(x_1, x_2) - h_1(x_1) - h_2(x_2)$. This expression matches a backfitting update. In isotonic modeling, we can instead fit an isotonic additive model to $h_{12}$ by minimizing $\rho(h_{12}(x_1, x_2), h_1(x_1) + h_2(x_2))$ subject to $h_1$ and $h_2$ being monotone (if monotonization is carried out along both $x_1$ and $x_2$). The same isotonizing algorithm is used; incidentally, this also resembles backfitting. When more than one interaction term involves $x_1$, say, the corresponding effects are added together to produce the main effect of $x_1$. Thus, as with the UA models and the parametric approach of Section 2, the fitting method itself is used on each higher order term to extract lower order terms.

We would like to mention here an important special case, where the domain of an interaction is a rectangle and the weights in matching $h_{12}$ with $h_1(x_1) + h_2(x_2)$ are the same over the domain (this parallels fitting an unweighted additive model). In this case, $h_1(x_1) = \bar{h}_{12}(x_1, \cdot)$ and $h_2(x_2) = \bar{h}_{12}(\cdot, x_2)$ assuming there is an intercept term in the model and so $\bar{h}_{12}(\cdot, \cdot) = 0$. Note that they are monotone because $h_{12}$ is as a limit of $h_{12}^{(m)}$.

We now turn to those predictors whose univariate effects are not included into the model. Now instead of extracting part of a univariate effect from each interaction term and adding these pieces together, our goal is to juggle around these effects in order to achieve the closest match with the representation $\hat{f}$. To understand, consider a simple example: $\hat{f} = \hat{f}_{12} + \hat{f}_{13}$, $\hat{h} = h_{12} + h_{13}$. Suppose now that we do not need to include the univariate effect for predictor one into the isotonic model. Therefore, we can arbitrarily add a univariate function $v(x_1)$ to $h_{12}$ and subtract it from $h_{13}$ without introducing any change $\hat{h}$. We would like to choose $v(x_1)$ as to minimize $\rho(h_{12} + v, \hat{g}_{12}) + \rho(h_{13} - v, \hat{g}_{13})$. For a general additive model, we are to minimize

$$\sum_{p<r} \rho(\hat{g}_{pr}, h_{pr} + v_p^{(r)} I_{\{p \notin U\}} + v_r^{(p)} I_{\{r \notin U\}}) =$$

$$\sum_{p<r} \sum_{z_p, z_r} (\hat{g}_{pr}(z_p, z_r) - h_{pr}(z_p, z_r) - v_p^{(r)}(z_p) I_{\{p \notin U\}} - v_r^{(p)}(z_r) I_{\{r \notin U\}})^2$$

over functions $v_p^{(r)}$ subject to $\sum_{r \in V_p} v_p^{(r)}(z_p) = 0$ for all $p$, $z_p$, where $V_p = \{r : p, r \in I\}$, those predictors $r$ whose interactions with $p$ enter the model. If each interaction is defined on a rectangle, we have the solution in closed form. It is presented in the appendix. Even if the interactions are not defined on a rectangle, this is still a standard optimization problem.

### 3.2. Removing Flat Spots

As can be seen in Figure 2 (dashed line), the monotone additive function $\hat{h}$ obtained at the end of the two-step procedure may have regions over which it is flat. As a simple example as to the origin of a flat region, the reader can quickly be convinced that the closest nonincreasing match to $\{5, 4, 3, 4\}$ is $\{5, 4, 3.5, 3.5\}$, thus having a flat spot on the right side. Such behavior may not be visually appealing. Then one could trade the proximity of $\hat{h}$ to $\hat{f}$ for a smoother $\hat{h}'$.

In many situations we would expect strictly monotone response, which would make flat spots infeasible. Since such an "expectation" is difficult to translate into a useful optimization constraint, we will instead propose a heuristic scheme that is likely to remove most of flat regions via additional smoothing.

We again alternate between smoothing and isotonizing steps. Given the closest IA function $\hat{h}$ to $\hat{f}$, $\hat{h}$ is smoothed to produce $f^{(1)}$. That brings in values from surrounding nonflat areas to smooth out a flat spot. Since $f^{(1)}$ need not be isotonic, it is monotonized to produce $h^{(1)}$. In our implementation, we used the same additive smoother as the one to fit the original data. Thus, $f^{(1)}$ and consequently $h^{(1)}$ retain the same functional form. Intuitively, if this additional smoothing were carried out by a projection operator, by idempotency, the fit should not be affected much in places where there was no monotone correction. In general, however, extra smoothing may worsen the fit. To alleviate this, small amount of smoothing may be given to $\hat{h}$ and if this does not deal with flat spots adequately, $h^{(1)}$ is further smoothed into $f^{(2)}$ and the latter is monotonized into $h^{(2)}$. One may also decide to stop early if the fit is altered in other places. In anticipation of this additional smoothing, the original model may be chosen to overfit the data to an extent. This overfitting will then be compensated here. Also, if the fitting procedure allows for different amount on smoothing in different regions, more smoothing may be carried out in the vicinity of flat spots. This process is then iterated until flat spots are removed. We deliberately remain vague as to the stopping criterion and other fine points as we are dealing with very much a heuristic process.

Another variation that [8] entertained here is to subtract further some of the correction produced by the isotonic transformation before further smoothing, thus moving away from $\hat{g}$. That is, instead of smoothing $\hat{h}$, one smooths $\hat{h} - u(\hat{f} - \hat{h})$, where $0 \leq u$ is a sort of a learning rate. This will accentuate a problem region. See Figure 3b for an illustration.

### 3.3. Discussion

The idea of using a combination of smoothing and a monotonic transformation goes back to [7]. In that work, given data pairs $(x_i, y_i)$ with univariate $x$, the authors first smoothed the data using the running mean smoother with the span chosen by cross-validation to produce smoothed response $s_i$. The $s_i$ were then monotonized using the PAV algorithm.

In [16], the author derived asymptotics for estimates $m_{SI}$ and $m_{IS}$ obtained by smoothing-isotonization and isotonization-smoothing combinations respectively, where kernel smoothing with an asymptotically optimal bandwidth was used.

The algorithm in Section 3.1 used these ideas in the context of fitting additive models. It was shown that the iterative application of PAV works for the isotonization step. It is implementationally and computationally simpler than solving a true penalized smoothing problem described in the introduction. Also, it was explained how to sensibly define the main effects and the interactions once the isotonic function is constructed.

Figure 2 is a quick illustration of the whole process in a one dimensional case. The data (the solid line at the integer arguments) are smoothed to produce the dotted fit. It is not monotone, so the PAV algorithm results in the dashed line. That has a flat spot towards the right end of the range. It can be removed thereby resulting in the fit whose fitted values are shown.

We begin the discussion by considering the population case. Suppose we have a random pair $(\underline{X}, Y)$ and the goal is to predict $Y$ from $\underline{X}$ under the squared error loss. As explained in detail in Section 5.2.1 of [10],

$$\hat{f} = \operatorname{argmin}_{f \in H_{add}} E(Y - f(\underline{X}))^2,$$

where $H_{add}$ is a linear subspace of additive functions $f$ acting on $\underline{X}$. Residual $Y - \hat{f}(\underline{X})$ is perpendicular to $H_{add}$. In particular, any additive function $h$ in $H_{add}$ satisfies

$$E(Y - h(\underline{X}))^2 = E(Y - \hat{f}(\underline{X}))^2 + E(\hat{f}(\underline{X}) - h(\underline{X}))^2 \tag{10}$$

Since any IA function is in $H_{add}$, from the population case perspective, minimizing $E(Y - h(\underline{X}))^2$ is equivalent to minimizing $E(\hat{f}(\underline{X}) - h(\underline{X}))^2$. This suggests that finding the closest isotonic approximant to the best additive fit is appropriate.

In the data case, carrying out the two-step procedure offers several advantages over the parametric method described and the approach of matching the data points themselves. Starting with an unconstrained additive model allows for more accurate model selection, in particular, mediating the problem of getting rugged functions and overfitting. Also, one is able to estimate the functional form and interpret the model more accurately. UA model also allows one to work with smoothers that do not admit the isotonic constraint, such as *loess* ([10], [4]). Smoothers like *loess* are particularly important in problems where extrapolation is required; they are easier to control than, say, a smoothing spline.

In case when the isotonicity assumption is in question, one may argue that starting with an unconstrained fit provides a natural setting for testing. [2] presents a test for monotonicity in a univariate setting that is based on resampling the residuals of the constrained fit and observing the amount of smoothness needed to remove all flat spots. The same idea may be used here. Additionally, one can use the difference between the constrained and the unconstrained fits as a test statistic. We are not aware of any relevant literature. Also possible is testing monotonicity along subsets of variables.

One technical concern is the discrete nature of (5). It is chosen for computational convenience, but gives rise to the following problem. Suppose two different overlapping sets of $\underline{z}_j$, $\underline{z}_j^{(1)}$ and $\underline{z}_j^{(2)}$ were used separately to carry out the analysis to obtain $\hat{h}^{(1)}$ defined over $\underline{z}_j^{(1)}$ and $\hat{h}^{(2)}$ defined over $\underline{z}_j^{(2)}$. Let $\underline{z}_j^{(0)} = \underline{z}_j^{(1)} \cap \underline{z}_j^{(2)}$. Then in general, $\hat{h}^{(1)}(\underline{z}_j^{(0)}) \neq \hat{h}^{(2)}(\underline{z}_j^{(0)})$, and that is not nice. What it means in practice is that one cannot in general carry out several analyses locally and expect them to be consistent together. A solution is of course for the $\underline{z}_j$ to represent a dense grid, essentially approximating a continuous problem. In particular, the penalty $\sum_j c(h(\underline{z}_j) - \hat{f}(\underline{z}_j)) \approx Z \int c(h(\underline{z}) - \hat{f}(\underline{z})) d\underline{z}$ for a normalizing constant $Z$. This results in greater computational burden, but note that the additive form only requires working with projections of the grid rather than the whole grid itself.

Last, the issue of "flat spots" is addressed. It was seen that the IA $\hat{h}$ obtained at the end of the two-step procedure may have regions over which it is flat (Figure 2, dashed line). While this $\hat{h}$ is the closest isotonic approximant of $\hat{f}$ and is thus justifiable from the population point of view as discussed above, such behavior may not be visually appealing. In such a case, one could trade the proximity of $\hat{h}$ to $\hat{f}$ for a smoother $\hat{h}'$. [8] proposed a heuristic iterative scheme where the smoothing and isotonizing steps alternate in order to remove such flat spots via additional smoothing.

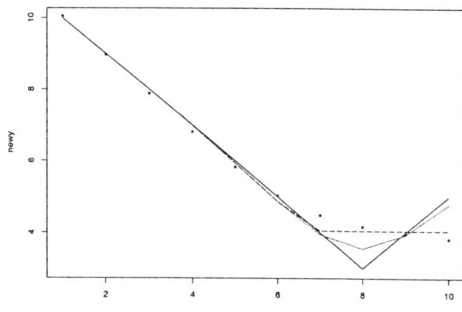

Figure 2. Data (solid), smooth (dotted), monotonized (dashed) and flat-spot-removed (points) curves.

## 4. Application: Computer Cache Rates

We applied the methodology of Section 3 to data obtained by running computer cache simulation for a variety of cache architectures. Here we present the construction of the model of the relationship between the total cache miss rate and the architectural parameters. The way caches work is described in [12]. Such model is important because due to technical and resource limitations ([14]), it is usually not possible to carry out cache simulations for all cache configurations of interest.

For the purposes of this study, $y = \log(\text{total miss rate})$ will be referred as response and the log-architectural parameters as predictors, which are:

- $x_1 = \log(\text{cache size})$
- $x_2 = \log(\text{cache sharing}/\#\text{CPUs})$
- $x_3 = \log(\#\text{CPUs})$
- $x_4 = \log(\text{cache associativity})$

These are described in detail in [12]. All variables are treated as continuous. 133 data points are given. The response is expected to be decreasing in predictors $x_1$ and $x_4$. Further details of these data and the problem in general can be found in [9].

The first modeling step is fitting an unconstrained additive model. The following model selection scheme is in the spirit of stepwise techniques in [10], Section 9.4. Each additive component has a regimen of options for the degree of smoothness, the equivalent degrees of freedom (df), including a zero for leaving the term out of the model. The model is fitted with the starting values for each term and the fit is evaluated based on some predefined criterion. We then vary df for each term, one at a time, and evaluate the fits. If any of them improves the criterion, we move the model to the greatest improvement and vary the dfs again. We continue until none of the changes improves the model. The process necessarily converges, since every step improves the criterion. The criterion used is 10-fold cross-validation ([11]). We use the following regimens of terms; exactly one term is to be chosen from each regimen:

- (skip, smooth$(x_1, x_2)$, rough$(x_1, x_2)$, smooth$(x_1, x_2, x_3)$)
- (skip, smooth$(x_1, x_3)$, rough$(x_1, x_3)$)
- (skip, smooth$(x_1, x_4)$, rough$(x_1, x_4)$)
- (skip, smooth$(x_2, x_3)$, rough$(x_2, x_3)$, smooth$(x_2, x_3, x_4)$)
- (skip, smooth$(x_2, x_4)$, rough$(x_2, x_4)$)
- (skip, smooth$(x_3, x_4)$, rough$(x_3, x_4)$)

Skip means not to include any of the terms in a regimen. The terms within each regimen are sorted from left to right in the ascending number of the dfs. That is, a fitted "rougher" term interpolates the data better, but has a larger estimation variance. We are seeking a balance between these two sources of error. We also include a limited number of trivariate interactions. Univariate effects may also be included for interpretation purposes, but are omitted in this study.

The smoothing method used to fit the additive components is loess, a running locally-weighted plane regression ([10], [4]).

Here we present the run of the stepwise technique described earlier. Each step it picked a regimen and then (though having a capability to move in either direction) went to the right, thus, increasing the df of the model. We also include the estimated error (using 10-fold cross-validation).

- Null model (intercept only); error = 0.784
- Step 1: Add smooth($x_1, x_2$); error = 0.243
- Step 2: Add smooth($x_3, x_4$); error = 0.107
- Step 3: Increase to rough($x_1, x_2$); error = 0.089
- Step 4: Add smooth($x_1, x_4$); error = 0.074

Roughly speaking, the final model explains 1 - 0.074/0.784 = 90.5% of the variabil-

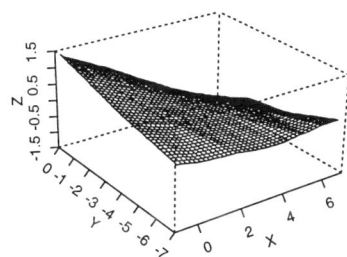

Figure 3. rough($x_1, x_2$)

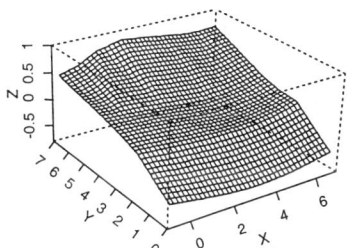

Figure 4. smooth($x_1, x_3$)

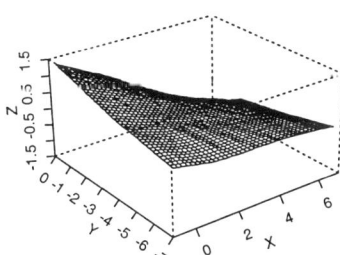

Figure 5. ($x_1, x_2$) - component of $\hat{h}$

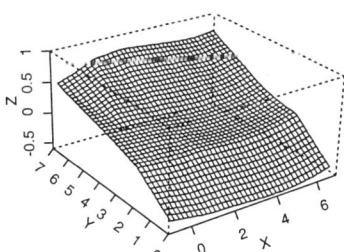

Figure 6. ($x_1, x_3$) - component of $\hat{h}$

ity in the data set. We also decided to add term smooth($x_1, x_3$) based on a 2-dimensional residual plot (not shown) indicating a moderate interaction effect, although the error

estimate went slightly up to 0.082. No trivariate interactions were included.
The final model is

$$f(x_1, x_2, x_3, x_4) = \text{rough}(x_1, x_2) + \text{smooth}(x_1, x_3) + \text{smooth}(x_1, x_4) + \text{smooth}(x_3, x_4) \quad (11)$$

Additive components $\text{rough}(x_1, x_2)$ and $\text{smooth}(x_1, x_3)$ are shown in Figures 3 and 4. The second modeling step is for (11) to undergo the isotonic transformation. The first two components of the IA function $\hat{h}$ are shown in Figures 5 and 6. The additive components of $\hat{h}$ are made to resemble those of $\hat{f}$ as closely as possible as described in Section 3.1. The overall average difference between $\hat{f}$ and $\hat{h}$, $\sqrt{\text{Ave}\left(\hat{f}(\underline{x}) - \hat{h}(\underline{x})\right)^2}/\text{Ave}(|\hat{f}(\underline{x})|)$, is about 8% of $|\hat{f}|$.

**Appendix. Solution to the optimization problem with rectangular domain interactions**

**Lemma 1.** $\sum_{\underline{z}} \hat{g}(\underline{z}) - \hat{h}(\underline{z}) = 0$

*Proof.* This is a direct implication of Theorem 1.4 of [3]

Therefore, let us adjust $h_{pr}$ so that $\sum_{z_p, z_r} h_{pr}(z_p, z_r) = \sum_{z_p, z_r} \hat{g}_{pr}(z_p, z_r)$. Then we have the following.

**Lemma 2.** There is a solution $\hat{v}_r^{(p)}(z_r)$ for which for any $p$ and $r$, $\sum_{z_r} \hat{v}_r^{(p)}(z_r) = 0$.

*Proof.* Suppose we have found a solution $v_p^{(r)}, p \notin U$. Since $\sum_{z_p, z_r} h_{pr}(z_p, z_r) - \sum_{z_p, z_r} \hat{g}_{pr}(z_p, z_r) = 0$, without the constraint, the solution can be improved by subtracting from $v_p^{(r)}$ their means, so that $\sum_{z_p, z_r} v_p^{(r)}(z_p) I_{\{p \notin U\}} + v_r^{(p)}(z_r) I_{\{r \notin U\}} = 0$ (unless it is already zero). But that is possible to do even under the constraint: since $\sum_r v_p^{(r)}(z_p) = 0, \sum \bar{v}_p^{(r)}(\cdot) = 0$ and so $\sum_{z_p} (v_p^{(r)}(z_p) - \bar{v}_p^{(r)}(\cdot)) = 0$.

For minimization, we use Lagrange multipliers. Taking the derivative with respect to $v_p^{(r)}(z_p), p \notin U$, obtain

$$-\sum_{z_r} (\hat{g}_{pr}(z_p, z_r) - h_{pr}(z_p, z_r) - v_p^{(r)}(z_p) - v_r^{(p)}(z_r) I_{\{r \notin U\}}) = \lambda_{p, z_p}$$

By Lemma 2, the last term sums to zero. We sum both sides over $r \in V_p$. We use the constraint $\sum_{r \in V_p} v_p^{(r)}(z_p) = 0$ to get

$$\lambda_{p, z_p} = -\sum_{r \in V_p, z_r} (\hat{g}_{pr}(z_p, z_r) - h_{pr}(z_p, z_r))/\#V_p$$

We plug this back to arrive at a solution:

$$v_p^{(r)}(z_p) = \sum_{z_r} (\hat{g}_{pr}(z_p, z_r) - h_{pr}(z_p, z_r))/N_r + \lambda_{p, z_p} \quad (12)$$

We note that to use (12), one first adjusts $\sum_{z_p, z_r} h_{pr}(z_p, z_r) = \sum_{z_p, z_r} \hat{g}_{pr}(z_p, z_r)$ as above.

# REFERENCES

1. P. Bacchetti, "Additive Isotonic Models," *JASA*, vol. 84, no. 405, pp. 289-294, 1989.
2. A. Bowman, M. Jones, I. Gijbels. "Testing Monotonicity of Regression," *Journal of Computational and Graphical Statistics*, vol. 7, no. 4, pp. 489-500, 1998.
3. R. E. Barlow, D. J. Bartholomew, J. M. Bremner and H. D. Brunk. *Statistical Inference under Order Restrictions*. John Wiley & Sons, 1972.
4. W. S. Cleveland and S. J. Devlin. "Locally-Weighted Regression: an Approach to Regression Analysis by Local Fitting," *JASA*, vol. 83, pp. 597-610, 1988.
5. R. L. Dykstra and T. Robertson. "An Algorithm for Isotonic Regreession for Two or More Independent Variables," *The Annals of Statistics*, vol. 10, no. 3, pp. 708-716, 1982.
6. J. H. Friedman. "Multivariate Adaptive Regression Splines," *Annals of Statistics*, 19(1), pp. 1-141, 1991.
7. J. H. Friedman and R. Tibshirani. "The Monotone Smoothing of Scatterplots," *Technometrics*, vol. 28, no. 3, pp. 243-250, 1984.
8. I. Gluhovsky. "Smooth Isotonic Additive Interaction Models with Application to Cache Architecture Design," submitted to *Technometrics*.
9. I. Gluhovsky and B. O'Krafka. "Interpolation and Extrapolation of Multiprocessor Cache Rates Using Multivariate Models," manuscript in preparation.
10. T. J. Hastie and R. J. Tibshirani. *Generalized Additive Models*. Chapman and Hall, 1990.
11. T. J. Hastie, R. J. Tibshirani and J. H. Friedman. *The Elements of Statistical Learning: Data Mining, Inference, and Prediction*. Springer-Verlag New York, 2001.
12. J. L. Hennessy, D. A. Patterson. *Computer Architecture: A Quantitative Approach. Third Edition*. Morgan Kaufmann Publishers, 2003.
13. C. Kelly and J. Rice. "Monotone Smoothing with Application to Dose-Response Curves and the Assessment of Synergism," *Biometrics*, vol. 46, pp. 1071-1085, 1990.
14. S. Kunkel, R. Eickemeyer, M. Lipasti, T. Mullins, B. O'Krafka, H. Rosenberg, S. VanderWeil, P. Vitale, L. Whitley. "A Performance Methodology for Commercial Servers," *IBM Journal of Research and Development*, vol. 44, no. 6, 2000.
15. D. G. Luenberger. *Linear and Nonlinear Programming* (2nd ed.) Reading, MA: Addison-Wesley, 1984.
16. E. Mammen. "Estimating a Smooth Monotone Regression Function," *The Annals of Statistics*, vol. 19, no. 2, pp. 724-740, 1991.
17. J. O. Ramsay. "Monotone Regression Splines in Action," *Statistical Science*, vol. 3, no. 4, pp. 425-461, 1988.
18. J. O. Ramsay. "Estimating Smooth Monotone Function," *Journal of The Royal Statistical Society*, vol. 60, part 2, pp. 365-375, 1998.
19. T. J. Hastie and R. J. Tibshirani. Discussion of J. O. Ramsay, "Monotone Regression Splines in Action," *Statistical Science*, vol. 3, no. 4, pp. 425-461, 1988.
20. Z. Wang. "An Algorithm for Generalized Monotonic Smoothing," *Journal of Applied Statistics*, vol. 27, no. 4, pp. 495-507, 2000.

*Recent Advances and Trends in Nonparametric Statistics*
Michael G. Akritas and Dimitris N. Politis (Editors)
© 2003 Elsevier Science B.V. All rights reserved.

# A Nonparametric Alternative to Analysis of Covariance

Arne Bathke [a] and Edgar Brunner [b]*

[a]Department of Statistics, University of Kentucky, 875 Patterson Office Tower, Lexington, KY 40506-0027, USA, email: arne@ms.uky.edu

[b]Department of Medical Statistics, University of Göttingen, Humboldtallee 32, D-37073 Göttingen, Germany, email: ebrunne1@gwdg.de

    The Analysis of Covariance (ANCOVA) is designed for the many practical situations in which factor effects are obscured by concomitant variables, or the main purpose of the investigation lies in assessing the effect of the concomitant variables. Not taking covariates into account may cause unprecise or biased results.

    However, if the response variable or the covariate are only measured on an ordinal scale (like typically psychological and other scores or grading scales), or if they show distinct nonnormal distributions, one would be reluctant to use parametric ANCOVA methods.

    In this paper, we consider a nonparametric model with covariates. The information contained in the covariates is used to minimize the variance of certain nonparametric estimators for the response variable.

    This model combines the power gain through introduction of covariates into a factorial design with the robustness of nonparametric procedures. We discuss asymptotic test procedures for inference about factor effects as well as for testing the effect of a covariate. To apply the suggested methods to real data, a SAS macro is provided and available for download. The use of this macro is briefly explained.

    The tests can be used for data with ties, and even for purely ordinal data, including ordinal covariates. The number of covariates that can be included into the model is not restricted. Simulations show extremely good small-sample performance. In many situations, the proposed tests only require sample sizes around 10.

## 1. Introduction

    In many biological experiments and ecological, psychological, or medical studies, the subjects are observed repeatedly under the same or under different conditions, described by a factorial design. Usually not only a response variable is measured, but there is also information available on other variables that are called covariates. These covariates can be nuisance variables that obscure the factor effects. Or it can be of its own interest to assess the effect of the covariates on the response variable. In the first case, not taking covariates into account may cause unnecessarily biased or unprecise results when analyzing the effect of the fixed factors.

---
*Supported in part by DFG/Br-655/12

There exist several more or less complicated models for the analysis of factorial designs with covariates. The parametric golden standard is the Analysis of Covariance (ANCOVA) which is described in many textbooks (see, e.g. [1]). However, there are also semi-parametric models as, e.g., the proportional odds model (refer to the monographs by Agresti [2,3] for a comprehensive treatment of this and related models), and there are completely nonparametric approaches [4–6]. Semiparametric models require assumptions that are sometimes not justifiable or hard to verify and it is difficult–if possible at all–to handle ordinal covariates within that framework. Moreover, the application of these models becomes difficult if the number of ordered categories is large. The above mentioned nonparametric approaches also do have their drawbacks. Quade [4] and Puri and Sen [5] were only considering the one-way model, and they were assuming continuous distribution functions. Thus, their methods are not applicable to ordinal data. The approach of Akritas, Arnold and Du [6] allows for ties, but is restricted to a maximum of three covariates.

In this paper, we consider a novel nonparametric analog to the parametric ANCOVA models that attempts to overcome the mentioned drawbacks. This model is based on the theory of rank methods for factorial designs introduced in [7] and [8]. It does not assume continuous distribution functions and can therefore be used for purely ordinal data and ordinal covariates. Also, the number of possible covariates to be included into the model is not limited. The resulting test statistics are asymptotically distribution-free, and they are invariant under any monotonic transformation of the data since they only use the ranks of the observations. As compared to other robust methods, the derived test statistics have the great advantage of an excellent small sample performance. The proposed methodology also works for dependent observations, but this paper will only treat the case of independent data. Some of the more technical details, as well as related material, can be found in [9–11].

The methods described in this paper are based on the nonparametric model for factorial designs that has been receiving considerable attention over the last decade. Akritas and Arnold [7] provided the simple but powerful idea to formulate the hypotheses in a two-way repeated measures design by means of the distribution functions. Based on this idea of the nonparametric hypotheses, Akritas, Arnold, and Brunner [8] developed rank tests for nonparametric main effects and interactions in an unbalanced two-factorial design without assuming continuous distribution functions. Thereby, the use of the normalized distribution function, as suggested in [12], allows for an elegant treatment of ties, since continuous and discontinuous distribution functions are handled in a unified form. Munzel and Brunner [13] were extending the results to multivariate factorial designs. For a review on nonparametric methods in factorial designs, see [14].

## 2. Basic Definitions

In the following, we introduce a nonparametric model with covariates. For simplicity, we introduce the model in a one-way layout. However, the approach is very general and the methods are not limited to a one-way model. They can be extended straightforward to higher factorial designs as described, e.g., in [15]. More detailed descriptions of this "alternative ANCOVA" can be found in [9–11].

Consider a one-factorial model with a fixed factor $A$ (levels $i = 1, \ldots, a$) and $n_i$ independent replications ($j = 1, \ldots, n_i$) for each level $i$ of factor $A$. The results can obviously be applied to other factorial models with independent observations just by imposing a special structure on the index $i$, i.e. $i = 1, \ldots, a$ is split into $i_1 = 1, \ldots, i_{a_1}$ and $i_2 = 1, \ldots, i_{a_2}$ and so forth. $N = \sum_{i=1}^{a} n_i$ is the total number of observations.

The response variables are denoted by $X_{ij}^{(0)}$, the corresponding covariates by $X_{ij}^{(r)}$, $r = 1, \ldots, d$, and $F_i^{(r)}(x)$ denotes the marginal distribution function of the random variables $X_{ij}^{(r)}$.

Table 1
Scheme for the $N$ responses variables and the covariates. For each factor level $i$ of factor $A$, $n_i$ responses are drawn from a marginal distribution $F_i^{(0)}$ (analogously for the covariates).

| Factor Level | Variables and Distribution Functions | | | | | |
|---|---|---|---|---|---|---|
| | Response | | Covariate 1 | | ... | Covariate $r$ |
| 1 | $X_{11}^{(0)}$ $\vdots$ $X_{1n_1}^{(0)}$ | $F_1^{(0)}$ | $X_{11}^{(1)}$ $\vdots$ $X_{1n_1}^{(1)}$ | $F_1^{(1)}$ | ... | $X_{11}^{(r)}$ $\vdots$ $X_{1n_1}^{(r)}$ $F_1^{(r)}$ |
| $\vdots$ | $\vdots$ | | $\vdots$ | | | $\vdots$ |
| $a$ | $X_{a1}^{(0)}$ $\vdots$ $X_{an_a}^{(0)}$ | $F_a^{(0)}$ | $X_{a1}^{(1)}$ $\vdots$ $X_{an_a}^{(1)}$ | $F_a^{(1)}$ | ... | $X_{a1}^{(r)}$ $\vdots$ $X_{an_a}^{(r)}$ $F_a^{(r)}$ |

Here, $F_i(x) = \frac{1}{2}(F_i^+(x) + F_i^-(x))$ denotes the normalized version of the $i$th marginal distribution function (see [12]), where $F_i^+(x) = P(X_{ij}^{(0)} \leq x)$ is the right continuous version and $F_i^-(x) = P(X_{ij}^{(0)} < x)$ is the left continuous version of the distribution function.

Finally, let $H^{(r)}(x) = N^{-1} \sum_{i=1}^{a} n_i F_i^{(r)}(x)$ denote the weighted average of the distribution functions. Note that the distribution functions $F_i^{(r)}$ need not necessarily be continuous. Even purely ordinal data is included within the framework of this model.

We assume that all observations $X_{ij}^{(0)}$, $i = 1, \ldots, a$, $j = 1, \ldots, n_i$, are independent. It is possible to include dependent replications for each subject into the model. However, the more complicated case of dependent observations will be considered in a separate paper.

The distribution functions $F_i^{(r)}$ are estimated by their empirical counterparts

$$\hat{F}_i^{(r)}(x) = n_i^{-1} \cdot \sum_{j=1}^{n_i} c(x - X_{ij}^{(r)}), \quad \text{where } c(t) = \begin{cases} 0, & t < 0 \\ 1/2, & t = 0 \\ 1, & t > 0, \end{cases}$$

and $H^{(r)}$ is estimated accordingly by $\hat{H}^{(r)}(x) = N^{-1} \cdot \sum_{i=1}^{a} \sum_{j=1}^{n_i} c(x - X_{ij}^{(r)})$. Also, the relative treatment effects

$$p_i^{(r)} = \int H^{(r)} dF_i^{(r)} \tag{1}$$

are estimated straightforward by

$$\hat{p}_i^{(r)} = \int \hat{H}^{(r)} d\hat{F}_i^{(r)}. \tag{2}$$

To derive a rank representation of the estimators $\hat{p}_i^{(r)}$, we define the asymptotic rank transforms (ART) of $X_{ij}^{(r)}$ by $Y_{ij}^{(r)} = H^{(r)}(X_{ij}^{(r)})$, as well as the rank transforms (RT)

$$\hat{Y}_{ij}^{(r)} = \hat{H}^{(r)}(X_{ij}^{(r)}) = N^{-1}(R_{ij}^{(r)} - \frac{1}{2}), \ r = 0, 1, \ldots, d, \tag{3}$$

where $R_{ij}^{(r)}$ denotes the (mid-)rank of the random variable $X_{ij}^{(r)}$ among all the $N$ observations $X_{11}^{(r)}, \ldots, X_{an_a}^{(r)}$. Here, $r = 0$ refers to the response variable while $r = 1, \ldots, d$ refers to the covariates. Note that the rankings $R_{11}^{(r)}, \ldots, R_{an_a}^{(r)}$ are performed separately within the response variable ($r = 0$) and within each covariate ($r = 1, \ldots, d$). Thus, $1 \leq R_{ij}^{(r)} \leq N$.

Let $\mathbf{Y}^{(r)} = (Y_{ij}^{(r)})_{i=1,\ldots,a;\ j=1,\ldots,n_i}$ and $\hat{\mathbf{Y}}^{(r)} = (\hat{Y}_{ij}^{(r)})_{i=1,\ldots,a;\ j=1,\ldots,n_i}$ denote the $N$-dimensional vectors of the ART and the RT, respectively. Means are denoted in the usual way by using the overline-dot notation, e.g., $\bar{Y}_{i.}^{(0)} = n_i^{-1} \cdot \sum_{j=1}^{n_i} Y_{ij}^{(0)}$. Finally, let $\mathbf{F}^{(0)} = (F_1^{(0)}, \ldots, F_a^{(0)})'$ denote the vector of the distribution functions of the response variable.

In order to derive tests on the factor effects, let

$$\bar{\mathbf{Y}} = \big((\bar{\mathbf{Y}}^{(0)})', \ldots, (\bar{\mathbf{Y}}^{(d)})'\big)' \tag{4}$$

denote the vector containing the cell means of the asymptotic rank transforms for the response variable and all covariates. Here, $\bar{\mathbf{Y}}^{(r)} = (\bar{Y}_{1.}^{(r)}, \ldots, \bar{Y}_{a.}^{(r)})'$ refers to the vector of the cell means of the ART of the $r$th variable.

When evaluating the effect of the covariate, we use the following adjustment for factor effects. Define $D_{ij}^{(r)} = Y_{ij}^{(r)} - \bar{Y}_{i.}^{(r)}$, and $\hat{D}_{ij}^{(r)}$ analogously. The respective vectors are denoted by $\mathbf{D}^{(r)}$ and $\hat{\mathbf{D}}^{(r)}$. Thus, $\mathbf{D}^{(r)}$ can be written as

$$\mathbf{D}^{(r)} = \Big(\bigoplus_{i=1}^{a} \mathbf{P}_{n_i}\Big) \mathbf{Y}^{(r)}, \tag{5}$$

where $\mathbf{P}_{n_i} = \mathbf{I}_{n_i} - n_i^{-1} \mathbf{J}_{n_i}$ denotes the so-called centering matrix and $\mathbf{J}_{n_i} = \mathbf{1}_{n_i} \mathbf{1}_{n_i}'$.

## 3. Variance Minimization for Treatment Effect Estimators in a Completely Randomized Design

In this section, we develop inferential tools for a "nonparametric ANCOVA" within the framework of completely randomized designs (CRD). CRD are very common in clinical trials, where patients are randomly allocated to different treatment groups. These different treatments groups should be as homogeneous as possible with regard to potentially confounding variables like age, gender, race, baseline disease status, and others. Otherwise, the results –regarding the variable of primary interest– might be biased due to the confounders. For the rest of this section, we will assume a CRD. That is, the marginal distributions of each covariate (representing a confounding variable) can be assumed to be equal for all different treatment levels. Yet, the additional information contained in observing them can be used to improve inference about the treatment effects. The case of unequal marginal distributions of the covariates will be discussed briefly in Section 4. Analogously to the parametric ANCOVA, we will utilize the covariates to minimize the variance of certain estimators for the variable of primary interest.

In the papers mentioned at the end of Section 1, nonparametric methods have been developed for factorial designs. These methods are based on the relative treatment effects $p_i$ and their empirical counterparts $\hat{p}_i$. The asymptotic multivariate distribution of the vector $\hat{\mathbf{p}}^{(0)} = (\hat{p}_1^{(0)}, \ldots, \hat{p}_a^{(0)})'$ under the nonparametric null hypothesis is used to construct test statistics.

The main goal of taking covariates into account is improving the precision of the estimators $\hat{p}_i^{(0)}$. Analogously to the parametric ANCOVA, this can be achieved by reducing the variance of these estimators through adjustment for covariates. The vector $\hat{\mathbf{p}}^{(0)}$ of estimated relative treatment effects (RTE) is in a certain sense asymptotically equivalent to the vector $\bar{\mathbf{Y}}^{(0)} = (\bar{Y}_{1.}^{(0)}, \ldots, \bar{Y}_{a.}^{(0)})'$ of asymptotic rank transforms (ART). Namely, if $\mathbf{CF}^{(0)} = \mathbf{0}$ for a contrast matrix $\mathbf{C}$, then $\sqrt{N}\mathbf{C}(\hat{\mathbf{p}}^{(0)} - \bar{\mathbf{Y}}^{(0)}) \xrightarrow{P} \mathbf{0}$ (see, e.g., [13], Corollary 4.4). However, the ART are technically more tractable than the RTE, due to the simpler covariance structure of $\bar{\mathbf{Y}}^{(0)}$ as compared to $\hat{\mathbf{p}}^{(0)}$. Therefore, we focus on minimizing the variance of the ART. Specifically, we attempt to minimize the variance $\sigma^2$ of a weighted sum of adjusted ART, i.e., $\sigma^2 = \mathrm{Var}\left(\frac{1}{N}\sum_{i=1}^{a} n_i \bar{Y}_i^*\right)$, where $\bar{Y}_i^* = \bar{Y}_{i.}^{(0)} - \sum_{r=1}^{d} \gamma^{(r)} \bar{Y}_{i.}^{(r)}$. Minimization is with respect to the coefficients $\gamma^{(r)}$ that combine covariates and response variable. If we assume partial differentiability of $\sigma^2$ w.r.t. $\gamma^{(r)}$, $r = 1, \ldots, d$, the minimizers are easily found as the solutions of the system of linear equations

$$\gamma^{(r)} = \frac{\sum_{i=1}^{a} n_i \left( \mathrm{Cov}(Y_{i1}^{(0)}, Y_{i1}^{(r)}) + \sum_{r \neq r'} \gamma^{(r')} \mathrm{Cov}(Y_{i1}^{(r)}, Y_{i1}^{(r')}) \right)}{\sum_{i=1}^{a} n_i \mathrm{Var}\, Y_{i1}^{(r)}}, \quad r = 1, \ldots, d.$$

Define

$$C^{rs} = \sum_{i=1}^{a} \sum_{j=1}^{n_i} (\hat{Y}_{ij}^{(r)} - \hat{Y}_{i.}^{(r)})(\hat{Y}_{ij}^{(s)} - \hat{Y}_{i.}^{(s)}), \quad r, s = 0, \ldots, d.$$

Then, a consistent estimator of $\boldsymbol{\gamma} = (\gamma^{(1)}, \ldots, \gamma^{(d)})'$ is given by $\hat{\boldsymbol{\gamma}} = (\hat{\gamma}^{(1)}, \ldots, \hat{\gamma}^{(d)})'$, where

$$\hat{\boldsymbol{\gamma}} = \begin{pmatrix} C^{11} & C^{12} & \cdots & C^{1d} \\ C^{21} & \ddots & & \vdots \\ \vdots & & & \\ C^{d1} & \cdots & & C^{dd} \end{pmatrix}^{-1} \begin{pmatrix} C^{01} \\ C^{02} \\ \vdots \\ C^{0d} \end{pmatrix}. \tag{6}$$

In the special case of $d = 1$ (one covariate), we obtain

$$\gamma^{(1)} = \frac{\sum_{i=1}^{a} n_i \mathrm{Cov}(Y_{i1}^{(0)}, Y_{i1}^{(1)})}{\sum_{i=1}^{a} n_i \mathrm{Var}\, Y_{i1}^{(1)}}.$$

and

$$\hat{\gamma}^{(1)} = \frac{\sum_{i=1}^{a}\sum_{j=1}^{n_i}(\hat{Y}_{ij}^{(0)} - \hat{Y}_{i.}^{(0)})(\hat{Y}_{ij}^{(1)} - \hat{Y}_{i.}^{(1)})}{\sum_{i=1}^{a}\sum_{j=1}^{n_i}(\hat{Y}_{ij}^{(1)} - \hat{Y}_{i.}^{(1)})^2} = (\hat{\mathbf{D}}^{(1)\prime}\hat{\mathbf{D}}^{(1)})^{-1}\hat{\mathbf{D}}^{(1)\prime}\hat{\mathbf{D}}^{(0)} \qquad (7)$$

For a discussion on variance reduction for RTE in a one-way CRD with one covariate, see also [17].

### 3.1. Testing the Factor Effects

With a slightly different motivation, Langer [9] has used the estimator $\hat{\gamma}$ as defined in (6) for deriving asymptotic inferential procedures regarding the factor effects in a CRD, while adjusting for the effect of a covariate. Hypotheses about factor effects are formulated using the marginal distribution functions. Thus, the hypotheses have the form $\mathbf{CF}^{(0)} = \mathbf{0}$. Here, $\mathbf{C}$ denotes a contrast matrix, as in parametric linear models. For example, in a one factorial design with vector of distribution functions $\mathbf{F}^{(0)} = (F_1^{(0)}, \ldots, F_a^{(0)})'$, the hypothesis of "no factor effect" can be formulated as $F_1^{(0)} = \ldots = F_a^{(0)}$, or equivalently as $\mathbf{CF}^{(0)} = \mathbf{0}$, where $\mathbf{C} = \mathbf{I}_a - \frac{1}{a}\mathbf{J}_a$.

The following theorem provides the asymptotic result on testing factor effects in a CRD. The proof is outlined in the appendix. It involves use of results on multivariate asymptotic normality of a vector of relative treatment effects (see, e.g., [13]).

**Theorem 1** *Let $X_{ij}^{(r)}$ be independent random variables with marginal distributions $F_i^{(r)}$. Let $\hat{\mathbf{p}}^* = (\hat{p}_1^*, \ldots, \hat{p}_a^*)'$, where $\hat{p}_i^* = \hat{p}_i^{(0)} - \sum_{r=1}^{d}\hat{\gamma}^{(r)}\hat{p}_i^{(r)}$ and where $\hat{\gamma}^{(r)}$ and $\hat{p}_i^{(r)}$ are defined in (6) and (2), respectively.*

*Then,*

1. *under the assumptions (A1)–(A5) given below, and under the null hypothesis $H_0 : \mathbf{CF}^{(0)} = \mathbf{0}$, the Wald type test statistic*

$$Q_N(\mathbf{C}) = N(\hat{\mathbf{p}}^*)'\mathbf{C}'(\mathbf{C}\hat{\mathbf{\Sigma}}_N^*\mathbf{C}')^+\mathbf{C}\hat{\mathbf{p}}^*$$

*is asymptotically $\chi_f^2$ distributed, where $f$ is estimated by $\hat{f} = \text{rank}(\mathbf{C}\hat{\mathbf{\Sigma}}_N^*)$, and $\hat{\mathbf{\Sigma}}_N^*$ is given in (8) below. Here, $(\mathbf{C}\hat{\mathbf{\Sigma}}_N^*\mathbf{C}')^+$ denotes the Moore-Penrose inverse of the matrix $\mathbf{C}\hat{\mathbf{\Sigma}}_N^*\mathbf{C}'$.*

2. *Under the assumptions (A1)–(A4) and (A6) given below, and under the null hypothesis $H_0 : \mathbf{CF}^{(0)} = \mathbf{0}$, the ANOVA type statistic (see [16] for an introduction of this term)*

$$A_N = \frac{Nf \cdot (\hat{\mathbf{p}}^*)'\mathbf{T}\hat{\mathbf{p}}^*}{\text{tr}(\mathbf{T}\hat{\mathbf{\Sigma}}_N^*)}$$

*is approximately $\chi_f^2$ distributed, where $f$ is estimated by*

$$\hat{f} = \frac{[\text{tr}(\mathbf{T}\hat{\mathbf{\Sigma}}_N^*)]^2}{\text{tr}(\mathbf{T}\hat{\mathbf{\Sigma}}_N^*\mathbf{T}\hat{\mathbf{\Sigma}}_N^*)}, \quad \text{and} \quad \mathbf{T} = \mathbf{C}'(\mathbf{CC}')^-\mathbf{C},$$

*where $(\mathbf{CC}')^-$ denotes a g-inverse of $\mathbf{CC}'$.*

The matrix $\hat{\Sigma}_N^*$ is given by

$$\hat{\Sigma}_N^* = \hat{\Gamma}'\hat{\Sigma}_N\hat{\Gamma}, \quad \text{where } \hat{\Gamma} = \begin{pmatrix} 1 \\ -\hat{\gamma} \end{pmatrix} \otimes \mathbf{I}_a, \tag{8}$$

and $\hat{\Sigma}_N$ is an estimator of the covariance matrix $\Sigma_N$ of $\sqrt{N}\bar{\mathbf{Y}}$, where $\bar{\mathbf{Y}}$ is defined in (4). Note that $\hat{\Sigma}_N$ is a partitioned matrix that can be written as

$$\hat{\Sigma}_N = \begin{pmatrix} \hat{\Sigma}_N^{(00)} & \cdots & \hat{\Sigma}_N^{(0d)} \\ \vdots & & \vdots \\ \hat{\Sigma}_N^{(0d)} & \cdots & \hat{\Sigma}_N^{(dd)} \end{pmatrix},$$

where the sub-matrices are diagonal matrices, i.e., $\hat{\Sigma}_N^{(rs)} = \mathrm{diag}(\hat{\sigma}_1^{(rs)}, \ldots, \hat{\sigma}_a^{(rs)})$. The components $\hat{\sigma}_i^{(rs)}$ of the diagonal matrix $\hat{\Sigma}_N^{(rs)}$ are given by

$$\hat{\sigma}_i^{(rs)} = \frac{N}{n_i(n_i-1)} \sum_{j=1}^{n_i} (\hat{Y}_{ij}^{(r)} - \hat{Y}_{i.}^{(r)})(\hat{Y}_{ij}^{(s)} - \hat{Y}_{i.}^{(s)}), \quad r,s = 0, \ldots, d; \ i = 1, \ldots, a.$$

**Proof.** See appendix for an outline of the proof.

These are the technical assumptions needed for Theorem 1. (A1) states the CRD assumption.

**(A1)** $F_i^{(r)} = F^{(r)}$ for $r = 1, \ldots, d; \ i = 1, \ldots, a$

**(A2)** $\min_{i=1,\ldots,a} n_i \to \infty$

**(A3)** $\exists N_0 : \forall i = 1, \ldots, a : N/n_i \leq N_0 < \infty$

**(A4)** $\Sigma_N \xrightarrow{(B2)} \Sigma$, where $\Sigma_N$ denotes the covariance matrix of $\sqrt{N}\bar{\mathbf{Y}}$, and where $\bar{\mathbf{Y}}$ is defined in (4).

**(A5)** $\exists \lambda_0 > 0 : \forall N \in \mathbb{N} : \lambda_{\min}(\Sigma_N) \geq \lambda_*$, where $\lambda_{\min}(\Sigma_N)$ is the smallest eigenvalue of the matrix $\Sigma_N$

**(A6)** $\exists \kappa > 0 : \forall N \in \mathbb{N} : \mathrm{tr}(\mathbf{T}\hat{\Sigma}_\mathbb{N}) \geq \kappa$.

### 3.2. Testing the Covariate Effect

Based on the estimator in (7), Bathke [10] has developed a test for the covariate effect in a nonparametric factorial model with covariates. The distribution of the test statistic has been derived conditional on the values taken by the covariate. This is technically equivalent to treating the covariate as a deterministic variable. Therefore, contrary to the previous theorem, this result remains valid even if the CRD assumption of equal covariate distribution for the different factor levels is violated. The main theoretical result is formulated in the following theorem. For the proof and some more details see [10], where the theorem is also formulated in the more general setting of a nonparametric mixed model, including dependent observations. Note that conditioning on the covariate implies that the function $H^{(r)}$ coincides with its empirical counterpart, $\hat{H}^{(r)}$, and thus $\mathbf{D}^{(1)} = \hat{\mathbf{D}}^{(1)}$.

**Theorem 2** Let $X_{ij}^{(0)}$ be independent random variables with marginal distributions $F_i^{(0)}$, and $x_{ij}^{(1)}$ some (non-stochastic) constants. Let $\hat{Y}_{ij}^{(0)}$ and $y_{ij}^{(1)}$ be the rank transforms as defined in (3). $\hat{\mathbf{D}}^{(0)}$, $\mathbf{D}^{(1)}$ denote vectors of adjusted rank transforms according to (5).

Then, under the assumptions (B1)–(B3) stated below, $\hat{v} = \mathbf{D}^{(1)'}\hat{\mathbf{D}}^{(0)}/\hat{s}$ is asymptotically distributed according to a standard normal distribution.

Hereby, $\hat{s}$ denotes a consistent estimator of the standard deviation of $\mathbf{D}^{(1)'}\hat{\mathbf{D}}^{(0)}$. This estimate is given by

$$\hat{s}^2 = \sum_{i=1}^{a} \hat{\sigma}_i^2 \sum_{j=1}^{n_i} D_{ij}^{(1)2}, \quad \text{where } \hat{\sigma}_i^2 = (n_i - 1)^{-1} \sum_{j=1}^{n_i} \hat{D}_{ij}^{(0)2}. \tag{9}$$

The following are the technical assumptions needed to establish this theorem. Essentially, some minimal variation is necessary among the values taken by the response variable and among the values taken by the covariate.

**(B1)** $\min_{i=1,...,a} n_i \to \infty$

**(B2)** $\exists \sigma_0^2 > 0: \sum_{i=1}^{a} \text{Var}(Y_{i1}) = \sum_{i=1}^{a} \sigma_i^2 \geq \sigma_0^2.$

**(B3)** $\sum_{i=1}^{a} \sum_{j=1}^{n_i} \text{Var}(D_{ij}^{(1)} Y_{ij}^{(0)}) \xrightarrow{(A1)} \infty.$

## 4. General Design

In many experiments, the CRD assumption is satisfied by randomization. However, there are situations in which we cannot assume equal marginal distributions of a covariate for the different treatments. "Unhappy randomization", or simply a covariate whose outcomes are not available to the experimenter when the treatments are assigned, are possible reasons. In such a case, Theorem 1 is not applicable. In this section, we briefly outline an approach that can be used also for non-randomized designs.

Motivated by the linear equation in Theorem 1, a regression model can be formulated that is linear in the relative treatment effects.

$$p_i^* = p_i^{(0)} - \sum_{r=1}^{d} \gamma^{(r)} \cdot p_i^{(r)}, \tag{10}$$

where $p_i^{(r)}$ denotes the "relative effect" of treatment $i$ as defined in (1). If $p_i^{(r)} > 1/2$ then the observations $X_{ij}^{(r)}$ in the treatment group $i$ tend to be larger than the others. The idea underlying the regression equation (10) is the following. If the response variable tends to be larger in group A than in group B, and a covariate also tends to be larger in group A than in group B, this could indicate a positive relation between the response variable and that covariate. The quantity $p_i^*$ is called "adjusted treatment effect".

In this model, the vector $\mathbf{p}^* = (p_1^*, \ldots, p_a^*)'$ of adjusted treatment effects is used to define the hypotheses instead of the vector $\mathbf{F}^{(0)}$ of distribution functions. Thus, the hypotheses are of the form $\mathbf{Cp^*} = \mathbf{0}$. Corresponding hypotheses in this model and in the model described in Section 3 are formulated using the same contrast matrices. However, formulating the null hypothesis in terms of adjusted treatment effects has two important implications. First, the hypothesis depends on the sample sizes. This problem can be avoided by using an unweighted average distribution function $G$ instead of the weighted average $H$ in (1), leading to the so-called harmonic ranks. Second, the covariance matrix of the vector of estimated effects is much more complicated under $H_0 : \mathbf{Cp^*} = \mathbf{0}$ than under $H_0 : \mathbf{CF}^{(0)} = \mathbf{0}$. Siemer [11] has tackled the mentioned problems and has derived asymptotically valid test statistics using equation (10) in a design without the CRD assumption. However, a more detailed elaboration on this approach would require a somewhat different notation and should be treated separately.

## 5. Computer Simulation Results and Application

All the aforementioned test statistics are asymptotically valid. In practice, however, the sample sizes are always finite. Therefore, extensive simulation studies have been carried out in order to determine for which sample size the actual $\alpha$-level is convincingly close enough to the nominal $\alpha$-level.

For these simulations, different underlying distributions have been used, including normal, lognormal, and different discrete distributions on the numbers $\{1,\ldots,5\}$. For each setting, 10,000 simulations have been performed.

In general, these simulations show that:

- The Wald type statistics require large sample sizes of at least $\sim 20$ per cell. For sample sizes $n_i \geq 20$, the simulated 5% $\alpha$-level ranges from 5% to 7%, while the simulated 1% $\alpha$-level is between 1% and 2%.

- When the number of factor levels is at least 4, then the ANOVA type statistics require $\sim 10$ replications per cell. When $n_i \geq 10$, the simulated 5% $\alpha$-level for the ANOVA type statistic is between 5% and 6.5%, while the simulated 1% $\alpha$-level ranges from 1% to 2%.

- The more samples, the better: If there are only two factor levels, then the ANOVA type statistics coincide with the Wald type statistics and are likely to be liberal, especially for small sample sizes. When the number of factor levels increases, smaller sample sizes per cell are sufficient for the ANOVA type statistic.

- The test for covariate effect only needs a minimum sample size of $\sim 7$ per cell. Contrary to the other test statistics, it has a tendency to be slightly conservative when the sample sizes are small. For $n_i \geq 7$, the simulated 5% $\alpha$-level ranges from 4.3% and 5.3%, while the simulated 1% $\alpha$-level is between 0.7% and 1.1%.

- Balanced is best: When comparing balanced to unbalanced designs, the performance is optimal for the balanced designs. However, the differences are only small.

There are very few alternative and competitive inferential procedures for nonnormal/ordinal data in the ANCOVA framework, especially in the case of small sample sizes. Therefore, in order to facilitate the use of the procedures described in this paper, we finish with a short explanation of how the methods can be applied in practice.

A SAS macro that performs all the mentioned tests is available at the following website:
www.ams.med.uni-goettingen.de/Projekte/makros
This website, provided by the Medical Statistics Department of Göttingen, also contains detailed descriptions on how to use the macros. The only system requirement is having a computer with the software package SAS installed or accessible.

After download, include the macro run_npar with the command
%include 'run_npar.sas';

Assuming that the dataset name.dat is recognizable to SAS and has a single fixed factor called "group", a response variable "x1" and a covariate "x2", the macro run_npar is evoked by the following statement.

```
%run_npar(data=name,out=_no_,file=_no_,var=x1,factora1=group,
         factora2=_no_,factorb=_no_,subject=_no_,cov1=x2 );
```

Note that it is possible to define up to 10 covariates, and the output can be specified to be an html-file. Also, it is possible to analyze data from a mixed model with this macro. The mixed model ANCOVA is, however, beyond the scope of this exposition.

### A. Outline of the Proof of Theorem 1

1. Let $\bar{\mathbf{Y}}^* = \mathbf{\Gamma}'\bar{\mathbf{Y}}$, where

$$\mathbf{\Gamma} = \begin{pmatrix} 1 \\ -\boldsymbol{\gamma} \end{pmatrix} \otimes \mathbf{I}_a, \quad \text{and } \bar{\mathbf{Y}} \text{ is as defined in (4).}$$

Under $H_0 : \mathbf{CF}^{(0)} = \mathbf{0}$, the random vector $\sqrt{N}\mathbf{C}\bar{\mathbf{Y}}^*$ has asymptotically a multivariate normal distribution with mean vector $\mathbf{0}$ and covariance matrix $\mathbf{C\Sigma^*C'}$, where $\mathbf{\Sigma}^* = \mathbf{\Gamma}'\mathbf{\Sigma}\mathbf{\Gamma}$, and $\mathbf{\Sigma}$ is defined in (A4). This follows with the same arguments as in the proof of Theorem 5.3 in [13].

2. Under $H_0 : \mathbf{CF}^{(0)} = \mathbf{0}$, $\sqrt{N}\mathbf{C}(\hat{\mathbf{p}}^* - \bar{\mathbf{Y}}^*) \xrightarrow{p} \mathbf{0}$. Thus, $\sqrt{N}\mathbf{C}\hat{\mathbf{p}}^*$ is also asymptotically multivariate normal with mean $\mathbf{0}$ and covariance matrix $\mathbf{C\Sigma^*C'}$. This result is established using the consistency of the estimator $\hat{\gamma}$, as well as applying, e.g., Corollary 4.4. from [13].

3. The asymptotic covariance matrix $\mathbf{\Sigma}^*$ of $\sqrt{N}\bar{\mathbf{Y}}^*$ is estimated by $\hat{\mathbf{\Sigma}}_N^*$ as defined in (8). Consistency of this estimator follows analogously to Theorem 5.5 in [13]. Thus, $\mathbf{C}\hat{\mathbf{\Sigma}}_N^*\mathbf{C'} - \mathbf{C\Sigma^*C'} \xrightarrow{p} 0$.

4. ¿From 2. and 3., we have that under $H_0 : \mathbf{CF}^{(0)} = \mathbf{0}$,

$$Q_N(\mathbf{C}) = (\sqrt{N}\mathbf{C}\hat{\mathbf{p}}^*)'(\mathbf{C}\hat{\mathbf{\Sigma}}_N^*\mathbf{C'})^+(\sqrt{N}\mathbf{C}\hat{\mathbf{p}}^*)$$

is asymptotically $\chi_f^2$ distributed. Here, $f = \text{rank}(\mathbf{C\Sigma^*})$, which is estimated by $\hat{f} = \text{rank}(\mathbf{C}\hat{\mathbf{\Sigma}}_N^*)$.

5. Let $\mathbf{T} = \mathbf{C}'(\mathbf{CC}')^{-}\mathbf{C}$. Then, $\mathbf{CF}^{(0)} = \mathbf{0}$ is equivalent to $\mathbf{TF}^{(0)} = \mathbf{0}$ since $\mathbf{C}'(\mathbf{CC}')^{-}$ is a $g$-inverse of $\mathbf{C}$. Under $H_0 : \mathbf{CF}^{(0)} = \mathbf{0}$, the quadratic form $N(\hat{\mathbf{p}}^*)'\mathbf{T}\hat{\mathbf{p}}^*$ is asymptotically distributed according to a weighted sum of $\chi_1^2$ random variables. This distribution is approximated by $g \cdot \chi_f^2$ (see, e.g., [16], Section 3.4) such that the first two moments coincide. Thus,

$$f = \frac{[\mathrm{tr}(\mathbf{T}\mathbf{\Sigma}^*)]^2}{\mathrm{tr}(\mathbf{T}\mathbf{\Sigma}^*\mathbf{T}\mathbf{\Sigma}^*)} \quad \text{and} \quad g = \frac{\mathrm{tr}(\mathbf{T}\mathbf{\Sigma}^*\mathbf{T}\mathbf{\Sigma}^*)}{\mathrm{tr}(\mathbf{T}\mathbf{\Sigma}^*)} = \frac{\mathrm{tr}(\mathbf{T}\mathbf{\Sigma}^*)}{f}, \quad \text{respectively.}$$

Finally, the constants $f$ and $g$ are estimated by replacing $\mathbf{\Sigma}^*$ with its consistent estimator $\hat{\mathbf{\Sigma}}_N^*$ given in (8), and the result about the ANOVA type statistic stated in Theorem 1 is obtained. □

## REFERENCES

1. B.E. Huitema (1980), *The Analysis of Covariance and Alternatives*, Wiley, New York.
2. A. Agresti (1990), *Categorical Data Analysis*, Wiley, New York.
3. A. Agresti (1996), *An Introduction to Categorical Data Analysis*, Wiley, New York.
4. D. Quade (1967), Rank Analysis of Covariance, *Journal of the American Statistical Association*, 62, 1187–1200.
5. M.L. Puri and P.K. Sen (1969), Analysis of Covariance Based on General Rank Scores. *Annals of Mathematical Statistics*, 40, 610-618.
6. M.G. Akritas, S.F. Arnold, and Y. Du (2000), Nonparametric Models and Methods for Nonlinear Analysis of Covariance, *Biometrica*, 87, 3, 507–526.
7. M.G. Akritas and S.F. Arnold (1994), Fully Nonparametric Hypotheses for Factorial Designs I: Multivariate Repeated Measures Designs, *Journal of the American Statistical Association*, 89, 336–343.
8. M.G. Akritas, S.F. Arnold, and E. Brunner (1997), Nonparametric Hypotheses and Rank Statistics for Unbalanced Factorial Designs, *Journal of the American Statistical Association*, 92, 258–265.
9. F. Langer (1998), *Berücksichtigung von Kovariablen in nichtparametrischen gemischten Modellen*, Ph.D. dissertation, Universität Göttingen, Institut für Mathematische Stochastik.
10. A. Bathke (2002), Testing the Effect of Covariates in Nonparametric Mixed Models, *Submitted*.
11. A. Siemer (2002), *Die statistische Auswertung von ordinalen Daten bei zwei Zeitpunkten und zwei Stichproben*, Ph.D. dissertation, Universität Göttingen, Institut für Mathematische Stochastik.
12. F.H. Ruymgaart (1980), A Unified Approach to the Asymptotic Distribution Theorey of Certain Midrank Statistics, in *Statistique non Parametrique Asymptotique*, 1–18, J.P. Raoult (Ed.), Lecture Notes on Mathematics, No. 821, Springer, Berlin.
13. U. Munzel and E. Brunner (2000), Nonparametric Methods in Multivariate Factorial Designs, *Journal of Statistical Planning and Inference*, 88, 1, 117–132.
14. E. Brunner and M.L. Puri (2001), Nonparametric Methods in Factorial Designs, *Statistical Papers* 42, 1–52.

15. M.G. Akritas and E. Brunner (1997), A Unified Approach to Rank Tests in Mixed Models, *Journal of Statistical Planning and Inference*, 61, 249–277.
16. E. Brunner, U. Munzel, and M.L. Puri (1999), Rank-Score Tests in Factorial Designs With Repeated Measures, *Journal of Multivariate Analysis*, 70, 286–317.
17. S. Domhof (2001), *Nichtparametrische relative Effekte*, Ph.D. dissertation, Universität Göttingen, Institut für Mathematische Stochastik.

# 4. Goodness of Fit

# Assessing structural relationships between distributions - a quantile process approach based on Mallows distance

G. Freitag[a], A. Munk[a] * and M. Vogt[b]

[a]Institut für Mathematische Stochastik, Georg-August-Universität Göttingen, 37083 Göttingen, Germany

[b]Fakultät für Mathematik, Ruhr-Universität Bochum, 44780 Bochum, Germany

We suggest diagnostic tools for the assessment of shape similarity of distributions, i.e. whether several distributions are equal up to a pre-specified semiparametric deviation, such as a simple shift in location or a scale deviation. This will be called a *structural relationship* between distributions, and it is of importance, for instance, in an initial analysis within the (nonparametric) analysis of variance. Our approach is based on a modification of a trimmed version of Mallows distance between distribution functions. Asymptotic theory for the corresponding test statistics is provided and a bootstrap limit theorem is shown. In particular, the case of *dependent samples* is covered as well. A simulation study reveals the bias corrected and accelerated bootstrap as an adequate method for the assessment of shape similar distributions. A medical application is discussed.

*Keywords*: $BC_a$-bootstrap, $\delta$-method, goodness of fit, Hadamard derivative, location-scale models, Mallows metric, model checking, $p$-value curves, rank tests, semiparametrics

## 1. INTRODUCTION

The aim of this paper is to provide a method for deciding whether a given set of distributions belongs to a semiparametric class, for instance to a location family. This problem arises e.g. in nonparametric analysis of variance, when the assumption of normality is not justified. [1] developed a general theory for obtaining locally most powerful rank tests in these models. For the two-sample case we refer to [2, chapter 3], where different rank tests for the location model are discussed. Statistics based on ranks are often attractive for practitioners, because of their several robustness and invariance properties. Initiated by the work of [3–7], this has been generalized to linear models, resulting in a 'nonparametric' analysis of variance (cf. [8–12] among many others). A survey can be found in [13].

However, ranking methods essentially require semiparametric assumptions (such as shift or scale alternatives) in order to be consistent and powerful, as highlighted by Proposition 2.1 in [13]. Therefore, any application of a ranking method should be accompanied by a proper

---

*corresponding author : Axel Munk, Institut für Mathematische Stochastik, Georg-August-Universität Göttingen, Lotzestr. 13, 37083 Göttingen, Germany. Tel. +49/551/397801, Fax +49/551/395997, email: munk@math.uni-goettingen.de; http:\\www.stochastik.math.uni-goettingen.de

'model check', e.g. whether a given set of cumulative distribution functions $F_i$ ($i = 1, \ldots, d$) is in a location model, say,

$$F_i(x) = F(x - \mu_i), \quad i = 1, \ldots, d, \tag{1}$$

for some unknown basis function $F$. Other important semiparametric models include scale families, location-scale families or Lehmann-alternatives (cf. e.g. [2,14,15] for a survey). We will term semiparametric relationships as (1) in general as structural relationships between a given set of cumulative distribution functions $F_i$ ($i = 1, \ldots, d$).

In order to investigate the presence of a structural relationship between two cumulative distribution functions, it is convenient to express the deviation between the distributions up to this relationship by a suitable measure of discrepancy. To this end we suggest as a measure of discrepancy the minimum $L^2$ distance between the quantile functions of two cumulative distribution functions $F$ and $G$ with existing second moment, respectively. This gives for the case of a location relationship

$$\Gamma_L^2(F,G) := \min_{\delta \in \mathbb{R}} \int_0^1 \left(F^{-1}(t) - G^{-1}(t) - \delta\right)^2 dt = \min_{\delta \in \mathbb{R}} \int_{\mathbb{R}} \left(x - F^{-1} \circ G(x) - \delta\right)^2 dG(x).$$

A main reason for defining our discrepancy in terms of quantile functions - and not of distributions - is that this allows to express differences in the original scale, which often yields a simpler interpretation. The adept reader will recognize $\Gamma_L$ as a generalization of Mallows [16] distance which is sometimes also called the Wasserstein distance [17]. Observe that presence of a *location model*

$$(F,G) \in \mathcal{F}_L := \{(F,G) : G(x) = F(x - \delta),\ x \in \mathbb{R},\ \text{for some } \delta \in \mathbb{R}\} \tag{2}$$

is equivalent to $x - F^{-1} \circ G(x) \equiv \delta$, hence $\Gamma_L$ is a measure for the deviation from the exact location model on the original scale.

Now we can provide - as a means for diagnostic model checks - tests for the hypotheses

$$H : \Gamma_L(F,G) > \Delta_0 \quad vs. \quad K : \Gamma_L(F,G) \leq \Delta_0. \tag{3}$$

Here $\Delta_0$ is a fixed tolerance bound summarizing the extent of discrepancy from a location model the experimenter is willing to tolerate. Observe that the hypotheses (3) will allow a valid assessment *in favor of* the location model $F(\cdot) = G(\cdot - \delta)$ at a controlled error rate $\alpha$ within the bound $\Delta_0$.

Hence, this type of hypotheses for model testing overcomes the drawback of the classical null hypothesis $H_0 : \Gamma_L(F,G) = 0$ (and variants of it) that acceptance of $H_0$ might be, for instance, due to a lack of power of the particular goodness-of-fit test for $H_0$ and not to the presence of the model. For a detailed methodological discussion of related testing problems see [18–20].

Hypotheses as in (3) adapt to this problem and are also directly related to the computation of one-sided confidence intervals for $\Gamma_L$. Moreover, as an additional graphical tool to quantify the discrepancy up to location shifts between $F$ and $G$ we will extend the asymptotic $p$-value curves introduced in [18] to the present problem of model checking. Finally, we would like to point out that the investigation of structural relationships between distributions is not only important from a point of view of subsequent application of efficient tests, it also allows to provide important direct insight into the data structure itself, as it becomes apparent in the next example.

In this paper we will also extend this approach to scale models and location-scale models. Here it is important to distinguish two types of 'scale differences' between distribution functions. For positive random variables, an *acceleration model* (cf. e.g. [21]) is given by the class of pairs of distributions

$$\mathcal{F}_A := \{(F, G) : G(x) = F(x/\sigma),\ x \geq 0,\ \text{for some}\ \sigma > 0\}, \tag{4}$$

whereas a *scale model* will be defined throughout this paper as

$$\mathcal{F}_S := \left\{(F, G) : G(x - G^{-1}(\tfrac{1}{2})) = F\left(\frac{x - F^{-1}(\tfrac{1}{2})}{\sigma}\right),\ G^{-1}(\tfrac{1}{2}) = F^{-1}(\tfrac{1}{2}),\ x \in \mathbb{R},\ \sigma > 0\right\}. \tag{5}$$

Accordingly, for the investigation of an acceleration difference we suggest as a measure of discrepancy

$$\Gamma_A^2(F, G) := \min_{\sigma > 0} \int_0^1 \left(F^{-1}(t) - \sigma G^{-1}(t)\right)^2 dt,$$

and for a scale model (with no location difference)

$$\Gamma_S^2(F, G) := \min_{\sigma > 0} \int_0^1 \left(F^{-1}(t) - F^{-1}(\tfrac{1}{2}) - \sigma(G^{-1}(t) - G^{-1}(\tfrac{1}{2}))\right)^2 dt + \left(F^{-1}(\tfrac{1}{2}) - G^{-1}(\tfrac{1}{2})\right)^2.$$

Finally, a *location-scale* model,

$$\mathcal{F}_{LS} := \{(F, G) : G(x) = F((x - \delta)/\sigma),\ x \in \mathbb{R},\ \delta \in \mathbb{R},\ \sigma > 0\},$$

can be evaluated with help of the discrepancy

$$\Gamma_{LS}^2(F, G) := \min_{\delta \in \mathbb{R},\ \sigma > 0} \int_0^1 \left(F^{-1}(t) - \sigma G^{-1}(t) - \delta\right)^2 dt.$$

**Remark 1.1** It has to be noted that sometimes it is more appropriate to use a definition for a scale model different from our model (5). For instance, in [2, p.176] a two-sample scale model is stated as $G(\cdot) \equiv F(\cdot/\tau),\ \tau > 0$, where $F \in \Omega_0$, the set of cumulative distribution functions $F_0$ s.t. $F_0$ is absolutely continuous and $F_0^{-1}(0.5) = 0$, uniquely. We denote this model as $\mathcal{F}_S^0$. Clearly, $\mathcal{F}_S$ from (5) is a generalization of $\mathcal{F}_S^0$, allowing for nonzero (but equal) medians. Our approach can be adapted to the assessment of $\mathcal{F}_S^0$ by using the discrepancy measure

$$\Gamma_{S0}^2(F, G) := \min_{\sigma > 0} \int_0^1 \left(F^{-1}(t) - \sigma G^{-1}(t)\right)^2 dt + \left(F^{-1}(\tfrac{1}{2})\right)^2.$$

**Example 1** (*Mortality data*). In [22] it is reported on a study from the general medical department in Oslo, Norway. There the relationship between the hospitalization (as a measure of morbidity) of patients and the mortality in the general population served by the medical department was analyzed. For this study, patient data were obtained for 367 consecutive admissions (176 males and 191 females) from 1980. The mortality data were provided by the Central Bureau of Statistics in Norway and consisted of 6140 deaths (2989 males and 3151 females). In Figure 1 estimated quantile functions from the hospitalization and mortality data of the females are displayed. Observe that Figure 1 suggests that a location model seems to be appropriate

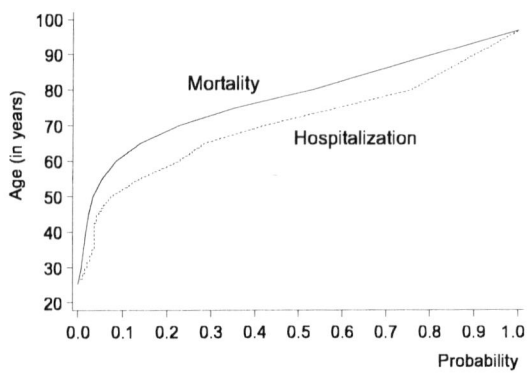

Figure 1. Mortality data: quantile functions of mortality and hospitalization

when the tails are neglected. Therefore, we will incorporate this in our investigation with help of trimmed versions of the Mallows distances for location, acceleration, scale and location-scale families. For ease of notation we use in the following *symmetric* trimming by the amount $\beta, 0 \leq \beta < \frac{1}{2}$, at both tails. For the location model this yields, e.g.,

$$\Gamma^2_{L,\beta}(F,G) := \min_{\delta \in \mathbb{R}} \|F^{-1} - G^{-1} - \delta\|^2_\beta,$$

where $\|f(\cdot)\|^2_\beta := \frac{1}{1-2\beta} \int_\beta^{1-\beta} f(\cdot)^2 dt$. With help of $\Gamma^2_{L,\beta}$ we can again formulate testing problems as in (3).

For the scale, acceleration, and location-scale models, it has to be taken into account that the corresponding discrepancies are not symmetric with respect to the distributions $F$ and $G$. Therefore, we suggest symmetrized versions

$$\tilde{\Gamma}^2_{(\cdot),\beta}(F,G) := \frac{1}{2}\left(\Gamma^2_{(\cdot),\beta}(F,G) + \Gamma^2_{(\cdot),\beta}(G,F)\right) \tag{6}$$

instead, where $(\cdot)$ refers to $A, S, LS$, respectively (cf. also Remark 2.8).

The paper is organized as follows. In the next section we provide the asymptotic theory for model checking of several two-sample models such as location or acceleration models. This will be done under a multivariate sampling scheme (for applications see e.g. [23–25]) as well as under independence. An asymptotic test based on a normal approximation with estimation of the limiting variance can be derived for each of the problems considered. It turns out that the estimation of the asymptotic variances is computationally rather complicated, hence in Section 3 we discuss the application of bootstrap versions of our tests. A simulation study in Section 4 investigates the accuracy of the bias corrected and accelerated bootstrap method with respect to level and power. In Section 5 we re-analyze the given data example with the proposed methods. Section 6 contains a short discussion of the results. In order to improve readability, all proofs are postponed to the Appendix. Finally, we mention that our results are based on the Hadamard differentiability of the proposed distances in the two-sample case, a result which might be of interest on its own.

## 2. ASYMPTOTIC THEORY

### 2.1. Notation

In the following we start with presenting the asymptotic theory for testing the functionals $\Gamma^2_{L,\beta}$, $\Gamma^2_{A,\beta}$ and $\Gamma^2_{LS,\beta}$. Results for the scale model can be obtained analogously. Throughout the following $\beta \in [0, \frac{1}{2})$ denotes a fixed trimming bound. We assume that $Z_i = (X_i, Y_i)$, $i = 1, \ldots, n$, are independent and identically distributed random variables according to an unknown cumulative distribution function $H$, which is supposed to be two times differentiable with continuous derivatives, $H \in C^2[\mathbb{R}^2]$, with marginals

$$F(\cdot) = H(\cdot, \infty), \quad G(\cdot) = H(\infty, \cdot), \text{ s.t. } F' = f \text{ and } G' = g \text{ are bounded away}$$
from zero on the intervals $[F^{-1}(\beta), F^{-1}(1-\beta)]$ and $[G^{-1}(\beta), G^{-1}(1-\beta)]$, resp. (7)

Furthermore, let in the sequel $H_n$ denote the empirical cumulative distribution function corresponding to the sample $(Z_i)_{i=1,\ldots,n}$ and let $F_n, G_n$ be its marginals. Let

$$F_n^{-1}(t) = \inf\{x : F_n(x) \geq t\} = \sum_{i=1}^{n} X_{(i)} \mathbf{I}_{\{(i-1)/n < t \leq i/n\}}, \quad t \in (0,1),$$

be the left continuous inverse of $F_n$, where $X_{(i)}$ denotes the i-th order statistic of $X_1, \ldots, X_n$.

### 2.2. Location model

In order to make ideas more transparent we treat the location model separately. Observe that

$$\Gamma^2_{L,\beta}(F, G) = \frac{1}{1 - 2\beta} \int_{\beta}^{1-\beta} \left(F^{-1}(t) - G^{-1}(t) - \delta_\beta(F, G)\right)^2 dt,$$

where

$$\delta_\beta = \delta_\beta(F, G) = \frac{1}{1-2\beta}\left(\int_{F^{-1}(\beta)}^{F^{-1}(1-\beta)} x\, dF(x) - \int_{G^{-1}(\beta)}^{G^{-1}(1-\beta)} x\, dG(x)\right). \quad (8)$$

Let $C[0,1]$ be the metric space of continuous functions on $[0,1]$ equipped with the sup-norm and the Borel $\sigma$-field. A Brownian Bridge will be denoted as $\mathcal{B}$, i.e. a centered Gaussian process in $C[0,1]$ with $\text{cov}[\mathcal{B}(s), \mathcal{B}(t)] = s \wedge t - st$, where $s \wedge t := \min\{s, t\}$. The following theorem treats the case of paired data.

**Theorem 2.1** Let $H \in C^2[\mathbb{R}^2]$ with marginals $F, G$ and $Z_1, \ldots, Z_n \stackrel{i.i.d.}{\sim} H(\cdot, \cdot)$. Then under the assumptions (7) we have for $\Gamma^2_{L,\beta} > 0$, as $n \to \infty$,

$$n^{\frac{1}{2}}\left\{\Gamma^2_{L,\beta}(F_n, G_n) - \Gamma^2_{L,\beta}(F, G)\right\} \stackrel{d}{\longrightarrow} X_{L1},$$

where $X_{L1}$ is given by

$$X_{L1} = \frac{2}{1 - 2\beta}\left\{\int_0^{1-\beta}\int_{\beta \vee s}^{1-\beta} \frac{F^{-1}(t) - G^{-1}(t) - \delta_\beta}{f(F^{-1}(t))} dt\, d\mathcal{B}_F(s) \right.$$
$$\left. + \int_0^{1-\beta}\int_{\beta \vee s}^{1-\beta} \frac{G^{-1}(t) - F^{-1}(t) + \delta_\beta}{g(G^{-1}(t))} dt\, d\mathcal{B}_G(s)\right\}.$$

Here $\mathcal{B}_F, \mathcal{B}_G$ are Brownian bridges with $\mathrm{cov}[\mathcal{B}_F(s), \mathcal{B}_G(t)] = H(F^{-1}(s), G^{-1}(t)) - st$. Thus, $X_{L1}$ is a centered normally distributed random variable with variance

$$\sigma_{L1,\beta}^2 = \frac{4}{(1-2\beta)^2}\left[\int_0^{1-\beta}\left(\int_{\beta\vee s}^{1-\beta}\frac{F^{-1}(t)-G^{-1}(t)-\delta_\beta}{f(F^{-1}(t))}dt\right)^2 ds - \left(\int_0^{1-\beta}\int_{\beta\vee s}^{1-\beta}\frac{F^{-1}(t)-G^{-1}(t)-\delta_\beta}{f(F^{-1}(t))}dtds\right)^2\right.$$

$$+ \int_0^{1-\beta}\left(\int_{\beta\vee s}^{1-\beta}\frac{G^{-1}(t)-F^{-1}(t)+\delta_\beta}{g(G^{-1}(t))}dt\right)^2 ds - \left(\int_0^{1-\beta}\int_{\beta\vee s}^{1-\beta}\frac{G^{-1}(t)-F^{-1}(t)+\delta_\beta}{g(G^{-1}(t))}dtds\right)^2$$

$$+ 2\int_0^{1-\beta}\int_0^{1-\beta}\left(\int_{\beta\vee r}^{1-\beta}\frac{F^{-1}(t)-G^{-1}(t)-\delta_\beta}{f(F^{-1}(t))}dt\right)\left(\int_{\beta\vee s}^{1-\beta}\frac{G^{-1}(t)-F^{-1}(t)+\delta_\beta}{g(G^{-1}(t))}dt\right)\frac{\partial H(F^{-1}(r), G^{-1}(s))}{\partial r \partial s}drds$$

$$\left. - 2\left(\int_0^{1-\beta}\int_{\beta\vee s}^{1-\beta}\frac{F^{-1}(t)-G^{-1}(t)-\delta_\beta}{f(F^{-1}(t))}dtds\right)\left(\int_0^{1-\beta}\int_{\beta\vee s}^{1-\beta}\frac{G^{-1}(t)-F^{-1}(t)+\delta_\beta}{g(G^{-1}(t))}dtds\right)\right].$$

Now we deal with the case of two independent samples as in the Example 1.

**Theorem 2.2** Assume that $X_1, \ldots, X_m \overset{i.i.d.}{\sim} F$, $Y_1, \ldots, Y_n \overset{i.i.d.}{\sim} G$ are independent samples, and $F, G$ satisfy the conditions in (7). If $n/(n+m) \to \lambda \in (0,1)$ as $m \wedge n \to \infty$, then we obtain

$$\left(\frac{nm}{n+m}\right)^{\frac{1}{2}}\left\{\Gamma_{L,\beta}^2(F_m, G_n) - \Gamma_{L,\beta}^2(F,G)\right\} \xrightarrow{d} X_{L2},$$

where $X_{L2}$ is given by

$$X_{L2} := \frac{2}{1-2\beta}\left\{\sqrt{\lambda}\int_0^{1-\beta}\int_{\beta\vee s}^{1-\beta}\frac{F^{-1}(t)-G^{-1}(t)-\delta_\beta}{f(F^{-1}(t))}dt\, d\mathcal{B}_F(s)\right.$$

$$\left. + \sqrt{1-\lambda}\int_0^{1-\beta}\int_{\beta\vee s}^{1-\beta}\frac{G^{-1}(t)-F^{-1}(t)+\delta_\beta}{g(G^{-1}(t))}dt\, d\mathcal{B}_G(s)\right\},$$

with two independent Brownian bridges $\mathcal{B}_F$ and $\mathcal{B}_G$. Furthermore, $X_{L2}$ is a centered normally distributed random variable with variance

$$\sigma_{L2,\beta}^2 = \frac{4}{(1-2\beta)^2}\left[\lambda\left(\int_0^{1-\beta}\left(\int_{\beta\vee s}^{1-\beta}\frac{F^{-1}(t)-G^{-1}(t)-\delta_\beta}{f(F^{-1}(t))}dt\right)^2 ds - \left(\int_0^{1-\beta}\int_{\beta\vee s}^{1-\beta}\frac{F^{-1}(t)-G^{-1}(t)-\delta_\beta}{f(F^{-1}(t))}dtds\right)^2\right)\right.$$

$$\left. + (1-\lambda)\left(\int_0^{1-\beta}\left(\int_{\beta\vee s}^{1-\beta}\frac{G^{-1}(t)-F^{-1}(t)+\delta_\beta}{g(G^{-1}(t))}dt\right)^2 ds - \left(\int_0^{1-\beta}\int_{\beta\vee s}^{1-\beta}\frac{G^{-1}(t)-F^{-1}(t)+\delta_\beta}{g(G^{-1}(t))}dtds\right)^2\right)\right].$$

### 2.3. Other models

Now we turn to the cases of acceleration and location-scale families. As for the location model, the minimizing values of the parameter vector can be determined explicitly. The resulting minimum distance for the acceleration model is

$$\Gamma_{A,\beta}^2 = \|F^{-1} - \sigma_{A,\beta} G^{-1}\|_\beta^2,$$

with

$$\sigma_{A,\beta} = \frac{\int_\beta^{1-\beta} F^{-1}(t) G^{-1}(t) dt}{\int_\beta^{1-\beta} (G^{-1}(t))^2 dt},$$

and for the location-scale model we obtain

$$\Gamma_{LS,\beta}^2 = \|F^{-1} - \sigma_{LS,\beta} G^{-1} - \delta_{LS,\beta}\|_\beta^2,$$

with

$$\delta_{LS,\beta} = \frac{1}{1-2\beta}\int_\beta^{1-\beta}\left(F^{-1}(t) - \sigma_{LS,\beta} G^{-1}(t)\right) dt,$$

$$\sigma_{LS,\beta} = \frac{\int_\beta^{1-\beta} F^{-1}(t) G^{-1}(t) dt - \frac{1}{1-2\beta}\int_\beta^{1-\beta} F^{-1}(t) dt \int_\beta^{1-\beta} G^{-1}(t) dt}{\int_\beta^{1-\beta}(G^{-1}(t))^2 dt - \frac{1}{1-2\beta}\left(\int_\beta^{1-\beta} G^{-1}(t) dt\right)^2}.$$

For a concise notation we will write in the following $\int = \int_\beta^{1-\beta}$ and $k_\beta = \frac{1}{1-2\beta}$.

**Theorem 2.3** Let $H \in C^2[\mathbb{R}^2]$ with marginals $F, G$, and let $Z_1, \ldots, Z_n \stackrel{i.i.d.}{\sim} H(\cdot, \cdot)$.

i) Then under the assumptions of Theorem 2.1 we have for $\Gamma^2_{A,\beta}(F, G) > 0$, as $n \to \infty$,
$n^{\frac{1}{2}}\left\{\Gamma^2_{A,\beta}(F_n, G_n) - \Gamma^2_{A,\beta}(F, G)\right\} \stackrel{d}{\longrightarrow} X_A$, where $X_A$ is a centered normal r.v. defined as

$$X_A = 2k_\beta \int \left(F^{-1}(u) - \sigma_{A,\beta}G^{-1}(u)\right)\left[\tilde{h}_1(u) - \sigma_{A,\beta}\tilde{h}_2(u)\right.$$
$$\left.-G^{-1}(u)\left(\frac{\int\left(\tilde{h}_1(t)G^{-1}(t)+F^{-1}(t)\tilde{h}_2(t)\right)dt}{\int(G^{-1}(t))^2 dt} - 2\frac{\int\left(F^{-1}(t)G^{-1}(t)\right)dt \int\left(G^{-1}(t)\tilde{h}_2(t)\right)dt}{\left[\int(G^{-1}(t))^2 dt\right]^2}\right)\right]du,$$

with $\tilde{h}_1(u) := \frac{\mathcal{B}_F(F^{-1}(u))}{f \circ F^{-1}(u)}$, $\tilde{h}_2(u) := \frac{\mathcal{B}_G(G^{-1}(u))}{g \circ G^{-1}(u)}$, and $\mathcal{B}_F, \mathcal{B}_G$ as defined in Theorem 2.1.

ii) Furthermore, if $\Gamma^2_{LS,\beta}(F, G) > 0$, we have $n^{\frac{1}{2}}\left\{\Gamma^2_{LS,\beta}(F_n, G_n) - \Gamma^2_{LS,\beta}(F, G)\right\} \stackrel{d}{\longrightarrow} X_{LS}$, where $X_{LS}$ is defined as

$$X_{LS} = 2k_\beta \int \left(F^{-1}(u) - \sigma_{LS,\beta}G^{-1}(u) - \delta_{LS,\beta}\right)\left[\tilde{h}_1(u) - \sigma_{LS}\tilde{h}_2(u)\right.$$
$$-G^{-1}(u)\left(\frac{\int\left(\tilde{h}_1(t)G^{-1}(t)+F^{-1}(t)\tilde{h}_2(t)\right)dt - k_\beta\left(\int \tilde{h}_2(t)dt \int F^{-1}(t)dt + \int \tilde{h}_1(t)dt \int G^{-1}(t)dt\right)}{\int(G^{-1}(t))^2 dt - k_\beta\left[\int G^{-1}(t)dt\right]^2}\right.$$
$$\left.-2\frac{\left[\int\left(F^{-1}(t)G^{-1}(t)\right)dt + k_\beta \int F^{-1}(t)dt \int G^{-1}(t)dt\right]\left[\int\left(G^{-1}(t)\tilde{h}_2(t)\right)dt - k_\beta \int G^{-1}(t)dt \int \tilde{h}_2(t)dt\right]}{\left[\int(G^{-1}(t))^2 dt - k_\beta\left(\int G^{-1}(t)dt\right)^2\right]^2}\right)$$
$$-k_\beta \int \left\{\tilde{h}_1(t) - \sigma_{LS,\beta}\tilde{h}_2(t)\right.$$
$$-G^{-1}(t)\left(\frac{\int\left(\tilde{h}_1(s)G^{-1}(s)+F^{-1}(s)\tilde{h}_2(s)\right)ds - k_\beta\left(\int \tilde{h}_2(s)ds \int F^{-1}(s)ds + \int \tilde{h}_1(s)ds \int G^{-1}(s)ds\right)}{\int(G^{-1}(s))^2 ds - k_\beta\left[\int G^{-1}(s)ds\right]^2}\right.$$
$$\left.\left.-2\frac{\left[\int\left(F^{-1}(s)G^{-1}(s)\right)ds + k_\beta \int F^{-1}(s)ds \int G^{-1}(s)ds\right]\left[\int\left(G^{-1}(s)\tilde{h}_2(s)\right)ds - k_\beta \int G^{-1}(s)ds \int \tilde{h}_2(s)ds\right]}{\left[\int(G^{-1}(s))^2 ds - k_\beta\left(\int G^{-1}(s)ds\right)^2\right]^2}\right)\right\}dt\right]du.$$

Integration by parts shows again that both random variables $X_S$ and $X_{LS}$ are normally distributed.

**Remark 2.4** (*Testing similarity of marginals / Equivalence test*). Testing equality of the marginals, $F = G$, in a bivariate setup has been treated rather extensively in the literature, mainly by methods based on generalized Chernoff-Savage [26] theorems, leading to tests based on rank statistics (cf. [8,24,27–29]). In our setting it would be of interest to show that

$$\Gamma_{E,\beta}(F, G) := \|F^{-1} - G^{-1}\|_\beta \le \Delta_0,$$

for a pre-specified tolerance margin $\Delta_0$. This leads to a statement similar to the above results (cf. also [18] for the case $H = F \otimes G$).

**Theorem 2.5** Let $H \in C^2[\mathbb{R}^2]$ with marginals $F, G$, and let $Z_1, \ldots, Z_n \stackrel{i.i.d.}{\sim} H(\cdot, \cdot)$. Then under the assumptions of Theorem 2.1 we have for $\Gamma^2_{E,\beta}(F, G) > 0$,

$$n^{\frac{1}{2}}\left\{\Gamma^2_{E,\beta}(F_n, G_n) - \Gamma^2_{E,\beta}(F, G)\right\} \stackrel{d}{\longrightarrow} X_E,$$

where $X_E$ is defined as

$$X_E := 2k_\beta \left\{ \int_0^{1-\beta} \int_{\beta \vee s}^{1-\beta} \frac{F^{-1}(t) - G^{-1}(t)}{f(F^{-1}(t))} dt dB_F(s) + \int_0^{1-\beta} \int_{\beta \vee s}^{1-\beta} \frac{G^{-1}(t) - F^{-1}(t)}{g(G^{-1}(t))} dt dB_G(s) \right\},$$

with Brownian bridges $B_F$ and $B_G$ on $[0,1]$ as in Theorem 2.1.

**Remark 2.6** (*Estimating the asymptotic variances*). In principle, consistent estimators $\hat{\sigma}^2_{(\cdot),\beta}$ for the asymptotic variances of the above test statistics can be constructed by plugging in the empirical estimators for the involved distribution functions. Thus, we obtain asymptotic level $\alpha$ tests for the hypotheses

$$H_{(\cdot)} : \Gamma_{(\cdot),\beta}(F,G) > \Delta_0 \quad \text{vs.} \quad K_{(\cdot)} : \Gamma_{(\cdot),\beta} \leq \Delta_0, \tag{9}$$

if $H_{(\cdot)}$ is rejected whenever $T_{n,(\cdot)} < -u_{1-\alpha}$, with $u_\alpha$ as the $\alpha$-quantile of the standard normal distribution, and $T_{n,(\cdot)} = n^{\frac{1}{2}} (\Gamma^2_{(\cdot),\beta}(F_n,G_n) - \Delta_0^2)/\hat{\sigma}_{(\cdot),\beta}$. However, the expressions for the estimators of the asymptotic variances are rather complicated, thus in the next section we will suggest bootstrap versions of our tests in order to circumvent these difficulties.

**Remark 2.7** (*Distributions under $\Gamma^2_{(\cdot),\beta}(F,G) = 0$*). Observe that under the 'classical null hypothesis' $H_0 : \Gamma^2_{(\cdot),\beta}(F,G) = 0$, we obtain degenerate limit laws from Theorems 2.1, 2.2, 2.3 and 2.5, i.e. all test statistics converge to zero in probability. In these cases limit theorems can be obtained after multiplying by the factor $n$ instead of $n^{\frac{1}{2}}$. For example, for the assessment of the location model we find that if $\Gamma^2_{L,\beta}(F,G) = 0$, i.e. $G^{-1}(\cdot) = F^{-1} - \delta_\beta(F,G)$ on $[\beta, 1-\beta]$,

$$n\Gamma^2_{L,\beta}(F_n,G_n) = n k_\beta \int \left\{ F_n^{-1}(t) - F^{-1}(t) - \left(G_n^{-1}(t) - G^{-1}(t)\right) - (\delta_\beta(F_n,G_n) - \delta_\beta(F,G)) \right\}^2 dt$$

$$\xrightarrow{d} k_\beta \int \left\{ B_F(t) - B_G(t) - \left( \int t \, dB_F(t) - \int t \, dB_G(t) \right) \right\}^2 dt,$$

where $B_F, B_G$ are defined as in Theorem 2.1. This is a complicated law of series of weighted $\chi_1^2$ random variables depending on the particular bivariate distribution $H$. For similar distributions occurring in goodness-of-fit testing for the one sample case we refer to [30–32].

**Remark 2.8** The above results can be immediately transferred to the symmetrized versions of the $\Gamma^2_{(\cdot),\beta}$ (cf. (6)).

**Remark 2.9** (*More than two samples*). It has to be noted that, of course, the suggested approach can be extended in a straight forward manner to the case of multivariate distribution functions $(F_1, ..., F_d) =: H$, for $d \geq 3$. Here, e.g., all pairwise distances $\Gamma^2_{(\cdot),\beta}(F_i, F_j) =: \gamma^{ij}_{(\cdot),\beta}$ can be summarized in the measure

$$\Gamma^2_{(\cdot),\beta}(F_1, ..., F_d) := \frac{1}{d(d-1)} \sum_{i \neq j} \gamma^{ij}_{(\cdot),\beta}. \tag{10}$$

The techniques used in the Appendix for proving the above results can immediately be transferred to $\Gamma^2_{(\cdot),\beta}(F_1, ..., F_d)$. Hence, limit theorems similar to Theorems 2.1, 2.2, 2.3 and 2.5 can be derived, provided that $\gamma^{ij}_{(\cdot),\beta} > 0$, $1 \leq i < j \leq d$, taking into account the invariance principle of the multivariate empirical process $n^{\frac{1}{2}}(H_n - H)(\cdot)$ as stated e.g. in [33], together with the Hadamard differentiability of (10).

## 3. BOOTSTRAP TESTS

As mentioned above, estimating the limiting variances of the test statistics under consideration is a very cumbersome task. Therefore, we suggest in the following a bootstrapped version of all tests presented in Section 2.

Using the weak convergence results given there, it can be concluded that a percentile bootstrap can be applied. More precisely, under the assumptions of Theorem 2.1, the weak consistency of the bootstrap method for the statistics considered here can be proved if $\Gamma^2_{(\cdot),\beta}(F,G) > 0$ along the lines of [34, Theorem 5], taking into account the Hadamard differentiability of the underlying functional (cf. the proof of Theorems 2.1, 2.2, 2.3 and 2.5 in the Appendix) together with the fact that the limit law of $n^{\frac{1}{2}}(\Gamma^2_{(\cdot),\beta}(F_n,G_n) - \Gamma^2_{(\cdot),\beta}(F,G))$ is normal with finite variance.

However, we found that the finite sample accuracy of the percentile bootstrap is not always satisfying. Following [35, p. 214], we therefore introduce for practical purposes the bias corrected and accelerated bootstrap ($BC_a$) in our context as follows. Let $H_n$ denote the empirical cumulative distribution function of the sample $\mathbf{Z_n} = (Z_1, \ldots, Z_n)$ with marginals $(F_n, G_n)$ and let

$$T := T_{\Delta_0,\beta}(\mathbf{Z_n}) := n^{\frac{1}{2}}\left(\Gamma^2_{(\cdot),\beta}(F_n, G_n) - \Delta_0^2\right). \quad (11)$$

For a set of B bootstrap samples $\mathbf{Z}^*_{n,1}, \ldots, \mathbf{Z}^*_{n,B}$ from $H_n$, define $T^b := T_{\Delta_0,\beta}(\mathbf{Z}^*_{n,b})$, $b = 1, \ldots, B$, and $\alpha_{up}(\alpha) = \Phi\left(\hat{z}_0 + \frac{\hat{z}_0 + u_{1-\alpha}}{1-\hat{a}(\hat{z}_0 + u_{1-\alpha})}\right)$, where $\Phi$ denotes the cumulative distribution function of a standard normal random variable and $\alpha$ is the chosen level of the test. Here $\hat{z}_0 = \Phi^{-1}\left(\frac{\#\{T^b \leq T : b=1,\ldots,B\}}{B}\right)$, and $\hat{a}$ denotes the correction $\hat{a} = \frac{\sum_{i=1}^n (\bar{T} - T_{(i)})^3}{6[\sum_{i=1}^n (\bar{T} - T_{(i)})^2]^{\frac{3}{2}}}$, with $T_{(i)}$ as the resulting test statistic obtained from (11) by leaving out the $i^{th}$ observation, and $\bar{T} = \frac{1}{n}\sum_{i=1}^n T_{(i)}$. Now we can define the bootstrapped p-value

$$p_{BC_a}(\Delta_0) = \alpha_{up}^{-1}\left(\frac{\sum_{b=1}^B I_{\{T^b \leq 0\}}}{B}\right). \quad (12)$$

## 4. SIMULATION RESULTS

Here we present a small selection of results from an extensive simulation study on the tests based on the $BC_a$ method for validating the models presented in Section 1. In particular, only some results for the location model and a summary of some simulations on the acceleration model are given. Further results can be obtained from the authors on request. The simulation study was designed to evaluate the performance of the tests with respect to different correlations, varying magnitudes of the sample variances, and the use of trimming. We used bivariate normal distributions with several parameter configurations, and for each setting equal sample sizes $n_1 = n_2 =: n$ were chosen. The bootstrap sample sizes were taken as $B = 500$, and 500 replications were generated for each scenario. All simulations were done using SAS V8.

### 4.1. Location model

We considered location-scale differences between the two marginal distributions, using $\Delta_0^2 = 1$, trimming bounds $\beta = 0, 0.05$, correlations $\rho = -0.5, 0, 0.5$, and sample sizes $n = 50, 100, 200$. In Tables 1 and 2, the observed power and significance levels are displayed, respectively. Table 1 shows the results for a small difference in location, $\mu_2 - \mu_1 = 1$, whereas in Table 2 the location difference was chosen as 5. Overall, the $BC_a$ method is slightly liberal for large variances.

Trimming improves the performance of the test. There is no clear effect of the correlation to be seen. Altogether it can be deduced that the $BC_a$ method yields quite satisfactory results.

Table 1
Location test: normal margins ($\mu_1 = 0$, $\mu_2 = 1$, $\sigma_1 = 1.5$, $\Delta_0^2 = 1$)

| $\beta$ | $\sigma_2$ | $\Gamma_{L,\beta}^2$ | n = 50 Correlation | | | n = 100 Correlation | | | n = 200 Correlation | | |
|---|---|---|---|---|---|---|---|---|---|---|---|
| | | | -0.5 | 0 | 0.5 | -0.5 | 0 | 0.5 | -0.5 | 0 | 0.5 |
| | | | | | $\alpha = 0.05$ | | | | | | |
| 0 | 0.5 | 1.00 | 0.072 | 0.068 | 0.086 | 0.072 | 0.062 | 0.060 | 0.074 | 0.040 | 0.052 |
| | 1.5 | 0.00 | 0.996 | 0.990 | 0.998 | 1.000 | 1.000 | 1.000 | 1.000 | 1.000 | 1.000 |
| | 2.5 | 1.00 | 0.084 | 0.094 | 0.102 | 0.050 | 0.074 | 0.056 | 0.054 | 0.052 | 0.058 |
| 0.05 | 0.233 | 1.00 | 0.034 | 0.050 | 0.038 | 0.044 | 0.064 | 0.060 | 0.058 | 0.042 | 0.044 |
| | 1.5 | 0.00 | 0.998 | 0.998 | 1.000 | 1.000 | 1.000 | 1.000 | 1.000 | 1.000 | 1.000 |
| | 2.767 | 1.00 | 0.088 | 0.076 | 0.062 | 0.056 | 0.054 | 0.046 | 0.050 | 0.058 | 0.062 |
| | | | | | $\alpha = 0.10$ | | | | | | |
| 0 | 0.5 | 1.00 | 0.120 | 0.128 | 0.154 | 0.120 | 0.108 | 0.106 | 0.132 | 0.092 | 0.116 |
| | 1.5 | 0.00 | 1.000 | 0.996 | 1.000 | 1.000 | 1.000 | 1.000 | 1.000 | 1.000 | 1.000 |
| | 2.5 | 1.00 | 0.118 | 0.144 | 0.152 | 0.108 | 0.128 | 0.104 | 0.094 | 0.084 | 0.104 |
| 0.05 | 0.233 | 1.00 | 0.068 | 0.106 | 0.086 | 0.094 | 0.114 | 0.108 | 0.110 | 0.086 | 0.104 |
| | 1.5 | 0.00 | 1.000 | 1.000 | 1.000 | 1.000 | 1.000 | 1.000 | 1.000 | 1.000 | 1.000 |
| | 2.767 | 1.00 | 0.150 | 0.122 | 0.108 | 0.100 | 0.108 | 0.106 | 0.104 | 0.096 | 0.118 |

Table 2
Location test: normal margins ($\mu_1 = 0$, $\mu_2 = 5$, $\sigma_1 = 1.5$, $\Delta_0^2 = 1$)

| $\beta$ | $\sigma_2$ | $\Gamma_{L,\beta}^2$ | n = 50 Correlation | | | n = 100 Correlation | | | n = 200 Correlation | | |
|---|---|---|---|---|---|---|---|---|---|---|---|
| | | | -0.5 | 0 | 0.5 | -0.5 | 0 | 0.5 | -0.5 | 0 | 0.5 |
| | | | | | $\alpha = 0.05$ | | | | | | |
| 0 | 0.5 | 1.00 | 0.056 | 0.090 | 0.094 | 0.078 | 0.054 | 0.060 | 0.054 | 0.064 | 0.078 |
| | 1.5 | 0.00 | 0.996 | 0.996 | 0.996 | 1.000 | 1.000 | 1.000 | 1.000 | 1.000 | 1.000 |
| | 2.5 | 1.00 | 0.068 | 0.118 | 0.116 | 0.084 | 0.096 | 0.080 | 0.054 | 0.056 | 0.030 |
| 0.05 | 0.233 | 1.00 | 0.044 | 0.046 | 0.056 | 0.054 | 0.042 | 0.044 | 0.034 | 0.064 | 0.048 |
| | 1.5 | 0.00 | 1.000 | 1.000 | 1.000 | 1.000 | 1.000 | 1.000 | 1.000 | 1.000 | 1.000 |
| | 2.767 | 1.00 | 0.080 | 0.098 | 0.060 | 0.054 | 0.054 | 0.056 | 0.056 | 0.070 | 0.060 |
| | | | | | $\alpha = 0.10$ | | | | | | |
| 0 | 0.5 | 1.00 | 0.114 | 0.138 | 0.140 | 0.144 | 0.106 | 0.096 | 0.110 | 0.108 | 0.118 |
| | 1.5 | 0.00 | 0.998 | 0.998 | 1.000 | 1.000 | 1.000 | 1.000 | 1.000 | 1.000 | 1.000 |
| | 2.5 | 1.00 | 0.112 | 0.172 | 0.170 | 0.140 | 0.142 | 0.128 | 0.098 | 0.098 | 0.068 |
| 0.05 | 0.233 | 1.00 | 0.090 | 0.096 | 0.130 | 0.128 | 0.092 | 0.082 | 0.098 | 0.106 | 0.088 |
| | 1.5 | 0.00 | 1.000 | 1.000 | 1.000 | 1.000 | 1.000 | 1.000 | 1.000 | 1.002 | 1.000 |
| | 2.767 | 1.00 | 0.138 | 0.128 | 0.106 | 0.098 | 0.102 | 0.058 | 0.114 | 0.122 | 0.108 |

### 4.2. Acceleration model

For this model we used the symmetrized version of the test statistics, as indicated in (6). Several bivariate normal distribution settings were considered, and each test was performed with $\Delta_0^2 = 1$. We found that in general trimming improves the performance of the $BC_a$ test. For smaller variances, it turns out to be liberal in case of no trimming and conservative in case of trimming. For larger variances, the $BC_a$ method is always liberal, although trimming reduces the liberality. It was seen that an increase in location difference (while keeping the true distance $\Gamma_{A,\beta}(F,G)$ constant) results in more liberal tests for $\beta = 0$, and in less liberal tests for $\beta = 0.05$.

## 5. EXAMPLE

In this chapter we analyze the data set from Example 1 with the suggested methodology from Sections 2 and 3. To this end we use the concept of p-value curves as it was introduced by [18] and transfer it to the present setting. These curves are based on the tests for the hypotheses (9) (similarly for the other structural relationship models), as developed in the preceding section. For given observations $(Z_1, \ldots, Z_n)$, this provides a simple graphical method to evaluate the evidence for the (approximate) presence of a structural relationship with respect to the (trimmed) Mallows distance. More specifically, we regard the $BC_a$-p-value curve given by the values of (12) for varying $\Delta_0$. This function summarizes different quantities. First, the $\alpha$ cut-off point gives the smallest bound $\Delta_{0,\alpha}$, for which the neighborhood hypothesis $H_{(\cdot)} : \Gamma_{(\cdot),\beta} > \Delta_{0,\alpha}$ can be rejected. Conversely, the $(1 - \alpha)$ cut-off point yields the largest bound $\Delta_{0,1-\alpha}$, for which the hypothesis $K_{(\cdot)} : \Gamma_{(\cdot),\beta} < \Delta_{0,1-\alpha}$ can be rejected. Finally, confidence intervals for the true distance $\Gamma_{(\cdot),\beta}$ can be deduced from $p_{BC_a}(\Delta_0)$ easily. For example, a one-sided $(1 - \alpha)$-confidence interval for $\Gamma_{(\cdot),\beta}$ is given by $[0, p_{BC_a}^{-1}(\alpha))$.

In [22] it was concluded that after trimming there is a pure shift effect between the cumulative distribution functions for hospitalization and mortality. The authors observed that in the center of the distributions the difference between the two populations was constant about 6.75, while at the tails there was no constant difference. We investigate the female patients only. For different trimming constants $\beta$, the estimated Mallows distances are displayed in Table 3. Figure 2 shows the corresponding $BC_a$-p-value curves with respect to $\Delta_0$. In accordance with the graphical analysis by [22], it can be seen that with stronger trimming a location model seems to be appropriate for the data.

## 6. DISCUSSION

We have proposed a method in order to quantify semiparametric distances between distribution functions based on a trimmed version of Mallows distance. Asymptotic normality of the corresponding test statistics has been shown, which allows the assessment whether two distribution functions are 'close' (up to a specific structural relationship) at a controlled error rate. Moreover, this overcomes the methodological difficulty encountered with classical null hypotheses, where it is not possible to decide whether acceptance of the null model is due to the presence of the null model or simply due to the lack of power of the particular test which has been used. Our approach requires the quantification of a maximal distance $\Delta_0$ regarded as 'tolerable', which can be driven by various aspects. This could be based on medical reasoning or on statistical grounds, such as a loss in efficiency when applying a 'wrong model' outside a certain $\Gamma_{(\cdot),\beta}$ neighborhood. We are aware that the problem of the 'choice of $\Delta_0$' leads to an

Table 3
Mortality data: estimated Mallows distances for the location model

| $\beta$ | $\hat{\Gamma}_{L,\beta}$ (in years) |
|---|---|
| 0.0 | 2.68 |
| 0.25 | 2.48 |
| 0.05 | 2.31 |
| 0.1 | 2.05 |
| 0.2 | 1.31 |
| 0.25 | 0.81 |

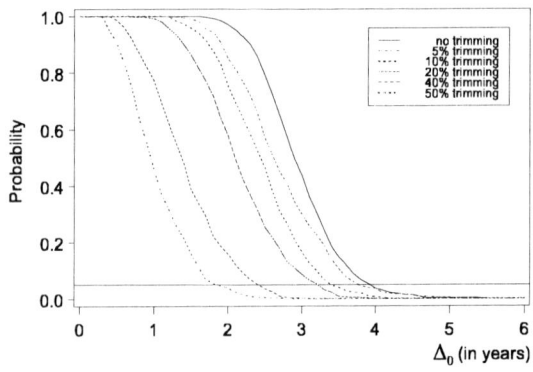

Figure 2. Mortality data: $BC_a$-p-value curves for the location model

increased complexity compared to classical testing procedures. However, the proposed p-value curves may serve as a supplementary graphical tool which helps to visualize quickly (and a posteriori) the evidence of a certain magnitude of model discrepancy. This includes in particular confidence intervals for the discrepancy measure $\Gamma_{(\cdot),\beta}$.

## 7. APPENDIX

For the proof of Theorems 2.1, 2.2, 2.3 and 2.5 we require the following lemma.

**Lemma 7.1** Let $X := \int p(t)d\mathcal{B}_1(t) + \int q(t)d\mathcal{B}_2(t)$, where $p, q \in L^2(\lambda)$ and $\mathcal{B}_1$ and $\mathcal{B}_2$ are Brownian Bridges on $C[0,1]$ with $cov[\mathcal{B}_1(s), \mathcal{B}_2(t)] = C(s,t) - st$. The function $C$ is supposed to be a two times differentiable copula with continuous second derivative, i.e. $C \in C^2[0,1]^2$. Then we have $X \sim \mathcal{N}(0, \sigma_{p,q}^2)$, where the variance is given as

$$\sigma_{p,q}^2 = \int p^2(t)dt - \left(\int p(t)dt\right)^2 + \int q^2(t)dt - \left(\int q(t)dt\right)^2 \\ + 2\int\int p(t)q(s)\frac{\partial^2 C(s,t)}{\partial s \partial t}dsdt - 2\int p(t)dt \int q(s)ds.$$

**Proof:** Let us first consider the case when $p$ and $q$ are step functions. Then we can write

$$X = \sum_{i=1}^{n} \left\{ (p_i(\mathcal{B}_1(t_{i+1}) - \mathcal{B}_1(t_i)) + q_i(\mathcal{B}_2(t_{i+1}) - \mathcal{B}_2(t_i)) \right\},$$

where $0 = t_1 < t_2 ... < t_{n+1} = 1$ is a partition of $[0,1]$. Hence X is normally distributed and $E(X) = 0$. Furthermore,

$$E(X^2) = E\left(\sum_{i=1}^{n} p_i(\mathcal{B}_1(t_{i+1}) - \mathcal{B}_1(t_i))\right)^2 + E\left(\sum_{i=1}^{n} q_i(\mathcal{B}_2(t_{i+1}) - \mathcal{B}_2(t_i))\right)^2 \\ + 2E\sum_{i=1}^{n}\sum_{j=1}^{n} p_i q_j (\mathcal{B}_1(t_{i+1}) - \mathcal{B}_1(t_i))(\mathcal{B}_2(t_{j+1}) - \mathcal{B}_2(t_j)) =: \text{I} + \text{II} + 2 \cdot \text{III}.$$

For the expressions I and II and the approximation of functions in $L^2(\lambda)$ with simple functions see [36], Chapter 2.2. Hence it remains to consider III. We have

$$\begin{aligned}
\text{III} &= \sum_{i=1}^{n} \sum_{j=1}^{n} p_i q_j \, E(\mathcal{B}_1(t_{i+1}) - \mathcal{B}_1(t_i)) \, E(\mathcal{B}_2(t_{j+1}) - \mathcal{B}_2(t_j)) \\
&= \sum_{i=1}^{n} \sum_{j=1}^{n} p_i q_j \left(C(t_{i+1}, t_{j+1}) - C(t_{i+1}, t_j) - C(t_i, t_{j+1}) + C(t_i, t_j)\right) \\
&\quad - \sum_{i=1}^{n} \sum_{j=1}^{n} p_i q_j \left(t_{i+1} t_{j+1} - t_{i+1} t_j - t_i t_{j+1} + t_i t_j\right) \\
&= \sum_{i=1}^{n} \sum_{j=1}^{n} p_i q_j \frac{C(t_{i+1},t_{j+1})-C(t_{i+1},t_j)-C(t_i,t_{j+1})+C(t_i,t_j)}{(t_{i+1}-t_i)(t_{j+1}-t_j)} (t_{i+1}-t_i)(t_{j+1}-t_j) \\
&\quad - \sum_{i=1}^{n} \sum_{j=1}^{n} p_i q_j (t_{i+1}-t_i)(t_{j+1}-t_j) \\
&= \int \int p(t) q(s) \frac{\partial^2 C(s,t)}{\partial s \partial t} ds \, dt - \int p(t) dt \int q(s) ds + o(1) \, . \qquad \square
\end{aligned}$$

In the following, $D[\mathbb{R}^p]$, $p \geq 1$, shall denote the space of functions on $\mathbb{R}^p$ with existing quadrant limits and continuous from above (cf. [37]), and equipped with the sup-norm and the open ball $\sigma$-field (cf. [38]).

**Proof of Theorems 2.1, 2.2, 2.3 and 2.5:** The main step in the proof is to show the Hadamard differentiability at $(F, G)$ tangentially to $C[\mathbb{R}]^2$ of the location-scale functional $T: D[\mathbb{R}]^2 \to \mathbb{R}$,

$$\begin{aligned}
T(F,G) &:= \min_{(\delta,\sigma) \in \mathbb{R} \times \mathbb{R}} \frac{1}{1-2\beta} \int_{\beta}^{1-\beta} \left(F^{-1}(u) - \sigma G^{-1}(u) - \delta\right)^2 du \\
&= k_\beta \int \left(F^{-1}(u) - \hat{\sigma} G^{-1}(u) - \hat{\delta}\right)^2 du,
\end{aligned}$$

where

$$\hat{\sigma} := \hat{\sigma}(F^{-1}, G^{-1}) := \frac{\int \left(F^{-1}(t) G^{-1}(t)\right) dt - k_\beta \int G^{-1}(t) dt \int F^{-1}(t) dt}{\int \left(G^{-1}(t)\right)^2 dt - k_\beta \left[\int G^{-1}(t) dt\right]^2},$$

$$\hat{\delta} := \hat{\delta}(F^{-1}, G^{-1}, \hat{\sigma}) := k_\beta \int \left(F^{-1}(t) - \hat{\sigma} G^{-1}(t)\right) dt \, .$$

This can be shown along the lines of [34] or [39]. The Hadamard derivative of $T$ is obtained as

$$\begin{aligned}
T'_{(F,G)}(h_1, h_2) = -2k_\beta \int & \left(F^{-1}(u) - \hat{\sigma} G^{-1}(u) - \hat{\delta}\right) \Bigg[ \tilde{h}_1(u) - \hat{\sigma} \tilde{h}_2(u) \\
& - G^{-1}(u) \Bigg( \frac{\int \left(\tilde{h}_1(t) G^{-1}(t) + F^{-1}(t) \tilde{h}_2(t)\right) dt - k_\beta \left(\int \tilde{h}_2(t) dt \int F^{-1}(t) dt + \int \tilde{h}_1(t) dt \int G^{-1}(t) dt\right)}{\int (G^{-1}(t))^2 dt - k_\beta \left[\int G^{-1}(t) dt\right]^2} \\
& -2 \frac{\left[\int \left(F^{-1}(t) G^{-1}(t)\right) dt + k_\beta \int F^{-1}(t) dt \int G^{-1}(t) dt\right]\left[\int \left(G^{-1}(t) \tilde{h}_2(t)\right) dt - k_\beta \int G^{-1}(t) dt \int \tilde{h}_2(t) dt\right]}{\left[\int (G^{-1}(t))^2 dt - k_\beta \left(\int G^{-1}(t) dt\right)^2\right]^2} \Bigg) \\
-k_\beta \int & \Bigg\{ \tilde{h}_1(t) - \hat{\sigma} \tilde{h}_2(t) \\
& - G^{-1}(t) \Bigg( \frac{\int \left(\tilde{h}_1(s) G^{-1}(s) + F^{-1}(s) \tilde{h}_2(s)\right) ds - k_\beta \left(\int \tilde{h}_2(s) ds \int F^{-1}(s) ds + \int \tilde{h}_1(s) ds \int G^{-1}(s) ds\right)}{\int (G^{-1}(s))^2 ds - k_\beta \left[\int G^{-1}(s) ds\right]^2} \\
& -2 \frac{\left[\int \left(F^{-1}(s) G^{-1}(s)\right) ds + k_\beta \int F^{-1}(s) ds \int G^{-1}(s) ds\right]\left[\int \left(G^{-1}(s) \tilde{h}_2(s)\right) ds - k_\beta \int G^{-1}(s) ds \int \tilde{h}_2(s) ds\right]}{\left[\int (G^{-1}(s))^2 ds - k_\beta \left(\int G^{-1}(s) ds\right)^2\right]^2} \Bigg) \Bigg\} dt \Bigg] du,
\end{aligned} \qquad (13)$$

where $\tilde{h}_1(u) = \frac{h_1(F^{-1}(u))}{f \circ F^{-1}(u)}$, $\tilde{h}_2(u) := \frac{h_2(G^{-1}(u))}{g \circ G^{-1}(u)}$. Thus, using the previous lemma we can apply the functional delta method (cf. e.g. Theorem 3.9.4 in [39]) together with the invariance principle for the multivariate empirical process (cf. [33]). The assertions of the three theorems follow immediately. □

**REFERENCES**

1. J. Hajek & Z. Sidak, Theory of Rank Tests, Academic Press, New York-London, 1967.
2. T. P. Hettmansperger, Statistical Inference Based on Ranks, Krieger Publishing Company, Malabar-Florida, 1991.
3. M. Friedman, The use of ranks to avoid the assumption of normality implicit in the analysis of variance, J. Am. Stat. Assoc. 32 (1937) 675–699.
4. W. H. Kruskal and W. A. Wallis, The use of ranks in one-criterion variance analysis, J. Am. Stat. Assoc. 47 (1952) 583–621.
5. W. H. Kruskal and W. A. Wallis, A nonparametric test for the several sample problem, Ann. Math. Stat. 23 (1952) 525–540.
6. W. H. Kruskal and W. A. Wallis, Errata in: The use of ranks in one-criterion variance analysis, J. Am. Stat. Assoc. 48 (1953) 907–909.
7. E. B. Page, Ordered hypotheses for multiple treatments: A significance test for linear ranks, J. Am. Stat. Assoc. 58 (1963) 216–230.
8. M. L. Puri and P. K. Sen, On a class of multivariate multisample rank-order tests, Sankhya Ser. A 28 (1966) 353–376.
9. M. L. Puri and P. K. Sen, A class of rank order tests for general linear hypotheses, Ann. Math. Stat. 40 (1969) 1325–1343.
10. G. L. Thompson, A unified approach to rank tests for multivariate and repeated measures designs, J. Am. Stat. Assoc. 86 (1991) 410–419.
11. M. G. Akritas and S. F. Arnold, Fully nonparametric hypotheses for factorial designs, I: Multivariate repeated measures designs, J. Am. Stat. Assoc. 89 (1994) 336–343.
12. M. G. Akritas and E. Brunner, A unified approach to rank tests in mixed models, J. Stat. Plann. Inf. 61 (1997) 249–277.
13. E. Brunner and M. L. Puri, Nonparametric Methods in Design and Analysis of Experiments, Handbook of Statistics 13, 631–703, 1996.
14. E. B. Manoukian, Mathematical Nonparametric Statistics, Gordon and Breach Science Publishers, New York, 1986.
15. D. Compagnone and M. Denker, Nonparametric tests for scale and location, J. Nonpar. Stat. 7, 1996, 123–154.
16. C. L. Mallows, A note on asymptotic joint normality, Ann. Math. Stat. 43 (1972) 508–515.
17. L. N. Wasserstein, Markovian processes on countable space product describing large systems of automata (Russian), Probl. Peredaci Inform. 5 (1969) 64–72.
18. A. Munk and C. Czado, Nonparametric validation of similar distributions and assessment of goodness of fit, J. R. Stat. Soc. Ser. B 60 (1998) 223–241.
19. H. Dette and A. Munk, Validation of linear regression models, Ann. Stat. 26 (1998) 778–800.
20. C. Goutis and C. P. Robert, Model choice in generalised linear models: A Bayesian approach via Kullback-Leibler projections, Biometrika 85 (1998) 29–37.

21. J. D. Kalbfleisch and R. L. Prentice, The Statistical Analysis of Failure Time Data, Wiley Series in Probability and Mathematical Statistics, Wiley & Sons, New York, 1980.
22. P. Laake, K. Laake and R. Aaberge, On the problem of measuring the distance between distribution functions: analysis of hospitalization versus mortality, Biometrics 41 (1985) 515–523.
23. M. J. Podgor and J. L. Gastwirth, Efficiency robust rank tests for stratified data, in: Research Developments in Probability and Statistics (Festschrift in honor of Madan L. Puri, eds.: E. Brunner and M. Denker). VSP International Science Publishers, 1996.
24. Z. Govindarajulu, A class of asymptotically distribution-free test procedures for equality of marginals under multivariate dependence, Am. J. Math. Manag. Sci. 15 (1995) 375–394.
25. Z. Govindarajulu, A class of asymptotically distribution-free tests for equality of marginals in multivariate populations, Math. Meth. in Stat. 6 (1997) 92–111.
26. H. Chernoff & I. R. Savage, Asymptotic normality and efficiency of certain nonparametric test statistics, Ann. Math. Stat. 29 (1958) 971–994.
27. Z. Govindarajulu, L. le Cam and I. M. Raghavachari, Generalizations of theorems of Chernoff and Savage on the asymptotic normality of test statistics, Proc. Fifth Berkeley Symp. Math. Stat. & Probab., 609–638, University of California Press, Berkeley, 1967.
28. F. Ruymgaart and M. C. A. van Zuijlen, Asymptotic normality of multi-variate linear rank statistics in the non i.i.d. case, Ann. Stat. 6 (1978) 588–602.
29. Z. Govindarajulu and T. L. Lai, A multivariate Chernoff-Savage Theorem with applications to rank statistics from multivariate population, in: Research Developments in Probability and Statistics (Festschrift in honor of Madan L. Puri, eds.: E. Brunner and M. Denker). VSP International Science Publishers, 1996.
30. T. de Wet and J. H. Venter, Asymptotic distributions of certain test criteria of normality, South Afr. Stat. J. 6 (1972) 135–149.
31. T. de Wet and J. H. Venter, Asymptotic distributions for quadratic forms with applications to tests of fit, Ann. Stat. 1 (1973) 380–387.
32. E. del Barrio, E. Gine and C. Matran, Central limit theorems for the Wasserstein distance between the empirical and the true distributions, Ann. Probab. 27 (1999) 1009–1071.
33. P. Gaenssler and W. Stute, A survey of results for independent and identically distributed random variables, Ann. Probab. 7 (1979) 193–243.
34. R. D. Gill, Non- and semi-parametric maximum likelihood estimators and the von Mises method. I, Scand. J. Stat. 16 (1989) 97–128.
35. B. Efron and R. J. Tibshirani, An Introduction to the Bootstrap, Monographs on Statistics and Applied Probabilities 57, Chapman & Hall, New York (1993).
36. M. Denker, Asymptotic Distribution Theory in Nonparametric Statistics, Advanced Lectures in Mathematics, Vieweg & Sohn, Braunschweig, 1985.
37. G. Neuhaus, On weak convergence of stochastic processes with multidimensional time parameter, Ann. Math. Stat. 42 (1971) 1285–1295.
38. R. M. Dudley, Weak convergence of probabilities on nonseparable metric spaces and empirical measures on Euclidean spaces, Illin. J. Math. 10 (1966) 109–126.
39. A. van der Vaart and J. A. Wellner, Weak Convergence and Empirical Processes – With Applications to Statistics, Springer Series in Statistics, Springer, New York, 1996.

## Almost sure representations in survival analysis under censoring and truncation. Applications to goodness-of-fit tests

R. Cao[a*], W. González Manteiga[b†] and C. Iglesias Pérez[c]

[a]Departamento de Matemáticas,
Universidade da Coruña, 15071 A Coruña, Spain

[b]Departamento de Estadística e Investigación Operativa,
Universidade de Santiago de Compostela,
15771 Santiago de Compostela, Spain

[c]Departamento de Estadística e Investigación Operativa,
Universidade de Vigo, Pontevedra, Spain

An almost sure representation is presented for a generalized product-limit estimator of the conditional distribution function when the data are subject to random left truncation and right censorship (LTRC). This result extends strong representations studied on conditional survival analysis for censored data as well as for truncated data. As consequences of this representation, goodness-of-fit tests in the LTRC setup are obtained.

## 1. INTRODUCTION.

Many biomedical studies are interested in predicting the survival time of a patient for a given vector of covariables of this individual (age, sex, cholesterol, etc.). Frequently, in works where the survival time is the variable of interest, two different problems appear: the first one, when a subject is not included in the study because its lifetime origin precedes the starting time of the study dying before this moment, (for instance a short period of illness), these subjects are referred to as left truncated (LT); on the other hand, when a patient is into the study but its lifetime may not be completely observed due to different causes (death for a reason unrelated to the study or change of address), these subjects are called right censored (RC). More specifically, let $(Y, T, C)$ be a random vector, where $Y$ is the lifetime, $T$ is the random left truncation time and $C$ denotes the random right censoring time. In addition $Y$ is assumed to be independent of $(T, C)$. In a random LTRC model one observes $(Z, T, \delta)$ if $Z \geq T$, where $Z = \min(Y, C)$ and $\delta = 1_{\{Y \leq C\}}$. When $Z < T$ nothing is observed. Take $\alpha = P(T \leq Z)$, then necessarily, we assume $\alpha > 0$. Let $(Z_i, T_i, \delta_i)$, for $i = 1, 2, ..., n$, be an i.i.d. random sample from $(Z, T, \delta)$ which one observes ( then $T_i \leq Z_i, \forall i$ ). If $F$ denotes the distribution function of $Y$, the product-limit

---

*Research partially supported by the MCyT Grant BFM2002-00265, European FEDER support included
†Research partially supported by the MCyT Grant BFM2002-03213, European FEDER support included

estimator (PLE), $\hat{F}_n$, of $F$ is defined in [20] as follows:

$$1 - \hat{F}_n(y) = \prod_{i=1}^{n}\left(1 - \frac{1_{\{Z_i \leq y,\, \delta_i = 1\}}}{nC_n(Z_i)}\right) \tag{1}$$

where $C_n(y) = \frac{1}{n}\sum_{i=1}^{n} 1_{\{T_i \leq y \leq Z_i\}}$ is the empirical estimator of $C(y) = P(T \leq y \leq Z | T \leq Z)$.

Note that $\hat{F}_n$ equals the Kaplan-Meier PLE (see [15]) when there is no left truncation ($T = 0$) and the Lynden-Bell PLE (see [17]) when there is no right censorship. For LTRC data, Gijbels and Wang (see [9]) provide a breakdown of the PLE, defined in (1), as a mean of i.i.d. random variables plus a remainder term of order $O(n^{-1}\ln n)$ a.s. uniformly over compact intervals, assuming some conditions on the support of the distribution functions of $Z$ and $T$. A similar result is obtained in [23] with a remainder term of order $O(n^{-1}(\ln n)^{1+\varepsilon})$ a.s. for any $\varepsilon > 1/2$ and extends the results in [9] for more general conditions about supports.

On conditional survival analysis, a PLE for RC data has been considered, (see [4] and [10]), and an almost sure asymptotic representation has been obtained with a remainder term of order $O\left(\left(\frac{\ln n}{nh}\right)^{3/4}\right)$, where $h$ is the smoothing parameter, for a variety of contexts; for instance, when a fixed design is considered with nonparametric weights of Gasser and Müller type (see [10] and [21]), or even for a random design with Nadaraya and Watson or $k$-nearest neighbours weights (see [7] and [2]). A general expression for these estimators is given by

$$1 - \hat{F}_n^{RC}(y|x) = \prod_{i=1}^{n}\left(1 - \frac{1_{\{Z_i \leq y,\, \delta_i = 1\}} B_{ni}(x)}{1 - \sum_{j=1}^{n} 1_{\{Z_j < Z_i\}} B_{nj}(x)}\right)$$

where $\{B_{ni}(x)\}_{i=1}^{n}$ is the sequence of the above commented nonparametric weights.

A conditional PLE for LT data has been considered in [16] where an almost sure asymptotic representation with a remainder term of order $O\left(\left(\frac{\ln n}{nh}\right)^{3/4}\right)$ is obtained. This estimator has the following expression:

$$1 - \hat{F}_n^{LT}(y|x) = \prod_{i=1}^{n}\left(1 - \frac{1_{\{Y_i \leq y\}} B_{ni}(x)}{\sum_{j=1}^{n} 1_{\{T_j \leq Y_i \leq Y_j\}} B_{nj}(x)}\right)$$

where $\{B_{ni}(x)\}_{i=1}^{n}$ is a sequence of nonparametric weights of $k$-nearest neighbours type.

In [14] a PLE for LTRC data when covariables are present is considered with an almost sure asymptotic representation for it. The latter representation will extend the results above. In order to describe the estimator, we introduce some notation.

Let $X$ be the covariable related with $Y$. Then denote by $(X_i, Z_i, T_i, \delta_i)$, $i = 1, 2, ..., n$ an i.i.d. sample from $(X, Z, T, \delta)$ which one observes ( $T_i \leq Z_i, \forall i$ ). We define the following curves:

(i) $M(x) = P(X \leq x)$, represents the distribution function of $X$.
(ii) $F(y|x) = P(Y \leq y|X=x)$, is the conditional distribution function of $Y$ when $X = x$.
(iii) $G(y|x) = P(C \leq y|X=x)$, is the conditional distribution function of $C$ at $X = x$.
(iv) $L(y|x) = P(T \leq y|X=x)$, is the conditional distribution function of $T$ at $X = x$.
(v) $H(y|x) = P(Z \leq y|X=x)$, is the conditional distribution function of $Z$ at $X = x$.

(vi) $H_1(y|x) = P(Z \leq y, \delta = 1|_{X=x})$, is the conditional subdistribution function of $Z$ in absence of censoring at $X = x$.

(vii) $\alpha(x) = P(T \leq Z|_{X=x})$, represents the conditional probability of absence of truncation at $X = x$.

(viii) the conditional cumulative hazard rate function of $Y$ at $X = x$

$$\Lambda(y|x) = \int_{-\infty}^{y} \frac{dF(t|x)}{1 - F(t^-|x)} = -\ln\left(1 - F(y^-|x)\right) \qquad (2)$$

Finally, for any distribution function $W(t) = P(\eta \leq t)$, we denote the left and right support endpoints by $a_W = \inf\{t/W(t) > 0\}$ and $b_W = \inf\{t/W(t) = 1\}$, respectively. Also, for $W$, we define $W^*(t) = P(\eta \leq t|_{T \leq Z})$. Then we will consider: $M^*(x) = P(X \leq x|_{T \leq Z})$, $L^*(y|x) = P(T \leq y|_{X=x, T \leq Z})$, $H^*(y|x) = P(Z \leq y|_{X=x, T \leq Z})$ and the subdistribution function $H_1^*(y|x) = P(Z \leq y, \delta = 1|_{X=x, T \leq Z})$.

It is supposed that:

(H1) $X, Y, T, C$ are absolutely continuous random variables.

(H2) Let $I$ a rectangle contained in the support of $m$ (density function of $X$). For all $x \in I$ the r.v. $Y, T, C$ are conditionally independent at $X = x$, and the support endpoints of the conditional functions of $T$ and $Z$ given $X = x$ hold $a_{L(\bullet|x)} \leq a_{H(\bullet|x)}$ and $b_{L(\bullet|x)} \leq b_{H(\bullet|x)}$ (Compare with Woodroofe's results in [22] about identifiability of $F$ for truncated data without covariables).

We will define a generalized product-limit estimator of $F(\cdot|x)$, $x \in I$, by defining a conditional cumulative hazard rate function estimator and making use of (2).

If $Y$ is conditionally independent of $(T, C)$ (which is true under H2) we have:

$$\begin{aligned} H_1^*(y|x) &= P(Z \leq y, \delta = 1|_{X=x, T \leq Z}) = \frac{P(Z \leq y, \delta = 1, T \leq Z|_{X=x})}{P(T \leq Z|_{X=x})} = \\ &= \frac{P(Y \leq y, Y \leq C, T \leq Y|_{X=x})}{P(T \leq Z|_{X=x})} = \alpha(x)^{-1} \int_{-\infty}^{y} P(T \leq t \leq C|_{X=x}) dF(t|x) \end{aligned}$$

By the latter relation and H2, we can write $\Lambda(\cdot|x)$ as a function of empirical estimable expressions:

$$\begin{aligned} \Lambda(y|x) &= \int_{-\infty}^{y} \frac{dF(t|x)}{1 - F(t^-|x)} = \int_{-\infty}^{y} \frac{\alpha(x)^{-1} P(T \leq t \leq C|_{X=x}) dF(t|x)}{\alpha(x)^{-1} P(T \leq t \leq C|_{X=x})(1 - F(t^-|x))} \\ &= \int_{-\infty}^{y} \frac{dH_1^*(t|x)}{C(t|x)} \end{aligned} \qquad (3)$$

for $y < b_{H(\bullet|x)}$, where

$$C(t|x) = \alpha(x)^{-1} P(T \leq t \leq C|_{X=x})\left(1 - F(t^-|x)\right) = P(T \leq t \leq Z|_{X=x, T \leq Z})$$

Under H1 the use of smooth nonparametric estimators for the conditional distribution functions seems to be reasonable. Let

$$\hat{H}_{1n}^*(y|x) = \sum_{i=1}^{n} 1_{\{Z_i \leq y, \delta_i = 1\}} B_{ni}(x)$$

and

$$\hat{C}_n(y|x) = \sum_{i=1}^{n} 1_{\{T_i \leq y \leq Z_i\}} B_{ni}(x)$$

be the nonparametric estimators of $H_1^*$ and $C$, respectively, where $\{B_{ni}(x)\}$ denotes a sequence of a suitable type of weights (Gasser-Müller, Nadaraya-Watson, ...). Substituting $H_1^*$ by $\hat{H}_{1n}^*$ and $C$ by $\hat{C}_n$ in (3), we define a natural estimator for $\Lambda(\bullet|x)$:

$$\hat{\Lambda}_n(y|x) = \int_{-\infty}^{y} \frac{d\hat{H}_{1n}^*(t|x)}{\hat{C}_n(t|x)} = \sum_{i=1}^{n} \frac{1_{\{Z_i \leq y, \delta_i = 1\}} B_{ni}(x)}{\sum_{j=1}^{n} 1_{\{T_j \leq Z_i \leq Z_j\}} B_{nj}(x)} \quad (4)$$

Finally, using (2) and applying a Taylor's expansion of first order to $\exp(-x)$ around $x = 0$, we define a generalized product-limit estimator, GPLE, of the conditional survival function of $Y$ for LTRC data, given by

$$1 - \hat{F}_n(y|x) = \prod_{i=1}^{n} \left[ 1 - \frac{1_{\{Z_i \leq y\}} \delta_i B_{ni}(x)}{\sum_{j=1}^{n} 1_{\{T_j \leq Z_i \leq Z_j\}} B_{nj}(x)} \right] \quad (5)$$

Note that the GPLE reduces to the estimator defined in [16] when there is no right censoring ($\delta = 1, Z = Y$), and to the estimator given in [4] when there is no left truncation ($T = 0$), (provided that the nonparametric weights are the same in each considered estimator). On the other hand, the GPLE reduces to the PLE estimator defined by (1) when one is in absence of covariables situation ( $B_{ni}(x) = 1/n$, $\forall i$ ).

## 2. AN ALMOST SURE REPRESENTATION.

Because of the nature of truncated data we will consider a random design. For the sake of notational simplicity we use only a real covariable. In this context, it is common to assume that the sequence of weights are of Nadaraya-Watson type:

$$B_{ni}(x) = \frac{K(\frac{x-x_i}{h})}{\sum_{j=1}^{n} K(\frac{x-x_j}{h})} \quad i = 1, 2, ..., n$$

where $K$ denotes a kernel function, and $h = h_n \geq 0$ is the bandwidth parameter.

Finally, and without loss of generality, we work with nonnegative variables, as usual in survival analysis.

Let $m$ denote the density of $X$, and $M^*$ the conditional distribution function of $X$ when $T \leq Z$ with density $m^*$, then

$$m^*(x) = \frac{\alpha(x)}{\alpha} m(x)$$

where $i(x) = \alpha(x)/\alpha$ is an index of truncation in $x$. We need to consider $x$ values with $i(x) \neq 0$.

In addition to H1 and H2, the following assumptions:

(H2') a) Let $I = [x_1, x_2]$ be an interval contained in the support of $m^*$, such that

$$0 < \gamma = \inf\,[m^*(x) : x \in I_\delta] < \sup\,[m^*(x) : x \in I_\delta] = \Gamma < \infty$$

for some $I_\delta = [x_1 - \delta, x_2 + \delta]$ with $\delta > 0$ and $0 < \delta\Gamma < 1$.
And for all $x \in I$ the r.v. $Y, T, C$ are conditionally independent at $X = x$,
b)Moreover:
i) $a_{L(\bullet|x)} \leq a_{H(\bullet|x)}$, $b_{L(\bullet|x)} \leq b_{H(\bullet|x)}$, for all $x \in I_\delta$
ii) Exist $a, b \in R$ with $a < b$, satisfying $\inf [\alpha^{-1}(x)(1 - H(b|x)) L(a|x) : x \in I_\delta] \geq \theta > 0$
(note that, if $a_{L(\bullet|x)} < y < b_{H(\bullet|x)}$ then $C(y|x) = \alpha^{-1}(x)(1 - H(y|x)) L(y|x) > 0$,
condition ii) say that $C(y|x) \geq \theta > 0$ in $[a, b] \times I_\delta$ ).
and some regularity conditions on $L, H, H_1$ and $K$, are used in [14] to prove:

**Theorem 2.1** *Under the assumptions in [14] and $h \to 0$, $nh \to \infty$, $\frac{\ln n}{nh} \to 0$ and $\frac{nh^5}{\ln n} = O(1)$,*
*a) The following representation holds*

$$[\hat{\Lambda}_n(y|x) - \Lambda(y|x)] - [\hat{\Lambda}_n(a|x) - \Lambda(a|x)] = \sum_{i=1}^n B_{ni}(x)\xi_a(Z_i, T_i, \delta_i, y, x) + R_{na}(y|x)$$

*for $x \in I$, $y \in [a, b]$, where*

$$\sup{}_{a \leq y \leq b, x \in I} | R_{na}(y|x) | = O\left(\left(\frac{\ln n}{nh}\right)^{3/4}\right) \quad a.s.$$

$$\xi(Z, T, \delta, y, x) = \frac{1_{\{Z \leq y, \delta=1\}}}{C(Z|x)} - \int_0^y \frac{1_{\{T \leq u \leq Z\}}}{C^2(u|x)} dH_1^*(u|x)$$

$$\xi_a(Z, T, \delta, y, x) = \xi(Z, T, \delta, y, x) - \xi(Z, T, \delta, a, x)$$

*b) If $a < a_{H(\bullet|x)}$ for all $x \in I$, then (a) is reduced to*

$$\hat{\Lambda}_n(y|x) - \Lambda(y|x) = \sum_{i=1}^n B_{ni}(x)\xi(Z_i, T_i, \delta_i, y, x) + R_n(y|x)$$

*for $x \in I$, $y \in [a, b]$, where*

$$\sup_{[a,b] \times I} | R_n(y|x) | = O\left(\left(\frac{\ln n}{nh}\right)^{3/4}\right) \quad a.s.$$

*c) If $a < a_{H(\bullet|x)}$ for all $x \in I$, then*

$$\hat{F}_n(y|x) - F(y|x) = (1 - F(y|x)) \sum_{i=1}^n B_{ni}(x)\xi(Z_i, T_i, \delta_i, y, x) + R'_n(y|x)$$

*for $x \in I$, $y \in [a, b]$, where*

$$\sup_{[a,b] \times I} | R'_n(y|x) | = O\left(\left(\frac{\ln n}{nh}\right)^{3/4}\right) \quad a.s.$$

## 3. GOODNESS-OF-FIT TESTS

One of the main aims of the survival analysis, in a conditional setup, is to analyze how an explanatory variable, $X$, influences the survival time, $Y$. This dependency may be modelled in a different number of ways but, typically, the key idea is to assume some kind of functional relationship for the conditional distribution function or some other conditional curve. In the following we make a short overview of some of the most popular conditional models in this setup. Models PH, AR and PO have been studied in detail by Grigoletto and Akritas [12].

### 3.1. General polynomial regression model (PR)

We assume that the conditional distribution function, $F(\bullet|x)$, satisfies the following equation:

$$T(F(\bullet|x)) = \beta_0 + \beta_1 x + \cdots + \beta_p x^p, \tag{6}$$

where $\beta_0, \beta_1, \ldots, \beta_p$ are unknown parameters, the functional $T$ is given by $T(N) = \int_0^1 N^{-1}(s) J(s) ds$, for any distribution function, $N$,

$$N^{-1}(s) = \inf\{u : N(u) \geq s\}$$

is the quantile function and $J$ is some nonnegative real function satisfying $\int_0^1 J(s) ds = 1$. In other words, $T$ is an $L$-functional in the terminology of Serfling [19], page 265. If $J$ is the density function of a $U[0,1]$ distribution, it is easy to check that $T(F(\bullet|x)) = E(Y|_{X=x})$, i.e. we are imposing a polynomial structure to the regression function of $Y$ given $X$. In survival analysis, where the survival time is typically asymmetric, it seems more reasonable to set $J(s) = \frac{1}{b-a} I_{\{a \leq s \leq b\}}$ for some $[a, b] \subset [0, 1]$, which gives rise to trimmed conditional means or, in an extreme case, the conditional median. For a detailed study of model (6) see [3].

### 3.2. Proportional hazards model (PH)

The idea behind this model is two split the conditional hazard rate as a product of two factors: one independent of the covariates and another covariate dependent. It assumes that, $\lambda(t|x)$, the conditional hazard rate of $Y$ given $X = x$ is of the form

$$\lambda(t|x) = \lambda_0(t) \exp(\beta_0 + \beta_1 x + \cdots + \beta_p x^p), \tag{7}$$

where $\lambda_0$ is the so-called baseline hazard rate and $\beta_0, \beta_1, \ldots, \beta_p$ are unknown real constants. This popular model, also known as Cox regression model, was proposed by Cox [6] and can also be expressed in terms of the cumulative hazard function. If we let the coefficients $\beta_j$, in equation (7), depend on the lifetime, we get the general Cox model with time depending coefficients:

$$\Lambda(t|x) = \Lambda_0(t) \exp(\beta_0(t) + \beta_1(t) x + \cdots + \beta_p(t) x^p). \tag{8}$$

This model has been studied in [18].

### 3.3. Additive risks model (AR)

An alternative approach to model the conditional hazard rate is considered now:
$$\lambda(t|x) = \lambda_0(t) + \beta_0 + \beta_1 x + \cdots + \beta_p x^p. \tag{9}$$

It is clear that some conditions on the polynomial $\beta_0 + \beta_1 x + \cdots + \beta_p x^p$ are needed now in order for $\lambda(t|x)$ to be a hazard rate. This model is a special case of the multiplicative intensity model suggested by Aalen [1].

### 3.4. Proportional odds model (PO)

The conditional cumulative hazard function of $Y$ given $X = x$, $\Lambda(t|x)$ is assumed to satisfy the following equation
$$\text{logit}(1 - \exp(-\Lambda(t|x))) = \alpha(t) + \beta_0 + \beta_1 x + \cdots + \beta_p x^p, \tag{10}$$

where $\alpha(t)$ is an increasing function and $\text{logit}(u) = \ln\left(\frac{u}{1-u}\right)$. Observe that this model is equivalent to
$$\ln\left(\frac{F(t|x)}{1 - F(t|x)}\right) = \alpha(t) + \beta_0 + \beta_1 x + \cdots + \beta_p x^p,$$

or, in other terms, to the following parametric model for the odds ratio:
$$\frac{P(Y \leq t|X=x)}{P(Y > t|X=x)} = \exp(\alpha(t) + \beta_0 + \beta_1 x + \cdots + \beta_p x^p).$$

### 3.5. Parameter estimation

In [3] some properties are studied for a general least squares estimator, $\hat{\beta}$, of the parameter vector $\beta = (\beta_0, \beta_1, \ldots, \beta_p)^t$ in the PR model for complete, censored or truncated data, but not for both situations simultaneously. This estimator is defined as
$$\hat{\beta} = \arg\min_\beta \hat{\psi}_n(\beta),$$

where
$$\hat{\psi}_n(\beta) = \frac{1}{n} \sum_{r=1}^n (\hat{m}_r - (\beta_0 + \beta_1 X_r + \cdots + \beta_p X_r^p))^2 \tag{11}$$

and $\hat{m}_r = T\left(\hat{F}_n(\bullet|X_r)\right)$ is an estimator of $m_r = T(F(\bullet|X_r))$ for $r = 1, 2, \ldots, n$. To avoid definiteness problems with $T\left(\hat{F}_n(\bullet|X_r)\right) = \int_0^1 \hat{F}_n^{-1}(s|X_r) J(s) ds$ we modify the estimator $\hat{F}_n(\bullet|X_r)$, if necessary, forcing it to attain the value 1 in the largest point with positive probability mass. This results in
$$\hat{\beta} = (\chi^t \chi)^{-1} \chi^t \hat{m}, \tag{12}$$

where $\hat{m} = (\hat{m}_1, \hat{m}_2, \ldots, \hat{m}_n)^t$ and $\chi = \left(X_i^{k-1}\right)_{1 \leq i \leq n; 1 \leq k \leq p+1}$ is the design matrix.

Grigoletto and Akritas [12] define $\hat{\beta}$, some version of the least squares estimator of the true parameter, $\beta$, in models PH, AR and PO:
$$\hat{\beta} = \arg\min_\beta \bar{\psi}_n(\beta) = \arg\min_\beta \frac{1}{n} \sum_{r=1}^n \left(\hat{\Omega}_r - (\beta_0 + \beta_1 X_r + \cdots + \beta_p X_r^p)\right)^2, \tag{13}$$

where $\hat{\Omega}_r$ is some estimation of $\Omega_r$, a suitable transformation of $\Lambda(\bullet|_{X_r})$ that will change from one model to other. An explicit expression for the estimator in (13) can be easily derived:

$$\hat{\beta} = (\chi^t \chi)^{-1} \chi^t \hat{\Omega},$$

with $\hat{\Omega} = \left(\hat{\Omega}_1, \hat{\Omega}_2, \ldots, \hat{\Omega}_n\right)^t$. These authors give some asymptotic properties of $\hat{\beta}$ under either censoring or truncation but not when the two mechanisms are present.

For model PH, straight forward integration of the terms in equation (7) leads to

$$\Lambda(t|_x) = \Lambda_0(t) \exp(\beta_0 + \beta_1 x + \cdots + \beta_p x^p)$$

and, taking logarithms,

$$\ln \Lambda(t|_x) = \ln \Lambda_0(t) + \beta_0 + \beta_1 x + \cdots + \beta_p x^p.$$

Finally, integrating both terms with respect to a weight function, $W$, satisfying $W(s) \geq 0$ for $s \in [0, \infty)$ and $\int_0^\infty dW(s) = 1$, we have

$$\int_0^\infty \ln \Lambda(s|_x) dW(s) = \int_0^\infty \ln \Lambda_0(s) dW(s) + \beta_0 + \beta_1 x + \cdots + \beta_p x^p$$
$$= \beta_0' + \beta_1 x + \cdots + \beta_p x^p,$$

with $\beta_0' = \beta_0 + \int_0^\infty \ln \Lambda_0(s) dW(s)$. Completely parallel calculations for model (8) give:

$$\int_0^\infty \ln \Lambda(s|_x) dW(s) = \tilde{\beta}_0 + \tilde{\beta}_1 x + \cdots + \tilde{\beta}_p x^p, \text{ with}$$

$$\tilde{\beta}_0 = \int_0^\infty \ln \Lambda_0(s) dW(s) + \int_0^\infty \beta_0(s) dW(s),$$

$$\tilde{\beta}_j = \int_0^\infty \beta_j(s) dW(s), \text{ for } j = 1, 2, \ldots, p,$$

which essentially tells that this model can also be transformed to a polynomial regression model. Although the functions $\beta_j(t)$ are not identifiable then, this idea can be used for testing (8).

Now, $\Omega_x = \int_0^\infty \ln \Lambda(s|_x) dW(s)$, $\Omega_r = \Omega_{X_r}$ and $\hat{\Omega}_r = \int_0^\infty \ln \hat{\Lambda}_n(s|_{X_r}) dW(s)$, where $\hat{\Lambda}_n(t|_x)$ is the conditional cumulative hazard function estimator connected to (5), i.e.

$$\hat{\Lambda}_n(t|_x) = \int_{-\infty}^t \frac{d\hat{H}_{1n}^*(s|_x)}{\hat{C}_n(s|_x)}, \qquad (14)$$

where

$$\hat{H}_{1n}^*(s|_x) = \sum_{i=1}^n 1_{\{Z_i \leq s, \delta_i = 1\}} B_{ni}(x)$$

is an estimator of the conditional subdistribution function

$$H_1^*(s|_x) = P(Z \leq s, \delta = 1|_{T \leq Z, X = x})$$

and

$$\hat{C}_n(s|x) = \sum_{i=1}^{n} 1_{\{T_i \leq s \leq Z_i\}} B_{ni}(x)$$

is an estimator of the populational function $C(s|x) = P(T \leq s \leq Z|_{T \leq Z, X=x})$.

For model AR things are a little bit easier since a direct integration of (9) leads to

$$\Lambda(t|x) = \Lambda_0(t) + (\beta_0 + \beta_1 x + \cdots + \beta_p x^p) t.$$

Another integration with respect to

$$\tilde{W} = \frac{W}{\int_0^\infty s \, dW(s)},$$

where $W$ is a weight function with the same conditions as above, transform this model in a kind of polynomial model:

$$\int_0^\infty \Lambda(s|x) \, d\tilde{W}(s) = \beta_0'' + \beta_1 x + \cdots + \beta_p x^p,$$

with

$$\beta_0'' = \beta_0 + \frac{\int_0^\infty \Lambda_0(s) \, dW(s)}{\int_0^\infty s \, dW(s)}.$$

As a consequence

$$\Omega_x = \int_0^\infty \Lambda(s|x) \, d\tilde{W}(s), \, \Omega_r = \Omega_{X_r} \text{ and } \hat{\Omega}_r = \int_0^\infty \hat{\Lambda}_n(s|X_r) \, d\tilde{W}(s),$$

where $\hat{\Lambda}_n(t|x)$ is the estimator defined in (14).

Under model PO, integrating both terms in (10) with respect to a weight function, $W$, this equation reduces to

$$\int_0^\infty \text{logit}(1 - \exp(-\Lambda(s|x))) \, dW(s) = \beta_0''' + \beta_1 x + \cdots + \beta_p x^p,$$

with $\beta_0''' = \beta_0 + \int_0^\infty \alpha(s) \, dW(s)$. Hence, $\Omega_r = \Omega_{X_r}$, with

$$\Omega_x = \int_0^\infty \text{logit}(1 - \exp(-\Lambda(s|x))) \, dW(s)$$

and

$$\hat{\Omega}_r = \int_0^\infty \text{logit}\left(1 - \exp\left(-\hat{\Lambda}_n(s|X_r)\right)\right) dW(s).$$

### 3.6. The test statistic

Let us first focus in model PR. The problem under study is to test

$H_0$ : $\exists \beta \in \mathbf{R}^{p+1}$ such that equation (6) holds

versus the alternative (15)

$H_1$ : equation (6) does not hold for any $\beta \in \mathbf{R}^{p+1}$.

An intuitive way of measuring the discrepancy between the hypothesized model and the data is to consider the stochastic process $\hat{\psi}_n(\beta)$ at $\beta = \hat{\beta}$:

$$\hat{\psi}_n\left(\hat{\beta}\right) = \frac{1}{n}\sum_{r=1}^{n}\left(\hat{m}_r - \left(\hat{\beta}_0 + \hat{\beta}_1 X_r + \cdots + \hat{\beta}_p X_r^p\right)\right)^2.$$

This is a kind of $L_2$-distance between the transformed data and the response predicted by the fitted model. Hence, it is reasonable to reject $H_0$ when $\hat{\psi}_n\left(\hat{\beta}\right)$ is large.

In models PH, AR and PO, the hypothesis testing is completely similar to (15) but replacing equation (6) by (7), (9) or (10), respectively. The definition of the statistic is also parallel for models PH, AR and PO:

$$\bar{\psi}_n\left(\hat{\beta}\right) = \frac{1}{n}\sum_{r=1}^{n}\left(\hat{\Omega}_r - \left(\hat{\beta}_0 + \hat{\beta}_1 X_r + \cdots + \hat{\beta}_p X_r^p\right)\right)^2.$$

These test statistics have to be multiplied by a normalizing sequence in order to have a limit distribution. This leads to the test statistics

$T_n^{(1)} = n\sqrt{h}\hat{\psi}_n\left(\hat{\beta}\right)$ for model PR,

$T_n^{(2)} = n\sqrt{h}\bar{\psi}_n\left(\hat{\beta}\right)$ for models PH, AR and PO.

In [5] the following result is presented assuming suitable conditions.

**Theorem 3.1** *Under* $H_0$,

$$\tilde{T}_n - b_{0h} \xrightarrow{d} N(0, V),$$

where $\tilde{T}_n = T_n^{(1)}$ for model PR and $\tilde{T}_n = T_n^{(2)}$ for models PH, AR and PO, with $b_{0h} = h^{-\frac{1}{2}}K^{(2)}(0)\int \sigma^2(x)\,dx$ and $V = 2K^{(4)}(0)\int \sigma^4(x)\,dx$. Here $K^{(2)}(u) = K*K(u)$ and $K^{(4)}(u) = K*K*K*K(u)$, where $*$ denotes convolution, $\sigma^2(x) = Var\left(\eta(Z,T,\delta,x)|_{X=x}\right)$,

$$\eta(Z,T,\delta,x) = \int (1 - F(y|_x))\,\xi(Z,T,\delta,y,x)\,J(F(y|_x))\,dy \text{ for model PR}$$

and

$$\eta(Z,T,\delta,x) = \begin{cases} \int \xi(Z,T,\delta,y,x)\frac{dW(y)}{\Lambda(y|_x)} & \text{for model PH} \\ \int \xi(Z,T,\delta,y,x)\,dW(y) & \text{for model AR} \\ \int \xi(Z,T,\delta,y,x)\frac{dW(y)}{F(y|_x)} & \text{for model PO} \end{cases}$$

**Remark 3.1** *If we choose $J$ to be the uniform density in $(0,1)$, the functional $T$ used in model PR becomes*

$$T(N) = \int_0^1 N^{-1}(s)\, ds = \int_0^\infty t\, dN(t).$$

*As a consequence, $T(F(\bullet|_x)) = E(Y|_{X=x})$. In the case of no censoring and no truncation the estimator (5) reduces to*

$$\hat{F}_n(y|_x) = \frac{\sum_{j=1}^n 1_{\{Z_j \leq y\}} B_{nj}(x)}{\sum_{j=i}^n B_{nj}(x)}$$

*and $T\left(\hat{F}_n(\bullet|_x)\right)$ is the classical Nadaraya-Watson kernel estimator of the regression function. Under these circumstances, the test statistic $T_n^{(1)}$ coincides with the test statistic $\Delta ASE$ proposed by González-Manteiga and Cao [11] except for the fixed design used in that paper. The limit distribution of $\Delta ASE$ in that paper is also a particular case of our Theorem 1 above, without censoring and truncation. In the same sense $T_n^{(1)}$ can be also viewed as a generalization of the test proposed in [13] to the case with censoring and truncation. However, these authors did not consider least squares parameter estimators with the smoothed responses.*

**Remark 3.2** *As soon as one transforms the goodness-of-fit problem of interest into the model check for linearity of the values $(X_r, \hat{m}_r)$ it is clear that plenty of the alternative approaches for constructing goodness-of-fit tests in regression can be directly used in this setup. For instance, the ideas in [8] can be translated to this setup by of constructing a test based on the difference between two estimators of the integrated conditional variance. One based on the linearity assumption of the $(X_r, \hat{m}_r)$ and the other that is purely non-parametric. Similar ideas could be exploited for models PH, AR and PO using the data $\left(X_r, \hat{\Omega}_r\right)$.*

**Remark 3.3** *Let us consider some local alternative to model PR of the following form*

$$T(F(\bullet|_x)) = \beta_0 + \beta_1 x + \cdots + \beta_p x^p + n^{-\frac{1}{2}} h^{-\frac{1}{4}} g(x),$$

*where $g(x)$ is a squared integrable function orthogonal to $\beta_0 + \beta_1 x + \cdots + \beta_p x^p$. A careful inspection of the proof of Theorem 1 gives*

$$T_n^{(1)} - b_{0h} \xrightarrow{d} N\left(\int g(x)^2 f(x)\, dx, V\right),$$

*under this local alternative model. This implies that the test is able to detect local alternatives that approach the hypothesized model at any rate slower than $n^{-\frac{1}{2}} h^{-\frac{1}{4}}$.*

*A similar property can be derived for the test $T_n^{(2)}$ for suitable local alternatives to models PH, AR and PO. It is clear that such local alternatives have to be formulated in terms of*

$$\Omega_x = \beta_0 + \beta_1 x + \cdots + \beta_p x^p + n^{-\frac{1}{2}} h^{-\frac{1}{4}} g(x),$$

*where $\Omega_x$ is the appropriate transformation introduced for each of these models early in this section.*

**Remark 3.4** *The quadratic form structure of the dominant terms in the test statistics $\tilde{T}_n$ gives a slow convergence rate to the normal limit. For this reason some bootstrap resampling plan has been proposed in [5] to approximate the null distrbiution of the test statistic. These authors propose a residual-based bootstrap resampling rather a model-based aproach, since some of the hypothesized models under $H_0$ do not uniquelly determine the underlying joint distribution.*

## REFERENCES

1. O.O. Aalen, A model for nonparametric regression analysis of counting processes, Lect. Notes Statist. No. 2 (1980) 1.
2. M.G. Akritas, Nearest neighbor estimation of a bivariate distribution under random censoring, Ann. Statist. No. 22 (1994) 1299.
3. M.G. Akritas, On the use of nonparametric regression techniques for fitting parametric regression models, Biometrics No. 52 (1996) 1342.
4. R. Beran, Nonparametric regression with randomly censored data, Technical report, University of California, Berkeley (1981).
5. R. Cao and W. González-Manteiga, Goodness-of-fit tests for conditional models under censoring and truncation, Unpublished manuscript (2003).
6. D.R. Cox, Regression models and life tables, Roy. Statist. Soc. B No. 34 (1972) 187.
7. D. Dabrowska, Uniform consistency of the kernel conditional Kaplan-Meier estimate, Ann. Statist. No. 17 (1989) 1157.
8. H. Dette, A consistent test for the functional form of a regression based on a difference of variance estimators, Ann. Statist. No. 27 (1999) 1012.
9. I. Gijbels and J.L. Wang, Strong representation of the survival function estimator for truncated and censored data with applications, J. Multivariate Anal. No. 47 (1993) 210.
10. W. González-Manteiga and C. Cadarso-Suárez, Asymptotic properties of a generalized Kaplan-Meier estimator with some applications, Nonparametric Statist. No. 4 (1994) 65.
11. W. González-Manteiga and R. Cao, Testing the hypothesis of a general linear model using nonparametric regression estimation, Test No. 2 (1993) 161.
12. M. Grigoletto and M.G. Akritas, Analysis of covariance with incomplete data via semiparametric model transformations, Biometrics No. 55 (1999) 1177.
13. W. Härdle and E. Mammen, Comparing nonparametric versus parametric regression fits, Ann. Statist. No. 21 (1993) 1926.
14. C. Iglesias-Pérez and W. González-Manteiga, Strong representation of a generalized product-limit estimator for truncated and censored data with some applications, J. Nonpar. Statist. No. 10 (1999) 213.
15. E.L. Kaplan and P. Meier, Nonparametric estimation from incomplete observations, J. Amer. Statist. Assoc. No. 53 (1958) 457.
16. M.P. LaValley and M.G. Akritas, Extensions of the Lynden-Bell-Woodroofe method for truncated data, Unpublished manuscript, (1994)
17. D. Lynden-Bell, A method of allowing for non observational selection in small samples applied to 3CR quasars, Mon. Not. R. Astr. Soc. No. 155 (1971) 95.

18. L. Marzec and P. Marzec, On fitting Cox's regression model with time-dependent coefficients, Biometrika No. 84 (1997) 901.
19. R.J. Serfling, Approximation theorems of mathematical statistics, John Wiley & Sons, New York, 1980.
20. W.Y. Tsai, N.P. Jewell and M.C. Wang, A note on the product-limit estimator under right censoring and left truncation, Biometrika No. 74 (1987) 883.
21. I. Van Keilegom and N. Veraverbeke, Estimation and bootstrap with censored data in fixed design nonparametric regression, Ann. Inst. Statist. Math. No. 49 (1997) 467.
22. M. Woodroofe, Estimating a distribution function with truncated data, Ann. Statist. No. 13 (1985) 163.
23. Y. Zhou, A note on the TJW product-limit estimator for truncated and censored data, Statist. and Probab. Letters. No. 26 (1996) 381.

# 5. High-dimensional Data and Visualization

*Recent Advances and Trends in Nonparametric Statistics*
Michael G. Akritas and Dimitris N. Politis (Editors)
© 2003 Elsevier Science B.V. All rights reserved.

# Data Depth: Center-Outward Ordering of Multivariate Data and Nonparametric Multivariate Statistics

Regina Y. Liu[a]*

[a]Department of Statistics, Rutgers University, Hill Center
Piscataway, NJ 08854-8019, USA

A *data depth* is a measure of depth or centrality of a given point with respect to a multivariate distribution. It gives rise to a new set of parameters which can easily quantify the many complex multivariate features of the underlying distribution. These parameters can be expressed by simple one-dimensional graphs which facilitate greatly the visualization and interpretation of multivariate features. Furthermore, a data depth also provides a natural center-outward ordering of data points in a given sample. This ordering leads to a systematic nonparametric multivariate inference scheme and many practical applications. This paper discusses some examples, including depth-based multivariate descriptive statistics, multivariate rank tests, $DD$-plots for testing in one- and two-sample problems, and multivariate process control.

## 1. INTRODUCTION

The recent advances in computing technology have made the collection of massive multivariate data a routine practice in many fields. The demand for effective multivariate analyses has never been greater, especially for the nonparametric ones. A nonparametric multivariate methodology based on the concept of data depth will be discussed here. The classical multivariate analysis is well developed, but its applicability is much restricted by its intrinsic elliptical nature, see for example [1].

Roughly speaking, a data depth is a measure of depth or centrality of a given point relative to the underlying distribution. It provides a description of the probability distribution starting from the center of the probability mass. This description is elaborated further in Section 2. In the context of sample data, the depth-based methodology may be viewed as a multivariate generalization of standard univariate rank methods, in the sense that it is also based on the idea of *ranking* (or *ordering*) of data points. For any point in a given multivariate sample, a data depth measures its depth with respect to the sample and eventually to the distribution from which the sample is drawn. Ordering the sample points according to their decreasing depth values immediately provides a center-outward ordering or ranking of these sample points. In the univariate setting, the depth-ranking reduces to the standard univariate rank method, although the standard

---

*The research is in part supported by *The Federal Aviation Administration*, *The National Science Foundation*, and *The National Security Agency*. The author thanks Andrew Cheng, Jesse Parelius, Kesar Singh, and Julie Teng for their collaborations on this subject.

ranking in the univariate case is a *linear ranking* from the minimum to the maximum, while the depth-induced one is a *center-outward ranking*.

Based on depth ordering, many qualitative and graphical methods have been introduced for the analysis of multivariate data. They range from distributional descriptive statistics to various statistical inference methods. Different notions of depth defined in [21] [25] [5] [22] [10] [4] [24] [27] have produced their respective "deepest" points, which have all been considered as multivariate medians (or location parameters) in the literature. To characterize the features of the underlying distribution, the paper [13] (LPS) has used depth ordering to define descriptive statistics such as scale, skewness and kurtosis. These descriptive statistics are presented in simple graphs in the plane, regardless of the dimension of the distribution. These simple graphs make it easy to visualize and interpret multivariate distributional features. For example, applying the scale curve idea proposed in LPS, we can compare the consistency of airline performance among the top ten airlines, as seen in Section 3.1.3. In LPS, $DD$-plots (depth vs. depth plots) are introduced as graphical inference tools for comparing multivariate samples. Multivariate rank tests based on different types of depth rankings have been studied in [7] [11] [15] [8] [6] [18]. Combining the depth ranking and the bootstrap, [26] proposed a nonparametric method for constructing confidence regions, and [16] provided several methods for determining $p$-values for general hypothesis testing. In [14], the idea of depth ordering was extended to the analysis of directional data as well. The paper [12] applied the depth ordering to constructing nonparametric control charts for monitoring multivariate processes. These control charts combined with properly chosen and justified false-alarm rates yield an attainable and meaningful threshold system in the context of monitoring multivariate process, as discussed in [3]. Finally, it should be mentioned that both [23] [19] have introduced very promising regression methods based on depth. These regression methods possess desirable robustness and invariance properties.

The goal of this paper is to provide a brief survey of some of the recent developments mentioned above. Data depth should *not* be viewed as just another way of deriving a multivariate median, which is sometimes misconceived to be. In fact, the subject of "data depth" is quite rich, and there are excellent prospects for rapid progress in depth-based statistical analysis.

The paper is organized as follows. Section 2 provides a general description of data depth, the depth induced ranking scheme, and the depth contours assciate with the underlying distribution. Section 3 lists existing depth based inference methods and highlights some of the less explored ones. Section 4 provides an account of nonparametric multivariate control charts using depth ranking. An airline performance dataset (collected by the FAA from 1993 to 1998) is used as a running example to show the usefulness of the proposed control charts and scale curves.

## 2. DATA DEPTH: DEFINITIONS & PROPERTIES

Several different notions of data depth have been introduced, e.g. [21] [25] [5] [22] [10] [4] [24] [27], and applied in various statistical contexts. A detailed review of various notions of depth can be found in LPS. For expository purposes, we recall here two depths to facilitate the descriptions of the depth-based rankings/orderings of sample points, descriptive statistics, inference methods, and some applications.

Let $F$ be an absolutely continuous distribution function in $\mathbb{R}^d$, $d \geq 1$, and let $\mathbf{X} \equiv \{X_1, \ldots, X_n\}$ be a random sample from $F$. Each sample point $X_i$ is viewed as a $1 \times d$ row vector. A data depth is a way of measuring how deep or how central a given point $x \in \mathbb{R}^d$ is w.r.t. $F$ or w.r.t. a given data cloud $\mathbf{X}$.

(I) The *Halfspace Depth (HD)* ([9] [25]) at $x$ w.r.t. $F$ is defined to be

$$HD(F; x) = \inf_H \{P(H) : H \text{ is a closed halfspace in } \mathbb{R}^d \text{ and } x \in H\} \tag{1}$$

The sample version of $HD(F; x)$ is $HD(F_n; x)$. Here $F_n$ denotes the empirical distribution of the sample $\{X_1, \ldots, X_n\}$.

(II) The *Simplicial Depth (SD)* ([10]) at $x$ w.r.t. $F$ is defined to be

$$SD(F; x) = P_F\{x \in S[X_1, \ldots, X_{d+1}]\}. \tag{2}$$

Here $S[X_1, \ldots, X_{d+1}]$ is a closed simplex formed by $(d+1)$ random observations from $F$. The sample version of $SD(F; x)$ is obtained by replacing $F$ in $SD(F; x)$ by $F_n$, or alternatively, by computing the fraction of the sample random simplices containing the point $x$, i.e.

$$SD(F_n; x) = \binom{n}{d+1}^{-1} \sum_{\star} I_{(x \in S[X_{i_1}, \ldots, X_{i_{d+1}}])} \tag{3}$$

where $I_{(\cdot)}$ is the indicator function.

For convenience, $D(\cdot; \cdot)$ or $D_F(\cdot)$ is used to indicate any given depth unless specified otherwise. The value of $D(F; x)$ may vary for different notions of data depth, but for each notion of depth, a larger value of $D(F; x)$ always implies a deeper $x$ w.r.t. $F$. For convenience, we also omit the underlying distribution $F$ and simply use $D(\cdot)$ and $D_n(\cdot)$ to denote respectively $D(F; \cdot)$ and $D(F_n; \cdot)$.

Intuitively, it is easy to see how $SD(F_n; x)$ defined in (3) can be a measure of depth. If $x$ is deep inside the data cloud $\mathbf{X}$, then it will be covered by many simplices generated from the data. On the other hand, if $x$ is relatively outlying in the data cloud, then it will be covered by relatively few such simplices. In general, a function $D(\cdot)$ can be viewed a measure of depth if it satisfies the following desirable properties. These properties were first introduced in [10], and later studied further in [28]. Note that both the halfspace depth and the simplicial depth are shown to satisfy these properties.

1. *Maximality at center*: If a distribution has a uniquely defined "center" (e.g. the center of a symmetric distribution), then the function $D(\cdot)$ attains the maximum value at this center.

2. *Monotonicity Relative to the Deepest point*: $D(x)$ decreases monotonically as the point $x \in \mathbb{R}^d$ moves away from the "deepest point" along any ray originated from the deepest point. (The deepest point here refers to the point at which the function $D(\cdot)$ attains the maximum value, such as the center of a symmetric distribution.)

3. *Vanishing at Infinity*: $D(x)$ approaches zero as $\|x\|$ approaches infinity.

4. *Affine Invariance*: $D(x)$ is invariant under affine transformations, i.e. it does not depend on the underlying coordinate system.

The first three properties are essential to ensure that the contours derived from the level sets of the depth function are nested within one another. This in turn implies that the depth ordering of the data points in a sample is indeed from the center outward. The last property ensures the affine invariance of the depth based inference methods or applications, which is often needed in practice. Note that the affine invariance property may not be needed in some depth related applications, such as in Euclidean-distance based classification or clustering approaches. Different notions of depth have different advantages in capturing different features of the distribution. If a specific feature is considered critical for a given application, then the depth notion which can capture best that feature ought to be used. For example, if the underlying distribution is close to elliptical, then it is more efficient to use the Mahalanobis depth (Cf. [21]). Otherwise, the more geometric type of depth such as the simplicial depth and the halfspace depth may be more desirable, since they reflect more accurately the underlying probabilistic geometry and do not even require moment conditions.

## 2.1. Ordering/Ranking Multivariate Data by Depth

Unlike for univariate data, there is generally no natural ordering of multivariate data, as noted in [2]. It was first observed in [10] that the measure of depth can provide a natural ordering for multivariate data from the center outward. Specifically, given a notion of data depth, we can compute the depth values of all the sample points $\{X_1, \ldots, X_n\}$ and order them according to decreasing depth values. This gives a ranking of the sample points from the center outward. Let $X_{[i]}$ denote the sample point associated with the $i$th highest depth value. We view $X_{[1]}, \ldots, X_{[n]}$ as the order statistics, with $X_{[1]}$ being the *deepest* or the *most central* point or simply the *center*, and $X_{[n]}$ the most outlying point. The implication is that *a larger rank is associated with a more outlying position w.r.t. the data cloud*. These ordered statistics are referred to as *depth order statistics*, and their ordering or ranking as *depth ordering* or *depth ranking*. When ties occur in the ordering, the corresponding sample points are viewed as depth-equivalent, and the set of these points is termed a *depth-equivalence class*. In case there are more than one sample point with the highest depth value, we refer to their average as the deepest point for convenience.

## 2.2. Contours Derived from the Center-Outward Ordering

To illustrate depth ordering and its ramifications, we use simplicial depth to order sample points and graph some representative contours. Applying simplicial depth ordering to a sample of 500 points drawn from the bivariate standard normal distribution, we obtain in Figure 1a) the sample $p$-th level depth contours for $p = .25, .5, .75$, and $.9$. For example, the smallest convex hull in Figure 1a) corresponds to the .25-th level contour which is the smallest convex set that contains the 25% most central data points, $\{X_{[1]}, \ldots, X_{[125]}\}$ in this case. The contours are nested within one another. As the $p$ value increases, the contour expands. The deepest point in the sample is marked by a cross. Figure 1b) shows the depth contours, for the same set of $p$ levels, for a sample drawn from the standard bivariate exponential distribution (i.e. with independent margins and each with marginal mean 1). As the $p$ value increases, the contours there also expand from the center outward. How the depth contour expands in terms of the support of the underlying distribution, the shape, the speed and direction, and rate of change, motivate the definitions of scale (dispersion), skewness and kurtosis in LPS.

The depth contours in Figures 1a) show an almost symmetric nested expansion which reflects the symmetry of the underlying normal distribution. The contours in Figures 1b) fan out up-right. They reflect clearly the asymmetric probabilistic geometry of the underlying exponential distribution. New definitions of skewness are proposed in LPS by examining the degree of asymmetry of the depth contours w.r.t. different notions of multivariate symmetry. The plot of the depth contours can also indicate scale or dispersion of the underlying distribution. For example, consider increasing the standard deviation in each marginal variable, say from 1 to 2, for the normal distribution used in Figure 1a). Then it is not hard to see that the depth contours for the same set of $p$ levels will be expanded farther out than the ones in Figure 1a). Therefore, measuring the areas (or volumes in the case of a higher dimension) of the expansions of contours immediately lead to a measure of multivariate scale, called a scale curve (see Section 3.1.3).

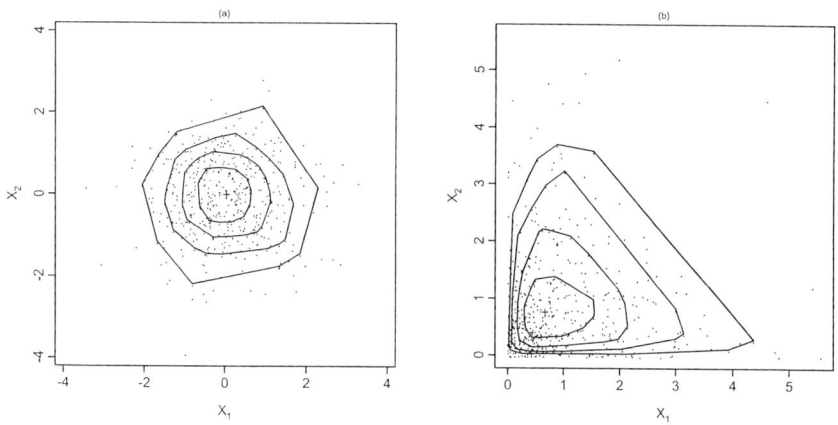

Figure 1. a) Normal Contours by $SD$. b) Exponential Contours by $SD$.

The above discussion and other useful depth-based statistics can be made more precise with the following definitions.

**Definition 1.** The set $\{x \in \mathbb{R}^d : D(x) = t\}$ is called the *level set* or *contour* of depth $t$.

**Definition 2.** The set $\{x \in \mathbb{R}^d : D(x) > t\}$ is called the *region enclosed by the contour of depth $t$*, and denoted by $R(t)$.

**Definition 3.** The set $C_p = \cap_t \{R(t) : P_F(R(t)) \geq p\}$. is called the *p-th central region*.

In other words, $C_p$ is the smallest region enclosed by depth contours to have amassed probability $p$. The contours shown in Figure 1 are sample versions of $\{C_{.25}, C_{.5}, C_{.75}, C_{.9}\}$. The boundary of $C_p$ is called the *p-th level contour*, and is denoted by $Q(p)$ (or $Q_F(p)$ when we need to stress that $F$ is the underlying distribution). In fact, if the density function $f$ is nonzero everywhere, $Q_F(p)$ is the contour $\{x \in \mathbb{R}^d : D(x) = t_p\}$ where $P\{x \in \mathbb{R}^d : D(x) \geq t_p\} = p$. To distinguish it from univariate quantiles, we shall call $Q_F(p)$, $0 \leq p \leq 1$ the *center-outward quantile surface*.

## 3. NONPARAMETRIC MULTIVARIATE ANALYSIS BY DATA DEPTH

The classical multivariate analysis is essentially elliptically structured. Its approach to obtaining multivariate distributional characteristics is a straightforward extension of the moment approach in the univariate case. More specifically, the location, scale, skewness, and kurtosis are defined respectively in terms of the first, second, third, and fourth moments. This leads to matrix or vector forms of outputs which are hard to grasp conceptually or graphically. Worse still, this approach would not even be applicable if the moments do not exist. Applying the idea of data depth, LPS introduced several quantitative and graphical multivariate distributional characteristics. These characteristics and their corresponding descriptive statistics are defined as functionals of data depth. They can be displayed as simple graphs in the plane, and can be easily visualized. Since these graphs are all based on the analysis of the contours derived from a data depth, they convey a more intuitive picture of the distributional properties. Furthermore, the depth approach is *moment-free* if the underlying data depth is, which in fact is the case for almost all known data depths.

### 3.1. Depth-Based Descriptive Statistics

#### 3.1.1. Location and $DL$-Statistics

The deepest point of a distribution derived from a given depth is viewed as the location parameter or multivariate median. The sample location estimator or sample median is then defined as the sample point with the highest sample depth value. More general location parameters are given in LPS as a class of depth-induced $L$-statistics, denoted by $DL$-statistics. Given the depth-order-statistics $X_{[1]}, \ldots, X_{[n]}$, we denote the stochastic process associated with these statistics by

$$\xi_n(t) = \begin{cases} X_{[i]}, & \text{for } \frac{i-1}{n} < t \leq \frac{i}{n}; \\ X_{[1]}, & \text{at } t = 0. \end{cases}$$

Let $\bar{\xi}_n(t)$ be the average of $\xi_n(t)$ over the class of sample points with the same depth value. The $DL$-statistic based on the weight function $\omega(t)$ is defined as $DL_n = \int \bar{\xi}_n(t)\omega(t)dt$. The population version of $DL_n$, denoted by $DL_F$, is defined as $DL_F = \int_0^1 \bar{Q}_F(t)\omega(t)dt$, where $\bar{Q}_F(t)$ is the population counterpart of $\bar{\xi}_n(t)$. More specifically, $Q_F(t)$ is the $t$-th center-outward quantile defined in Definition 3, and $\bar{Q}_F(t)$ is the mean along $Q_F(t)$.

Different choices of the weight function $\omega(\cdot)$ yields different alternatives for the location parameter and the associated estimators. The multivariate trimmed center (or median) is such an example and it provides a more robust location parameter.

### 3.1.2. Multivariate Quantiles

If $F$ is absolutely continuous and its density function $f$ is nonzero everywhere, then we have $C_p = R(t_p)$, where $R(t_p)$ is characterized by the requirement that $P_F(R(t_p)) = p$. In theory, the empirical versions of $C_p$ and $Q_F(p)$ should be defined by replacing $F$ and $D(\cdot)$ in (1) with $F_n$ and $D_n(\cdot)$ respectively. However, for computational and graphical convenience, we shall focus only on the $D_n(\cdot)$ values computed on the sample $\{X_1, \ldots, X_n\}$ and view the convex hull containing the most central fraction $p$ sample points as the sample estimate of $C_p$. More precisely, we set $C_{n,p} = $ convex hull $\{X_{[1]}, \ldots, X_{[\lceil np \rceil]}\}$, where $\lceil np \rceil = np$ if $np$ is an integer, and $(1 +$ the integer part of $np)$ otherwise. This simplification can also be justified by the fact that $C_p$ is typically a convex region. We call $C_{n,p}$ the *sample p-th central hull*, and its boundary, denoted by $Q_n(p)$, the *sample p-th level contour* or the *empirical p-th center-outward quantile surface*. Here, $Q_n(p)$ may be viewed as an estimate of the quantile $Q_F(p)$.

These center-outward quantiles can provide a natural theoretical foundation for multivariate inferences such as constructing confidence regions or computing p-values for testing hypotheses.

### 3.1.3. Scales

The scale curve and its estimate are defined in LPS respectively as

$$S(p) = volume\{C_p\} \quad \text{and} \quad S_n(p) = volume\{C_{n,p}\}, \tag{4}$$

for $0 \le p \le 1$. Here $S_n(p)$ is simply the volume of the convex hull containing the $\lceil np \rceil$ most central points. This plot of $S(p)$ vs. $p$ shows the scale of the distribution as a simple curve in the plane, which is easily visualized and interpreted. It is used in LPS as a simple visual tool for comparing the efficiency of different estimators when they are all used to estimate the same multivariate parameter, e.g. using the sample mean, the sample componentwise median and simplicial median to estimate the mean of a bivariate normal distribution. Figure 2 shows the scale curves derived from bivariate aviation safety data (collected by the FAA from 1993 to 1998) from some air carriers. The scale curve of Carrier 4 lies above that of Carrier 1 throughout which suggests that the distribution of Carrier 4 has a larger scale than Carrier 1. This indicates that the sample points from Carrier 4 are more scattered than those from Carrier 1. Therefore, the performance of Carrier 1 is more consistent (or less erratic) than that of Carrier 4. Clearly, from the point of view of aviation safety, the more consistent or stable performance of a carrier is more desirable and thus Carrier 1 outperforms Carrier 4. In this example, the plot of scale curves gives a convenient visual tool for the FAA to compare the consistency of performance of air carriers. Needless to say, performance consistency alone is not a sufficient measure for rating the overall performance of carriers. In Section 4, control charts derived from data depth ($r$-, $Q$- and DDMA-charts) are constructed and applied to the same set of airlines data to provide a more detailed comparison of the carriers.

LPS introduced four ways of measuring multivariate skewness: they are in terms of the departure of four different types of symmetry, namely spherical, elliptical, antipodal and angular symmetry. Kurtosis is also defined in LPS using Lorenz curves.

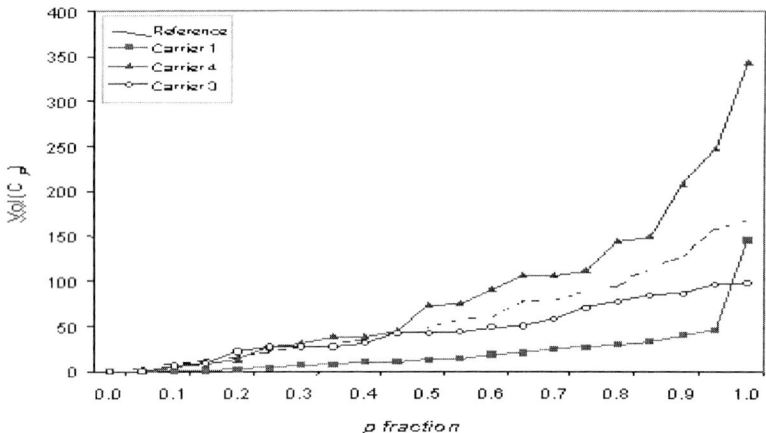

Figure 2. Scale curve comparisons for four air carriers.

## 3.2. Depth-Based Inference Methods

The fact that depth ranking is from the center outward enables us to identify central or tail regions in a given data cloud. Thus it immediately facilitates the statistical inference procedures such as constructing confidence regions and determining $p$-values for hypothesis testing. The specific details of these two aspects are lengthy. They both use extensively the bootstrap to generate enough many bootstrap estimates, and then assess the relative depth of the original estimate w.r.t. the cloud of bootstrap estimates. The details can be found in [26] [16]. We discuss here a few less explored depth rank tests.

### 3.2.1. One- and Two-Sample Tests Based on $DD$-Plots

$DD$-plots were introduced in LPS as simple graphical tools for a one- or two-sample test. Let $F$ and $G$ be two distributions on $\mathbb{R}^d$, and $D(\cdot)$ be an affine-invariant depth. We define $DD(F, G)$ as

$$DD(F,G) = \{(D_F(x), D_G(x)), \text{ for all } x \in \mathbb{R}^d \}. \tag{5}$$

Note that $DD(F,G)$ is a subset of $\mathbb{R}^2$. Clearly, if $F$ and $G$ are identical, then $D_F(x) = D_G(x)$ and thus $DD(F,G)$ coincides with the 45° line in the first quadrant. Any $DD(F,G)$ that deviates from this 45° line would indicate that $F$ and $G$ are not identical. When the distributions are unknown, we may construct the empirical versions of $DD$-plots. Based on a sample $\mathbf{X} \equiv \{\mathbf{X_1}, \ldots, \mathbf{X_n}\}$ from $F$, we may determine whether or not $F$ is the specified distribution $G$, by examining the following $DD$-plot,

$$DD(F_n, G) = \{(D_{F_n}(x), D_G(x)), \text{ for all } x \in \mathbf{X}\}. \tag{6}$$

If $F$ and $G$ are the distributions for the samples $\{X_1, \ldots, X_n\}(= \{\mathbf{X}\})$ and $\{Y_1, \ldots, Y_m\}(= \{\mathbf{Y}\})$, then the $DD$-plot below can be used to determine whether or not $F$ and $G$ are identical.

$$DD(F_n, G_m) = \{(D_{F_n}(x), D_{G_m}(x)), \ x \in \{\mathbf{X} \cup \mathbf{Y}\}\}, \tag{7}$$

where $F_n$ and $G_m$ are the empirical distributions of $\mathbf{X}$ and $\mathbf{Y}$ respectively. Since $DD(F,G)$ is a subset of $\mathbb{R}^2$, its plot can be easily visualized. It can be shown that a specific departure pattern of a $DD$-plot from the $45^o$ line corresponds to a specific type of difference between the two distributions in terms of location, scale, skewness or kurtosis. As expected, Figure 3a) shows a $DD$-plot clustering tightly along the $45^o$ line, since the two samples are drawn from the same distribution, either bivariate normal or bivariate exponential. The heart-shape $DD$-plot in Figure 3b) suggests a location difference, the half-moon shape in Figure 4a) suggests a scale difference, and the wedge-shape in Figure 4b) suggests a skewness difference.

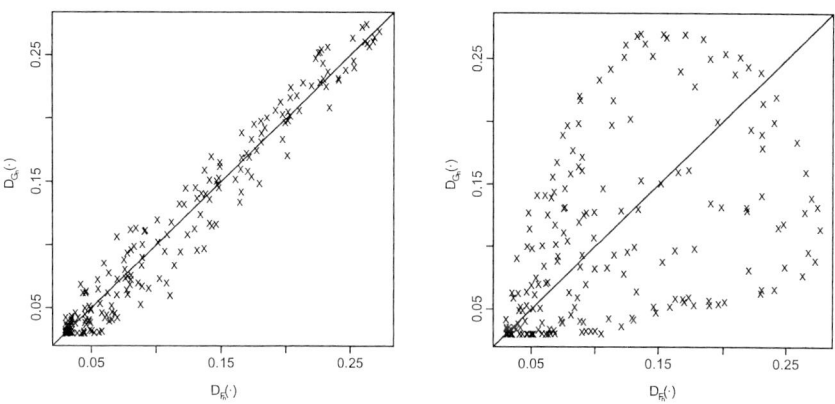

Figure 3. a) $DD$-Plot for Identical Distributions. b) $DD$-Plot with Location Difference.

### 3.2.2. Multivariate Rank Tests

The depth ranking has motivated many rank tests for testing the equality of multivariate locations or scales. Details can be found in [7] [11] [15] [8] [6] [18]. We briefly mention the rank tests developed recently in [18] for comparing multivariate scales in two or multiple samples.

Consider two random samples drawn from two multivariate distributions which have the same center but possibly different scales (or dispersions). If we combine the two samples, then the sample with the smaller scale would tend to cluster more tightly around the center of the combined sample, while the sample with the larger scale would tend to scatter at more outlying positions. The center-outward ranking induced by a data depth thus gives rise to a nonparametric rank sum test for the scale comparison of two multivariate

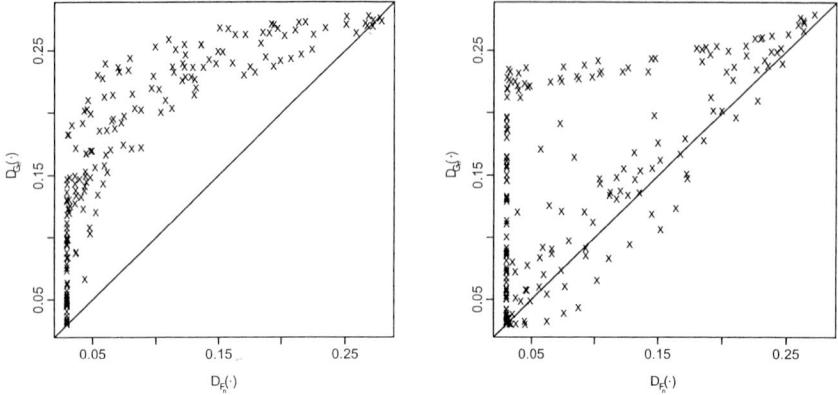

Figure 4. a) $DD$-Plot with Scale Difference. b) $DD$-Plot with Skewness Difference.

samples. More specifically, we apply depth ranking to the combined sample, and use the sum of the ranks of one sample as the test statistic. Since the more outlying points in the combined sample would receive smaller depth values and thus larger ranks, the sample with the larger scale would tend to receive larger ranks and consequently yield a greater sum of ranks. The critical value or $p$-value of this rank sum test can be obtained exactly in the same way as the well known Wilcoxon rank-sum test. This rank test is proposed in [18] for testing scale expansion or contraction when comparing multivariate samples. Furthermore, [18] also generalizes this test to the testing of scale difference in the setting of multiple multivariate samples, and proposes ways to address the issue of large numbers of ties.

In the univariate setting, the above multivariate rank sum test essentially coincides with the Siegel-Tukey or the Ansari-Bradley rank tests for testing the equality of variance for two univariate samples.

## 4. MULTIVARIATE CONTROL CHARTS: $r$- $Q$- & DDMA-CHARTS

Thanks to powerful computer facilities, most process monitoring procedures now routinely collect large amounts of measurements of multiple quality characteristics. Therefore, simple nonparametric multivariate control charts are highly desirable. In [12], the following $r$-chart was introduced: Let $\mathbf{Y} \equiv \{Y_1, \cdots, Y_N\}$ be a given reference sample (or the observations from an *accepted lot*), and $\{X_1, X_2, X_3, \cdots\}$ be the new observations coming from the process. The relative depth rank of $X_i$ w.r.t. to $\mathbf{Y}$ is

$$r(X_i) = \sum_{j=1}^{N} I\{D_N(Y_j) < D_N(X_i)\}/N \tag{8}$$

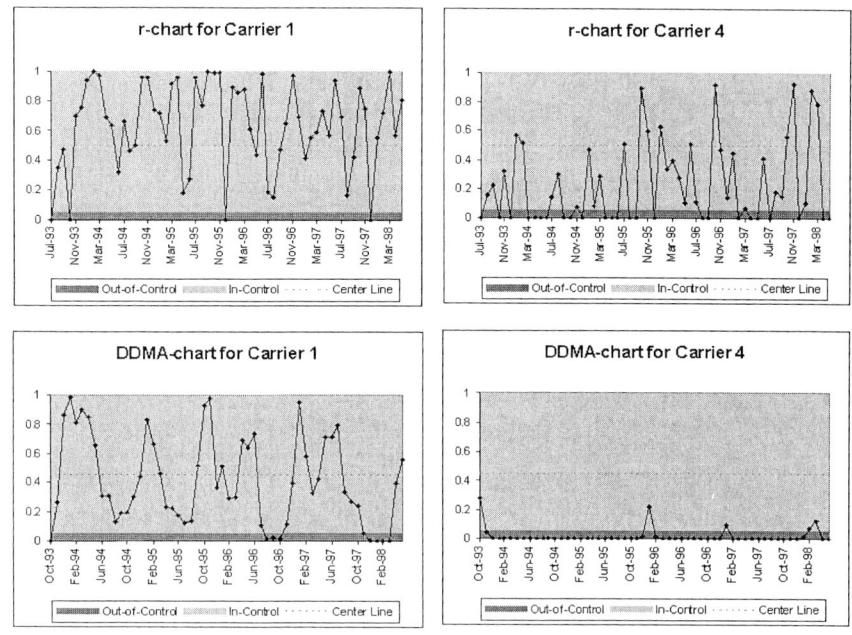

Figure 5. $r$-chart and DDMA-chart for Carriers 1 and 4.

where $I\{\cdot\}$ is the indicator function. The smaller $r(X_i)$, the more outlying $X_i$ is w.r.t. the data cloud $\mathbf{Y}$. We have shown in [15] that if the underlying distribution of the distribution is the same as that for $\mathbf{Y}$, then $r(X_i)$ follows a uniform distribution on the interval [0,1]. Therefore, the preset false alarm rate $\alpha$ is exactly the control limit for the $r$-chart. Regardless of the dimension of the observations, the $r$-chart is a simple graph in the plane, and is as easily visualized and interpreted as the standard univariate Shewhart chart. The advantage of the $r$-chart over the traditional Hoteling-$T^2$ chart is that the latter requires a normality assumption for the process while the earlier is completely nonparametric. Moreover, the $r$-chart allows simultaneous detections of location shifts and scale increases.

In [3], $r$-charts are applied to the monitoring of FAA multivariate aviation safety data. Properly chosen false alarm rates ($\alpha$-levels) in this charting technique can be used by FAA for the purpose of rating airline performance. For example, it provides a threshold system, labeled *concern, advisory, informational* and *expected*, as a concrete performance measure in terms of the severity of potential flaws, and serve as a guideline for the general rating of inspection results.

In the same setting as of the $r$-chart, the average relative ranks of $r(X_i)$'s leads to the $Q$-chart studied in [12], while the moving average of $r(X_i)$'s leads to the DDMA-chart studied in [20]. All three charts can detect simultaneously the location and scale process changes. The DDMA-chart improves upon $r$- and $Q$-chars in the efficiency for detecting location changes. Based on the same FAA aviation safety data leading to scale curves in Figure 2, Figure 5 shows the $r$- and DDMA-charts for Carriers 1 and 4. The out-of-control limit is set at $\alpha = .05$, and any observations falling below should be considered out-of-control. As it turns out, Carrier 1 outperforms Carrier 4 on all accounts. Not only does Carrier 4 perform more erratically or inconsistently (as shown by a larger scale in Figure 4), but it also has out-of-control performance more often throughout the five years inspection period (as seen in the control charts in Figure 5). Note that the $r$-chart for Carrier 4 in Figure 5 with many out-of-control observations indicates that there are possible location changes and scale increases in Carrier 4. The DDMA-chart for Carrier 4 further accentuates the location change by showing more out-of-control points throughout.

## REFERENCES

1. T. W. Anderson (1994) *An Introduction to Multivariate Analysis.* 3rd ed., John Wiley & Sons, New York.
2. V. Barnett (1976) The ordering of multivariate data. *Journal of the Royal Statistical Society,* Ser. **A**, **139**, 319-354.
3. A. Cheng, R. Liu and J. Luxhøj (2000) Monitoring multivariate aviation safety data by data depth: control charts and threshold systems. *IIE Transactions on Operations Engineering,* **32**, 861-872. (*Best Paper Award for Feature Applications for 2000-2001* by *Institute of Industrial Engineers*).
4. D. Donoho and M. Gasko (1992) Breakdown properties of location estimates based on halfspace depth and projected outlyingness. *The Annals of Statistics,* **20**, 1803-1827.
5. W. Eddy (1982) Convex hull peeling. *COMPSTAT,* 42-47. ed. Caussinus, H. et. al., Physica-Verlag, Wien.
6. T. Hettmansperger, J. Mottonen and H. Oja (1997) Affine-invariant multivariate one-sample signed-rank tests. *Journal of the American Statistical Association,* **92**, 1591-1600.
7. T. Hettmansperger, J. Nyblom and H. Oja (1992) On multivariate notions of sign and rank. *L-1 Statistical and Related Methods.* 267-278. ed. Y. Dodge, Notch-Holland, Amsterdam.
8. T. Hettmansperger and H. Oja (1994) Affine invariant multivariate multisample sign tests. *Journal of Royal Statistical Society,* Ser. *B*, **56**, 235-249.
9. J. Hodges (1955) A bivariate sign test. *The Annals of Mathematical Statistics,* **26**, 523-527.
10. R. Liu (1990) On a notion of data depth based on random simplices. *The Annals of Statistics,* **18**, 405-414.
11. R. Liu (1992) Data depth and multivariate rank tests. $L_1-Statistical$ *Analysis and Related Methods,* pp. 279-294, ed. Y. Dodge, Elsevier, Amsterdam.
12. R. Liu (1995) Control charts for multivariate processes. *Journal of the American Statistical Association,* **90**, 1380-1388.
13. R. Liu, J. Parelius and K. Singh (1999) Multivariate analysis by data depth: descriptive statistics, graphics and inference (with discussions). *Annals of Statistics,* **27**,

783-858.
14. R. Liu and K. Singh (1992) Ordering directional data: concepts of data depth on circles and spheres. *The Annals of Statistics*, **20**, 1468-1484.
15. R. Liu and K. Singh (1993) A quality index based on data depth and multivariate rank tests. *Journal of the American Statistical Association*, **88**, 257-260.
16. R. Liu and K. Singh (1997) Notions of limiting P-values based on data depth and bootstrap. *Journal of the American Statistical Association*, **91**, 266-277.
17. R. Liu and K. Singh (2002) DDMA-charts: nonparametric multivariate moving average control charts based on data depth. Technical report, Rutgers University.
18. R. Liu and K. Singh (2003) Rank tests for comparing multivariate scale using data depth: testing for expansion or contraction. Technical report, Rutgers University.
19. R. Liu, K. Singh and J. Teng (2001) Linear fitting by simplicial intercept depth: reflection invariance and robustness. Technical report, Rutgers University.
20. R. Liu, K. Singh and J. Teng (2002) DDMA-Chart: nonparametric multivariate moving average charts based on data depth. Technical report, Rutgers University.
21. P. C. Mahalanobis (1936) On the generalized distance in statistics. *Proceedings of the National Academy India*, **12**, 49-55.
22. H. Oja (1983) Descriptive statistics for multivariate distributions, *Statistics and Probability Letters*, **1**, 327-332.
23. P. Rousseeuw and M. Hubert (1999) Regression depth (with discussion). *Journal of the American Statistical Association*, **94**, 388-433.
24. K. Singh (1992) Majorty depth. Technical report, Rutgers University.
25. J. Tukey (1975) Mathematics and picturing data. *Proceedings of the 1975 International Congress of Mathematics*, **2**, 523-531.
26. A. Yeh and K. Singh (1997) Balanced confidence regions based on Tukey's depth and bootstrap. *Journal of the Royal Statistical Society, Ser. B*, **59**, 639-652.
27. Y. Zuo (2001) Projection based depth functions and associated medians. To appear in *The Annals of Statistics*.
28. Y. Zuo and R. Serfling (2000a) General notions of statistical depth function. *The Annals of Statistics*, **28**, 461-482.
29. Y. Zuo and R. Serfling (2000b) Structural properties and convergence results for contours of sample statistical depth functions. *The Annals of Statistics*, *28*, 483-499.

# Visual exploration of data through their graph representations

George Michailidis[a] *

[a]Department of Statistics
The University of Michigan,
Ann Arbor, MI 48109-1092

We consider the problem of visually exploring structured and unstructured data sets. The representation of the data as the adjacency matrix of an appropriately defined graph transforms the problem into one of graph drawing. The necessary mathematical framework for producing fast and informative graph layouts is introduced and the associated optimization issues discussed. Finally, the improvements that may be obtained by introducing negative weights in the underlying graph structure are investigated.

## 1. INTRODUCTION

Advances in hardware and software technologies have enabled today's computer systems to collect and store vast amounts of data. Scientific and commercial data are most often automatically recorded via sensors and monitoring systems resulting in data sets comprised of millions of objects. Furthermore, many attributes are also recorded, resulting in very high dimensional data sets. Finally, new applications areas such as biochemical pathways, gene and protein functions, and web documents, to name a few, produce data with inherent *structure* that can not simply be captured by numbers.

In order to extract useful information from such large data sets, a first step is to be able to *visualize* their global structure and identify patterns, trends and outliers. For structured data the goal is to capture visually the underlying relationships, hierarchies and taxonomies between the objects under study. The basic idea of visual data exploration is to present the data in some visual form, allowing the human eye to gain insight from the representation, so that the user can draw conclusions and interact with the data in order to enhance her understanding. Visual data mining techniques have proven to be of high value in exploratory data analysis, as the increased research interest on the topic [1,2] and the number of new data visualization products attest. Visual data exploration is especially useful when little is known about the data and the direct involvement of the user in the exploration process helps define and adjust the data mining goals [3].

In this paper we focus on the visual exploration of unstructured and structured data through their graph representations. More specifically, we show how the commonly objects by attributes data sets encountered in statistics can be represented by graphs and ways to

---
*Work supported in part by NSF under grant IIS-9988095. I would like to thank Jan de Leeuw for many fruitful discussions and suggestions on the topic.

visualize the resulting graph structures. Special attention is paid to categorical data. We also provide examples of visualizing structured data, for which graphs represent a natural way for capturing their underlying structure.

## 2. DATA AND GRAPHS

Graphs are useful entities since they can represent relationships between a set of objects. Areas of applications include VLSI and Web layouts, subroutine-call diagrams, semantic and social networks, molecular and genetic maps, evolutionary and phylogenetic trees, just to name a few. In statistics we usually encounter them as dendrograms in cluster analysis and as path diagrams in structural equations models and Bayesian belief networks. In Figure 1, an example of the graph representation of a structured data set that shows the protein interaction network implicated in the membrane fusion process of vesicular transport is shown [4].

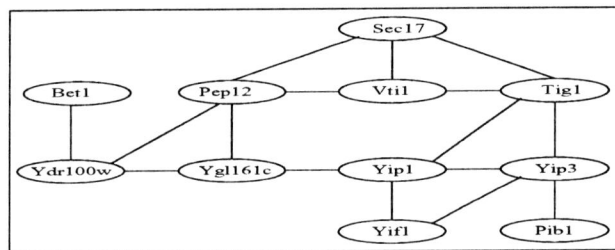

Figure 1. The graph representation of a structured data set of protein interactions. Nodes correspond to proteins and edges to the relationships amongst them.

However, graphs are also capable of representing aspects of unstructured data as well as the following three examples show. In the first two examples a contingency table and a similarity matrix are shown as graphs, while in the third example the graph representation of a categorical multivariate data set is given.

| Contingency table | | | Similarity matrix | | | | | |
|---|---|---|---|---|---|---|---|---|
| 36 | 0 | 7 | 0 | | | | | |
| 16 | 0 | 0 | 7 | 0 | | | | |
| 0 | 2 | 0 | 2 | 3 | 0 | | | |
| 0 | 22 | 42 | 8 | 7.5 | 1 | 0 | | |
| | | | 5 | 6.4 | 3.2 | 3 | 0 | |
| | | | 3 | 2.7 | 12.1 | 1.2 | 12 | 0 |

Table 1
The data for a contingency table and a similarity matrix.

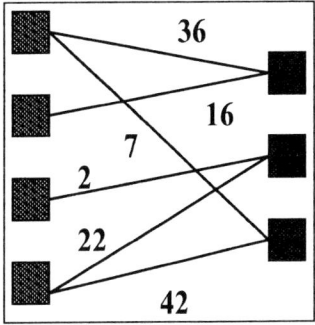 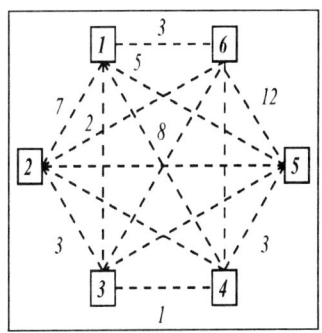

Figure 2. The graph representation of a contingency table. The four nodes on the left correspond to the four categories of the first variable and similarly for the three nodes on the right. The edge weights represent the counts in the corresponding cell of the table.

Figure 3. The graph representation of a similarity matrix. The numbered squares correspond to the objects, while the weights on certain edges correspond to the dissimilarities.

|  |  | \multicolumn{3}{c}{Price} | \multicolumn{2}{c}{Fiber} | \multicolumn{3}{c}{Quality} |
| --- | --- | --- | --- | --- | --- | --- | --- | --- | --- |
|  | Sleeping Bag | cheap | not expensive | expensive | down fibers | synthetic fibers | good | acceptable | bad |
| 1 | One Kilo Bag | 1 | 0 | 0 | 0 | 1 | 1 | 0 | 0 |
| 2 | Sund | 1 | 0 | 0 | 0 | 1 | 0 | 0 | 1 |
| 3 | Kompakt Basic | 1 | 0 | 0 | 0 | 1 | 1 | 0 | 0 |
| 4 | Finmark Tour | 1 | 0 | 0 | 0 | 1 | 0 | 0 | 1 |
| 5 | Interlight Lyx | 1 | 0 | 0 | 0 | 1 | 0 | 0 | 1 |
| 6 | Kompakt | 0 | 1 | 0 | 0 | 1 | 0 | 1 | 0 |
| 7 | Touch the Cloud | 0 | 1 | 0 | 0 | 1 | 0 | 1 | 0 |
| 8 | Cat's Meow | 0 | 1 | 0 | 0 | 1 | 1 | 0 | 0 |
| 9 | Igloo Super | 0 | 1 | 0 | 0 | 1 | 0 | 0 | 1 |
| 10 | Donna | 0 | 1 | 0 | 0 | 1 | 0 | 1 | 0 |
| 11 | Tyin | 0 | 1 | 0 | 0 | 1 | 0 | 1 | 0 |
| 11 | Travellers Dream | 0 | 1 | 0 | 1 | 0 | 1 | 0 | 0 |
| 13 | Yeti Light | 0 | 1 | 0 | 1 | 0 | 1 | 0 | 0 |
| 14 | Climber | 0 | 1 | 0 | 1 | 0 | 0 | 1 | 0 |
| 15 | Viking | 0 | 1 | 0 | 1 | 0 | 1 | 0 | 0 |
| 16 | Eiger | 0 | 0 | 1 | 1 | 0 | 0 | 1 | 0 |
| 17 | Climber light | 0 | 1 | 0 | 1 | 0 | 1 | 0 | 0 |
| 18 | Cobra | 0 | 0 | 1 | 1 | 0 | 1 | 0 | 0 |
| 19 | Cobra Comfort | 0 | 1 | 0 | 1 | 0 | 0 | 1 | 0 |
| 20 | Foxfire | 0 | 0 | 1 | 1 | 0 | 1 | 0 | 0 |
| 21 | Mont Blanc | 0 | 0 | 1 | 1 | 0 | 1 | 0 | 0 |

Table 2
A categorical data set analyzed in [5] shown as a super-indicator matrix [6].

These three examples show that graphs are capable of representing some types of unstructured data. The graphs contain the same information as the original data. However,

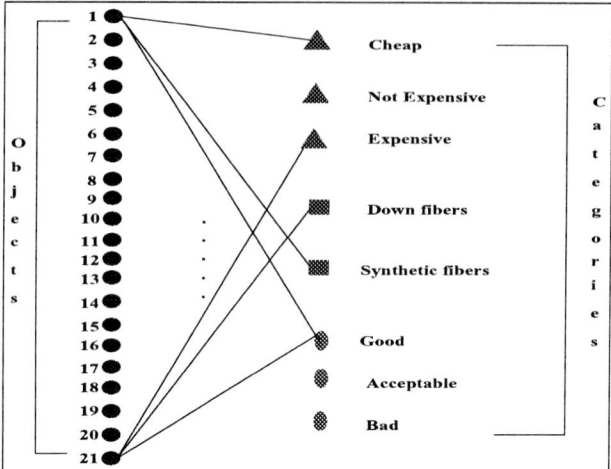

Figure 4. The graph representation of the sleeping bag data set presented in Table 2, with one set of nodes corresponding to the objects (sleeping bags) another set of nodes to the categories of the three attributes (price, fiber, quality) and the edges capturing the relationship between objects and categories.

these representations do not reveal 'interesting' features in the data; the same fact holds true for the protein interaction data set as well. What is required is a *drawing* of the graph that would show the overall structure of the data, as well as reveal interesting patterns and unexpected relationships.

To formalize our setting, we consider an undirected graph $G = (V, E)$, where $V = \{v_1, v_2, \cdots, v_n\}$ is the set of the $n$ vertices (nodes) and $E \subset V \times V$ the set of edges. It is assumed that the graph $G$ does not contain either self-loops or multiple edges between any pair of vertices. The set of edges can be represented in matrix form through the *adjacency matrix* $A = \{a_{ij}|, i, j = 1, \cdots, n\}$. Thus, vertices $i, j \in G$ are connected if and only if $a_{ij} > 0$, otherwise $a_{ij} = 0$. If $a_{ij} \in \{0, 1\}$ we are dealing with a *simple* graph, otherwise with a *weighted* graph. The contingency table and the dissimilarity matrix correspond to weighted graphs, while the protein interaction and the sleeping bags data sets to a simple graph. We note that for two of the examples considered above, namely the contingency table and the sleeping bags, the adjacency matrix has a special form given by

$$A = \begin{bmatrix} 0 & W \\ W' & 0 \end{bmatrix}$$

where $W$ corresponds either to the contingency table itself or to the super-indicator matrix familiar from multiple correspondence analysis [6]. This is because the node set can be partitioned in two subsets (e.g. objects and categories) and they do not exist edges that connect nodes within the subsets, but only across subsets.

The goal is to make a 'meaningful' picture of the graph. More specifically, our objective is to to represent the nodes of a graph as points in $\mathbf{R}^s$ and the vertices as lines connecting the points. Graph drawing is an active area in computer science, and it is very ably reviewed in the recent book by [7]. The choice of $\mathbf{R}^s$ is due to its attractive underlying geometry and the fact that it renders the necessary computations more manageable. In practice, $s$ is usually chosen equal to 2 or 3, in order to be able to visualize the resulting representation.

## 3. GRAPH LAYOUT

The problem of graph drawing/layout has received a lot of attention from various scientific communities. Simply put the problem is defined as: given a set of nodes connected by a set of edges, calculate the position of the nodes and the curve to be drawn for each edge. This simple description also reveals the intricacy of the problem; which space to use for the positions and what type of curves. For example, grid layouts position the nodes at points with integer coordinates, while hyperbolic layouts [8] embed the points on a sphere. Most layouts use straight lines for drawing the edges but some use curves of a certain degree [7]. Many layouts algorithms try to impose a set of *aesthetic* rules on the final drawing. For example, nodes and edges must be evenly distributed, edges should have all the same length, edge-crossings should be kept at a minimum, etc. Some of these rules clearly apply to certain graphs and/or are important in certain applications, while others have a more 'absolute' character [9]. Furthermore, each of the rules defines an associated optimization problem and some of them such as the edge-crossing minimization are computationally intractable but for very small graphs [7]. In addition in graph visualization, a major problem that needs to be addressed is the *size* of the graph. Few layout algorithms can deal effectively with thousands of nodes, although graphs with such size appear in a wide variety of applications. Some systems that use a combination of optimization algorithms and heuristics to handle such large graphs are NicheWorks [10], GVF [11] and H3Viewer [12].

A popular multivariate analysis technique that can be used for graph drawing purposes is Multidimensional Scaling (MDS) [13]. Its objective is to embed the vertices in a Euclidean space of appropriate dimensionality so that the Euclidean distances between the points that represent the nodes approximate well the path-length distances defined between the vertices in the original graph. The quality of the resulting representation is measured by an appropriate fit (loss) function.

In our approach, we adopt primarily the *adjacency model*, i.e. we do not emphasize graph-theoretical distances, but we pay special attention to which vertices are adjacent and which vertices are not. Obviously, this is related to distance, but the emphasis is different. We also use a fit (loss) function to measure the *quality* of the resulting embedding.

Define the fit function

$$\sigma(Z) = \sum_{i=1}^{n} \sum_{j=1}^{n} \alpha_{ij} d_{ij}(Z) \tag{1}$$

where $Z$ is a $n \times s$ matrix that would contain the coordinates of the $n$ nodes in the s-dimensional Euclidean space and $d_{ij}(Z)$ denotes the distance between the $s$-dimensional points $z_i$ and $z_j$. We are interested in minimizing (1), which in the case of a simple graph the end result would be that nodes sharing lots of interconnections should end up close together in the layout, while nodes without too many connections are expected to be located on the periphery of the drawing. For a weighted graph, the larger the weights are, the stronger the bond between the connected nodes and hence the closer together they should be drawn. Thus, in this formulation, unlike the MDS one, the number and strength of bonds is the main factor that determines the final layout.

In this study, we restrict attention to squared Euclidean distances; i.e. $d_{ij}(Z) = (z_i - z_j)^2$. Then, some straightforward algebra shows that (1) can be written as

$$\sigma(Z) = \operatorname{trace}(Z'LZ), \tag{2}$$

where $L$ a $n \times n$ given by $L = D - A$ with $D$ containing the row sums of the adjacency matrix $A$. The matrix $L$ is known in the graph theory literature as the Laplacian matrix of a graph [14]. In [5], the graph drawing problem under other distance functions such as the $\ell_h$, $h \in [1, 2)$ is considered.

In order to avoid the trivial solution $Z = 0$ (all nodes collapse to the same point in $\mathbf{R}^s$) we impose the following normalization constraint

$$Z'DZ = I_s \text{ and } u'Z = 0 \tag{3}$$

where $u$ is a column vector comprised of ones. The second constraint will center the layout of the graph, while the first one would provide non-trivial solutions. Other possible normalizations of the general $\phi$ fit function are discussed in [5]. Some routine algebra shows that minimizing $\sigma(Z)$ subject to (3) is equivalent to *maximizing* [2]

$$\psi(Z) = \operatorname{trace}(Z'AZ) \text{ subject to (3).} \tag{4}$$

The solution to this new problem is known to be given by the $s+1$ largest generalized eigenvectors of $A$, which can be computed in $\mathcal{O}(n^{3/2})$ operations [15], since the largest one corresponding to an eigenvalue $\lambda = 1$ is a constant one and must be discarded. Notice that the solution is not orthogonal in Euclidean space but in the weighted (by $D$) Euclidean space. The spectrum of the problem at hand is $1 = \lambda_1 \geq \lambda_2 \geq ... \geq \lambda_n \geq -1$, since the matrix $D^{-1}A$ that possesses the same spectrum is a stochastic matrix. Other properties of the solution that provide information about the underlying graph (e.g connected, bipartite, etc) as well as approximate colorings based on the the eigenvectors corresponding to the largest negative eigenvalues are discussed in [16].

In Figures 5 and 6 the graph layouts based on the above described method of the protein interactions structured data set and of a correlation matrix of protein expression data from three different experimental conditions are shown. It can be seen that the

---

[2] Another possible normalization is $Z'Z = I_s$, which also leads to an optimization problem whose solution is given by the $s$ *smallest* eigenvalues of $L$. However, computing the smallest eigenvalues of a matrix is a somewhat harder computational problem than computing the largest ones [15]. Experience with this algorithm indicates that the nodes of large graphs tend to form tight clusters, thus impairing interpretation.

proteins PIB1 and BET1 that have very few interactions are located on the periphery of the drawing as expected. Moreover, the solution places the 'hub' proteins TLG1 and YIP1 close to the center of the plot and reveals a clustering structure in the data. The layout of the correlation matrix reveals a strong clustering pattern between the three experimental conditions, but also more variability (smaller correlations and thus looser bonds) within one set of conditions (the one depicted on the upper left corner of the plot).

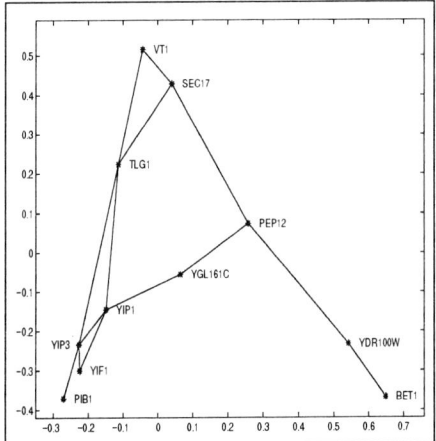
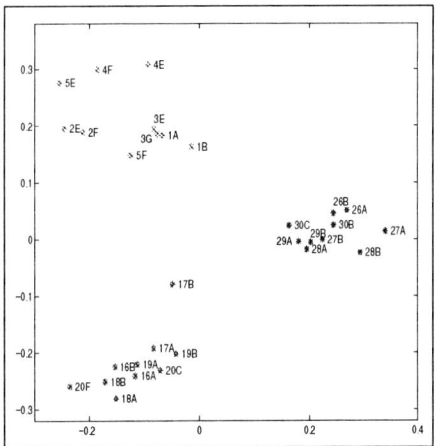

Figure 5. The graph layout of the protein interactions data set.

Figure 6. The graph layout of a correlation matrix. The edges have been removed, since we deal with a complete graph (every node is connected to every other node) and thus increase the interpretability of the representation.

### 3.1. The special case of bipartite graphs

As noted earlier, the graph representation of a contingency table and of a categorical data set have some special features, namely that the set of nodes can be naturally partitioned in two subsets; the categories of the each variable for the former case and the subset of nodes representing the objects (sleeping bags) and the subset of nodes representing the categories of all the variables for the latter one, thus giving rise to a *bipartite* graph. Let $Z = [X'\ Y']'$, where $X$ contains the coordinates of one subset of the nodes (e.g. the objects) and $Y$ the coordinates of the remaining subset (e.g. the categories of the variables). The fit function can then be written as (given the special structure of the adjacency matrix $A$)

$$\sigma(X, Y) = \text{trace}(X'D_X X + Y'D_Y Y - 2Y'WX) \tag{5}$$

where $D_Y$ is a diagonal matrix containing the column sums of $W$ and $D_X$ another diagonal matrix containing the row sums of $W$. In the case of a contingency table both $D_Y$ and $D_X$ contain the marginal frequencies of the two variables, while for a categorical data set $D_Y = \text{diag}(W'W)$ contains the univariate marginals of the categories of all the variables and $D_X = JI$ is a constant multiple of the identity matrix with $J$ being the number of variables in the data set. By adopting an analogous normalization constraint, namely $X'D_XX = I_s$, we can solve the resulting optimization problem through a block relaxation algorithm [17] as follows:
Step 1: $\hat{Y} = D^{-1}W'X$;
Step 2: $\hat{X} = \frac{1}{J}WY$;
Step 3: Orthonormalize $X$ using the Gram-Schmidt procedure [6].
It can be seen that the solution at the optimum satisfies $\hat{Y} = D^{-1}W'X$; i.e. category points are in the center of gravity of the objects belonging to a particular category. This is known in the literature as the *centroid principle* [6,17].

In Figure 7 the graph layout of the sleeping bags data set is shown. The solution captures the presence in the data set of good, expensive sleeping bags filled with down fibers and cheap, bad quality sleeping bags filled with synthetic fibers and the absence of bad, expensive sleeping bags. It also shows that there are some intermediate sleeping bags in terms of quality and price filled either with down or synthetic fibers. The centroid principle proves useful in the interpretation of the graph layout.

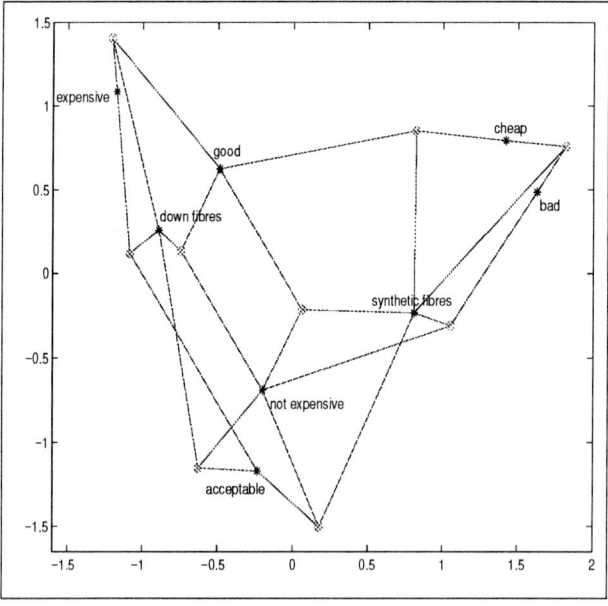

Figure 7. The graph layout of the sleeping bags data set.

## 4. GRAPH DRAWING WITH NEGATIVE WEIGHTS

The graph drawing model posited in this paper takes into consideration only the links between the nodes. Nodes not connected do not make any contribution. However, it is reasonable to look at a model where not connected vertices push each other apart, while connected ones pull each other closer together. This is the main idea behind a large class of graph drawing techniques, collectively known as *force-directed* techniques ([7] Chapter 10). Such techniques treat the graph nodes as bodies that attract and repel each other, for instance because the edges are springs or because the vertices have electric charges. A popular choice for the pulling force is Hooke's law [18], i.e. the force is proportional to the difference between the distance of the vertices and the zero-energy length of the spring, while the repelling force follows an inverse square law. A Bayesian framework for dynamic graph drawing by using force-directed methods is provided in [19]. One of the issues with this class of techniques is that different choices of the pulling and pushing forces would lead to very different and in many cases rather unappealing solutions as the layout in Figure 8 illustrates. The other problem is that the whole approach is dominated by heuristics, thus not allowing one to study properties of the optimal solution and obtain the necessary insight of when a particular technique would work on a specific type of problem.

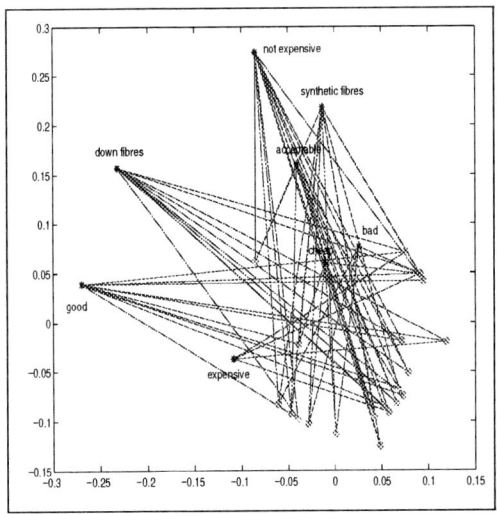

Figure 8. The graph layout of the sleeping bags data set using a force-directed technique with the wrong choice of a pushing force.

In this paper the pushing force is introduced through negative entries in the adjacency matrix. One crucial question turns out to be what is the magnitude of the negative weights (entries) allowed. Going back to the formulation of our problem given in (2), it is seen that we want to minimize $\sigma(Z) = \sum_{i,j=1}^{n} \alpha_{ij}(z_i - z_j)^2$. The non-negative weights assumed so far, prevent the fit function from becoming negative. By allowing $\alpha_{ij} < 0$ we can get $\sigma(Z) < 0$, which in turn implies that we can find a $Z$ such that $\sigma(Z) \to -\infty$; a totally uninteresting solution from a graph drawing point of view. The following constraint prevents this from happening.

*Necessary Constraint:* Let $G = (V, E)$ be a weighted graph with adjacency matrix $A$ with $\alpha_{ij} \in \mathbf{R}$ and corresponding Laplacian matrix $L$. For graph drawing purposes $L$ has to be positive semi-definite.

The positive semi-definiteness constraint on $L$ guarantees that its diagonal elements $\ell_{ii}$ are also non-negative [20]. We examine next in more detail the presence of negative weights in the adjacency matrix of a graph, which also leads to a sufficient condition for the semi-positiveness of $L$.

Write the adjacency matrix $A = A_P - A_N$, where $A_P$ is a $n \times n$ matrix containing non-negative weights and $A_N$ a $n \times n$ matrix containing the non-positive weights. In our previous formulation we simply have $A_N = 0$. Let $D_P$ be a diagonal matrix containing the row sums of $A_P$ and analogously define $D_N$. We then have that

$$L = D - A = (D_P - A_P) - (D_N - A_N) = L_P - L_N, \tag{6}$$

where $L_N$ is a positive semi-definite matrix.

A sufficient condition for $L$ to be positive semi-definite is that $\rho(L_N L_P^{-1}) < 1$; the spectrum of the product to be smaller than 1 ([20] Chapter 7). This condition tivially holds if, for example, the smallest eigenvalue of $L_P$ is larger than the largest eigenvalue of $L_N$. Although the above condition is easy to check when assigning negative weights to a graph, it does not lead to rules of thumb of how to choose the magnitude of the negative weights. The optimization problem remains the same as defined by (2) and (3).

The effects of introducing negative weights are shown in Figures 9-11. It can be seen that graphs with more nodes offer more opportunities for producing considerably different layouts. The choice of negative weights (as explained in the captions of the corresponding graphs) has been such that it emphasizes the clustering effects in the graph layouts. Experience shows that a choice of uniform negative weights makes a difference only in very large graphs. The judicious choice of negative weights for improving the interpretability of the graph layout and thus aid the visual exploration of structured and unstructured data sets is a topic of current research.

## 5. SOME CONCLUDING REMARKS

In this study the problem of visualizing data sets through their graph representations is considered. The connection with the problem of graph drawing is made and a mathematical framework for dealing with the latter is introduced. The optimality and properties of the solution to the posed graph drawing problem are examined and illustrated through several examples. Finally, the introduction of negative weights seems a promising way for expanding the capabilities of the proposed framework.

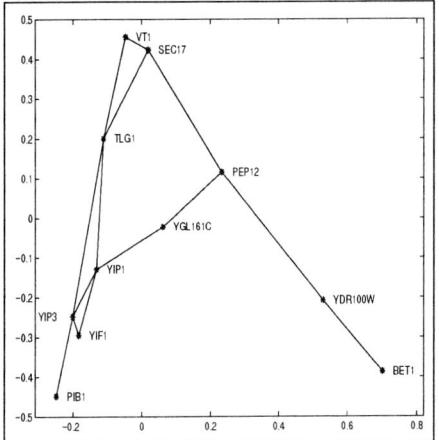
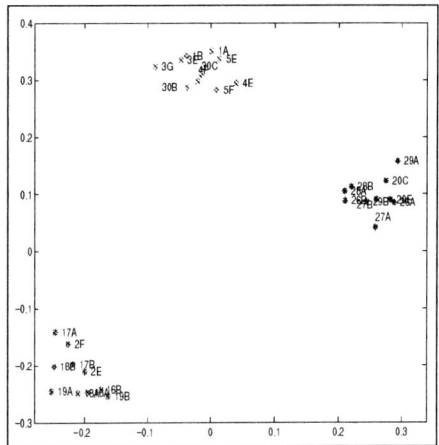

Figure 9. The graph layout of the protein interactions data set with negative weights applied *only* to the highly connected nodes.

Figure 10. The graph layout of a correlation matrix with negative weights applied to the nodes not belonging to the originally detected clusters.

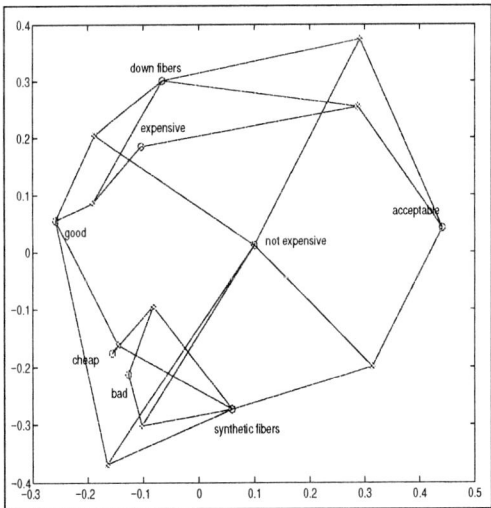

Figure 11. The graph layout of the sleeping bags data set using a force-directed technique with negatived weights applied only to the objects.

## REFERENCES

1. S.G. Eick and G.J. Wills, High interaction graphics, European J. of Oper. Research 84 (1995) 445-459,
2. A.F.X. Wilhelm, A.R. Unwin and M. Theus, Software for interactive statistical graphics, Proc. Int'l Softstat 95 Conf. (1995).
3. A. Buja, D.F. Swayne and D. Cook., Interactive high-dimensional data visualization, J. Comp. and Graphical Statistics, 5 (1996) 78-99.
4. T. Ito et al., Toward a protein-protein interaction map of the budding yeast: a comprehensive system to examine two-hybrid interactions in all possible combinations between the yeast proteins, PNAS, 97 (2000) 1143-1147.
5. G. Michailidis and J. de Leeuw, Data visualization through graph drawing, Comp. Statistics, 16 (2001) 435-450.
6. A. Gifi, Nonlinear Multivariate Analysis, Wiley, Chichester, 1990
7. Di Battista, G., Eades, P., Tammasia, R. and Tollis, I., Graph Drawing: Algorithms for the Geometric Representation of Graphs, Prentice Hall, NY, 1998
8. T. Muenzer and P. Buchard, Visualizing the structure of the World Wide Web in 3D hyperbolic space, Proc. VRML Conf. (1995).
9. H.C. Purchase, Which aesthetic has the greatest effect on human understanding?, Proc. Symp. Graph Drawing GD 97, (1998) 248-261.
10. G.J. Wills, NicheWorks - Interactive visualization of very large graphs, Proc. Symp. Graph Drawing GD 97, (1998) 403-415.
11. I. Herman, M.S. Marshall and G. Melancon, An object oriented design for graph visualization, Software Practice and Experience, 31 (2001) 739-756.
12. T. Muenzer, Drawing large graphs with H3Viewer and site manager, Proc. Symp. Graph Drawing GD 98, (1999) 384-393.
13. A. Buja and D.F. Swayne, Visualization methodology for multidimensional scaling, J. of Classification, 19 (2002) 7-43.
14. F.R.K. Chung, Spectral Graph Theory, AMS, Providence, 1997.
15. G.H. Golub and C.F. van Loan, Matrix Computations, Johns Hopkins University Press, Baltimore, 1993
16. G. Michailidis, Clustering and drawing of graphs through eigenvalue decomposition, Technical Report, Dept. of Stats, The U. of Michigan, 2001.
17. G. Michailidis and Jan de Leeuw, The Gifi system of descriptive multivariate analysis, Stat. Science, 13 (1998) 307-336.
18. P. Eades, A heuristic for graph drawing, Congressus Numerantium, 42 (1980) 149-160.
19. U. Brandes and D. Wagner, A Bayesian paradigm for dynamic graph layout, Proc. Symp. Graph Drawing GD 96, (1997) 236-247.
20. R.A. Horn and C.R. Johnson, Matrix Analysis, Cambridge Univ. Press, Cambridge, 1985.

# 6. Nonparametric Regression

# Inference for nonsmooth regression curves and surfaces using kernel-based methods

I. Gijbels

Institut de Statistique, Université catholique de Louvain,
Voie du Roman Pays 20, B-1348 Louvain-la-Neuve, Belgium.*

In this paper we review kernel-based methods for detecting discontinuities in an otherwise smooth regression function or surface. In case of a possible discontinuous curve the interest might be in detecting the discontinuities, their jump sizes and finally to estimate the discontinuous curve. Alternatively, one might be uniquely interested in estimating directly the discontinuous curve preserving the jumps. A brief discussion on available kernel-based methods for testing for a continuous versus a discontinuous regression function, and for detecting discontinuities in regression surfaces is also provided.

## 1. INTRODUCTION

When estimating a regression curve it is often assumed that this function is smooth, and standard smoothing techniques will result in an estimator that smooths away discontinuities. As a consequence these estimators are not consistent at points of discontinuity of the domain. To obtain consistent estimators over the whole domain, one needs to adapt the standard techniques. Apart from discontinuities a curve may show other irregularities, such as for example cusps, discontinuities in a higher order derivative, etc. In this (non-exhaustive) survey we mainly focus on simple jump discontinuities in the regression function itself.

Several approaches have been proposed to estimate such discontinuous curves. A first approach is to estimate first the locations of the discontinuities, as well as their jump sizes. The estimated jump locations divide the domain into several segments and within these segments the underlying regression curve is continuous. The estimation of the regression function can then be done by simply using standard smoothing techniques on each of the segments. A second approach focuses on the estimation of the regression function directly and aims not at estimating separately the jump locations nor their sizes. We will refer to these as direct estimation methods.

Standard smoothing techniques include kernel-based methods, local polynomial fitting, wavelet-based methods, splines etc. It would take us too far to discuss the adaptations to estimating discontinuous curves for all these techniques. We will therefore restrict to a survey of kernel-based methods and local polynomial fitting methods for estimating

---

*Financial support from the contract 'Projet d'Actions de Recherche Concertes' nr 98/03-217 from the Belgian government, and from the IAP research network nr P5/24 of the Belgian State (Federal Office for Scientific, Technical and Cultural Affairs) is gratefully acknowledged.

curves with possible jump discontinuities. Among the papers dealing with spline-based methods are Laurent and Utreras (1986), Girard (1990), Potier and Vercken (1994), Shiau (1988), Koo (1997) and Lee (2002). Semiparametric methods for change-point analysis have been dealt with by Speckman (1995) and Cline, Eubank and Speckman (1995). Wavelet methods for change-point detection have been studied by Mallat and Hwang (1992), Wang (1995, 1998, 1999), Raimondo (1998), Oudshoorn (1998) and Antoniadis and Gijbels (2002), among others.

The paper is organized as follows. In Section 2 we recall a standard jump detection method. Section 3 then provides a brief survey of available methods all based in some way on the standard method. In this section we also discuss some identification problem when detecting jump discontinuities, and discuss the case of a finite (unknown) number of jump points. Section 4 focuses on estimation of a discontinuous regression curve and Section 5 provides some references for tests for a continuous versus a discontinuous curve. We end by giving a few basic references for generalizations of the methods to the two-dimensional case.

## 2. A STANDARD JUMP DETECTION METHOD

Suppose that we have a sample of $n$ data pairs $(X_1, Y_1), \cdots, (X_n, Y_n)$ generated from the model

$$Y_i = g(X_i) + \sigma(X_i)\varepsilon_i, \qquad 1 \leq i \leq n, \tag{1}$$

where $\varepsilon_i$ are i.i.d. variables, independent of the $X_i$, with mean zero and variance one. The quantity $\sigma^2(\cdot)$ is the conditional variance function. Without loss of generality we assume that the design points are in the interval $[0, 1]$. The regression function $g$ is smooth except at a finite number $k$ of points where the function is discontinuous. For simplicity we will first restrict attention to the case of one discontinuity point (i.e. $k = 1$), located at $\tau \in ]0, 1[$. The case of more than one discontinuity point will be discussed in Section 3.4.

We can write

$$g(x) = g_0(x) + \gamma \mathbf{1}_{[\tau,1]}(x) = \begin{cases} g_0(x) & \text{if } x \in [0, \tau[ \\ g_1(x) \equiv g_0(x) + \gamma & \text{if } x \in [\tau, 1] \end{cases}, \tag{2}$$

where the function $g_0$ is a continuous function and where $\gamma$ denotes the size of the jump of the regression function $g$ at location $\tau$. Clearly, at the point $\tau$ the left-hand limit of the function $g$ (denoted by $g_-$; $g_-(x) = \lim_{\substack{t \to x \\ <}} g(t)$) differs (for $\gamma \neq 0$) from the right-hand limit (denoted by $g_+$; $g_+(x) = \lim_{\substack{t \to x \\ >}} g(t)$), whereas at all other points they coincide. Therefore, a reasonable estimator of the unknown point $\tau$ is provided by the point at which the difference between estimates of the right-hand limit and the left-hand limit of $g$ is maximal in absolute value. An estimate $\widehat{g}_+(x)$ of $g_+(x)$ is obtained via a nonparametric kernel estimate using all observations located to the right of $x$. Similarly, an estimate $\widehat{g}_-(x)$ of $g_-(x)$ is based on all observations located to the left of $x$. An estimate of $\tau$ is then

$$\widehat{\tau} = \arg\max_x |\widehat{g}_+(x) - \widehat{g}_-(x)|. \tag{3}$$

For the jump size of $g$ at the point $\tau$ we have

$$\gamma = \gamma(\tau) = g_+(\tau) - g_-(\tau), \tag{4}$$

and an estimate of this quantity is

$$\hat{\gamma} = \hat{\gamma}(\hat{\tau}) = \hat{g}_+(\hat{\tau}) - \hat{g}_-(\hat{\tau}). \tag{5}$$

An estimate of the unknown regression function $g$ is obtained by using a standard smoothing technique on all observations falling into the interval $[0, \hat{\tau}[$ and obtaining as such $\hat{g}_0$, as well as on all observations falling into the interval $[\hat{\tau}, 1]$ for obtaining $\hat{g}_1$. In correspondence with (2) an estimate for the regression curve $g$ is then

$$\hat{g}(x) = \begin{cases} \hat{g}_0(x) & \text{if } x \in [0, \hat{\tau}[ \\ \hat{g}_1(x) & \text{if } x \in [\hat{\tau}, 1]. \end{cases} \tag{6}$$

This approach of estimating $\tau$, $\gamma$ and finally the regression curve $g$ has been followed in many papers. A first difference between the various kernel-based methods available in the literature is related to the use of different smoothers for estimating $g_+$, $g_-$ and the final estimate of $g$, such as, for example, the Nadaraya-Watson kernel estimator, the Gasser-Müller estimator, a local linear estimator or more generally a local polynomial estimator. (See for example Wand and Jones (1995) and Fan and Gijbels (1996).) A second difference is in the choice of the kernel functions. Thirdly, some procedures involve several steps in the estimation in order to achieve better rates of convergence as well as a better performance for the resulting estimators.

The minimax optimal rate of convergence for detecting $\tau$ is $O(n^{-1})$, for either regularly-spaced or stochastic design. Indeed, the 'average' spacings between design points are of order $n^{-1}$, and so jump discontinuities cannot be determined with a greater accuracy. See also Korostelov and Tsybakov (1993) for optimal rates mainly in parametric settings.

In the next section we briefly discuss the different methods, and indicate how to deal with the case of more than one discontinuity point.

The generalization of this approach for detecting discontinuities in the $\nu$th ($\nu \geq 0$) derivative of a regression function $g$ is straightforward. One then looks at differences of estimates of the $\nu$th left-hand and right-hand derivative of the regression function, denoted by $\hat{g}_+^{(\nu)}(\cdot)$ and $\hat{g}_-^{(\nu)}(\cdot)$ respectively. The location of a discontinuity point $\tau$ of the $\nu$th derivative of the function $g$ is then estimated by

$$\hat{\tau}_\nu = \arg\max_x \left| \hat{g}_+^{(\nu)}(x) - \hat{g}_-^{(\nu)}(x) \right|,$$

and an estimate of the jump size $\gamma$ of the derivative function $g^{(\nu)}$ at the point $\tau$ is

$$\hat{\gamma}_\nu = \hat{\gamma}_\nu(\hat{\tau}) = \hat{g}_+^{(\nu)}(\hat{\tau}) - \hat{g}_-^{(\nu)}(\hat{\tau}).$$

## 3. DETECTION OF DISCONTINUITIES IN REGRESSION CURVES

### 3.1. One-step procedures

We first discuss the one-step procedures and then provide a brief survey on the more-step procedures.

Among the first papers using kernel-based smoothing techniques, there is the work by Canny (1986).

Müller (1992) considered a fixed-design regression model with homoscedastic errors and studied estimation of a single discontinuity point in the $\nu$th derivative of a regression function. The approach is as the one explained in Section 2. As a smoother he used the Gasser-Müller estimator with one-sided kernels. A one-sided kernel $K_+$ supported on $[-1, 0]$ is used to estimate the right-hand limit and a one-sided kernel $K_-$ defined via $K_-(x) = K_+(-x)$, supported on $[0, 1]$, is used for estimating the left-hand limit. Furthermore it is assumed that $K_+(-1) = K_+(0) = 0$. The resulting estimator for $\tau$ can achieve the rate $O(n^{-1+\delta})$ for some $\delta > 0$. See also Eubank and Speckman (1994) for similar results obtained via a least-squares type of estimator.

Wu and Chu (1993a) consider a diagnostic function

$$J(x) = \hat{g}_2(x) - \hat{g}_3(x) , \tag{7}$$

where $\hat{g}_2(\cdot)$ and $\hat{g}_3(\cdot)$ are Gasser-Müller estimators, using the same bandwidth $h$, and different kernel functions $K_2$ and $K_3$ respectively, with $K_2$ and $K_3$ compactly supported on $[-1, 1]$ and satisfying $K_2(x) = K_3(-x)$ and $\int_0^1 K_2 \neq \int_0^1 K_3$. Discontinuity points are then detected by searching for locations where $|J(x)|$ is large. Indeed, for points $x$ where the regression function $g$ is continuous both estimates $\hat{g}_2(x)$ and $\hat{g}_3(x)$ are consistent estimates for $g(x)$ and hence $J(x)$ will be close to zero; one can show that $E[J(x)] = O(h)$. For a point $x$ which is a discontinuity point of $g$ though, it can be shown that the expectation of $J(x)$ is given by

$$E[J(x)] = \gamma \left[ \int_0^1 K_3(u)du - \int_0^1 K_2(u)du \right] + O\left(h + (nh)^{-1}\right) , \tag{8}$$

where $\gamma$ is the jump size at the discontinuity point $x$. An estimate for $\tau$ in model (2) is then the point where $|J(x)|$ is maximal. To estimate the jump size $\gamma$ Wu and Chu (1993a) consider the quantity

$$S(x) = \hat{g}_4(x) - \hat{g}_5(x) ,$$

where $\hat{g}_4(\cdot)$ and $\hat{g}_5(\cdot)$ are Gasser-Müller estimators, using a same bandwidth $g$, and different kernel functions $K_4$ and $K_5$, satisfying conditions similar to these imposed on $K_2$ and $K_3$. An expression similar to (8) can be proved for $S(x)$. Therefore, an appropriate estimate for $\gamma$ is given by

$$\hat{\gamma} = S(\hat{\tau}) \left[ \int_0^1 K_5(u)du - \int_0^1 K_4(u)du \right]^{-1} .$$

The bandwidth $g$ is taken to be larger than the bandwidth $h$, such that in the neighbourhood of a discontinuity point the quantity $S(x)$ is expected to be of order $O(g)$ and hence could be of a better performance then an estimate based on $J(\hat{\tau})$ for $\hat{\tau}$ in a neighbourhood of the true $\tau$.

Modifications of kernel regression estimates for points near the boundary of a support and discontinuity points are proposed in Wu and Chu (1993b). The proposed modifications built further on the methodology of Hall and Wehrly (1991) proposed for regression estimation at boundary points.

Loader (1996) considers regression model (1) with $g$ as in (2), but restricts to equispaced fixed design points and to Gaussian errors. The estimator for $\tau$ is as in (3) where the estimates $\hat{g}_+(\cdot)$ and $\hat{g}_-(\cdot)$ are obtained by local polynomial fitting of degree $p$. Essential assumptions on the kernel functions $K_+$ and $K_-$ are that $K_+(x) = K_-(-x)$ and $K_-$ is defined on $[0,1]$ with $K_-(1) = 0$ and $K_-(0) > 0$, and integrates to one. Loader (1996) shows that the estimate of $\tau$ achieves rate of convergence $O(n^{-1})$ and also derives the asymptotic law for $\hat{\tau}$ in this restrictive setup of normal errors. The assumption that $K_-(0) > 0$ is essential for obtaining the $O(n^{-1})$ rate of convergence of the estimator.

Grégoire and Hamrouni (2002a) consider the more general random regression model (1) and use local linear smoothing with (3) for detecting jump points. They use a kernel $K_+(\cdot)$ that is supported on $[-1,0]$ with $K_-(x) = K_+(-x)$ and make the basic assumption that $K_+(0) > 0$. They show that the rate of convergence $O(n^{-1})$ is obtained for all choices of the bandwidth parameter $h = h_n$ satisfying the usual conditions $h_n \to 0$ and $nh_n \to \infty$. This result is along the same lines as the result in Loader (1996) obtained under the particular setup. The use of local linear smoothing is motivated by the fact that local linear smoothers have the nice property of adapting automatically to the boundary of the support. Note that estimation of the left-hand and right-hand limits $g_+$ and $g_-$ is indeed assimilated to estimation at 'boundary' points. Grégoire and Hamrouni (2002a) study the jump estimate $\hat{\gamma}(x)$ as a process in $x$, and establish convergence of the process. From this they deduce the asymptotic normality of the estimate $\hat{\tau}$ as well as of the estimate $\hat{\gamma}(\hat{\tau})$ of the jump size $\gamma$.

Extensions of these methods to the case of a jump discontinuity in the $\nu$th derivative of the regresion function is rather straightforward, by relying on local polynomial fitting of degree $p$ with $p \geq \nu$.

A key issue in these estimation methods is the choice of the bandwidth parameter $h$. This choice is rather crucial as was pointed out by many authors. Among the proposals available in the literature there are those of Grégoire and Hamrouni (2002a) who suggest to carry out the estimation procedure for an extended grid of $h$-values. If the estimates show a certain stability than the bandwidth used is a reasonable one. A more involved approach would be to obtain $\hat{\tau} = \hat{\tau}(h)$, for a fixed value of $h$, and from this derive the regression estimate $\hat{g}$ as defined in (6), using the same bandwidth. Consider then a measure of performance of the regression estimate, such as for example, an Integrated Squared Error, or an Average Squared Error, and get an estimate $I(h)$ of this measure of error. The idea is then to choose the bandwidth, among a grid of bandwidth values, for which this estimated error is minimal. See also Wu and Chu (1993c) for such a procedure. Note that such a method focuses on final estimation of the regression function $g$. An important question here is whether the bandwidth for estimating the location of the jump point on one hand and the estimation for the regression curve on the other hand should be taken the same, and whether an appropriate error criterion is one that focuses on the estimation of the regression function.

Horváth and Kokoszka (2002) consider regression model (1) with fixed equidistant design $x_i = i/n$, $1 \leq i \leq n$, and fit locally polynomials of degree $\nu$ to the left and to the right of points $x$ of the domain of definition of the regression function. Consistency of the resulting estimates for the change point in the $\nu$th derivative is shown under different conditions on the kernel function. The issue of choice of the bandwidth parameters is not addressed.

The above estimation procedures are commony-based on looking at differences of estimates of left-hand and right-hand limits (see (3) and (5)) or on a diagnostic function which is the difference between two kernel type estimates of the regression function as in (7).

## 3.2. Two-step procedures

Other kernel-based procedures have been proposed. Among these are several two-step procedures. A first two-step procedure was introduced in Korostelev (1987) who considered estimation of jump points in a Gaussian white noise model.

Müller and Song (1997) consider a fixed design regression model and propose a two-step estimator for the case of a single jump discontinuity. Instead of (2) they consider

$$g(x) = g_0(x) + \gamma_n \mathbf{1}_{[\tau,1]}(x),$$

with $|\gamma_n| = a_n^{-1}$, where $a_n$ is a positive sequence of real numbers. The 'contiguous' jump size case is characterized by $a_n \to \infty$ as $n \to \infty$ whereas the fixed jump case, described above, corresponds to $a_n = $ constant. In this contiguous jump size case Müller and Song (1997) propose a two-step estimator for $\tau$ and establish its rate of convergence as well as its limiting distribution. In step one of the procedure a pilot estimator $\tilde{\tau}$ of $\tau$ is obtained via (3) using a bandwidth $h_n$, and kernels $K_+$ and $K_-$ as indicated before, without requiring that $K_+(0) > 0$. Müller and Song (1997) showed that, under suitable conditions, $P(|\tilde{\tau} - \tau| \leq h_n) \to 1$ as $n \to \infty$. In a second step of the procedure random intervals

$$[r_0, r_1] = [\tilde{\tau} - h_n, \tilde{\tau} + h_n] \quad \text{and} \quad [s_0, s_1] = [\tilde{\tau} - 2h_n, \tilde{\tau} + 2h_n],$$

are considered, where $h_n > 0$ should satisfy $a_n h_n \to 0$, $a_n^2 (nh_n)^{-1} \to 0$ and $nh_n^2 a_n^{-1} \to 0$. Note that this implies that $h_n \to 0$ and $nh_n \to \infty$. Note also that the interval $[r_0, r_1]$ contains the true unknown location point $\tau$ with probability tending to one, as shown in the theoretical property mentioned above. In this second step one calculates for each point $x$ in $[r_0, r_1]$, a weighted difference between averages of $Y$-observations to the left and to the right, contained in the bigger interval $[s_0, s_1]$. The final estimate is then the point $x$ for which this weighted difference is maximal. More precisely, the final estimation of $\tau$

is based on the diagnostic function

$$T_n(x) = [(x-s_0)(s_1-x)]^{1/2} \left\{ \frac{1}{\lambda_n([x,s_1])} \sum_{i:x_i \in [x,s_1]} Y_i - \frac{1}{\lambda_n([s_0,x))} \sum_{i:x_i \in [s_0,x)} Y_i \right\},$$

where $\lambda_n(A)$, with $A$ a set, is the counting measure counting the number of design points $x_i = i/n$ in the set $A$. The final estimator of $\tau$ is defined as

$$\hat{\tau} = \arg\max_{x \in [r_0, r_1]} |T_n(x)|.$$

Müller and Song (1997) prove that this final estimator has rate of convergence $O_P(n^{-1}a_n^2)$ in the contiguous jump case, and rate $O_P(n^{-1})$ in the fixed jump case (i.e. $a_n$ is a constant). Practical choices of the bandwidth parameter are not discussed.

Gijbels, Hall and Kneip (1999) introduced another two-step procedure under the regression model (1). The proposed estimator has desired rate $n^{-1}$. In addition practical choices of the bandwidth parameters involved have been proposed, resulting into a fully data-driven estimation procedure.

The two-step procedure of Gijbels, Hall and Kneip (1999), consists of obtaining a rough estimate of $\tau$, via maximization of an appropriate diagnostic function, and then to do a refined search by fitting a piecewise constant function via least squares in an interval surrounding the rough estimate. The first step involves a diagnostic function, which is taken to be an estimate of the 'derivative' of the unknown regression function. The motivation behind this is simple: the estimated 'continuous' regression curve will typically show a large slope at the location of the discontinuity point of $g$. Therefore a good pilot estimator is obtained by taking the location where an estimate of the first derivative is large. Examples of estimates of derivative curves are: the derivative of the Nadaraya-Watson or Gasser-Müller estimate, or the direct estimator of the derivative obtained via local linear (or polynomial) fitting. For simplicity of presentation we take the derivative of the Nadaraya-Watson estimator, and define the diagnostic function $D$ by

$$D(x, h_1) = \frac{\partial}{\partial x} \left( \frac{\sum_{i=1}^n K\{(x-X_i)/h_1\} Y_i}{\sum_{i=1}^n K\{(x-X_i)/h_1\}} \right), \tag{9}$$

where $K$ is a kernel function, supported on $[-1,1]$, with two Hölder continuous derivatives on $[0,1]$, $K(0) \geq K(x) > 0$ for all $x \neq 0$ and $K''(0) > 0$. The parameter $h_1 = h_{1,n} > 0$ is a bandwidth parameter. A first rough estimator for $\tau$ is

$$\tilde{\tau} = \arg\max_{x \in ]h_1, 1-h_1[} |D(x, h_1)|.$$

In the second step one considers an interval around $\tilde{\tau}$ that contains the unknown point $\tau$ with high probability (justified by theoretical results). Consider the interval $[\tilde{\tau}-h_2, \tilde{\tau}+h_2]$, with $h_2 = h_{2,n}$ bandwidth parameter (possibly different from $h_1$). Denote by $\{i_1, i_1, \cdots, i_2\}$ the set of integers $i$ for which $X_i \in [\tilde{\tau}-h_2, \tilde{\tau}+h_2]$. Then fit a step function on the interval

$[\tilde{\tau} - h_2, \tilde{\tau} + h_2]$ using least-squares. The jump point is estimated to occur between design points $X_{i_0}$ and $X_{i_0+1}$, where $i_0$ is chosen to minimize the (residual) sum of squares

$$\sum_{i=i_1}^{i_0} \left\{ Y_i - (i_0 - i_1 + 1)^{-1} \sum_{j=i_1}^{i_0} Y_j \right\}^2 + \sum_{i=i_0+1}^{i_2} \left\{ Y_i - (i_2 - i_0)^{-1} \sum_{j=i_0+1}^{i_2} Y_j \right\}^2.$$

Denoting by $\hat{i}_0$ the minimizer of this sum of squares, the final estimator of $\tau$ is given by the midpoint between $X_{i_0}$ and $X_{i_0+1}$, i.e.

$$\hat{\tau} = \frac{1}{2}(X_{i_0} + X_{i_0+1}).$$

Gijbels, Hall and Kneip (1999) proved that this estimator achieves the optimal $n^{-1}$ rate of convergence. Moreover, their extensive simulation study shows that the introduction of a second step really improves the estimation: if the pilot estimation is bad, the second step permits an improvement of it.

The performance of the estimator $\hat{\tau}$ heavily depends on the choice of the bandwidth parameters $h_1$ and $h_2$. Data-driven choices for $h_1$ and $h_2$, based on a bootstrap procedure relying on an appropriate error criterion for the estimation of $\tau$, have been worked out by Gijbels and Goderniaux (2002a). The fully data-driven procedure has been tested extensively via simulation studies and shows a very good performance of the method. Its performance has also been compared, via simulations with other $n^{-1}$-rate estimation procedures, such as the one-step procedure of Grégoire and Hamrouni (2002a). See Goderniaux (2001) and Gijbels and Goderniaux (2002a,c). One of the findings is that the data-driven two-step method tends to be less variable than the one-step procedure of Grégoire and Hamrouni (2002a), although both achieve, theoretically, the optimal rate of convergence.

Several remarks can be made here. Recalling the definition of a derivative one can see that the diagnostic functions in (7) and (9) are similar in spirit. The diagnostic function (7) is focusing on the numerator in the definition of the derivative. Note also that in the case of fixed equally spaced design, one could, instead of relying on (9), simply look at the derivative of the numerator of the Nadaraya-Watson estimator, as this is a consistent estimator of $g' \cdot f$, with $f$ the design density. The estimation can as such be based on $(nh_1)^{-1} \sum_{i=1}^{n} K'((x - x_i)/h_1) Y_i$. Bias and variance of the corresponding estimator $\hat{\tau} = \arg\max_x (nh_1)^{-1} \sum_{i=1}^{n} K'((x - x_i)/h_1) Y_i$ of $\tau$ are derived in Koch and Pope (1997).

Further, the two-step procedures by Müller and Song (1997) and Gijbels, Hall and Kneip (1999) are somewhat similar in spirit, but a major difference is in the error criterion used in the second step, and in the fact that the latter is a fully data-driven estimation procedure.

Another kernel-based algorithm for detecting jumps in the $v$th derivative of a regression function is provided in Qiu and Yandell (1998a). For detecting a jump in the regression function $g$ itself, a local linear fit is carried out in a neighbourhood of design points, and the estimates of the derivatives of the regression function at these design points are used to detect discontinuity points. This basically relies on ideas similar to the justification of (9). For determining whether a point is a jump point or not, Qiu and Yandell (1998a) rely on asymptotic theory for local linear (polynomial) smoothers.

### 3.3. Identification problem

A possible problem that might appear in all of the above methods is the distinction between a discontinuity point of the regression function $g$ and a location where $g$ shows a high derivative in absolute value. For convenience, we explain this problem in the context of the procedure of Gijbels, Hall and Kneip (1999). A simple search for the maximum in the diagnostic function $|D(x,h_1)|$, defined in (9), might give some problems, as was already pointed out in Gijbels, Hall and Kneip (1999). As an example, consider the cosinus function

$$g(x) = \cos\{8\pi(0.5 - x)\} - 2\cos\{8\pi(0.5 - x)\}\mathbf{1}(x > 0.5) , \qquad (10)$$

which has a discontinuity point at 0.5 of size $-2$, and shows many steep decreasing and increasing parts. As a consequence the diagnostic function exhibits many local maxima and the largest among these may correspond to a steep gradient instead of to the jump discontinuity. This is illustrated in Figure 1 (a) where we have represented, for a simulated dataset of size $n = 100$ from (1) with $\sigma^2 = \text{constant} = 0.1$, $\varepsilon \sim N(0;1)$ and equidistant design, the diagnostic function for $h_1 = h_{1,0} = 0.1$. We can see that the diagnostic function shows 8 local maxima and the largest of them does not correspond to the location of the jump discontinuity, 0.5. So the problem that is highlighted with this example is one of identifying that local maximum of the diagnostic function that corresponds to the jump discontinuity. This identification problem can be dealt with in a fairly easy way, via a three-step algorithm proposed by Gijbels, Hall and Kneip (1999). The algorithm also includes an automatic choice of the bandwidth $h_1$. Let $h_{1,i} = h_0 r^i$, for $i = 0, 1, 2, \cdots$, be a range of decreasing values of the bandwidth $h_1$ (starting with $h_{1,0} = h_0$), with $h_0 > 0$ and $0 < r < 1$. The three-step algorithm reads as follows:

*Step 1: Initialization.*
Let $M$ denote the number of local maxima of $|D(.,h_{1,0})|$ on $]h_{1,0}, 1 - h_{1,0}[$. Let $\{\xi_{0,j}, 1 \leq j \leq M\}$ be the set of points at which the maxima are achieved.

*Step 2: Iteration.*
Given a set $\{\xi_{i,j}, 1 \leq j \leq M\}$ of local maxima of $|D(.,h_{1,i})|$ on $]h_{1,0}, 1 - h_{1,0}[$, let $\xi_{i+1,j}$ denote the local maximum of $|D(.,h_{1,i+1})|$ that is nearest to $\xi_{i,j}$.

*Step 3: Termination.*
Stop the algorithm at iteration $i = \tilde{\imath}$, when the number of data values in some interval $]x - h_{1,i}, x + h_{1,i}[ \subseteq ]h_{1,0}, 1 - h_{1,0}[$ first falls below a predetermined value.
Take the preliminary estimate $\tilde{\tau}$ of the jump discontinuity $x_0$ to be this value of $\xi_{\tilde{\imath},j}$ for which $||D(\xi_{\tilde{\imath},j}, h_{1,\tilde{\imath}})| - |D(\xi_{0,j}, h_{1,0})||$ is the largest.

The reason for choosing the preliminary estimator of $\tau$ by maximizing $||D(\xi_{\tilde{\imath},j}, h_{1,\tilde{\imath}})| - |D(\xi_{0,j}, h_{1,0})||$ is the following: when the bandwidth $h_{1,i}$ decreases, the slope of the Nadaraya-Watson estimator increases around the jump, i.e. the value taken by the diagnostic function increases. Therefore, it is expected that the most noticeable differences in the diagnostic function, when decreasing $h$, are encountered around jump points. Figure 1 illustrates the three steps of the algorithm for four selected iterations corresponding to $i = 0, 4, 8$ and $\tilde{\imath} = 11$. Figure 1 (a) represents the initial iteration ($i = 0$) and the diagnostic function shows $M = 8$ local maxima at that iteration. For each iteration $i$ the vertical dotted lines present the 8 local maxima, $\xi_{i,j}$ for $j = 1, ..., 8$ that are closest to those obtained in the previous iteration, $\xi_{i-1,j}$ $j = 1, ..., 8$. In the last iteration we then have 8 (considered) local maxima, $\xi_{\tilde{\imath},j}$ for $j = 1, ..., 8$ and we must choose among them by maximizing the difference in magnitude between the maximal values of the diagnostic function achieved in the first and in the last iteration for the 8 local maxima, i.e. $||D(\xi_{\tilde{\imath},j}, h_{1,\tilde{\imath}})| - |D(\xi_{0,j}, h_{1,0})||$, $j = 1, 2, ..., 8$. These values are reported in Table 1 with the values of $\xi_{\tilde{\imath},j}$, $j = 1, ..., 8$. This leads to the conclusion that the preliminary estimator of the jump point is 0.50.

Table 1
Tabulated $||D(\xi_{\tilde{\imath},j}, h_{1,\tilde{\imath}})| - |D(\xi_{0,j}, h_{1,0})||$ and $\xi_{\tilde{\imath},j}$ values for a simulated data set for (10).

| $j$ | $||D(\xi_{\tilde{\imath},j}, h_{1,\tilde{\imath}})| - |D(\xi_{0,j}, h_{1,0})||$ | $\xi_{\tilde{\imath},j}$ |
|---|---|---|
| 1 | 395.94 | 0.134286 |
| 2 | 379.04 | 0.305714 |
| 3 | 683.21 | 0.420000 |
| 4 | **887.88** | **0.500000** |
| 5 | 677.06 | 0.580000 |
| 6 | 546.55 | 0.694286 |
| 7 | 418.44 | 0.785714 |
| 8 | 314.07 | 0.854285 |

Note that two parameters $h_0$ and $r$ appear in the automatic choice of the bandwidth $h_1$. These choices however are of little importance. They represent respectively the starting value of the range of $h_1$-values and the multiplicative decreasing step in the sequence of $h_1$ values. The only precaution is to choose $h_0$ and $r$ not too small. In other words, the grid of $h_1$ values should be fine enough and should be of a reasonable range. Safe choices are $h_0$ large (for example half of the length of the domain of the regression function) and $r$ close to 1.

### 3.4. More than one discontinuity

In case of several discontinuity points $\tau_1, \tau_2, \cdots, \tau_k$ one has, instead of (2),

$$g(x) = g_0(x) + \sum_{j=1}^{k} \gamma_j \mathbf{1}_{[\tau_j, 1]}(x),$$

where $\gamma_j$ denotes the size of the jump of the regression function $g$ at location $\tau_j$ and where it is assumed, for simplicity of presentation, that $|\gamma_j| > |\gamma_{j+1}|$ for $j = 1, \cdots, k$ with $\gamma_{k+1} = 0$. Furthermore one needs to assume that the discontinuity points $\tau_j$ are such that $\tau_j \in [\delta, 1 - \delta]$ for some $\delta > 0$, i.e. are located in the interior of $[0, 1]$. Furthermore, it is assumed that the jumps points are at a least a distance $\delta$ located from each other.

For $k$ known, Wu and Chu (1993a) suggest the following procedure for detecting all jump points.

STEP 1: Estimate $\tau_1$ by $\arg\max_x |J(x)|$;

STEP 2: For $j = 2, \cdots, k$, estimate $\tau_j$ by searching for the location of the maximum over the set

$$A_j = [\delta, 1-\delta] - \bigcup_{k=1}^{j-1} [\hat{\tau}_k - 2h, \hat{\tau}_k + 2h].$$

In other words, for each estimated discontinuity point, remove a symmetric interval of length $4h$, centered at the estimated point, from the domain and search for the next discontuity point in the remaining set.

Yin (1988) proposed a method to estimate the number of discontinuity points, by introducing a (pre-specified) threshold value. Wu and Chu (1993a) propose tests to test for the number of jump points. See also Qiu (1994). Other procedures for determining the number of change-points using kernel-based estimation methods can be found in Beunen (1998) and Bunt, Koch and Pope (1998).

As a second example, we describe the generalization of the two-step procedure of Gijbels, Hall and Kneip (1999) to the case of general $k$. Suppose first that $k$ is known. One can then proceed in the same way: identify the $k$ positions using a kernel method and improve the rate of convergence by a least-squares step on an interval around each pilot estimator. It is convenient to use the three-step algorithm of Section 3.3 which generalizes easily to the case of $k$ jump discontinuities $\tau_1, \tau_2, \cdots, \tau_k$ by a slight modification of step 3.

*Step 3: Termination* (addition):
Take the preliminary estimates $\tilde{\tau}_1, \ldots, \tilde{\tau}_k$ of the $k$ jump discontinuities $\tau_1, \tau_2, \cdots, \tau_k$ to be those values of $\xi_{\tilde{i},j}$ for which $||D(\xi_{\tilde{i},j}, h_{1,\tilde{i}})| - |D(\xi_{0,j}, h_{1,0})||$ is one of the $k$ largest.

For refining these preliminary estimators, we then fit via least-squares piecewise constant functions within the intervals $[\tilde{\tau}_j - h_2, \tilde{\tau}_j + h_2]$ for $j = 1, \ldots, k$. We then obtain the final estimators $\hat{\tau}_1, \ldots, \hat{\tau}_k$ of $\tau_1, \ldots, \tau_k$.

Also here data-driven choices of the bandwidth parameters involved have been proposed. See Gijbels and Goderniaux (2002a).

When the number of discontinuity points is unknown it can be estimated simply, by for example using a cross-validation method. For a fixed value of $k$ ($k = 0, 1, 2, 3, \cdots$) locate the $k$ discontinuity points, and estimate the regression curve $g$ by using a standard smoothing method with a good data-driven bandwidth selector on the segments, as indicated in (6) in the case of one jump point. Denote this estimate by $\hat{g}_k(\cdot)$. Consider the cross-validation sum of squares:

$$\mathrm{CV}(k) = \sum_{i=0}^{n} (Y_i - g_k^{-i}(X_i))^2,$$

where $g_k^{-i}(X_i)$ is the fit obtained at $X_i$, assuming $k$ discontinuity points and excluding the data point $(X_i, Y_i)$ when constructing the fit. This approach has been proposed by

Müller and Stadtmüller (1999). The number of discontinuities $k$ is estimated by

$$\widehat{k} = \arg\min_{k\in\{0,1,\ldots\}} \text{CV}(k) ,$$

and the estimated curve is then the one associated with this number of discontinuities.

## 4. ESTIMATION OF DISCONTINUOUS REGRESSION CURVES

### 4.1. Indirect methods

As explained in the introduction, once estimates for the locations of the jump discontinuity point are obtained, say $\widehat{\tau}_1, \cdots, \widehat{\tau}_k$, the interval of estimation is divided into $k+1$ intervals $[0, \widehat{\tau}_1[, [\widehat{\tau}_1, \widehat{\tau}_2[, \cdots, [\widehat{\tau}_k, 1]$ and on each segment the regression function is continuous and estimated using standard smoothing techniques (such as for example a local linear smoother with cross-validation bandwidth).

An alternative indirect method has been proposed by Kang, Koo and Park (2000) in the fixed design regression case. They suggest to estimate the location $\tau$ and $\gamma$ via (3) and (4) respectively, to obtain suitably adjusted data

$$Z_i = Y_i - \widehat{\gamma}\mathbf{1}_{[\widehat{\tau},1]}(x_i) , \quad i = 1, \cdots, n$$

and then to use standard kernel regression methods to estimate the function $g_0$ in (2).

### 4.2. Direct methods

Several estimation methods have been proposed for estimating directly the unsmooth regression curve, without estimating first the locations and sizes of the jumps separately.

McDonald and Owen (1986) suggest to obtain for any given point three smoothed estimates of the regression function at that point, based on data to the right, to the left and on both sides of the given point. They used least squares fits to obtain, locally, the three linear fits. These three local linear fits are then used to construct a 'split linear fit' as a weighted average of these three estimates with weights determined by the goodness-of-fit values of the estimates.

Hall and Titterington (1992) proposed an alternative to the smoothed linear fitting method of McDonald and Owen (1986). Let $m \in \{0, 1, 2, \cdots\}$ be a window size. At all design points $x_i = i/n$ three smooths, the central smooth, the right and the left smooth, are computed using a linear conbination of the $(2m+1)$st observations in the neighbourhood of the fixed design point. The central smooth is based on the observation at the design point itself and on those at its $m$ nearest neighbours to the left and to the right. The left (right) smooth is based on the observation at the design point itself and on those at the $2m$ nearest design points to the left (right) of the given design point. More precisely, the central, left and right smooths at the design point $x_i = i/n$ are defined as

$$\widehat{g}_c(x_i) = \sum_{j=-m}^{m} c_j Y_{i+j} \qquad \widehat{g}_l(x_i) = \sum_{j=-2m}^{0} r_{-j} Y_{i+j} \qquad \widehat{g}_r(x_i) = \sum_{j=0}^{2m} r_j Y_{i+j} ,$$

where the constants $c_j$ and $r_j$ are chosen such that the leading terms in Taylor expansions of the expected values of $\hat{g}_c(x_i)$ and $\hat{g}_r(x_i)$ coincide. The method of Hall and Titterington (1992) is in fact an indirect method, since it searches for discontinuity points and then constructs the estimate for the unsmooth curve. To find the discontinuity points the central, left and right smooths are compared with one another. If there is no discontinuity and the regression function is locally linear then the three smooths should be very similar. At discontinuity points the three smooths behave quite different, and various diagnostics can be used to detect discontinuity points based on comparisons of the three smooths. These diagnostics involve the choice of a threshold value to determine whether a point is a discontinuity point or not. Values of a threshold are determined by referring to a linear regression function and Gaussian errors.

Recently, Qui (2003) proposed a jump-preserving curve fitting procedure based on local piecewise-linear kernel estimation. For each $x$ in the interval $[h_n/2, 1-h_n/2]$, with $h_n > 0$ a bandwidth, obtain a local linear fit to the left and to the right, using respectively a kernel function $K_+$, defined on $[-1/2, 0)$ and $K_-$, defined on $[0, 1/2]$. These fits estimate the left and right-hand limits of the function $g$ at the given point $x$. Clearly, if $x$ is a point of continuity of $g$, then both estimates are very close. If $x$ is not a jump point but a jump point is near to it, then one of the two estimates will be a good estimate of $g$. The idea is now to choose for each $x \in [h_n/2, 1-h_n/2]$ between one of these estimates, according to the residual sum of squares of the two fits. Denote the left-hand and right-hand estimates by $\hat{g}_+(x)$ and $\hat{g}_-(x)$ respectively. Further let $\mathrm{RSS}_+(x)$ and $\mathrm{RSS}_-(x)$ denote the residual sum of squares of these fits. Then the proposed estimate for $g$ at the given point $x$ is

$$\hat{g}(x) = \begin{cases} \hat{g}_-(x) & \text{if } \mathrm{RSS}_+(x) > \mathrm{RSS}_-(x) \\ \frac{1}{2}(\hat{g}_+(x) + \hat{g}_-(x)) & \text{if } \mathrm{RSS}_+(x) = \mathrm{RSS}_-(x) \\ \hat{g}_+(x) & \text{if } \mathrm{RSS}_+(x) < \mathrm{RSS}_-(x) \,. \end{cases}$$

An adaptive estimation procedure that simultaneously adapts to the smoothness of the curve and is sensitive to discontinuities of the curve has been proposed by Spokoiny (1998). The basic idea behind the method is to try to find by data the maximal interval such that $g$ within that interval can be well approximated by a polynomial. The fits are evaluated via least-squares analysis of the residuals. The intervals are allowed to be asymmetric. Intuitively: when $x$ is a point near a jump discontinuity $\tau$, the automatic procedure will come up with an asymmetric window excluding the near jump discontinuity.

Other direct estimation methods involve M-smoothers. Chu, Glad, Godtliebsen and Marron (1998) estimate $g$ in (1) by minimizing

$$S(a; x_i) = (-1) \sum_{j=1}^{n} K_h(x_i - x_j) L_g(Y_j - a) \,,$$

with respect to $a$, and choosing among the local minima the one that is closest to $Y_i$. Here $K_h$ and $L_g$ are rescaled versions of kernels $K$ and $L$. For example $K_h(\cdot) = h^{-1}K(\cdot/h)$. This results into a so-called local constant M-smoother.

An improvement of this is a local linear M-smoother defined as the local minimizer of

$$S(a,b;x_i) = (-1)\sum_{j=1}^{n} K_h(x_i - x_j)L_g(Y_j - a - b(x_i - x_j)),$$

with respect to $a$ and $b$, and choosing among the local minima $(\widehat{a},\widehat{b})$ the one for which $\widehat{a}$ is closest to $Y_i$. This estimator is introduced and studied by Rue, Chu, Godtliebsen and Marron (2002). Both the constant and linear M-smoothers have been designed for signals showing abrupt changes.

## 5. TESTING PROBLEMS AND CONFIDENCE INTERVALS AND BANDS

Several testing problems have been dealt with in the literature. Denote by $x_0$ a point in the interval $]0,1[$, and by $\kappa = \sum_{j=1}^{k} \gamma_j^2$ the sum of squared jump sizes of all jump points. Among the testing problems of interest are:

$$\begin{cases} H_0 : g \text{ is continuous in the point } x_0 \\ H_1 : g \text{ has a jump discontinuity at the point } x_0 \end{cases} \quad (11)$$

$$\begin{cases} H_0 : g \text{ is continuous on the interval } ]0,1[ \\ H_1 : g \text{ is discontinuous in at least one point on the interval } ]0,1[ \end{cases} \quad (12)$$

$$\begin{cases} H_0 : \kappa = 0 \\ H_1 : \kappa > 0 \end{cases} \quad (13)$$

In testing problem (11), called a local testing problem, one focuses on testing whether the regression function $g$ has a jump discontinuity at a certain fixed (pre-specified) point $x_0$. In contrast, testing problem (12), called a global testing problem, aims at finding out whether the regression function is continuous or not. Similarly for problem (13).

Grégoire and Hamrouni (2002b) studied tests for testing problems (11) and (12) based on asymptotic theory for the estimates developed in Grégoire and Hamrouni (2002a). The tests involve the choice of a bandwidth parameter. Gijbels and Goderniaux (2002b) proposed data-driven bootstrap based tests for the same problems, as well as for similar testing problems regarding the $\nu$th derivative of a regression function. Horváth and Kokoszka (2002) also study problem (11) but in general for the $\nu$th derivative function $g^{(\nu)}$ in the case of fixed regression.

Müller and Stadtmüller (1999) and Dubowik and Stadtmüller (2000) consider testing problem (13) and propose tests based on asymptotic theory. An important practical issue here is also the choice of a span-parameter.

Apart from testing it is also interesting to derive confidence bands (and intervals) for curves with jumps. This has been dealt with in Gijbels, Hall and Kneip (2003) using the estimation procedure introduced by Gijbels, Hall and Kneip (1999) and relying on non-standard bootstrap procedures.

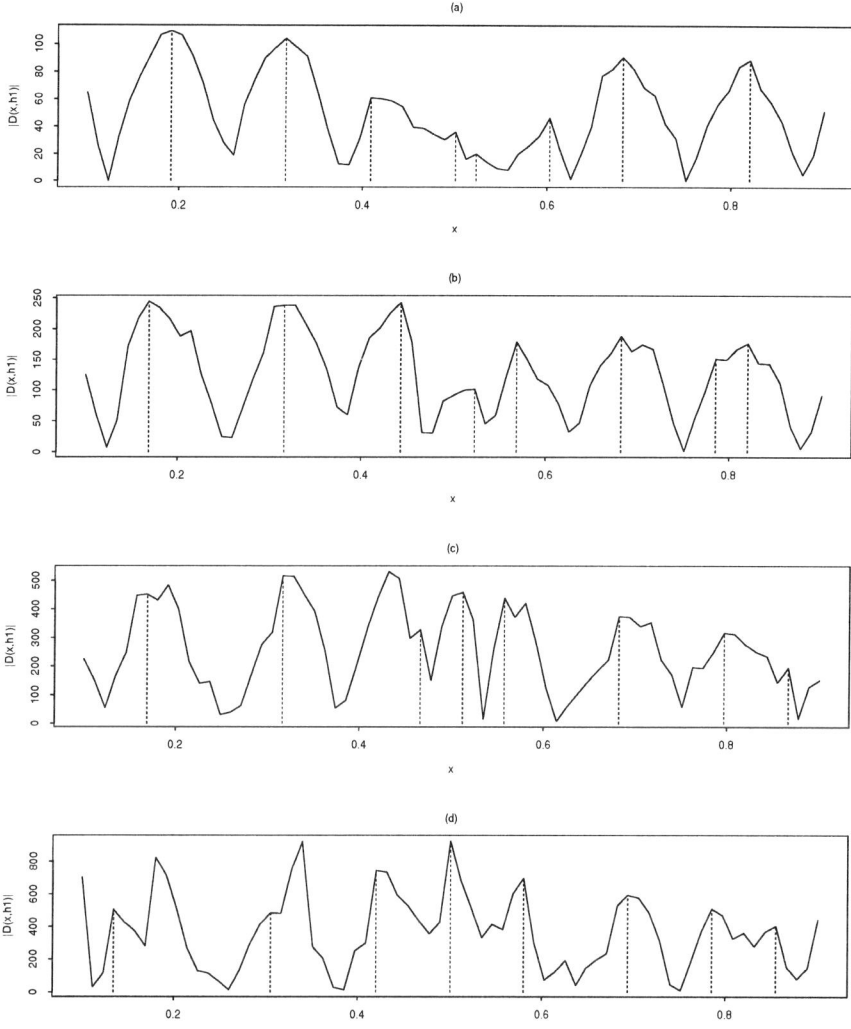

Figure 1. Performance of the diagnostic function for function (10), with $n = 100$ and $\sigma^2 = 0.1$ and using the three-step algorithm. Panels (a)-(d) illustrate plots of $|D(x, h_{1,i})|$ for $i = 0, 4, 8$ and $i = \tilde{i} = 11$ respectively.

## 6. DISCONTINUOUS BIVARIATE REGRESSION FUNCTIONS

Given space limitation we will only provide a few basic references for kernel-based methods for detecting jump curves in a regression surface, that generalize in some way the univariate methods discussed in Sections 2 and 3.

Detection of jump curves in regression surfaces by looking at differences of kernel estimates is exploited in Qiu (1997, 2002). Methods relying on partial derivatives of the regression surface can be found in Qiu (1998, 2000). A related approach using local least squares fitting of a plane is used in Qiu and Yandell (1996, 1997).

A generalization to the two-dimensional case of the estimation method of Gijbels, Hall and Kneip (1999) has been worked out by Goderniaux (2001), including automatic selection of the bandwidth parameters involved using bootstrap methods.

# References

Antoniadis, A. and Gijbels, I. (2002). Detecting abrupt changes by wavelet methods. *Journal of Nonparametric Statistics*, **14**, 7–29.

Beunen, J.A. (1998). Implementation of a method for detection and location of discontinuities. *Thesis* for Master of Scientific Studies (Statistics). University of Newcastle, Australia.

Bunt, M., Koch, I. and Pope, A. (1998). Counting discontinuities in nonparametric regression. *Research Report 98-2*, University of Newcastle, Australia.

Canny, J. (1986). A computational approach to edge detection. *IEEE Trans. Pattern Analysis and Machine Intelligence*, **8**, 679–698.

Chu, C.-K., Glad, I.K., Godtliebsen, F. and Marron, J.S. (1998). Edge preserving smoothers for image processing (with discussion). *Journal of the American Statistical Association*, **93**, 526–556.

Cline, D.B.H., Eubank, R.L. and Speckman, P.L. (1995). Nonparametric estimation of regression curves with discontinuous derivatives. *Journal of Statistical Research*, **29**, 17–30.

Dubowik, Ch. and Stadtmüller, U. (2000). Detecting jumps in nonparametric regression. In *Asymptotics in Statistics and Probability*, M.L. Puri (editors), pp. 171–184.

Eubank, R.L. and Speckman, P.L. (1994). Nonparametric estimation of functions with jump discontinuities. In *Change-point Problems*, IMS–Lecture Notes, Eds. E. Carlstein, H.-G. Müller and D. Siegmund, **23**, 130–143.

Fan, J. and Gijbels, I. (1996). *Local Polynomial Modelling and its Applications*. Chapman and Hall, New York.

Gijbels, I. and Goderniaux, A.-C. (2002a). Bandwidth selection for change point estimation in nonparametric regression. Institut de Statistique, Université catholique de Louvain, *Discussion Paper # 0024. Revised version.*

Gijbels, I, and Goderniaux, A.-C. (2002b). Bootstrap test for detecting change points in nonparametric regression. Institut de Statistique, Université catholique de Louvain, *Discussion Paper # 0144.* http://www.stat.ucl.ac.be

Gijbels, I, and Goderniaux, A.-C. (2002c). Discontinuity detection in derivatives of a regression function. Institut de Statistique, Université catholique de Louvain, *Discussion Paper # 0130.* http://www.stat.ucl.ac.be

Gijbels, I., Hall, P. and Kneip, A. (1999). On the estimation of jump points in smooth curves. *The Annals of the Institute of Statistical Mathematics,* **51,** 231–251.

Gijbels, I., Hall, P. and Kneip, A. (2003). Interval and band estimation for curves with jumps. *Journal of Applied Probability,* Special Issue in honour of Chris Heyde, to appear.

Girard, D. (1990). Détection de discontinuités dans un signal par inf-convolution spline et validation croisée. *Technical Report RR 702-I-M,* University of Grenoble, France.

Goderniaux, A.-C. (2001). Automatic detection of change-points in nonparametric regression. *Doctoral dissertation.* Institute of Statistics, Catholic University of Louvain, Louvain-la-Neuve, Belgium.

Grégoire, G. and Hamrouni, Z. (2002a). Change-point estimation by local linear smoothing. *Journal of Multivariate Analysis,* **83,** 56–83.

Grégoire, G. and Hamrouni, Z. (2002b). Two nonparametric tests for change-point problem. *Journal of Nonparametric Statistics,* **14,** 87–112.

Hall, P. and Titterington, D.M. (1992). Edge preserving and peak-preserving smoothing. *Technometrics,* **34,** 429–440.

Hall, P. and Wehrly, T.E. (1991). A geometrical method for removing edge effects from kernel-type nonparametric regression estimators. *Journal of the American Statistical Association,* **86,** 665–672.

Horváth, L. and Kokoszka, P. (2002). Change-point detection with non-parametric regression. *Statistics,* **36,** 9–31.

Kang, K.-H., Koo, J.-Y. and Park, C.-W. (2000). Kernel estimation of discontinuous regression functions. *Statistics & Probability Letters,* **47,** 277–285.

Koch, I. and Pope, A. (1997). Asymptotic bias and variance of a kernel-based estimator for the location of a discontinuity. *Journal of Nonparametric Statistics,* **8,** 45–64.

Koo, J.-Y. (1997). Spline estimation of discontinuous regression function. *Journal of Computational and Graphical Statistics,* **6,** 266–284.

Korostelev, A.P. (1987). On minimax estimation of a discontinuous signal. *Theory of Probability and its Applications*, **32**, 727–730.

Korostelev, A.P. and Tsybakov, A.B. (1993). *Minimax Theory of Image Reconstruction.* Lecture Notes in Statistics, **82**. Springer-Verlag, New York.

Laurent, P.J. and Utreras, F. (1986). Optimal smoothing of noisy broken data with spline functions. *Journal of Approximation Theory and its Applications*, **2**, 71–94.

Lee, Thomas C.M. (2002). Automatic smoothing for discontinuous regression functions. *Statistica Sinica*, **12**, 823–842.

Loader, C.R. (1996). Change point estimation using nonparametric regression. *The Annals of Statistics*, **24**, 1667–1678.

Mallat, S. and Hwang, W.L. (1992). Singularity detection and processing with wavelets. *IEEE Trans. Information Theory*, **2**, 617–643.

McDonald, J.A. and Owen, A.B. (1986). Smoothing with split linear fits. *Technometrics*, **28**, 195–208.

Müller, H.-G. (1992). Change-points in nonparametric regression analysis. *The Annals of Statistics*, **20**, 737–761.

Müller, H.-G. and Song, K.-S. (1997). Two-stage change-point estimators in smooth regression models. *Statistics & Probability Letters*, **34**, 323–335.

Müller, H.-G. and Stadtmüller, U. (1999). Discontinuous versus smooth regression. *The Annals of Statistics*, **27**, 299–337.

Oudshoorn, C.G.M. (1998). Asymptotically minimax estimation of a function with jumps. *Bernoulli*, **4**, 15–33.

Potier, C. and Vercken, C. (1994). Spline fitting numerous noisy data with discontinuities. In *Curves and Surfaces*, Eds Laurent, P.J. et al., 477–480, Academic Press.

Qiu, P. (1994). Estimation of the number of jumps of the jump regression functions. *Communications in Statistics, Theory and Methods*, **23**, 2141–2155.

Qiu, P. (1997). Nonparametric estimation of jump surface. *Sankhyā, Series A*, **59**, 268–294.

Qiu, P. (1998). Discontinuous regression surfaces fitting. *The Annals of Statistics*, **26**, 2218–2245.

Qiu, P. (2000). Some recent developments on edge detection and image reconstruction based on local smoothing and nonparametric regression. In *Recent Research Developments in Pattern Recognition*, **1**, 41–49.

Qiu, P. (2002). A nonparametric procedure to detect jumps in regression surfaces. *Journal of Computational and Graphical Statistics*, **11**, 799–822.

Qiu, P. (2003). A jump-preserving curve fitting procedure based on local piecewise-linear kernel estimation. *Journal of Nonparametric Statistics*, to appear.

Qiu, P. and Yandell, B. (1996). Discontinuity detection in regression surfaces. In *ASA Proc. Statist. Comput. Sect.*, 85–90.

Qiu, P. and Yandell, B. (1997). Jump detection in regression surfaces. *Journal of Computational and Graphical Statistics*, **6**, 332–354.

Qiu, P. and Yandell, B. (1998a). A local polynomial jump-detection algorithm in nonparametric regression. *Technometrics*, **40**, 141–152.

Qiu, P. and Yandell, B. (1998b). Discontinuous regression surfaces fitting. *The Annals of Statistics*, **26**, 2218–2245.

Raimondo, M. (1998). Minimax estimation of sharp change points. *The Annals of Statistics*, **26**, 1379–1397.

Rue, H., Chu, C.-K., Godtliebsen, F. and Marron, J.S. (2002). M-smoother with local linear fit. *Journal of Nonparametric Statistics*, **14**, 155–168.

Shiau, Jyh-Jen Horng (1988). Efficient algorithms for smoothing spline estimation of functions with or without discontinuities. In *Comput. Sci. Statist.*: Proc. 20th Symp. Interface, Wegman, Edward J., ed., 260–265.

Speckman, P.L. (1995). Fitting curves with features: semiparametric change-point methods. *Computing Science and Statistics*, **26**, 257–264.

Spokoiny, V.G. (1998). Estimation of a function with discontinuities via local polynomial fit with an adaptive window choice. *The Annals of Statistics*, **26**, 1356–1378.

Wand, M.P. and Jones, M.C. (1995). *Kernel smoothing*. Chapman and Hall, London.

Wang, Y. (1995). Jump and sharp cusp detection by wavelets. *Biometrika*, **82**, 385–397.

Wang, Y. (1998). Change curve estimation via wavelets. *Journal of the American Statistical Association*, **93**, 163–172.

Wang, Y. (1999). Change-point analysis via wavelets for indirect data. *Statitica Sinica*, **9**, 103–118.

Wu, J.S. and Chu, C.K. (1993a). Kernel type estimation of jump points and values of regression function. *The Annals of Statistics*, **21**, 1545–1566.

Wu, J.S. and Chu, C.K. (1993b). Modification for boundary effects and jump points in nonparametric regression. *Nonparametric Statistics*, **2**, 341–354.

Wu, J.S. and Chu, C.K. (1993c). Nonparametric function estimation and bandwidth selection for discontinuous regression functions. *Statistica Sinica*, **3**, 557–576.

Yin, Y.Q. (1988). Detection of the number, locations and magnitudes of jumps. *Communications in Statistics, Stochastic Models*, **4**, 445–455.

# Nonparametric smoothing methods for a class of non-standard curve estimation problems

O. Linton [a]* and E. Mammen [b]†

[a]Department of Economics, London School of Economics,
Houghton Street, London WC2A 2AE, United Kingdom.
E-mail address: lintono@lse.ac.uk

[b]Institut für Angewandte Mathematik, Ruprecht–Karls-Universität Heidelberg,
Im Neuenheimer Feld 294, 69120 Heidelberg, Germany.
E-mail address: enno@statlab.uni-heidelberg.de

We will introduce a class of models where an unknown function $m$ is defined as solution of an integral equation. Intercept and kernel of the integral equation are unknown but they can be directly estimated by application of classical smoothing methods. Estimates of $m$ will be given as solutions of the empirical integral equation. We will outline how an asymptotic distribution theory can be developed for the estimate of $m$. A series of examples will be given that motivate our class of models.

## 1. CLASS OF MODELS

We discuss a class of nonparametric/semiparametric models where the quantity of interest is a function $m(.)$ that is only implicitly defined but is known to satisfy a linear integral equation of the second kind in the space $L_2(p)$

$$m(x) = m^*(x) + \int \mathcal{H}(x,y) m(y) p(y) dy, \tag{1}$$

where the function $m^*(x)$ and the operator $\mathcal{H}(x,y)$ are defined explicitly in terms of the distribution of some observable quantities. We write this equation in short hand

$$m = m^* + \mathcal{H}m. \tag{2}$$

This sort of structure arises in many statistical problems and we shall give some examples below. A key question is whether there exists a solution to (1) and whether that solution is continuous in some sense. Even if the answer to these questions is affirmative, the implicit definition of $m$ leads to some challenging problems in statistical inference when noisy observations $\widehat{m}^*(x)$ and $\widehat{\mathcal{H}}(x,y)$ are available on $m^*(x)$ and $\mathcal{H}(x,y)$. We shall discuss general estimation strategies for this class of problems.

---

*Research was supported by the National Science Foundation, the Economic and Social Science Research Council of the United Kingdom, and the Danish Social Science Research Council.
†Research was supported by the the Deutsche Forschungsgemeinschaft, Project MA1026/8-1. and the Danish Social Science Research Council.

## 2. EXAMPLES

### 2.1. Example 1. Additive Regression
Suppose that we have

$$Y^i = m_0 + m_1(X_1^i) + \ldots + m_d(X_d^i) + \varepsilon^i,$$

with $(X^i, Y^i)$ stationary (e.g. $X_j^i = Y^{i-j}$) and $E[\varepsilon^i|X^i] = 0$ a.s.. It is convenient to normalize the functions $m_j$ by $E[m_j(X_j^i)] = 0$ in which case $E[Y^i] = m_0$. This additive model has been extensively studied by inter alia [1]. We have:

$$\begin{aligned} E\left[Y^i|X_j^i = x_j\right] &= m_0 + m_j(x_j) + \sum_{r \neq j} E\left[m_r(X_r^i)|X_j^i = x_j\right] \\ &= m_0 + m_j(x_j) + \sum_{r \neq j} \int m_r(x_r) \frac{p_{j,r}(x_j, x_r)}{p_j(x_j)} dx_r \end{aligned}$$

with $p_{j,r}$ ($p_j$) density of $(X_j^i, X_r^i)$ (or $X_j^i$, resp.). Write this as

$$\begin{pmatrix} m_1^* - m_0 \\ \vdots \\ m_d^* - m_0 \end{pmatrix} + \begin{pmatrix} \mathcal{H}_{11} & \cdots & \mathcal{H}_{1d} \\ \vdots & \ddots & \vdots \\ \mathcal{H}_{d1} & \cdots & \mathcal{H}_{dd} \end{pmatrix} \begin{pmatrix} m_1 \\ \vdots \\ m_d \end{pmatrix} = \begin{pmatrix} m_1 \\ \vdots \\ m_d \end{pmatrix}, \qquad (3)$$

where

$$m_j^* = E\left[Y^i|X_j^i = x_j\right],$$

$$m_0 = E\left[Y^i\right],$$

$$\mathcal{H}_{jr}(x_j, x_r) = \begin{cases} -\frac{p_{j,r}(x_j,x_r)}{p_j(x_j)} & \text{if } j \neq r \\ 0 & \text{else.} \end{cases}$$

With $m^* = (m_1^* - m_0, \ldots, m_d^* - m_0)^T$ this can be written in the form

$$m^* + \mathcal{H}m = m. \qquad (4)$$

Note that (4) is the first order condition of the least squares minimization problem

$$\min_{m_0, m_1(\cdot), \ldots, m_d(\cdot)} E\left[\{Y - m_0 - m_1(X_1) - \ldots - m_d(X_d)\}^2\right]$$

whose solution is the projection of $Y$ [or $E[Y|X]$] onto the space of additive functions. For an estimation approach that makes use of the representation of additive regression functions as solutions of integral equations see [2].

## 2.2. Example 2. Correlated Errors

Suppose that

$$Z_t = m(X_t) + u_t,$$

where $\{(X_t, u_t)\}$ is a stationary stochastic process with the process $\{u_t\}$ independent of the process $\{X_t\}$ and in particular

$$u_t = \rho u_{t-1} + \varepsilon_t,$$

where $\varepsilon_t$ is iid with mean zero and finite variance. Write $Y_t = Z_t - \rho Z_{t-1}$ then

$$Y_t = m(X_t) - \rho m(X_{t-1}) + \varepsilon_t.$$

This transformed model now has uncorrelated errors, but has an additive structure. Claim that

$$m = m^* + \mathcal{H}m,$$

where

$$m^*(x) = \frac{\int y \left( p_{Y_0, X_0}(y, x) - \rho p_{Y_0, X_{-1}}(y, x) \right) dy}{p_{X_0}(x)(1 - \rho^2)}$$

$$\mathcal{H}(u, v) = \frac{\rho}{(1 - \rho^2)} \frac{\left( p_{X_0, X_{-1}}(u, v) + p_{X_0, X_{-1}}(v, u) \right)}{p_{X_0}(u)}.$$

In this case it is more reasonable to suppose that the parameter $\rho$ is unknown along with the regression function $m$, and so both $m^*$ and $\mathcal{H}$ are parameter dependent quantities. We take this issue up further in our next example. Discussions of regression models with correlated errors can be found in [3,4].

## 2.3. Example 3. Semiparametric GARCH Model

We shall suppose that the observed process $\{y_t\}_{t=-\infty}^{\infty}$ is a (weakly and strongly) stationary martingale difference sequence, i.e., $E(y_t | \mathcal{F}_{t-1}) = 0$, where $\mathcal{F}_{t-1}$ is the sigma field generated by the entire past history of the $y$ process, and satisfies

$$E(y_t^2 | \mathcal{F}_{t-1}) \equiv \sigma_t^2 = \theta \sigma_{t-1}^2 + m(Y_{t-1}) = \sum_{j=1}^{\infty} \theta^{j-1} m(y_{t-j}). \qquad (5)$$

The parameters $\theta_0$ and the function $m(.)$ are unknown and to be estimated. Here one has $m^* + \mathcal{H}m = m$, where

$$m^*(y) = (1 - \theta^2) \sum_{j=1}^{\infty} \theta^{j-1} \, E\left[ Y_0^2 | Y_{-j} \right],$$

$$\mathcal{H}_\theta(y, x) = -\sum_{j=\pm 1}^{\pm \infty} \theta^{|j|} \frac{p_{0,j}(y, x)}{p(y)},$$

where $p_{0,j}$ ($p$) density of $(Y_0, Y_j)$ ($Y_0$, respectively.) This model is of interest in financial applications, and is a natural generalization of the GARCH(1,1) model of [5]. The generalization allows for flexibility in the 'news impact curve', i.e., the function $m$, which is interpreted as the channel through which news affects volatility in financial markets. For a discussion of this model see [6]. For a related model see also [7].

We can interpret the integral equation as the first order condition (with respect to $m$) from the following population criterion function

$$E\left[\{y_t^2 - \sum_{j=1}^{\infty} \theta^{j-1} m(y_{t-j})\}^2\right],$$

which is well defined when $E(y_t^4) < \infty$ for any $\theta \in \Theta$ and $m \in \mathcal{M}$, where $\Theta$ is a compact subset of $(-1, 1)$, $\mathcal{M} = \{m: \text{measurable}\}$ are parameter sets.

### 2.4. Example 4. Panels of Time Series

Let $Y^{it}$ ($i = 1, ..., n; t = 1, ..., T$) be stochastic variables representing observations on a set of $n$ individuals over $T$ time periods. A nonparametric autoregressive model for such data is given by

$$Y^{it} = m(Y^{i,t-1}) + \theta_t + \lambda_i + \varepsilon^{it}.$$

Here $\theta_t$ and $\lambda_i$ are nuisance parameters representing temporary or individual effects. Again, the function $m$ can be characterised as solution of an integral equation. For a discussion of this model and more general settings see [8].

### 2.5. Example 5. Estimation of Yield Curves

Consider the problem of extracting the yield curve from a sample of government coupon bonds. Coupon bonds generate several payments at future dates, and in an efficient bond market, the present value of these future payments should, apart from a small error, be equal to the trading price, $p_i$. Let the payments to the owner of bond $i$ at time $\tau_{ij}$ be $b_i(\tau_{ij})$, where for bond $i$, $\tau_{i1} < \ldots < \tau_{im_i}$ are the possible payment dates. Note that the time to maturity $\tau_{im_i}$ varies across bonds. The statistical model we adopt is

$$p_i = \sum_{j=1}^{m_i} b_i(\tau_{ij}) d(\tau_{ij}) + \varepsilon_i, \quad i = 1, \ldots, n, \tag{6}$$

where $\varepsilon_i$ is a random sequence satisfying $E[\varepsilon_i] = 0$, $i = 1, \ldots, n$. The problem is to extract the unknown, but smooth, discount function $d(.)$ from information $\{p_i, m_i, \tau_{i1}, \ldots, \tau_{im_i}, b_i(\tau_{i1}), \ldots, b_i(\tau_{im_i}), i = 1, \ldots, n\}$ on a sample of bonds. The issue is that $d$ is only implicitly defined and is not representable as a conditional expectation except in the special case that $m_i = 1$ for all $i$. Again, it can be shown that $d$ can be characterised as the solution of an integral equation, for details see [9].

### 2.6. Example 6: Nonparametric Simultaneous Equations

Suppose that we have a nonparametric model

$$Y = m(Z) + \varepsilon,$$

but that the error term satisfies

$$E(\varepsilon|Z) \neq 0.$$

This model can arise when there are mismeasured covariates or in a model of supply and demand in economics, where say $Y$ is price and $Z$ is quantity in a given market. In that case we can expect the price and quantity to be related through the same unobservables. Now suppose however that there are some instruments $X$ that satisfy

$$E(\varepsilon|X) = 0.$$

In the above example, these are variables that affect either supply or demand but not both. We have

$$m^*(x) = E(Y|X = x) = E[m(Z)|X = x] = \int m(y) \frac{f_{Z,X}(y,x)}{f_X(x)} dy,$$

i.e.,

$$m^* + \mathcal{H}m = 0, \quad \text{where } \mathcal{H}(x,y) = -\frac{f_{Z,X}(y,x)}{f_X(x)}.$$

This model has been studied by inter alia [10–12]. This is a Type I linear integral equation, and its solution requires a different approach to those we study.

### 2.7. Generalized Additive Models

Suppose that

$$Y^i = F\{m_0 + m_1(X_1^i) + \ldots + m_d(X_d^i)\} + \varepsilon^i,$$

where $F$ is some known transformation, e.g., the normal c.d.f. This model does not directly fit into our framework. The function $m$ can be characterised as solution of an integral equation but where now the kernel of the integral operator depends on $m$. Linear approximations of this integral equation can be used to be back into our framework, see [13]. For another approach see also [14].

## 3. SOME KEY PROPERTIES

The key properties are the nature of the operator or familiy of operators $\mathcal{H}$ that we define the integral equations. We will make use of the following condition.

ASSUMPTION A1. The operator $\mathcal{H}(x,y)$ is Hilbert-Schmidt

$$\int\int \mathcal{H}(x,y)^2 p(x) p(y) dx dy < \infty.$$

Under Assumption A1, $\mathcal{H}$ is a self-adjoint bounded linear operator on the Hilbert space of functions $L_2(p)$. Also this condition implies that $\mathcal{H}$ is a compact operator and therefore has a countable number of eigenvalues:

$$\infty > |\lambda_1| \geq |\lambda_2| \geq \ldots,$$

with

$$\sum_{j=1}^{\infty} \lambda_j^2 < \infty. \qquad (7)$$

This condition is satisfied in most cases under quite weak conditions.
Another key condition is that for a constant $0 < \gamma < 1$

$$\lambda_j < \gamma \qquad (8)$$

for $j \geq 1$. This condition requires some special arguments. If (8) is true we get that $I - \mathcal{H}$ has eigenvalues bounded from below by $1 - \gamma > 0$. Therefore $I - \mathcal{H}$ is invertible and $(I - \mathcal{H})^{-1}$ has only positive eigenvalues that are bounded by $(1 - \gamma)^{-1}$. So we can directly solve the integral equation and write

$$m = (I - \mathcal{H})^{-1} m^*.$$

If also

$$|\lambda_1| < 1, \text{ then } m = \sum_{j=0}^{\infty} \mathcal{H}^j m^*.$$

In this case, the sequence of successive approximations

$$m^n = m^* + \mathcal{H} m^{n-1}, n = 1, 2, \ldots$$

converges to the truth from any starting point.

There are some situations where $|\lambda_1| \geq 1$, and so the conditions that guarantee convergence of the successive approximations method are not satisfied. In that case, one has to transform the integral equation in order to obtain an equation which is more regular. Define

$$\nu = \min\{j : |\lambda_j| < 1\},$$
$$\pi_\nu = L_2 \text{ projection onto span}(e_1, \ldots, e_{\nu-1}),$$

where $e_j$ is the eigenfunction corresponding to $\lambda_j$. Then

$$m = m^* + \mathcal{H}(I - \pi_\nu)m + \mathcal{H}\pi_\nu m,$$

which is equivalent to

$$m = m_\pi^* + \mathcal{H}_\pi m, \qquad (9)$$

where $m_\pi^* = (I - \mathcal{H}\pi_\nu)^{-1} m^*$ and $\mathcal{H}_\pi = (I - \mathcal{H}\pi_\nu)^{-1} \mathcal{H}(I - \pi_\nu)$. It is easy to check that $\|\mathcal{H}_\pi\| < 1$, and so the method of successive approximations for example can be applied to the transformed equation. Here for an operator $A$ the norm $\|A\|$ denotes the absolute largest eigenvalue of the operator $A$.

## 4. ESTIMATION OF $m$

In all five models the curve $m$ can be estimated by using kernel smoothers of $m^*$ and $\mathcal{H}$ and by solving the empirical integral equation

$$\widehat{m}^* + \widehat{\mathcal{H}}\widehat{m} = \widehat{m}, \tag{10}$$

where $\widehat{m}^*, \widehat{\mathcal{H}}$ are estimates of $m^*, \mathcal{H}$. In the examples we have seen, $m^*$ involves some conditional expectations, while $\mathcal{H}$ depends on bivariate and univariate density functions. Therefore, we require nonparametric estimation of these quantities.

For the numerical solution of (10) we propose two approaches. In the first approach one uses discrete approximations of the integral equation $\widehat{m} = \widehat{m}^* + \widehat{\mathcal{H}}\widehat{m}$ by large system of linear equations and then one uses some appropriate numerical method to solve this system. Another approach can be based on successive approximations described in the last section. In this iterative procedure one starts with an initial value $\widehat{m}^{(0)}$ for $\widehat{m}$ and uses the following iteration steps:

$$\widehat{m}^{NEW} = \widehat{m}^* + \widehat{\mathcal{H}}\widehat{m}^{OLD}.$$

We call this approach also smooth backfitting. It will be further discussed below for additive models. Convergence of this algorithm requires $\|\widehat{\mathcal{H}}\| < 1$. If this condition does not hold the modified successive approximation method can be used that has been described at the end of the last section.

## 5. ASYMPTOTIC THEORY

There are some results on the asymptotic properties of the estimated functions: pointwise normal distribution; uniform stochastic expansions.

The asymptotic analysis starts by noting the following basic fact. The stochastic part of $\widehat{m}$ is asymptotically equivalent to stochastic part of

$$\widehat{m}^* - (\widehat{\mathcal{H}} - \mathcal{H})m.$$

We now give a heuristic motivation for this fact. By taking the differences of the left and right hand sides of (10) and (2) we get

$$\widehat{m} - m = \widehat{m}^* - m^* + \mathcal{H}(\widehat{m} - m) + (\widehat{\mathcal{H}} - \mathcal{H})m.$$

We will now argue that the stochastic part of $\mathcal{H}(\widehat{m} - m)$ is of lower order. This implies that the stochastic part of $\widehat{m} - m$ is asymptotically equivalent to the stochastic part of $\widehat{m}^* - m^* + (\widehat{\mathcal{H}} - \mathcal{H})m$. This immediately shows our claim.

So it remains to check that the stochastic part of $\mathcal{H}\widehat{m}$ is of lower order. The basic argument for this is the following. If $\widehat{g}$ is (the stochastic part of) a local smoother (i.e. a local weighted average $\widehat{g}(x) = \sum_{i=1}^{n} w_i(x)\varepsilon_i$ with weights fullfilling $\sum_{i=1}^{n} w_i(x) = 1$ and $w_i(x) = 0$ for $|X_i - x|$ larger a bandwidth $h \to 0$), then $\mathcal{H}\widehat{g}$ is a global weighted average and in particular, it holds that $\mathcal{H}\widehat{g}$ is of order $O_P(n^{-1/2})$. This can be easily checked because $\mathcal{H}\widehat{g}(x) = \sum_{i=1}^{n} [\int \mathcal{H}(x, y) w_i(y) \, dy] \varepsilon_i$ and the weights $\int \mathcal{H}(x, y) w_i(y) \, dy$ are of order $n^{-1}$ under weak assumptions.

Now our argument would go through if $\widehat{m}$ would be a local smoother. Unfortunately this does not hold. But it can be shown that in first order it is equivalent to a local smoother. This suffices for finishing our argumentation but this step requires some refined argumets. We donot want to discuss this here but refer to e.g. [2] for a detailed discussion.

The fact that the stochastic part of $\widehat{m}$ is asymptotically equivalent to stochastic part of $\widehat{m}^* - (\widehat{\mathcal{H}} - \mathcal{H})m$, simplifies the asymptotic discussion essentially. Both terms $\widehat{m}^*$ and $(\widehat{\mathcal{H}} - \mathcal{H})m$ can be treated analytically by standard asymptotic smoothing theory. Both are local weighted averages and asymptotic normality results can be easily derived. So immediately we get an asymptotic normality result for $\widehat{m}$ together with an expression for the asymptotic variance. It remains to calculate the asymptotic bias. This requires different methods for the mentioned examples and will not be discussed here.

In some cases, the stochastic part of $(\widehat{\mathcal{H}} - \mathcal{H})m = 0$ is of lower order and so the stochastic part of $\widehat{m}$ is the same as the stochastic part of $\widehat{m}^*$. An example is the additive model which we will discuss again in the next section. In other cases, the operator part enters.

## 6. ADDITIVE REGRESSION, REVISITED

We now give more details in the additive model:

$$Y^i = m_0 + m_1(X_1^i) + \ldots + m_d(X_d^i) + \varepsilon^i.$$

The empirical version of (3) is equivalent to

$$\widehat{m}_j(x_j) = \widehat{m}_j^*(x_j) - \widehat{m}_0 - \sum_{r \neq j} \int \widehat{m}_r(x_r) \frac{\widehat{p}_{j,r}(x_j, x_r)}{\widehat{p}_j(x_j)} dx_r. \tag{11}$$

Let $\widehat{m}_j^*(x_j)$ be the Nadaraya-Watson estimate of $E[Y^i|X_j^i = x_j]$, $\widehat{m}_0 = \bar{Y}$ and $\widehat{p}_{j,r}$ ($\widehat{p}_j$) be kernel density estimates of density of $(X_j^i, X_r^i)$ (or of $X_j^i$, resp.). We propose the following algorithmic calculation of $\widehat{m}$.

$$\widehat{m}_j^{NEW}(x_j) = \widehat{m}_j^*(x_j) - \widehat{m}_0 - \sum_{r \neq j} \int \widehat{m}_r^{OLD}(x_r) \frac{\widehat{p}_{j,r}(x_j, x_r)}{\widehat{p}_j(x_j)} dx_r.$$

We call this algorithm smooth backfitting. Smooth backfitting is a modification of the classical backfitting algorithm. In the classical backfitting one uses the iteration

$$\begin{aligned}\widehat{m}_j^{NEW}(x_j) &= \frac{\frac{1}{n}\sum_{i=1}^n K_{h_j}(X_j^i - x_j)\left[Y^i - \widehat{m}_0 - \sum_{r \neq j} \widehat{m}_r^{OLD}(X_r^i)\right]}{\widehat{p}_j(x_j)} \\ &= \widehat{m}_j^*(x_j) - \widehat{m}_0 - \sum_{r \neq j} \frac{\frac{1}{n}\sum_{i=1}^n K_{h_j}(X_j^i - x_j)\widehat{m}_r^{OLD}(X_r^i)}{\widehat{p}_j(x_j)},\end{aligned}$$

where $K_h(u) = h^{-1}K(h^{-1}u)$ is the localized version of a kernel function $K$. Thus, in the smooth backfitting we replace $\frac{1}{n}\sum_{i=1}^n K_{h_j}(X_j^i - x_j)\widehat{m}_r^{OLD}(X_r^i)$ by

$$\int \widehat{m}_r^{OLD}(x_r)\widehat{p}_{j,r}(x_j, x_r)dx_r = \int \frac{1}{n}\sum_{i=1}^n K_{h_r}(X_r^i - x_r)K_{h_j}(X_j^i - x_j)\widehat{m}_r^{OLD}(x_r)\, dx_r.$$

For $h_r \to 0$ for $r \neq j$ the latter term converges to the first one. Thus backfitting differs from smooth backfitting by putting the bandwidths to zero for all components that are not updated. This is the reason why we call our approach *smooth* backfitting.

Smooth backfitting allows an easier asymptotic mathematical analysis. In contrast to backfitting the estimate is explicitly defined by a minimisation criterion (see also below) whereas backfitting is only defined as limit of an iterative procedure. In an appropriate function space smooth backfitting can also be interpreted as an orthogonal projection of the data on the class of additive functions, see [15]. Simulations suggest that smooth backfitting works stable under weaker assumptions on the design and for quite larger number of additive components, see [16].

For additive models the algorithm of successive approximations, i.e. smooth backfitting converges at the geometric rate. This holds because

$$\|\widehat{\mathcal{H}}\| + o_P(1) = \|\mathcal{H}\| < 1$$

with $\|\ldots\|$ operator norm (absolutely largest eigenvalue). The norm $\|\mathcal{H}\|$ is smaller than 1 because $\mathcal{H}$ consists in iterative projections onto linear spaces that are not orthogonal.

We now introduce a version of smooth backfitting that uses local linear smoothing instead of local constant smoothing. It can be easily checked that for local constant smoothing the backfitting estimate $\widehat{m}$ minimizes the following smooth least squares criterion: minimize

$$\int \sum_{i=1}^{n} \left\{ Y^i - m_0 - m_1(x_1) - \ldots - m_d(x_d) \right\}^2 \prod_{j=1}^{d} K_h(X_j^i - x_j) \, dx$$

over real values $m_0$ and over all functions $m_1, \ldots, m_d$ with $\frac{1}{n} \sum_{i=1}^{n} m_k(X_k^i) = 0$. For $d = 1$ this gives the usual Nadaraya-Watson regression smoother. This minimisation criterion can be easily generalized to local linear smoothing.

Define

$$\left( \widehat{c}, \widehat{m}_1^{LL}, \ldots, \widehat{m}_d^{LL}, \widehat{m}_1^{LL,SL}, \ldots, \widehat{m}_d^{LL,SL} \right)$$

as minimizer of

$$\int \sum_{i=1}^{n} \Big\{ Y^i - c - m_1(x_1) - \ldots - m_d(x_d)$$
$$- m_1^{SL}(x_1)[X_1^i - x_1] - \ldots - m_d^{SL}(x_d)[X_d^i - x_d] \Big\}^2 \prod_{j=1}^{d} K_h(X_j^i - x_j) \, dx,$$

where the minimization runs over all real values $c$ and over all functions $m_1, \ldots, m_d$, $m_1^{SL}, \ldots, m_d^{SL}$ with $\frac{1}{n} \sum_{i=1}^{n} m_k(X_k^i) = 0$. The functions $m_1, \ldots, m_d$ are now the smooth backfitting estimates of the additive components. The functions $m_1^{SL}, \ldots, m_d^{SL}$ are estimates of their derivatives. Note that for $d = 1$ this gives the classical local linear estimate. Then with appropriate definitions of

$$\widehat{m}^{*,LL} = \left( \widehat{m}_1^{*,LL}, \ldots, \widehat{m}_d^{*,LL}, \widehat{m}_1^{*,LL,SL}, \ldots, \widehat{m}_d^{*,LL,SL} \right)^T$$

and $\widehat{\mathcal{H}}^{LL}$ one gets for
$$\widehat{m}^{LL} = \left(\widehat{m}_1^{LL}, \ldots, \widehat{m}_d^{LL}, \widehat{m}_1^{LL,SL}, \ldots, \widehat{m}_d^{LL,SL}\right)^T$$
the integral equation
$$\widehat{m}^{*,LL} + \widehat{\mathcal{H}}^{LL}\widehat{m}^{LL} = \widehat{m}^{LL}.$$
Again one can show that
$$\|\widehat{\mathcal{H}}^{LL}\| + o_P(1) = \|\mathcal{H}^{LL}\| < 1.$$

Thus $\widehat{m}^{LL}$ can be again calculated by the iterative algorithm of smooth backfitting.

A closed asymptotic theory is available both for local constant and for local linear smoothing, see [2]. Both smooth backfitting estimates have the same asymptotic variance as the estimate of the intercept $\widehat{m}^*$. So we are in a much more simpler situation than in the general case where also the stochastic part of $(\widehat{\mathcal{H}} - \mathcal{H})m$ enters in the asymptotics, see the discussion in the last section. We will now give a heuristic explanation why here this term is asymptotically negligible. We will do this for the local constant version of smooth backfitting. For this we make use of the following decomposition of the intercept of the empirical integral equation:

$$\begin{aligned}\widehat{m}_j^*(x_j) - \widehat{m}_0 &= \frac{\frac{1}{n}\sum_{i=1}^n K_{h_j}(X_j^i - x_j)Y^i}{\widehat{p}_j(x_j)} - \bar{Y} \\ &= \widehat{m}_j^{*,A}(x_j) + \widehat{m}_j^{*,B}(x_j) + \widehat{m}_j^{*,C}(x_j),\end{aligned}$$

where

$$\widehat{m}_j^{*,A}(x_j) = [m_j(x_j) - \bar{m}_j] + \sum_{r \neq j}\int \frac{\widehat{p}_{j,r}(x_j,x_r)}{\widehat{p}_j(x_j)}[m_r(x_r) - \bar{m}_r]\ dx_r,$$

$$\widehat{m}_j^{*,B}(x_j) = \frac{\frac{1}{n}\sum_{i=1}^n K_{h_j}(X_j^i - x_j)[m_j(X_j^i) - m_j(x_j)]}{\widehat{p}_j(x_j)}$$
$$+ \sum_{r \neq j}\int \frac{\frac{1}{n}\sum_{i=1}^n K_{h_j}(X_j^i - x_j)K_{h_r}(X_r^i - x_r)[m_r(X_r^i) - m_r(x_r)]}{\widehat{p}_j(x_j)}\ dx_r,$$

$$\widehat{m}_j^{*,C}(x_j) = \frac{\frac{1}{n}\sum_{i=1}^n K_{h_j}(X_j^i - x_j)\varepsilon^i}{\widehat{p}_j(x_j)} - \bar{\varepsilon}$$

with $\bar{m}_r = n^{-1}\sum_{i=1}^n m_r(X_r^i)$ for $r = 1,\ldots,d$ and $\bar{\varepsilon} = \sum_{i=1}^n \varepsilon^i$. By inverting our integral equation we have
$$\widehat{m} = (I - \widehat{\mathcal{H}})^{-1}\widehat{m}^* = \widehat{m}^A + \widehat{m}^B + \widehat{m}^C$$
with $\widehat{m}^j(x) = (I - \widehat{\mathcal{H}})^{-1}\widehat{m}^{*,j}(x)$ and $\widehat{m}^{*,j}(x) = [\widehat{m}_1^{*,j}(x_1),..,\widehat{m}_d^{*,j}(x_d)]^T$ for $j = A, B, C$. Recall that $\widehat{m}^*(x) = [\widehat{m}_1^*(x_1) - \widehat{m}_0, .., \widehat{m}_d^*(x_d) - \widehat{m}_0]^T$. We now discuss the components $\widehat{m}^A$, $\widehat{m}^B$ and $\widehat{m}^C$.

We start by a discussion of the last two components. Let us suppose that all bandwidths are chosen of order $n^{-1/5}$. The term $\widehat{m}_j^{*,C}(x_j)$ is a stochastic mean zero variable that is of order $O_P(n^{-2/5})$. The same holds for $\widehat{m}^C$. It can be also shown that

$$\widehat{m}^C(x) = (I - \widehat{\mathcal{H}})^{-1}\widehat{m}^{*,C}(x) = (I - \mathcal{H})^{-1}\widehat{m}^{*,C}(x) + o_P[n^{-2/5}] = \widehat{m}^{*,C}(x) + o_P[n^{-2/5}].$$

For the proof of the last equality one makes use of the fact that $\mathcal{H}^s\widehat{m}^{*,C}(x)$ is a global smoother for $s \geq 1$ and for this reason of lower order, see the discussion in the last section.

The component $\widehat{m}_j^{*,B}(x_j)$ can be expanded leading to a deterministic quantity of order $O(n^{-2/5})$ and lower order terms. This expansion can be used to get an expansion for $\widehat{m}^B$ again with leading deterministic term of order $O(n^{-2/5})$. The first order term consists in bias terms.

We now consider the first component $\widehat{m}^A$. We will argue that

$$\widehat{m}^A = [m_1(x_1) - \bar{m}_1, .., m_d(x_d) - \bar{m}_d]^T, \tag{12}$$

i.e. up to a (random) shift all functions $\widehat{m}_r^A$ are equal to the true functions $m_r$, $r = 1, ..., d$. Because of our norming assumption $Em_r(X_r^i) = 0$ we have

$$\bar{m}_r = Em_r(X_r^i) + o_P(n^{-2/5}) = o_P(n^{-2/5})$$

under weak assumptions on the design variables $X^i$. This gives $\widehat{m}^A = m + o_P(n^{-2/5})$. For the proof of (12) it suffices to show that

$$m = \widehat{m}^{*,A} + \widehat{\mathcal{H}}m,$$

i.e. one has to show that $m$ is a solution of the empirical integral equation with intercept $\widehat{m}^{*,A}$. This equation can be easily checked.

We now summarize our heuristic discussion of $\widehat{m}$. We have shown the following asymptotic decomposition of $\widehat{m}$. It consists of $m$ (see the discussion of $\widehat{m}^A$) + bias terms of order $O(n^{-2/5})$ (see the discussion of $\widehat{m}^B$) + a stochastic term that is asymptotically equivalent to the stochastic part of the intercept $\widehat{m}^*$ (see the discussion of $\widehat{m}^C$). In particular, we have shown our claim that the stochastic part of $\widehat{m}$ is asymptotically equivalent to the stochastic part of the intercept $\widehat{m}^*$. Remember that $\widehat{m}^*$ is a classical Nadaraya-Watson estimate in a standard nonparametric regression model with only one component. And the asymptotic equivalence does hold no matter how many additive components the additive model includes. In particular this implies that the asymptotic variance does not depend on the number of additive components and is the same as in the classical regression model where only $d = 1$ component is present.

With respect to the bias local linear smoothing outperforms local constant smoothing. The asymptotic bias of the local linear estimate does not depend on number or shape of other components. Heuristically, this may be explained by the form of the bias for local linear estimation of a classical full dimensional unconstrained estimate that does not make use of the additivity. For such an unconstrained estimate the bias is an additive function. The additive components are the bias terms of one-dimensional local linear estimates. Now, in the smooth backfitting the bias function is projected onto the space of additive functions. Because the bias function is already additive smooth backfitting lets the bias function unchanged. In particular, we get the same asymptotic bias for the additive components in the additive model as for a classical one-dimensional local linear estimate. For local constant estimates, i.e. Nadaraya-Watson estimates this does not hold because a nonadditive term depending on the design density enters in the bias expression. Asymptotic theory for smooth backfitting works under much weaker conditions on the design density than for backfitting. This may be reflected by the above mentioned simulations where smooth backfitting outperforms backfitting.

For issues of practical implementations of smooth backfitting see [16,17]. For a two step procedure that uses smooth backfitting as a pilot estimator see [18]. There it has been shown that the asymptotic performance of each type of a smoother for $d = 1$ can be achieved by this two step procedure for $d > 1$. As for local linear smooth backfitting this can be interpreted as achieving an oracle inequality.

## 7. OUTLOOK

In this paper we have discussed a large class of models where one or more nonparametric components enter linearly into the model. We have used characterisations of the nonparametric functions by integral equations to develop estimation strategies. It is now challenging to carry over this approach to more general models where the nonparametric components enter in a more complicated nonlinear form:

$$Y_i = G(m_1, \ldots, m_d, \theta, X^i) + S(g_1, \ldots, g_p, \beta, Z^i)\varepsilon_i.$$

These models have been called generalised structured models by [13]. One promising estimation strategie is to linearize the model and apply smooth backfitting in the linearised model.

## REFERENCES

1. T. J. Hastie and R. J. Tibshirani (1990). Generalized additive models. Chapman and Hall, London.
2. E. Mammen, O. Linton and J. Nielsen. The existence and asymptotic properties of a backfitting projection algorithm under weak conditions. Ann. Statist. 27 (1999), 1443 - 1490.
3. Z. Xiao, O. Linton, R. J. Carroll and E. Mammen. More efficient kernel estimation in nonparametric regression with autocorrelated errors. preprint (2003).
4. R. J. Carroll, X. Lin, O. B. Linton and E. Mammen. Accounting for correlation in marginal longitudinal nonparametric regression. *Second Seattle Symposium on Biostatistics*, editor D. Lin, to appear (2003).
5. T. Bollerslev. Generalized autoregressive conditional heteroscedasticity. J. Economet. 31 (1986), 307-327.
6. O. Linton and E. Mammen. Estimating semiparametric ARCH($\infty$) models by kernel smoothing methods. preprint (2003).
7. R. J. Carroll and W. Härdle and E. Mammen. Estimation in an additive model when the parameters are linked parametrically. Econometric Theory 18 (2002), 886-912.
8. E. Mammen and D. Tjøstheim. Nonparametric additive models for panels of time series. preprint (2003).
9. O. Linton, E. Mammen, J. Nielsen and C. Tanggaard. Estimating yield curves by kernel smoothing methods. Journal of Econometrics 105/1 (2001), 185-223.
10. W. K. Newey and J. L. Powell. Instrumental variables estimation for nonparametric regression models. (1989, revised 2002). Forthcoming in Econometrica.
11. Darolles, S., Florens. J.P., and E. Renault (2000). Nonparametric instrumental regression. Preprint, Toulouse.

12. Florens, J.P., J. Heckman, C. Meghir, and E. Vytlacil (2001). Instrumental variables, local instrumental variables, and control functions. Preprint, Toulouse.
13. E. Mammen and J. P. Nielsen. Generalised structured models. Biometrika (2003), to appear.
14. J. Horowitz and E. Mammen. Nonparametric estimation of an additive model with a link function. preprint (2002).
15. E. Mammen, J. S. Marron, B. A. Turlach and M. P. Wand. A general framework for constrained smoothing. Statistical Science 16 (2001), 232-248
16. J.P. Nielsen and S. Sperlich. Smooth backfitting in practice. preprint (2003).
17. E. Mammen and B. Park. Bandwidth choice for smooth backfitting in additive models. preprint (2003).
18. J. Horowitz, J. Klemelä and E. Mammen. Optimal estimation in additive regression models. preprint (2002).

# Weighted Local Linear Approach to Censored Nonparametric Regression

Zongwu Cai[a]*

[a]Department of Mathematics, University of North Carolina, Charlotte, NC 28223, USA,
E-mail: zcai@uncc.edu

This article proposes a unified and easily implemented nonparametric regression method for estimating the regression function for censored data under both iid and time series contexts. The basic idea of the method is to use a constructed weighted local likelihood by combining the benefits of the local polynomial fitting. The estimation procedure is implemented and a bandwidth selection criterion, based on the generalized cross-validation criterion, is proposed. Further, the finite sample operating characteristics of the proposed method is examined through simulations, and its usefulness is also illustrated on two real examples. Finally, the consistency and the asymptotic normality of the proposed estimator are established, which provide an insight into the large sample behavior of the proposed estimator. In particular, the explicit expression for the asymptotic variance of the resulting estimator is given and its consistent estimate is provided.

**Key Words:** Adaptive bandwidth; asymptotic theory; censored data; nonlinear time series; weighted local likelihood.

## 1. INTRODUCTION

Consider the nonparametric regression model

$$Y_i = m(\mathbf{X}_i) + \varepsilon_i, \tag{1}$$

where $Y$ is the survival time of an individual taking part in a clinical trial or other experimental study, $\mathbf{X}$ is the associated covariate vector in $\Re^d$ of covariates such as age, sex, blood pressure, cholesterol level, etc, $m(\cdot)$ is an unknown regression function, $E(\varepsilon_i \,|\, \mathbf{X}_i) = 0$ and $\text{var}(\varepsilon_i \,|\, \mathbf{X}_i) = \sigma^2(\mathbf{X}_i)$. Since $Y_i$ is not always available, standard methods which require knowledge of all $Y$'s are not directly applicable. Under the censored scheme, rather than the survival time $Y_i$, one observes $Z_i$ together with an indicator $\delta_i$ such that $Z_i = Y_i$ if and only if $\delta_i = 1$. Let $Y_i$ and $C_i$ be conditionally independent given $\mathbf{X}_i$, where $C_i$ is the censoring time associated with the survival time $Y_i$. The standard assumption in the literature is $Z_i = \min\{Y_i, C_i\}$ and $\delta_i = I(Y_i \leq C_i)$. The observations are $\{(\mathbf{X}_i, Z_i, \delta_i)\}_{i=1}^n$, which is a random sample from the population $(\mathbf{X}, Z, \delta)$.

---

*This work was supported, in part, by the National Science Foundation grant DMS 0072400 and funds provided by the University of North Carolina at Charlotte.

In regression analysis, in particular in survival analysis, (1) provides an adequate approximation to the true relationship between $Y$ and $\mathbf{X}$ after an appropriate transformation of the survival time $Y$ and it also might need a transformation on the covariates. With a known monotone transformation $g(Y)$, if $m(\cdot)$ is linear and $\sigma(\cdot)$ is a constant, (1) becomes the Cox's proportional hazards model when the distribution $F_\varepsilon(x) = 1 - \exp(-\exp(x))$ of $\varepsilon$ is the extreme value distribution and it becomes the proportional odds model ([1,19,20,31]) when $F_\varepsilon(x) = \exp(x)/\{1 + \exp(x)\}$ is the standard logistic distribution. Also, it reduces to the accelerated failure time (AFT) model ([11,13,26,32]) if $g(y) = \log(y)$. Further, if the errors $\{\varepsilon_t\}$ arise from a stationary process, (1) is the regression model with censored time series data; see, for example, [8,27,34] for the case that $m(\cdot)$ is linear.

In the absence of the right censoring, to estimate the nonparametric regression function $m(\cdot)$ for both independent and dependent data, some linear smoothers have been proposed in the literature, such as kernel, spline, local polynomial, orthogonal series methods, among others. For the available methods and results on both theory and applications, see the book by [10], among others. Among the aforementioned linear smoothers, the local polynomial method has become particularly popular in recent years due to its attractive mathematical efficiency, bias reduction and adaptation of edge effects; see, for example, the book by [10] for the detailed methods and results.

For censoring data, when $m(\cdot)$ is a linear function of unknown parameters, [17] applied a least squares method to estimate the parameters with the error distribution estimated by the Kaplan-Meier estimator. [3] initiated the study of a modified least squares estimator of the parameters by transforming the data in an unbiased way to account for censoring. However, their transformation required estimation of the error distribution via Kaplan-Meier and it led to an iterative scheme. To overcome this drawback, [15] considered a transformation depending only on the censoring distribution. [35] proposed a more general class of transformations of this type. In each of the aforementioned papers, inferences were made under the assumption that $Y_t$s are iid. [8,27,34] considered the situation that $\{Y_t\}$ are not independent. Recent developments include those of [16,26,29,32] based on linear rank tests and [30,33] based on median regression approach. Recently, [24] proposed a weighted least-squares method to estimate the unknown parameters in the nonlinear parametric models. Under the nonparametric setting, the nonparametric techniques such as kernel and local linear fitting were applied by [2,7,9,10,14,36]. The common approach for the above transformation techniques is to first transform the censored data and then to apply a variety of statistical methods to analyze the transformed data as if they were uncensored. As pointed out by [10], such a transformation increases the variability because it does not involve the distribution of the response variable. To remedy this problem, [10] studied systematically a more general class of transformations and proposed a data-driven method to transfer the censored data. However, as criticized by [14], the approach proposed by [10] inherited the drawback of the transformation.

This work is motivated from an analysis of the Stanford heart transplant data, benchmark status in the literature of survival analysis. The main interest is to examine the relationship between the survival time after transplant and the age at transplant. There have been several methods proposed to analyze this famous dataset, such as linear and quadratic Cox's regression models ([18,30]), and linear or nonlinear median and mean regression models ([10,14]). Among the existing results, the quadratic Cox's regression models overestimated the survival time for the middle ages and the methods by ([10,14]) overestimated the survival time for the younger ages. The proposed method here combines the advantages from the earlier analyses together. In another context, one is interested in the comparison of two standard therapies A and B for patients with small cell lung cancer. The data are taken from [33] and they have been analyzed by [14,30,33]. [33] and [30] claimed that A works always better than B. As pointed out by [14], the linear Cox models and the estimates of [33] completely missed the low-high feature in the data. The method proposed here gives a more precise description of the data structure and concludes that A works always better than B only for older ages but younger ages. The detailed analyses of these two real examples are reported in Section 3.

In this article I propose a novel nonparametric regression method, a unified and easily implemented approach for estimating the regression function. The method constructs a weighted local likelihood based on the idea of weighted likelihood, which may be regarded a generalization of the weighted least-squares method initialized by [37] for linear models and extended by [24] for nonlinear parametric models. The basic idea is as follows. The weighted local likelihood is constructed by expressing the local log-likelihood for uncensored data via an integral over the empirical function and replacing the empirical function for uncensored data by its counterpart for censored data (Kaplan-Meier estimator). In such a way, no transformation is needed so that it avoids the selection of smoothing parameter twice. More importantly, the idea can be extended to a variety of models; see [6] for more details.

The article is organized as follows. Section 2 describes in detail the proposed procedure for least-squares situation for the iid censored data and the generalized cross-validation for the smoothing parameter selection is proposed. In the same section, the asymptotic behavior of the proposed estimator is formulated and a consistent estimate of the asymptotic variance is proposed, followed by a brief remark on a comparison of asymptotic results with those for uncensored situation. An extension to the censored time series data is included. Section 3 presents the simulation results, followed by an application of the methodology to two real examples. Section 4 provides some concluding remarks.

## 2. METHODOLOGY

### 2.1. Weighted Local Linear Estimate

Although I pay my attention to the univariate case, I would like to mention that the methodology proposed continues to hold for the situations of two or more than two covariates. The local linear modeling scheme is used here, although general local polynomial methods are also applicable. Assume that the regression function $m(\cdot)$ has a second continuous derivative, so that it can be approximated by a linear function at a given grid point $x_0$, $m(x) \approx \beta_1 + \beta_2(x - x_0)$, where $\beta_1$ and $\beta_2$ depend on $x_0$, and the local linear fitting technique can be applied to the estimation of the regression function $m(\cdot)$. Namely, by assuming that the available uncensored observations are $\{(X_i, Y_i)\}_{i=1}^n$, the local linear estimator of $m(\cdot)$ is $\widetilde{m}(x_0) = \hat{a}_1$, where $\hat{a}_1$ and $\hat{a}_2$ minimize the locally weighted least squares

$$\frac{1}{n} \sum_{i=1}^n \{Y_i - a_1 - a_2(X_i - x_0)\}^2 K_h(X_i - x_0) \qquad (2)$$

with $K_h(\cdot) = K(\cdot/h)/h$ and $h = h_n > 0$ the bandwidth. It is clear that (2) can be expressed as the following form

$$\int \{y - a_1 - a_2(x - x_0)\}^2 K_h(x - x_0) F_n^*(dx, dy), \tag{3}$$

where $F_n^*(x, y)$ is the empirical distribution of $\{(X_i, Y_i)\}_{i=1}^n$, a consistent estimator of the joint distribution function $F^*(x, y) = P(X \leq x, Y \leq y)$ of $(X, Y)$.

Now, based on the censored data $\{(X_i, Z_i, \delta_i)\}_{i=1}^n$, I derive the local linear estimator for $m(\cdot)$. To this end, let $W_{in} = \delta_{[i:n]} (n - i + 1)^{-1} \prod_{j=1}^{i-1} \{(n - j)/(n - j + 1)\}^{\delta_{[j:n]}}$, where $Z_{1:n} \leq \cdots \leq Z_{n:n}$ are the order statistics of $Z_1, \cdots, Z_n$ and $\delta_{[i:n]}$ is the $\delta$ associated with $Z_{i:n}$. Clearly, $W_{in}$ is the mass attached to $Z_{i:n}$ under the Kaplan-Meier estimator $\widehat{F}_n(y)$. Therefore, $\widehat{F}_n(y)$ can be expressed as a weighted empirical function of $\{Z_{i:n}\}$ with weights $\{W_{in}\}$, namely, $\widehat{F}_n(y) = \sum_{i=1}^n W_{in} I(Z_{i:n} \leq y)$ (see [25]). In the presence of covariates, [22] extended the Kaplan-Meier estimator to include the covariates and proposed an estimator to the joint distribution function $F^*(x, y)$, namely, $\widehat{F}_n^*(x, y) = \sum_{i=1}^n W_{in} I(X_{[i:n]} \leq x, Z_{i:n} \leq y)$, where $X_{[i:n]}$ is the $X$ associated with $Z_{i:n}$. It follows from Corollary 1.5 of [22] that $\widehat{F}_n^*(x, y)$ is consistent. Motivated by the Miller's ([17]) method, I replace $F_n^*(x, y)$ in (3) by $\widehat{F}_n^*(x, y)$ to obtain the following *weighted local least squares*

$$\int \{y - \beta_1 - \beta_2(x - x_0)\}^2 K_h(x - x_0) \widehat{F}_n^*(dx, dy)$$
$$= \sum_{i=1}^n W_{in}^* \{Z_i - \beta_1 - \beta_2(X_i - x_0)\}^2 K_h(X_i - x_0), \tag{4}$$

where $W_{in}^*$ is the $W_{in}$-value associated with the natural order of $Z_i$. Note that there are two weights in the last expression of (4). $W_{in}^*$ accounts for the censoring and $K(\cdot)$ controls the amount of smoothing used in estimation. In contrast to the data-driven method of transformation in [10], no smoothing is needed here to account for the censoring so that the bandwidth selection becomes much simpler. When there is no censoring at all, one may formally set $W_{in} = 1/n$ and $Z_i = Y_i$, then, (4) becomes to (2), which is the ordinary local least squares. By minimizing (4) with respect to $\beta_1$ and $\beta_2$, one obtains the *weighted local linear estimator* of $m(x_0)$, $\widehat{m}(x_0) = \widehat{\beta}_1$, where $\widehat{\beta}_1$ and $\widehat{\beta}_2$ minimize (4). It is clear that the weighted local linear estimator for $m(x_0)$ can be expressed as

$$\widehat{m}(x_0) = \frac{S_{n,2}(x_0) T_{n,0}(x_0) - S_{n,1}(x_0) T_{n,1}(x_0)}{S_{n,2}(x_0) S_{n,0}(x_0) - S_{n,1}^2(x_0)}, \tag{5}$$

where, for $0 \leq l \leq 2$,

$$S_{n,l}(x_0) = \sum_{i=1}^n W_{in}^* K_h(X_i - x_0)(X_i - x_0)^l = \sum_{i=1}^n W_{in} K_h(X_{[i:n]} - x_0)(X_{[i:n]} - x_0)^l, \tag{6}$$

and

$$T_{n,l}(x_0) = \sum_{i=1}^n W_{in}^* K_h(X_i - x_0)(X_i - x_0)^l Z_i = \sum_{i=1}^n W_{in} K_h(X_{[i:n]} - x_0)(X_{[i:n]} - x_0)^l Z_{i:n}. \tag{7}$$

By a comparison (5) - (7) with the non-censored case (see [10]), one only needs to compute the weights $\{W_{in}\}$ and then can apply any existing software such as S-Plus to compute estimates. Note that similar to (4), the weighted local least squares for the transformed model can be derived just by replacing $Z_{i:n}$ in (4) by $g(Z)_{i:n}$.

## 2.2. Bandwidth Selection

As is the case for all nonparametric curve estimation, the smoothing parameter or bandwidth plays an essential role in the trade-off between reducing bias and variance. Here, I apply the generalized cross-validation (GCV) proposed by [28] to choose the bandwidth $h$, described as follows. It follows from (5) that the fitted values can be expressed as $(\widehat{m}(X_1), \cdots, \widehat{m}(X_n))^\tau = \mathbf{H}(Z_1, \cdots, Z_n)^\tau$, where $\mathbf{H} = \mathbf{H}(h)$ is the $n \times n$ hat matrix, depending only on the $X$-covariate and the $\delta$-censoring indicator, and the superscript $\tau$ denotes the transpose. The GCV selects the optimal global bandwidth $h$ to minimize

$$\text{GCV}(h) = \left\{1 - \frac{\text{tr}(\mathbf{H})}{n}\right\}^2 \sum_{i=1}^{n} W_{in} \left\{Z_{i:n} - \widehat{m}(X_{[i:n]})\right\}^2. \qquad (8)$$

Here, $\text{tr}(\mathbf{H}) = K_h(0) \sum_{i=1}^{n} W_{in} S_{n,2}(X_{[i:n]}) \left\{S_{n,2}(X_{[i:n]}) S_{n,0}(X_{[i:n]}) - S_{n,1}^2(X_{[i:n]})\right\}^{-1}$. Note that the GCV might have a tendency of undersmoothing when $\text{tr}(\mathbf{H})/n$ is large, particularly for small sample size. To alleviate this shortcoming, [12] proposed using the corrected version of Akaike information criterion

$$\text{AICC}(h) = \log\left[\sum_{i=1}^{n} W_{in}\left\{Z_{i:n} - \widehat{m}(X_{[i:n]})\right\}^2\right] + \frac{2(\text{tr}(\mathbf{H}) + 1)}{(n-1) - \text{tr}(\mathbf{H}) - 2},$$

which intends to give a better finite sample performance and to overcome some deficiencies of GCV. However, it is clear that the GCV and AICC are equivalent to the first order when $\text{tr}(\mathbf{H})/n$ is small, which is typically the case in the nonparametric context for large sample size.

## 2.3. Asymptotic Results

In this section, I provide an insight into the large sample behavior of the proposed estimator. To this end, I introduce some notation. Let $H(\cdot)$ denote the distribution function of $Z$ and let $F(\cdot)$ and $G(\cdot)$ be the distribution functions of $Y$ and $C$, respectively. Define $\tau_H = \inf\{x, H(x) = 1\}$. Similarly, define $\tau_F$ and $\tau_G$. Then, $\tau_H = \min\{\tau_F, \tau_G\}$. Let $f_x(\cdot)$ denote the marginal density of $X$. Finally, set $\mu_2 = \int u^2 K(u)\,du$ and $\nu_0 = \int K^2(u)\,du$. I have the following theorem but the detailed proof is not presented here. It is available from the author upon request; or see the sketch proof in [6]. First, the conditions are listed.

**Conditions:**

A1. Assume that $\{(X_i, \delta_i, Z_i)\}$ are iid.
A2. Assume that $f_x(x)$ is continuous at $x = x_0$ and $f_x(x_0) > 0$.
A3. The second derivative of $m(x)$ exists and is continuous in a neighborhood of $x_0$.
A4. $P(Y \leq C \mid X, Y) = P(Y \leq C \mid Y)$.
A5. The kernel function $K(\cdot)$ is symmetric and has a compact support.
A6. For each $y$ and $k = 0$ and $1$, functions

$$a_k(x, y) = E\left[Y^k I(Y > y) \mid X = x\right] \text{ and } b_{2k}(x) = E\left[Y^{2k}\left\{1 - G(Y)\right\}^{-1} \mid X = x_0\right]$$

are continuous in a neighborhood of $x = x_0$.
A7. $E\left[|Y|^{2+\alpha}/(1 - G(Y))^{1+\alpha} \mid X = x_0\right]$ is continuous in a neighborhood of $x = x_0$ for some $\alpha > 0$.

**Remark 1.** Conditions imposed here are standard in the literature except A4. Note that condition A4 was also imposed in [22-24] for parametric models. For its detailed interpretation, see the aforementioned three papers. It may be interpreted as a Markov property of the vector $(X, Y, \delta)$, where covariate vector $X$ is the "initial state" being always known to the observer. The model is flexible enough to allow for a dependence structure between $X$ and $C$. In terms of the language in the survival literature, A4 says that given the time of death, the covariates do not provide any further information as to whether censoring will take place or not. However, it is conjectured that condition A4 may not be necessary but I would not be able to dispense with it.

**THEOREM 1.** *Assume that $H(\cdot)$ is continuous and $\tau_F = \tau_H$ and conditions A1 - A7 hold true, then,*

$$\sqrt{nh}\left[\widehat{m}(x_0) - m(x_0) - \frac{h^2}{2}\mu_2 \, m''(x_0) + o_p(h^2)\right] \longrightarrow N\{0, \Sigma(x_0)\} \quad \text{in distribution,}$$

*where $\Sigma(x_0) = \nu_0 f_x^{-1}(x_0) b_2(x_0)$ with $b_2(x_0) = E\left[Y^2\{1 - G(Y)\}^{-1} \mid X = x_0\right]$.*

**Remark 2.** First, the continuity assumption imposed in the theorem is just for convenience in the proof since under continuity, identification of the involved quantities is little simpler. Secondly, as a consequence of the theorem, $\widehat{m}(x_0)$ is a consistent estimator of $m(x_0)$, and its asymptotic mean squares error (AMSE) is provided by AMSE = $\frac{h^4}{4}\mu_2^2 \{m''(x_0)\}^2 + \frac{\nu_0 b_2(x_0)}{n h f_x(x_0)}$. Therefore, the optimal bandwidth, which minimizes the AMSE, is given by $h_{opt} = \left[\frac{\nu_0 b_2(x_0)}{n \mu_2^2 f_x(x_0) \{m''(x_0)\}^2}\right]^{1/5}$, so that the optimal convergent rate of the AMSE is of order $n^{-4/5}$. Further, in comparison of the results in the theorem with the results without censoring (see [10]), it is surprising to note that the asymptotic bias for the censored case is exactly same as the one for the uncensored context but the asymptotic variance for the censored case is larger than its counterpart for the uncensored situation. In other words, the asymptotic bias does not depend on the censoring scheme but the asymptotic variance does rely on the censoring structure. This is not surprising because the asymptotic bias comes from the linear approximation. Finally, when there is no censoring at all, one may formally set $G = \delta_\infty$, Dirac at infinity. In this case, the asymptotic variance $\Sigma(x_0)$ reduces to $\nu_0 f_x^{-1}(x_0) E[Y^2 \mid X = x_0]$, which is the asymptotic variance without censoring (see [10]).

### 2.4. Standard Errors

In practice, a simple and quick way to estimate the asymptotic variance given in the theorem is desirable. Now, I propose an easily implemented estimate of $\Sigma(x_0)$, described as follows. It can be showed that $\lim_{n \to \infty} \text{var}\{\xi_{n,i}(x_0)\} = \nu_0 f_x(x_0) b_2(x_0)$, where $\xi_{n,i}(x_0) = \sqrt{h}\, K_h(X_i - x_0)\{1 - G(Z_i)\}^{-1}\delta_i$, which involves the unknown censoring distribution $G(\cdot)$. Clearly, $\{\xi_{n,i}(x_0)\}_{i=1}^n$ are iid. If $G(\cdot)$ were known, a consistent estimate of $\text{var}\{\xi_{n,i}(x_0)\}$ would be the sample variance of $\{\xi_{n,i}(x_0)\}_{i=1}^n$. In practice, of course, it is unknown. First, I need to estimate $G(\cdot)$. In many applications, censoring is caused by a determination of a study. Then, it is reasonable to assume that the distribution of censoring variable does not depend on covariates. Therefore, the Kaplan-Meier product limit estimator can be applied to estimating $G(z)$, defined by $1 - \widehat{G}(z) = \prod_{j: Z_{j:n} \leq z} \{(n-j)/(n-j+1)\}^{1-\delta_{[j:n]}}$. Alternative methods are also available in the literature, such as the empirical approach of [23] and the Beran's method ([2]). This leads to the estimate of $\xi_{n,i}(x_0)$, given by $\widehat{\xi}_{n,i}(x_0) = \sqrt{h}\, K_h(X_i - x_0) \{1 - \widehat{G}(Z_i)\}^{-1} \delta_i$. It is easy to show that $S_{n,0}(x_0)$ is the consistent estimator of $f_x(x_0)$. Therefore, the consistent estimator of $\Sigma(x_0)$ is defined as $\widehat{\Sigma}(x_0) = \frac{1}{n} \sum_{i=1}^{n} \{\widehat{\xi}_{n,i}(x_0) - \overline{\widehat{\xi}}_n(x_0)\}^2 S_{n,0}^{-2}(x_0)$, where $\overline{\widehat{\xi}}_n(x_0) = n^{-1} \sum_{i=1}^{n} \widehat{\xi}_{n,i}(x_0)$. Note that this consistent estimate of the asymptotic variance is implemented in Examples 3 and 4 in Section 3.

### 2.5. Censored Time Series Data

In some applications, censored observations may arise naturally in time series if there is an upper or lower limit of detection, for example, when one is monitoring levels of an airborne contaminant or recording daily bioassays of hormone levels in a patient. For this aspect, see the articles by [21,34]. In contrast to the well-studied conventional regression methods for complete time series data, there remains open for the nonparametric regression methods for the censored time series data. Under some regularity conditions, [4,5] showed that the Kaplan-Meier estimator still enjoys the consistency and asymptotic normality for censored time series data. Therefore, similar to the conventional regression methods for complete time series data, the weighted local linear method defined in (4) can be applied to dealing with the nonparametric regression estimation for the censored time series data and the bandwidth selection criterion described in (8) is still applicable. The details are omitted since they are parallel to those for the iid case.

## 3. NUMERICAL STUDY

In this section, I illustrate the proposed estimation method through five examples. In the first three examples, I use simulated data for both iid and time series cases and three sample sizes ($n = 100, 200,$ and $400$) for each case to evaluate the performance of the weighted local linear estimator. The method is then applied to two real examples. In all the presented examples, the kernel $K(\cdot)$ is taken to be the standard normal density function and the bandwidth is automatically selected by the procedure outlined in Section 2.2.

### 3.1. Simulated Examples

**Example 1:** I use the model considered in [10] to simulate $n$ observations from the following model:

$$Y_i = m(X_i) + 0.25\,\varepsilon_i, \qquad X_i \stackrel{iid}{\sim} \text{uniform }[0,1], \quad \text{and} \quad \varepsilon_i \stackrel{iid}{\sim} N(0,1), \qquad (9)$$

where $m(x) = 4.5 - 64\,x^2\,(1-x)^2 - 16\,(x-0.5)^2$ and $\{X_i\}$ and $\{\varepsilon_i\}$ are independent. The censoring time $C_i$ is conditional independent of the survival time $Y_i$ given $X_i$ and is distributed as $(C_i \mid X_i = x) \sim \text{exponential }(c(x))$, where $c(x)$ is the mean conditional censoring time defined by $c(x) = 3\,(1.25 - |4\,x - 1|)$, if $0 \leq x \leq 0.5$, and otherwise, $c(x) = 3\,(1.25 - |4\,x - 3|)$. Figures 1(a) and 1(b) give the scatter-plots of unobserved ($Y$ vs $X$) and observed ($Z$ vs $X$) simulated 200 observations. In this example, the mean censoring rates are 37.9% (5.2%) for $n = 100$, 37.7% (3.5%) for $n = 200$, and 37.6% (2.4%) for $n = 400$, where the number in the parentheses stands for the standard deviation based on 500 replications.

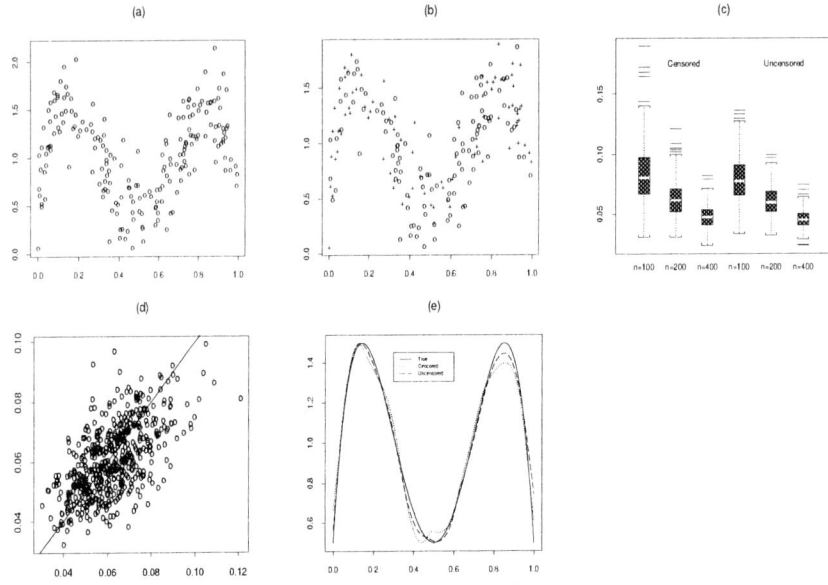

Figure 1. Simulated Data for the iid Case. (a)-(b) Uncensored observations are denoted by "o" and censored observations are indicated by "+". (c) Boxplots of 500 values of MADE of $\widehat{m}(\cdot)$ (the left three) and $\widetilde{m}(\cdot)$ (the right three). (d) Scatter plot of the $\mathcal{E}_2$ for $\widetilde{m}(\cdot)$ versus the $\mathcal{E}_1$ for $\widehat{m}(\cdot)$; the straight line marks the position where the two $\mathcal{E}$'s are equal. (e) $m(\cdot)$ (solid line), $\widehat{m}(\cdot)$ (dotted line), and $\widetilde{m}(\cdot)$ (dashed line).

To evaluate the finite sample performance of the proposed estimators, the mean absolute deviation error (MADE) is defined by $\mathcal{E}_1 = \sum_{j=1}^{n_0} |\widehat{m}(x_j) - m(x_j)|/n_0$ and $\mathcal{E}_2 = \sum_{j=1}^{n_0} |\widetilde{m}(x_j) - m(x_j)|/n_0$, where $x_1, \cdots, x_{n_0}$ are the grid points on $[0, 1]$ with $n_0 = 101$. Figure 1(c) displays the boxplots of 500 values of $\mathcal{E}_1$ (the first three) and $\mathcal{E}_2$ (the last three) for $n = 100, 200$, and 400. It shows clearly that $\mathcal{E}$ decreases when the sample size $n$ increases and $\mathcal{E}_2$ is not much smaller than $\mathcal{E}_1$, which indicates that two estimators perform almost same, reflecting the effectiveness of the procedure. To gain further insight, Figure 1(d) plots the MADEs $\mathcal{E}_2$ versus $\mathcal{E}_1$, using the same sample data for $n = 200$. Clearly, there is about almost equal chance that each estimator beats the other although $\widetilde{m}(\cdot)$ performs slightly better. Figure 1(e) depicts the estimated curves $\widehat{m}(\cdot)$ (dotted line) and $\widetilde{m}(\cdot)$ (dashed line) of the regression function $m(\cdot)$ (solid line) based on a typical example, which is chosen such that the corresponding $\mathcal{E}_1$ is equal to its median of the 500 $\mathcal{E}_1$-values.

**Example 2:** To demonstrate that the proposed procedure works for the censored time series data, I simulate the censored time series data as follows. I use the model (9) but $\varepsilon_i$ is generated from AR(1) model with the correlation coefficient $\rho = 0.85$. Similar to the iid case, the simulation is repeated 500 times. Approximately 37% of the observations are censored for all three sample sizes $n = 100, 200,$ and 400. Figures 2(a) and 2(b) depict the scatter-plots of unobserved ($Y$ vs $X$) and observed ($Z$ vs $X$) simulated 200 observations, Figure 2(c) presents the boxplots of 500 values of $\mathcal{E}_1$ (the left three) and $\mathcal{E}_2$ (the right

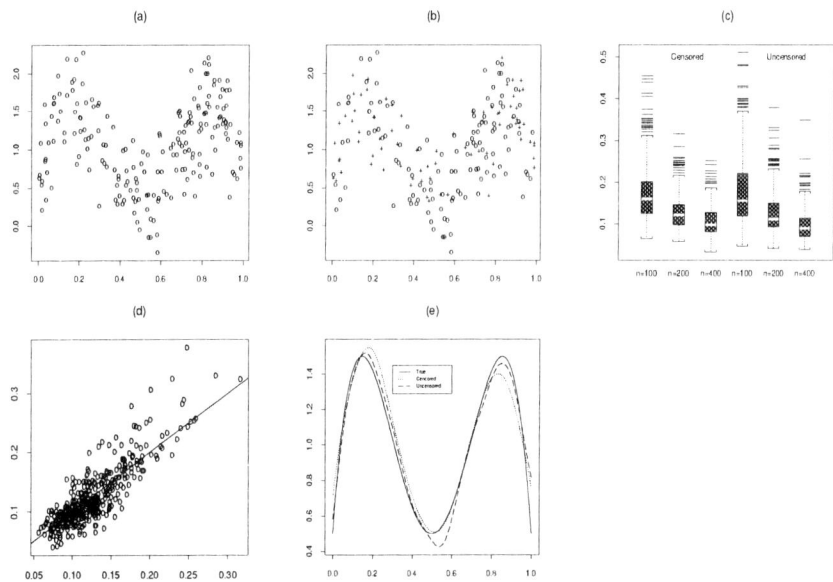

Figure 2. The caption is the same as that in Figure 1.

three) for $n = 100, 200,$ and 400, Figure 2(d) compares the weighted local linear estimator for censored data with the local linear estimator for uncensored data for $n = 200$, and Figure 2(e) summarizes the estimated curves $\widehat{m}(\cdot)$ (dotted line) and $\widetilde{m}(\cdot)$ (dashed line) of the regression function $m(\cdot)$ (solid line) based on a typical example. From those graphs, it concludes that the proposed estimator performs reasonably well for the censored time series data.

## 3.2. Real Examples

**Example 3:** Now, the proposed method is illustrated via the data from the Stanford heart transplant program. From October 1967 to February 1980, 184 of the 249 patients had received heart transplants. The focus here is on only two variables, with the survival time (days) as the response and age (years) at transplant as the covariate. Patients alive beyond February 1980 were censored. See [18] for more details about the dataset and some related work in the literature. The data are taken from [18] and Figure 3 gives the scatterplot of the $\log_{10}$ survival time versus age for 157 patients who had complete tissue typing. Among 152 patients with complete record and survival time exceeding 10 days, 55 were censored and the censoring proportion is 0.36. The purpose of this analysis is to use the proposed method to compare with the work of [9,10,18,14,30]. In various analyses of this dataset in the literature, the quadratic age models were considered by [18,30], and the nonparametric median and mean regression models were explored by [10,14,18,30]. The

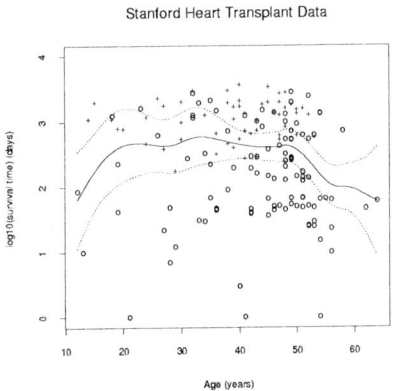

Figure 3. Stanford Heart Transplant data with log-survival time plotted against age for 157 patients with complete record and the estimated mean regression function (solid line) plus/minus twice estimated standard errors (dotted lines). The character "+" indicates the censored observations, and the uncensored observations are presented by "o".

proposed mean regression estimate (solid line) relating to survival time to age on the base 10 logarithmic scale is presented in Figure 3, together with plus/minus twice estimated standard errors (dotted lines), which give an idea of the pointwise confidence interval with bias ignored. The bandwidth $h = 4$ is used for the proposed estimate. The proposed estimate suggests that survival time increases linearly when age is less than 20, remains relatively constant with age between 20 and 48, and then decreases linearly from age 48 on. This reflects that in the middle ages, the log-survival time is nearly independent of age, but it increases at the extreme younger ages and it decreases at the extreme older ages.

The estimates for the extreme younger and older ages are more reasonable, which are on the same line, to a lesser extent, as those in [18,30] by using the quadratic models. Note that the flat behavior of the estimated curve for age between 20 and 48 is similar to that of [10,14]. Therefore, in comparison with previous studies, for example, [9,10,14,18,30], this analysis combines all advantages from the earlier studies and gives a more precise description of the data structure.

**Example 4:** Finally, I use the small cell lung cancer (SCLC) data study to illustrate the censored regression model. A combination of cisplatin (P) and etoposide (E) had been a standard therapy for patients with SCLC, of which the optimal sequencing and administration schedule needed to be established. The data are taken from Table 1 of [33] from a clinical study which was designed to evaluate two groups: Arm A (P followed by E) and Arm B (E followed by P). 121 patients were randomly designed to one of

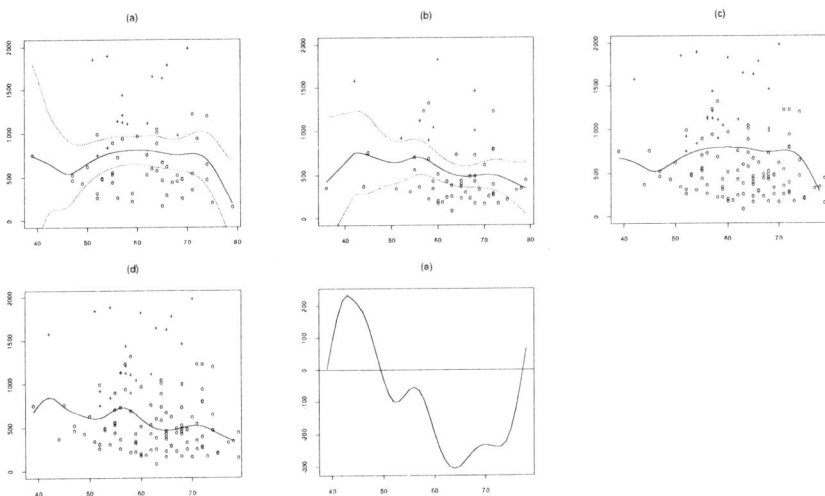

Figure 4. The character "+" indicates the censored observations and the uncensored observations are presented by "o". (a) Scatterplot for Arm A with $\widehat{m}(\cdot)$ (solid line) plus/minus twice estimated standard errors (dotted lines). (b) Scatterplot for Arm B with $\widehat{m}(\cdot)$ (solid line) plus/minus twice estimated standard errors (dotted lines). (c) Scatterplot for the whole data with $\widehat{a}_1(\cdot)$. (d) Scatterplot for total 121 patients with $\widehat{a}_1(\cdot) + \widehat{a}_2(\cdot)$. (e) Estimated coefficient function $\widehat{a}_2(\cdot)$.

these two groups: 62 patients to Arm A and 59 patients to Arm B. At the time of the analysis, there was no loss to follow-up. Each death time was either observed or censored. Therefore, the censoring variable does not depend on the covariate, the patient's entry age. Among 121 patients, 23 had their survival time censored. Thus, the censoring rate is 0.19. Figures 4(a) and 4(b) depict the scatter-plots of the survival time in days versus age for 62 patients for Arm A and 59 patients for Arm B, respectively, and total 121 patients for both Arm A and Arm B are plotted together in Figures 4(c) and 4(d). This dataset had been analyzed by [14,30,33] by using the median regression techniques. Note that [30,33] fitted only one model for the whole dataset and [14] fitted two models for Arm A and Arm B, respectively.

Here, I use the mean regression method to analyze this dataset and I first fit two separate models for Arm A and Arm B, respectively. The bandwidths $h = 3$ in Arm A and $h = 4$ in Arm B are used for the proposed estimates. The proposed estimates (solid line) of regression mean functions are displayed in Figure 4(a) for Arm A and Figure 4(b) for Arm B, respectively, together with plus/minus twice estimated standard errors. They suggest that for Arm A, the mean survival time decreases for ages less than 47, increase for ages from 47 to 55, keeps constant until 72, and decreases sharply from 72 on. For Arm B, the highest mean survival time occurred at age 42 and then it tends to decline with the exception of the ages from 51 to 62 since the second highest peak occurred at 56. This indicates that the relationship between survival time and age in Arm A differs from that in Arm B. Also, the proposed estimates shows that the mean survival time in Arm A is shorter than in Arm B for younger patients under age 50 but longer than in Arm B for patients over age 50. Therefore, one might conclude that Arm A works better than Arm B for older age patients (over 50). Because there are only few data points for the younger patients (6.6% of patients under age 50), it should be cautious to exercise the extrapolation when using the linearity assumption in this data.

Now, I use the functional-coefficient model $Y = a_1(X) + a_2(X)U + \varepsilon$ with $X$ being age and $U$ being group code ($U = 0$ for Arm A and $U = 1$ for Arm B) to fit the whole dataset. Figures 4(c) and 4(d) give the estimated coefficient functions $\hat{a}_1(\cdot)$ (corresponding to $U = 0$) and $\hat{a}_1(\cdot) + \hat{a}_2(\cdot)$ (corresponding to $U = 1$), respectively. By a comparison of Figures 4(a) with 4(c) and 4(b) with 4(c), it is clear that the estimated mean survival times based on two separate models and functional-coefficient model are almost same. Figure 4(e) depicts the estimated curve $\hat{a}_2(\cdot)$, measuring the difference of mean survival time between the two groups, which shows clearly that it changes over age. It is striking that $\hat{a}_2(\cdot)$ is positive when age is under 49 and it then becomes negative over 50. This indicates that Arm B performs better than Arm A for ages under 49 but Arm A does better than Arm B from age 50 on. This supports the above conclusion based on two separate models fitting. Such a finding appears to be new for this dataset.

In comparison with the results from the median regression methods, it is noted that the estimates by [30,33] completely missed the low-high feature in the data. For this aspect, to a lesser extent, the proposed estimates are close to those by [14,30,33] concluded that Arm A works always better than Arm B. However, the proposed estimates along with those in [14] confirm the assertion by [30,33] for age over 50, but not for age under 50. Whether Arm A works better than Arm B for younger patients remains to be confirmed in the future investigations.

## 4. CONCLUDING

The motivation of this research is to develop a unified and easily implemented method for estimating the mean regression function for both iid and time series data in the presence of the right censoring. This can be done by using the proposed weighted local least squares approach, in contrast to the parametric or semi-parametric procedures, which assume a linear or proportional hazard model to construct the relationship between the survival time and covariates, together with a new bandwidth selector, based on the generalized cross-validation criterion. The proposed estimator is expected to be shown to be consistent and asymptotically normal and a consistent estimator of the asymptotic variance is constructed. It is surprising to note that the asymptotic bias for the censoring case is exactly same as the one for the uncensored situation but the asymptotic variance is larger. The numerical studies based on the simulations and real examples show clearly that the proposed method works reasonably well. In particular, in the simulations, the censoring variable was generated to depend conditionally on the covariate. However, for the sake of simplicity in the proof, a technical assumption is imposed on the dependence structure between the censoring variable and covariates, see condition A4, which, unfortunately, was also imposed in [22-24] for parametric models. It is conjectured that condition A4 may not be necessary but I would not be able to dispense with it (see Remark 2).

## REFERENCES

1. Bennett, S. (1983), Log-logistic regression models for survival data, *Applied Statistics* **32**, 165-171.
2. Beran, R. (1981), Nonparametric regression with randomly censored survival data, *Technical Report*, Department of Statistics, University of California at Berkeley.
3. Buckley, J. and I.R. James (1979), Linear regression with censored data, *Biometrika* **66**, 429-436.
4. Cai, Z. (1998), Asymptotic properties of Kaplan-Meier estimator for censored dependent data, *Statistics and Probability Letters* **37**, 381-389.
5. Cai, Z. (2001), Estimating a distribution function for censored time series data, *Journal of Multivariate Analysis* **78**, 299-318.
6. Cai, Z. (2003), Weighted local linear approach to censored nonparametric regression, *Techincal Report*, Deaprtment of mathematics, University of North Carloina at Charlotte.
7. Dabrowska, D.M. (1987), Nonparametric regression with censored data, *Scandinavian Journal of Statistics* **14**, 181-197.
8. Dagenais, M. (1991). Comments on S.L. Zeger and R. Brookmeyer: "Regression analysis with censored autocorrelated data" [*Journal of American Statistical Association* **81**, 721-729], *Journal of the American Statistical Association* **86**, 255.
9. Doksum, K.A. and B.S. Yandell (1982), Properties of regression estimates based on censored survival data, In *A Festschrift for Erich Lehmann* (P.J. Bickel, K.A. Dosum, and J.L. Hpdges, Jr., eds.), Belmont CA: Wadsworth Internation Group, pp. 140-156.
10. Fan, J. and I. Gijbels (1996), *Local Polynomial Modeling and Its Applications*, Chapman and Hall, London.

11. Gray. R.J. (2000), Estimation of regression parameters and the hazard function in transformed linear survival data, *Biometrics* **56**, 571-576.
12. Hurvich, C.M., J.S. Simonoff and C.-L. Tsai (1998), Smoothing parameter selection in nonparametric regression using an improved Akaike information criterion, *Journal of the Royal Statistical Society* **B 60**, 271-293.
13. Kalbfleisch, J.D. and R.L. Prentice (1980), *The Statistical Analysis of Failure Time Data*, John Wiley, New York.
14. Kim, H.T. and Y.K. Truong (1998), Nonparametric regression estimates with censored data: Local linear smoothers and their applications, *Biometrics* **54**, 1434-1444.
15. Koul, H., V. Susarla and J. Van Ryzin (1981), Regression analysis with randomly right-censored data, *The Annals of Statistics* **9**, 1276-1288.
16. Lai, T.L. and Z. Ying (1990), Linear rank statistics in regression analysis with censored or truncated data, *Journal of Multivariate Analysis* **19**, 531-536.
17. Miller, R.G. (1976), Least squares regression with censored data, *Biometrika* **63**, 449-464.
18. Miller, R. and J. Halpern (1982), Regression with censored data, *Biometrika*, **69**, 521-531.
19. Murphy, S.A., A.J. Rossini and A.W. Van der Vaart (1996), Maximum likelihood estimation in the proportional odds model, *Journal of the American Statistical Association* **92**, 968-976.
20. Shen, X. (1998), Proportional odds regression and sieve maximum likelihood estimation, *Biometrika* **85**, 165-177.
21. Shumway, R.H., A.S. Azari and P. Johnson (1988), Estimating mean concentrations under transformation for environmental data with detection limits, *Technometrics* **31**, 347-356.
22. Stute, W. (1993), Consistent estimation under random censorship when covariates are present, *Journal of Multivariate Analysis* **45**, 89-103.
23. Stute, W. (1996), Distributional convergence under random censorship when covariates are present, *Scandinavian Journal of Statistics* **23**, 461-471.
24. Stute, W. (1999), Nonlinear censored regression, *Statistica Sinica* **9**, 1089-1102.
25. Stute, W. and J.L. Wang (1993), The strong law under random censorship, *The Annals of Statistics* **21**, 1591-1607.
26. Tsiatis, A.A. (1990), Estimating regression parameters using linear rank tests for censored data, *The Annals of Statistics* **18**, 354-372.
27. Vasudaven, M., M.G. Nair and M.M. Sithole (1996), On estimation for censored autoregressive data, *Statistics and Probability Letters* **31**, 97-105.
28. Wahba, G. (1977), A survey of some smoothing problems and the method of generalized cross-validation for solving them, In *Applications of Statistics* (P.R. Krisnaiah, ed.), 507-523. Amsterdam, North Holland.
29. Wei, L.J., Z. Ying and D.Y. Lin (1990), Linear regression analysis of censored survival data based on rank tests, *Biometrika* **77**, 845-891.
30. Yang, S. (1999), Censored median regression using weighted empirical survival and hazard functions, *Journal of the American Statistical Association* **94**, 137-145.
31. Yang, S. and R.L. Prentice (1999), Semiparametric inference in the proportional odds regression model, *Journal of the American Statistical Association* **94**, 125-136.

32. Ying, Z. (1993), A large sample study of rank estimation for censored regression data, *The Annals of Statistics* **21**, 76-99.
33. Ying, Z., S.H. Jung and L.J. Wei (1995), Survival analysis with median regression models, *Journal of the American Statistical Association* **90**, 178-184.
34. Zeger, S.L. and R. Brookmeyer (1986), Regression analysis with censored autocorrelated data, *Journal of the American Statistical Association* **81**, 721-729.
35. Zheng, Z. (1987), A class of estimators of the parameters in linear regression with censored data, *Acta Mathematicae Applicatae Sinica* **3**, 231-241.
36. Zheng, Z. (1988), Strong consistency of nonparametric regression estimates with censored data, *Journal of Mathematical Research and Exposition* **8**, 307-313.
37. Zhou, M. (1992), M-estimation in censored linear models, *Biometrika* **79**, 837-841.

# 7. Topics in Nonparametrics

Recent Advances and Trends in Nonparametric Statistics
Michael G. Akritas and Dimitris N. Politis (Editors)
© 2003 Elsevier Science B.V. All rights reserved.

## Adaptive quantile regression

S.A. van de Geer[a]

[a]Mathematical Institute, University of Leiden,
P.O. Box 9512, 2300 RA Leiden, The Netherlands

In this paper, the estimation of a regression quantile function is studied. We consider a linear regression model with dimension $m$, where $m$ may be as large as the number of observations $n$. As penalty on the quantile regression estimator, the sum of the absolute value if its coefficients is used. We show that the estimator adapts to the unknown smoothness of the underlying quantile regression function, as well as to unknown identifiablity properties.

### 1. INTRODUCTION

Consider $n \geq 2$ independent observations on a response variable $Y_i \in \mathbf{R}$ and a covariable $x_i \in \mathcal{X}$, $i = 1, \ldots, n$. We study the estimation of the $\beta$-quantile $g_0(x_i)$ of the distribution of $Y_i$ given the covariable $x_i$, for $i = 1, \ldots, n$. Here, $0 < \beta < 1$ is a given number, chosen by the statistician. The quantile regression problem was introduced by [6]. There, $g_0(x)$ is modelled as a linear function of a fixed (small) number of parameters. In this paper, we propose a linear model with many parameters, say $m$, where $m$ may be as large as the number of observations $n$. Moreover, we do not require strong assumptions on the distribution of the observations, in particular on the degree of identifiabililty of the unknown regression function $g_0$. Our aim is to construct an estimator that adapts to the amount of parameters needed, as well as to identifiability properties of $g_0$.

Let $\gamma = \gamma_\beta$ be the quantile regression loss function

$$\gamma(y) = \beta|y|\mathbf{1}\{y < 0\} + (1-\beta)|y|\mathbf{1}\{y \geq 0\}. \tag{1}$$

Consider the empirical loss function

$$\Gamma_n(g) = \frac{1}{n}\sum_{i=1}^{n}\gamma(Y_i - g(x_i)), \tag{2}$$

and let

$$\Gamma(g) = \mathbf{E}\Gamma_n(g) \tag{3}$$

be the theoretical loss function. Here, and throughout, we regard the covariables $x_1, \ldots, x_n$ as fixed (i.e., we work conditionally on the observed values of the covariables). It is then easily seen that

$$g_0 = \arg\min_{g} \Gamma(g), \tag{4}$$

where the minimum is taken over the class of all functions $g : \mathcal{X} \to \mathbf{R}$. More specifically, let for $i = 1, \ldots, n$,

$$F_i(y) = \mathbf{P}(Y_i \leq y), \ y \in \mathbf{R}, \tag{5}$$

be the distribution function of $Y_i$. Then, assuming the inverse of $F_i$ at $\beta$ exists, we have

$$g_0(x_i) = F_i^{-1}(\beta) = \arg\min_{c \in \mathbf{R}} \mathbf{E}\gamma(Y_i - c), \ i = 1, \ldots, n. \tag{6}$$

Now, the idea is to estimate $g_0$ by using the empirical counterpart, i.e., the minimizer of $\Gamma_n$. We then need to somehow control complexity, since the overall minimizer $\arg\min_g \Gamma_n(g)$ just reproduces the data $(Y_i)_{i=1}^n$. To avoid overfitting, one can add a penalty pen$(g)$ to the empirical loss function. The penalized quantile regression estimator is

$$\hat{g}_n = \arg\min_{g \in \mathcal{G}} \{\Gamma_n(g) + \text{pen}(g)\}. \tag{7}$$

Here, $\mathcal{G}$ is an a priori model class, which may be the class of all functions $g : \mathcal{X} \to \mathbf{R}$.

We now come to the description of the linear model, and the penalty. Let for $j = 1, \ldots, m$, $\psi_j : \mathcal{X} \to \mathbf{R}$ be given functions. We assume that $\psi_1, \ldots, \psi_m$ are linearly independent in $L_2(Q_n)$, with $Q_n = \sum_{i=1}^n \delta_{x_i}/n$ the empirical measure of the covariables. (Thus in particular, we assume that $m \leq n$). Take $\mathcal{G}$ as (a subset of) the linear functions

$$g = g_\alpha = \sum_{j=1}^m \alpha_j \psi_j, \ \alpha \in \mathbf{R}^d. \tag{8}$$

We let

$$\Sigma_n = \int \psi \psi^T dQ_n, \tag{9}$$

where $\psi = (\psi_1, \ldots, \psi_m)^T : \mathcal{X} \to \mathbf{R}^m$. Denote the smallest eigenvalue of $\Sigma_n$ as $\lambda_{\min}^2$, and its largest eigenvalue as $\lambda_{\max}^2$. Invoke the normalization

$$\lambda_{\max} = 1. \tag{10}$$

and choose the penalty

$$\text{pen}(g_\alpha) = \lambda_n \sum_{j=1}^m |\alpha_j|, \tag{11}$$

with $\lambda_n$ a regularization parameter, to be specified (see Theorem 1). We call (11) an $L_1$-penalty. It is closely related to soft-thresholding ([3]), and to the LASSO ([16,5]). Note that with this penalty, the computation of the penalized quantile estimator is quite feasible (using e.g. interior point methods, see also [14]).

In [7,8], total variation type penalties on the function $g$ or on (first or higher order) derivates of $g$ are used. Asymptotic theory for such estimators can be found in [13]. These estimators have local adaptive properties, but they may not be globally adaptive. Aiming at a globally adaptive procedure (see Section 2 for more details), one may propose to use model selection among a collection of linear (high-dimensional) models, and use a penalty proportional to model dimension (AIC ([1]), BIC ([15]), etc.). However, it is not at all clear whether such an approach is robust against violations of regularity conditions on the distribution of $Y_i$, $i = 1, \ldots, n$ (for instance, the assumption of existence of Lebesgue densities which are strictly positive near the $\beta$-quantile (assumption (i) in Theorem 4.2 of [6])). Such regularity conditions ensure quadratic behavior of the loss function near its minimum. In our setup, the behavior near the minimum may be unknown.

Below, a general condition (13) on the identifiability of $g_0$ is presented. Here, and in the sequel, we use the notation

$$\|g\|^2 = \sum_{i=1}^n g^2(x_i), \quad g : \mathcal{X} \to \mathbf{R}. \tag{12}$$

Thus $\|\cdot\|$ is the $L_2(Q_n)$-norm.

**Identifiability condition.** There exists a strictly increasing function $J : [0, \infty) \to [0, \infty)$ such that for all $g \in \mathcal{G}$,

$$\Gamma(g) - \Gamma(g_0) \geq \|g - g_0\| J(\|g - g_0\|). \tag{13}$$

In the regular case, $J$ is linear, so that $\Gamma$ is quadratic. If e.g., $J(\xi)$ decreases faster than linear as $\xi \to 0$, the regression function $g_0$ is harder to identify. We call $J$ the *margin function*: it describes the sensitivity of $\Gamma(\cdot)$ to deviations from its minimum. Condition (13) will be referred to as the *margin condition*.

An important point is that the function $J$ will generally be unknown, and that it will also be hard to estimate it with sufficient accuracy. Thus, we need a procedure which adapts to all possible $J$. Note also that the amount of identifiability may depend on possible a priori model assumptions, i.e., on the model class $\mathcal{G}$ for the regression function. We moreover make the rather trivial observation that one may replace $\mathcal{G}$ by an appropriate smaller class, but that one then has to take into account the (hopefully negligible) probability that $\hat{g}_n$ is not in the smaller class.

**Example 1.** Write $c_i = g_0(x_i)$ (so $F_i(c_i) = \beta$), and suppose that there exists constants, $0 < \epsilon < 1$, $\sigma > 0$ and $\rho > 0$ such that

$$|F_i(c) - F_i(c_i)| \geq |c - c_i|^\rho / \sigma^\rho, \tag{14}$$

for all $|c - c_i| \leq \epsilon$, $i = 1, \ldots, n$. Then, for $|c - c_i| \leq \epsilon$

$$\mathbf{E}\gamma(Y_i - c) - \mathbf{E}\gamma(Y_i - c_i) = (c - c_i)(F_i(c) - F_i(c_i))$$
$$\geq |c - c_i|^{1+\rho} / \sigma^\rho, \quad i = 1, \ldots, n. \tag{15}$$

So then, when $|g(x_i) - g_0(x_i)| \leq \epsilon$, for all $i = 1, \ldots, n$, we have

$$\Gamma(g) - \Gamma(g_0) \geq \|g - g_0\|^{1+\tilde{\rho}} / \sigma^\rho, \tag{16}$$

where $\tilde{\rho} = \max(\rho, 1)$. Thus, then the margin condition (13) is met on the set

$$\mathcal{G}_0 = \{g : \max_{i=1,\ldots,n} |g(x_i) - g_0(x_i)| \leq \epsilon\}, \tag{17}$$

with $J$ the function $J(\xi) = \xi^{1+\tilde{\rho}} / \sigma^\rho$, $\xi \geq 0$.

We will show in Theorem 1 of Section 3 that the quantile estimator with $L_1$-penalty adapts to the smoothness of $g_0$ as well as to the margin function $J$. In Section 2, we will explain in some detail what we mean by adaptation. Theorem 2 of Section 3 shows moreover that results can be extended to hold uniformly in $\beta$. In Section 4, we take a brief look at the concept of smoothness, considered in general terms of approximation theory. Section 5 presents the proof of Theorem 1 and Theorem 2.

## 2. THE ORACLE

Let the "dimension" of $g = g_\alpha$ be

$$d_g = \#\{\alpha_j \neq 0\}. \tag{18}$$

An "oracle" $g_c^*$ is

$$g_c^* = \arg\min_{g \in \mathcal{G}} \tau_c^2(g|g_0). \tag{19}$$

where

$$\tau_c^2(g|g_0) = \left\{\Gamma(g) - \Gamma(g_0) + 4c(\lambda_n/\lambda_{\min})\sqrt{d_g}J^{-1}(4c(\lambda_n/\lambda_{\min})\sqrt{d_g})\right\}. \tag{20}$$

Here $\lambda_n$ is the regularization parameter, and $c > 0$ is a constant. Furthermore, we let $\tau^2(g|g_0) = \tau_1^2(g|g_0)$, and we let $g^* = g_1^*$ be the oracle with constant $c = 1$. Its dimension is denoted by $d^* = d_{g^*}$. Note that $g_c^*$ depends on the regularization parameter $\lambda_n$. We regard this parameter as to be chosen by the statistician, and hence it may not depend on unknown quantities. In fact we will choose it as

$$\lambda_n = u\sqrt{\frac{\log n}{n}}, \tag{21}$$

where $u$ is equal, or larger than, some universal constant (see Theorem 1). The constant $c$ however is allowed to depend on unknown quantities, i.e., on the distribution of $Y_i$, $i = 1, \ldots, n$. The constant 4 in our definition of $\tau_c^2$ is only there because it comes out of our rough calculations in the proof of Theorem 1. It has no intrinsic meaning.

Now, we come to the question why we consider $g^*$ (more generally $g_c^*$) as an oracle. The reason is that it represents the best trade-off, over all linear submodels, between "bias" and "variance", where we take the terminology in a very loose sense ("bias" corresponding to "approximation error" and "variance" to "estimation error"). Of course, $g^*$ is non-random, and in fact, the "estimation error" comes rather from the estimator one could use if the set $\mathcal{J}^*$ (with cardinality $d^*$) of the non-zero coefficients of $g^*$ were known. Let us give the heuristics here, without going into details.

Consider a $d$-dimensional model, with only the $d$ coefficients in the index set $\mathcal{J}$, with cardinality $d$, possibly non-zero. Write

$$\mathcal{G}_d = \{g = \sum_{j \in \mathcal{J}} \alpha_j \psi_j\} \cap \mathcal{G} \tag{22}$$

for this model class and let $\hat{g}_{n,d}$ be the quantile estimator over this class, that is

$$\hat{g}_{n,d} = \arg\min_{g \in \mathcal{G}_d} \Gamma_n(g). \tag{23}$$

Let

$$g_{*,d} = \arg\min_{g \in \mathcal{G}_d} \Gamma(g) \tag{24}$$

be the best approximation of $g_0$ within the class $\mathcal{G}_d$. Then rewriting

$$\Gamma_n(\hat{g}_{n,d}) \leq \Gamma_n(g_{*,d}) \tag{25}$$

gives

$$\Gamma(\hat{g}_{n,d}) - \Gamma(g_0) \leq -\nu_{n,d} + \Gamma(g_{*,d}) - \Gamma(g_0), \tag{26}$$

where

$$\nu_{n,d} = [\Gamma_n(\hat{g}_{n,d}) - \Gamma_n(g_{*,d})] - [\Gamma(\hat{g}_{n,d}) - \Gamma(g_{*,d})] \tag{27}$$

is the "random part" of the problem. Now, empirical process theory (see e.g., [19,17], and their references) gives that $\nu_{n,d}$ behaves like $\sqrt{d/n}\|\hat{g}_{n,d} - g_{*,d}\|$, i.e. except on a set $\bar{\mathbf{A}}$, with small probability $\mathbf{P}(\bar{\mathbf{A}})$,

$$|\nu_{n,d}| \leq C\sqrt{d/n}\|\hat{g}_{n,d} - g_{*,d}\|, \tag{28}$$

where $C$ is a constant depending on the distribution of the observations. After some straightforward calculations, we see that except on the set $\bar{\mathbf{A}}$,

$$\Gamma(\hat{g}_{n,d}) - \Gamma(g_0) \leq 2\tilde{\tau}^2(g_{*,d}|g_0) \tag{29}$$

where

$$\tilde{\tau}^2(g|g_0) = \Gamma(g) - \Gamma(g_0) + C\sqrt{\frac{d}{n}} J^{-1}(2C\sqrt{\frac{d}{n}}) \tag{30}$$

(Similar arguments are used in the proof of Theorem 1, where more details are presented.) So the best result (up to constants) are obtained for the class $\mathcal{J}_*$ with cardinality $d_*$ corresponding to the non-zero coefficients of

$$g_* = \arg\min \tilde{\tau}^2(g|g_0). \tag{31}$$

Thus we see that apart from constants and the $\sqrt{\log n}$ term, the function $g^*$ represents the ideal approximation of $g_0$ by a linear model with smallest possible dimension, taking into account the "bias" term $\Gamma(g) - \Gamma(g_0)$ as well as the "variance" term $C\sqrt{d_g/n}J^{-1}(2C\sqrt{d_g/n})$.

## 3. ADAPTATION

This section provides our main results. We give explicit constants, so that the dependencies on $n$, $g_0$ and other quantities, is evident. We have however not attempted to optimize these constants.

Theorem 1 below formulates an oracle inequality for the $L_1$-penalized quantile estimator

$$\hat{g}_n = \arg\min_{g \in \mathcal{G}} \{\Gamma_n(g) + \text{pen}(g_\alpha)\}. \tag{32}$$

We took the terminology *oracle inequality* from [2], where a Gaussian sequence space model is considered (and in that context stronger results are obtained). We recall that

$$\mathcal{G} \subseteq \{g_\alpha = \sum_{j=1}^{m} \alpha_j \psi_j : \alpha \in \mathbf{R}^m\}, \tag{33}$$

and that

$$\Gamma_n(g) = \frac{1}{n} \sum_{i=1}^{n} \gamma(Y_i - g(x_i)) \tag{34}$$

is the quantile loss function, and

$$\text{pen}(g_\alpha) = \lambda_n \sum_{j=1}^{m} |\alpha_j| \tag{35}$$

is the $L_1$-penalty.

**Theorem 1.** *Assume that $\mathcal{G}$ is a convex set and that the margin condition (13) holds for each $g \in \mathcal{G}$, and some strictly increasing function $J$. Take $\lambda_n = (12+u)2\sqrt{\log n/n}$, with $u \geq 6$. Then for $n \geq N$, where $N$ is such that $\lambda_n/n + \tau^2(g^*|g_0) \leq 1/8$, and for any $0 < \delta \leq 1$,*

$$\mathbf{P}\left(\Gamma(\hat{g}_n) - \Gamma(g_0) \geq (1+\delta)^2 \left[\frac{\lambda_n}{n} + \tau_n^*(\delta)^2\right]\right) \leq 64R^2 \exp[-\frac{u^2 \log n}{1024 R^2}], \tag{36}$$

*where*

$$\tau_n^*(\delta)^2 = \left[\Gamma(g^*) - \Gamma(g_0) + \frac{2}{\delta^2} \frac{\lambda_n}{\lambda_{\min}} \sqrt{d^*} J^{-1}(\frac{4}{\delta} \frac{\lambda_n}{\lambda_{\min}} \sqrt{d^*})\right], \tag{37}$$

*and where $R = 1 + \|g^* - g_0\|$.*

The estimator can be computed for a whole path of values of $\lambda_n$ and $\beta$ simultaneously (see [8] for a related problem). Note that the theorem allows one to immediatly obtain simultaneous inequalities for all $\beta$ in (say) the finite grid $\beta \in \{k/n, (k+1)/n, \ldots, (n-k)/n\}$, $k = n\lfloor t \rfloor$, $t \in (0,1)$. It is in fact possible to extend Theorem 1 to hold uniformly for all $\beta$ in some domain of interest, say $\beta \in \mathcal{B} \subset (0,1)$. This will be shown in Theorem 2.

Note that almost everything we have defined so far depends on $\beta$. For example, $g_0 = g_{0,\beta}$, $\Gamma_n = \Gamma_{n,\beta}$ and $\Gamma = \Gamma_\beta$. In the margin condition (13), $J = J_\beta$ will generally depend on $\beta$ as well. We will assume however that the *same* functions $\psi_1, \ldots, \psi_m$ are used in the linear model, and for simplicity also the *same* model class $\mathcal{G}$. The oracle $g_\beta^*$ which minimizes

$$\tau_\beta^2(g|g_{0,\beta}) = \{\Gamma_\beta(g) - \Gamma_\beta(g_{0,\beta}) + 4(\lambda_n/\lambda_{\min})\sqrt{d_g} J_\beta^{-1}(4c(\lambda_n/\lambda_{\min})\sqrt{d_g})\}, \tag{38}$$

clearly also depends on $\beta$, as well as its dimension $d_\beta^*$. (Here, we have a ambiguity in notation as $\tau_c^2$ and $g_c^*$ were used in Section 2 with a different meaning, but we believe the risk of confusion is small.) The $L_1$-penalized $\beta$-quantile estimator will be denoted by $\hat{g}_{n,\beta}$.

**Theorem 2.** Assume that $\mathcal{G}$ is a convex set and that the margin condition (13) holds for strictly increasing functions $J_\beta$, and for each $g \in \mathcal{G}$ and $\beta \in \mathcal{B}$. Take $\lambda_n = (12 + u)2\sqrt{\log n/n}$, with $u \geq 6$. Then for $n \geq N$, where $N$ is such that $\lambda_n/n + \sup_{\beta \in \mathcal{B}} \tau_\beta^2(g_\beta^*|g_{0,\beta}) \leq 1/8$, and for any $0 < \delta \leq 1$,

$$\mathbf{P}\left(\Gamma_\beta(g_{n,\beta}) - \Gamma_\beta(g_{0,\beta}) \geq (1+\delta)^2 \left[\frac{\lambda_n}{n} + \tau_{n,\beta}^*(\delta)^2\right]\right) \leq 64R^2 \exp[-\frac{u^2 \log n}{1024 R^2}], \quad (39)$$

where

$$\tau_{n,\beta}^*(\delta)^2 = \left[\Gamma_\beta(g_\beta^*) - \Gamma_\beta(g_{0,\beta}) + \frac{2}{\delta^2}\frac{\lambda_n}{\lambda_{\min}}\sqrt{d_\beta^*}J_\beta^{-1}(\frac{4}{\delta}\frac{\lambda_n}{\lambda_{\min}}\sqrt{d_\beta^*})\right] \quad (40)$$

and where $R = 1 + \sup_{\beta \in \mathcal{B}} \|g_\beta^* - g_{0,\beta}\|$.

## 4. A TYPICAL EXAMPLE

In this section, we address the question what Theorem 1 has to say on adaptation to smoothness. We again fix $\beta \in (0,1)$ and drop subscripts $\beta$ in the notation. Let us introduce a smoothness parameter $s$. For example, if $g_0$ is a function of one variable $x$ in a bounded interval, say $x \in [0,1]$, one could describe the amount of smoothness by the value of the squared Sobolev pseudo-norm,

$$\int_0^1 |g_0^{(s)}(x)|^2 dx, \quad (41)$$

where $g_0^{(s)}$ is the $s$-th derivative of $g_0$. More generally, in higher dimensional space $\mathcal{X}$. the smoothness $s$ of a function $g_0 : \mathcal{X} \to \mathbf{R}$ can be the *effective* smoothness (such as (roughly) the number of derivatives divided by the dimension of $\mathcal{X}$). The parameter $s$ may also be the smoothness parameter appearing in the definition of the Besov pseudo-norm, etc. Here, we will not provide any approximation theory nor define the concept of smoothness in any precise way. We simply *assume* that for some $s$, and $c$, and all $d \leq m$,

$$\inf_{\mathcal{G}_d} \inf_{g \in \mathcal{G}_d} \|g - g_0\| \leq cd^{-s}. \quad (42)$$

Here, the infimum is taken over the $\binom{m}{d}$ $d$-dimensional spaces

$$\mathcal{G}_d = \{\sum_{j \in \mathcal{J}} \alpha_j \psi_j\} \cap \mathcal{G}, \ |\mathcal{J}| = d. \quad (43)$$

For smoothness classes where (42) is met, we refer to general work in approximation theory, for example [12], or [4].

Let us now assume that the margin condition (13) is met with $J(\xi) = \xi^r/\sigma^r$, for $\xi \leq \epsilon$, where $r > 0$, $\sigma > 0$ and $\epsilon > 0$ (see also Example 1). Thus

$$\Gamma(g) - \Gamma(g_0) \geq \|g - g_0\|^{r+1}/\sigma^r, \ g \in \mathcal{G}, \ \|g - g_0\| \leq \epsilon. \quad (44)$$

Assume also the reverse holds for an appropriate constant $\eta$, i.e.,

$$\Gamma(g) - \Gamma(g_0) \leq \|g - g_0\|^{r+1}/\eta^r, \ g \in \mathcal{G}, \ \|g - g_0\| \leq \epsilon. \quad (45)$$

We may assume that eventually $\|g_n - g_0\| \leq \epsilon$ by convexity arguments, provided also $\|g^* - g_0\|$ is small enough. The convexity argument is explained in [18] (it is also used in the proof of Theorem 1).

The relevant expression appearing in Theorem 1 is the value $\tau_n^*(1)^2$ at $g^*$ of the quantity

$$\Gamma(g) - \Gamma(g_0) + 2\frac{\lambda_n}{\lambda_{\min}}\sqrt{d_g}J^{-1}(4\frac{\lambda_n}{\lambda_{\min}}\sqrt{d_g}) \tag{46}$$

which can be bounded by

$$\frac{c^{r+1}}{\eta^r}\left(\|g - g_0\|^{-s(r+1)} + (\frac{4\lambda_n}{\lambda_{\min}})^{\frac{r+1}{r}}\frac{\sigma\eta^r}{c^{r+1}}d_g^{\frac{r+1}{2r}}\right). \tag{47}$$

Taking (47) as starting point, The optimal trade-off has solution

$$d_0^* = \frac{c^{r+1}}{\eta^r}\arg\min\{d^{-s(r+1)} + \delta_n d^{\frac{r+1}{2r}}\}, \tag{48}$$

where

$$\delta_n = (\frac{4\lambda_n}{\lambda_{\min}})^{\frac{r+1}{r}}\frac{\sigma\eta^r}{c^{r+1}}. \tag{49}$$

This optimal value is

$$d_0^* = (\frac{2rs}{\delta_n})^{\frac{2r}{(2rs+1)(r+1)}}, \tag{50}$$

where we tacitly assume that this is an integer. Inserting that value in (47) gives

$$\Gamma(g^*) - \Gamma(g_0) + 2\lambda_n\sqrt{d^*}J^{-1}(\frac{4\lambda_n}{\lambda_{\min}}\sqrt{d^*}) \leq \frac{c^{r+1}}{\eta^r}c_{r,s}\delta_n^{\frac{2rs}{2rs+1}}$$

$$= \frac{c^{r+1}}{\eta^r}c_{r,s}(\frac{4\lambda_n}{\lambda_{\min}})^{\frac{2s(r+1)}{2rs+1}}(\frac{\sigma\eta^r}{c^{r+1}})^{\frac{2rs}{2rs+1}}. \tag{51}$$

Here,

$$c_{r,s} = (2rs)^{-\frac{2rs}{2rs+1}} + (2rs)^{\frac{1}{2rs+1}}. \tag{52}$$

Thus, taking $\lambda_n$ of order $\sqrt{\frac{\log n}{n}}$ gives a rate of convergence of order

$$(\frac{\log n}{n})^{\frac{s(r+1)}{2rs+1}}. \tag{53}$$

For $r = 1$ this corresponds, up to the factor $(\log n)^{\frac{2s}{2s+1}}$, to the usual rate $n^{-2s/(2s+1)}$ for a model of smoothness $s$.

## 5. PROOFS

The proof of Theorem 1 is a modification of arguments used in [11]. Throughout the proofs, we use the notation

$$\nu_n(\alpha) = [\Gamma_n(g_\alpha) - \Gamma_n(g^*)] - [\Gamma(g_\alpha) - \Gamma(g^*)], \ \alpha \in \mathbf{R}^m, \tag{54}$$

for the empirical process, and

$$I(\alpha) = \sum_{j=1}^{m} |\alpha_j|, \ \alpha \in \mathbf{R}^m, \tag{55}$$

for the $L_1$-norm. We furthermore write, for $M > 0$, $R > 0$,

$$\mathcal{A}_{M,R} = \{\alpha \in \mathbf{R}^m : I(\alpha - \alpha^*) \leq M, \ \|g_\alpha - g^*\| \leq R\}. \tag{56}$$

Lemma 3 and Lemma 4 study the "random part" of the problem: they provide a probability inequality for the empirical process. Theorem 1 uses these lemmas to arrive at its result. The proof Theorem 2 is established in a similar fashion, at the end of this section.

**Lemma 3.** *For all $M > 0$, $R > 0$, and $u > 0$, the following upper bound holds*

$$\mathbf{P}\left(\sup_{\alpha \in \mathcal{A}_{M,R}} |\nu_n(\alpha)| \geq (12+u)M\sqrt{\frac{\log n}{n}}\right)$$
$$\leq \exp[-\frac{u^2((M^2/R^2) \vee 1)\log n}{32}]. \tag{57}$$

Here $a \vee b = \max(a, b)$.

**Proof.** Define the random variable

$$Z = \sup_{\alpha \in \mathcal{A}_{M,R}} |\nu_n(\alpha)|. \tag{58}$$

Set

$$U_i(\alpha) = \gamma(Y_i - g_\alpha(x_i)) - \gamma(Y_i - g^*(x_i)), \ i = 1, \ldots, n. \tag{59}$$

Then (see [9]),

$$\mathbf{P}(Z \geq \mathbf{E}(Z) + u) \leq \exp[-\frac{n^2 u^2}{8 b_n^2}], \tag{60}$$

where $b_n^2$ is assumed to satisfy

$$b_n^2 \geq \sup_{\alpha \in \mathcal{A}_{M,R}} \sum_{i=1}^{n} |U_i(\alpha) - \mathbf{E} U_i(\alpha)|^2. \tag{61}$$

But since $\gamma$ is 1-Lipschitz,

$$|U_i(\alpha)| \leq |g_\alpha(x_i) - g^*(x_i)|, \tag{62}$$

so we may take

$$b_n^2 \leq \sup_{\alpha \in \mathcal{A}_{M,R}} 4n\|g_\alpha - g^*\|^2 \leq 4n(M^2 \wedge R^2), \tag{63}$$

where $a \wedge b = \min(a,b)$, and where we used the normalization $\lambda_{\max} = 1$, so that

$$\|g_\alpha - g^*\|^2 \leq \sum_{j=1}^m (\alpha_j - \alpha_j^*)^2, \tag{64}$$

and moreover,

$$\sum_{j=1}^m (\alpha_j - \alpha_j^*)^2 \leq I^2(\alpha - \alpha^*). \tag{65}$$

A symmetrization procedure (see e.g., [10] or [19]), yields that

$$\mathbf{E}(Z) \leq \mathbf{E}\left(\sup_{\alpha \in \mathcal{A}_{M,R}} |\frac{1}{n}\sum_{i=1}^n \epsilon_i U_i(\alpha)|\right), \tag{66}$$

where $\epsilon_1, \ldots, \epsilon_n$ are Rademacher random variables. Because $\gamma$ is 1-Lipschitz, we can apply the contraction principle ([10], Theorem 4.12)), which gives

$$\mathbf{E}\left(\sup_{\alpha \in \mathcal{A}_{M,R}} |\frac{1}{n}\sum_{i=1}^n \epsilon_i U_i(\alpha)|\right) \leq 2\mathbf{E}\left(\sup_{\alpha \in \mathcal{A}_{M,R}} |\frac{1}{n}\sum_{i=1}^n \epsilon_i(g_\alpha(x_i) - g^*(x_i))|\right)$$
$$\leq 2M\mathbf{E}\left(\max_{j=1,\ldots,m} |\frac{1}{n}\sum_{j=1}^m \epsilon_j \psi_j(x_i)|\right). \tag{67}$$

Now, $\|\psi_j\| \leq 1$ for all $j = 1, \ldots, m$, because $\lambda_{\max} = 1$. Applying results in [19] (Chapter 2.2), we arrive at the bound

$$\mathbf{E}\left(\max_{j=1,\ldots,m} |\frac{1}{n}\sum_{j=1}^m \epsilon_j \psi_j(x_i)|\right) \leq 6\sqrt{\frac{\log n}{n}}. \tag{68}$$

As a consequence,

$$\mathbf{P}(Z \geq 12M\sqrt{\frac{\log n}{n}} + u) \leq \exp[-\frac{nu^2}{32(M^2 \wedge R^2)}]. \tag{69}$$

Replacing $u$ by $uM\sqrt{\log n/n}$ completes the proof. □

**Lemma 4.** *We have for all $R \geq 1$ and $u \geq 6$*

$$\mathbf{P}\left(\sup_{\|g_\alpha - g^*\| \leq R} \frac{|\nu_n(\alpha)|}{I(\alpha - \alpha^*) + \frac{1}{n}} > (12+u)2\sqrt{\frac{\log n}{n}}\right)$$
$$\leq 4R^2 \exp[-\frac{u^2 \log n}{64R^2}]. \tag{70}$$

**Proof.** First we consider the set where $I(\alpha - \alpha^*) \leq \frac{1}{n}$. By Lemma 3,

$$\mathbf{P}\left(\sup_{\alpha \in \mathcal{A}_{\frac{1}{n},R}} |\nu_n(\alpha)| > (12+u)\frac{2}{n}\sqrt{\frac{\log n}{n}}\right)$$
$$\leq \exp[-\frac{u^2 \log n}{32}]. \tag{71}$$

Next, we consider $(1/n) < I(\alpha - \alpha^*) \leq 1$. Take $j_0$ as the smallest integer such that $j_0 + 1 > \log_2 n$. We find from Lemma 3,

$$\mathbf{P}\left(\sup_{1/n < I(\alpha-\alpha^*) \leq 1,\ \|g_\alpha - g^*\| \leq R} \frac{|\nu_n(\alpha)|}{I(\alpha - \alpha^*)} > (12+u)2\sqrt{\frac{\log n}{n}}\right)$$
$$\leq \sum_{j=0}^{j_0} \mathbf{P}\left(\sup_{\alpha \in \mathcal{A}_{2^{-j},R}} |\nu_n(\alpha)| > (12+u)2^{-j}\sqrt{\frac{\log n}{n}}\right)$$
$$\leq (n+1)\exp[-\frac{u^2 \log n}{32}] \leq \exp[-\frac{u^2 \log n}{64}]. \tag{72}$$

Next, we consider the set where $I(\alpha - \alpha^*) > 1$. There

$$\mathbf{P}\left(\sup_{I(\alpha-\alpha^*)>1,\ \|g_\alpha - g^*\| \leq R} \frac{|\nu_n(\alpha)|}{I(\alpha - \alpha^*)} > (12+u)2\sqrt{\frac{\log n}{n}}\right)$$
$$\leq \sum_{j=0}^{\infty} \mathbf{P}\left(\sup_{\alpha \in \mathcal{A}_{2^{j+1},R}} |\nu_n(\alpha)| > (12+u)2^{j+1}\sqrt{\frac{\log n}{n}}\right)$$
$$\leq \sum_{j=0}^{\infty} \exp[-\frac{u^2 2^{2(j+1)} \log n}{32R^2}]$$
$$\leq \sum_{j=0}^{\infty} \exp[-\frac{u^2(j+2) \log n}{32R^2}] \leq 2R^2 \exp[-\frac{u^2 \log n}{32R^2}]. \tag{73}$$

$\square$

**Proof of Theorem 1.** First, suppose we already know that $\|\hat{g}_n - g^*\| \leq \bar{R}$, where we take $\bar{R} = 2 + 4\|g^* - g_0\|$. Let $\mathbf{A}$ be the set

$$\mathbf{A} = \{|\nu_n(\alpha)| \leq \lambda_n I(\alpha - \alpha^*) + \lambda_n/n,\ \text{for all}\ \|g_\alpha - g^*\| \leq \bar{R}\}. \tag{74}$$

Then by Lemma 4,

$$\mathbf{P}(\mathbf{A}) \geq 1 - 4\bar{R}^2 \exp[-\frac{u^2 \log n}{64\bar{R}^2}]. \tag{75}$$

So let us consider what happens on the set $\mathbf{A}$.

Clearly, the inequality

$$\Gamma_n(\hat{g}_n) + \lambda_n I(\hat{\alpha}_n) \leq \Gamma_n(g^*) + \lambda_n I(\alpha^*) \tag{76}$$

may be rewritten in the form

$$\Gamma(\hat{g}_n) - \Gamma(g_0) \leq -\nu_n(\hat{\alpha}_n) - \lambda_n[I(\hat{\alpha}_n) - I(\alpha_*)] + \Gamma(g^*) - \Gamma(g_0). \tag{77}$$

So on **A**,

$$\Gamma(\hat{g}_n) - \Gamma(g_0) \leq \lambda_n I(\hat{\alpha}_n - \alpha^*) + \frac{\lambda_n}{n} - \lambda_n[I(\hat{\alpha}_n) - I(\alpha_*)] + \Gamma(g^*) - \Gamma(g_0). \quad (78)$$

Let $\mathcal{J}^* = \{j: \alpha_j^* \neq 0\}$, and let for any $\alpha$, $I_1(\alpha) = \sum_{j \in \mathcal{J}^*} |\alpha_j|$ and $I_2(\alpha) = \sum_{j \notin \mathcal{J}^*} |\alpha_j|$. Since $I_2(\alpha - \alpha^*) = I_2(\alpha)$, we now find

$$\Gamma(\hat{g}_n) - \Gamma(g_0) \leq \lambda_n I_1(\hat{\alpha}_n - \alpha^*) + \lambda_n I_2(\hat{\alpha}_n)$$
$$-\lambda_n[I_1(\hat{\alpha}_n) - I_1(\alpha^*)] - \lambda_n I_2(\hat{\alpha}_n) + \Gamma(g^*) - \Gamma(g_0)$$
$$= \lambda_n I_1(\hat{\alpha}_n - \alpha^*) - \lambda_n[I_1(\hat{\alpha}_n) - I_1(\alpha^*)] + \Gamma(g^*) - \Gamma(g_0). \quad (79)$$

Since for any $a, b \in \mathbf{R}$, $|a| - |b| \leq |a - b|$, we arrive at

$$\Gamma(\hat{\alpha}_n) - \Gamma(\alpha_0) \leq 2\lambda_n I_1(\hat{\alpha}_n - \alpha^*) + \frac{\lambda_n}{n} + \Gamma(\alpha^*) - \Gamma(\alpha_0). \quad (80)$$

Application of first the Cauchy-Schwarz inequality and the inequality $\sum_{j=1}^{m} (\alpha_j - \alpha_j^*)^2 \leq \|g_\alpha - g^*\|^2 / \lambda_{\min}^2$, and then the triangle inequality yields

$$\Gamma(\hat{g}_n) - \Gamma(g_0) \leq 2 \frac{\lambda_n}{\lambda_{\min}} \sqrt{d^*} \|\hat{g}_n - g^*\| + \frac{\lambda_n}{n} + \Gamma(g^*) - \Gamma(g_0)$$
$$\leq 2 \frac{\lambda_n}{\lambda_{\min}} \sqrt{d_*} \|\hat{g}_n - g_0\| + 2 \frac{\lambda_n}{\lambda_{\min}} \sqrt{d^*} \|g^* - g_0\| + \frac{\lambda_n}{n} + \Gamma(g^*) - \Gamma(g_0)$$
$$= I + II + III, \quad (81)$$

where

$$I = 2 \frac{\lambda_n}{\lambda_{\min}} \sqrt{d^*} \|\hat{g}_n - g_0\|, \quad (82)$$

$$II = 2 \frac{\lambda_n}{\lambda_{\min}} \sqrt{d^*} \|g^* - g_0\|, \quad (83)$$

and

$$III = \frac{\lambda_n}{n} + \Gamma(g^*) - \Gamma(g_0). \quad (84)$$

If $I \geq \delta(II + III)$ and invoking margin condition (13), we now obtain

$$\|\hat{g}_n - g_0\| J(\|\hat{g}_n - g_0\|) \leq \Gamma(\hat{g}_n) - \Gamma(g_0) \leq 2(1 + \frac{1}{\delta}) \frac{\lambda_n}{\lambda_{\min}} \sqrt{d^*} \|\hat{g}_n - g_0\|. \quad (85)$$

This implies

$$\|\hat{g}_n - g_0\| \leq J^{-1}(2(\frac{1+\delta}{\delta}) \frac{\lambda_n}{\lambda_{\min}} \sqrt{d^*}) \quad (86)$$

and hence

$$\Gamma(\hat{\alpha}_n) - \Gamma(\alpha_0) \leq 2(\frac{1+\delta}{\delta}) \frac{\lambda_n}{\lambda_{\min}} \sqrt{d^*} J^{-1}(2(\frac{1+\delta}{\delta}) \frac{\lambda_n}{\lambda_{\min}} \sqrt{d^*}). \quad (87)$$

If $II \geq \delta(III)$, we get (by condition (13))

$$\|g^* - g_0\|J(\|g^* - g_0\|) \leq \Gamma(g^*) - \Gamma(g_0) \leq \frac{2}{\delta}\frac{\lambda_n}{\lambda_{\min}}\sqrt{d^*}\|g^* - g_0\|, \qquad (88)$$

which gives

$$\|g^* - g_0\| \leq J^{-1}(\frac{2}{\delta}\frac{\lambda_n}{\lambda_{\min}}\sqrt{d^*}). \qquad (89)$$

So then

$$II \leq \frac{2}{\delta}\frac{\lambda_n}{\lambda_{\min}}\sqrt{d^*}J^{-1}(\frac{2}{\delta}\frac{\lambda_n}{\lambda_{\min}}\sqrt{d_*}). \qquad (90)$$

So if $I \leq \delta(II + III)$ and $II \geq \delta(III)$,

$$\Gamma(\hat{g}_n) - \Gamma(g_0) \leq (1+\delta)(II + III)$$
$$\leq (1+\delta)(1+\frac{1}{\delta})\frac{2}{\delta}\frac{\lambda_n}{\lambda_{\min}}\sqrt{d^**}J^{-1}(\frac{2}{\delta}\frac{\lambda_n}{\lambda_{\min}}\sqrt{d^*}). \qquad (91)$$

Finally, if $I \leq \delta(II + III)$ and $II \leq \delta(III)$, we clearly have

$$\Gamma(\hat{g}_n) - \Gamma(g_0) \leq (1+\delta)(II + III) \leq (1+\delta)^2(III). \qquad (92)$$

We now come back to our starting point, namely the assumption $\|\hat{g}_n - g^*\| \leq \bar{R}$, with $\bar{R} = 2 + 4\|g^* - g_0\|$. We may replace in the proof we have so far, the function $\hat{g}_n$ by the convex combination $\hat{g}_{n,t} = t\hat{g}_n + (1-t)g^*$, where $t = 1/(1+\|\hat{g}_n - g^*\|/\bar{R})$. This is because clearly,

$$\|\hat{g}_{n,t} - g^*\| \leq \bar{R}. \qquad (93)$$

and because (77) is also true when $\hat{g}_n$ is replaced by $\hat{g}_{n,t}$. So, so far, we have shown that (36) holds for $\hat{g}_{n,t}$ instead of $\hat{g}_n$.

But now, for $\hat{g}_{n,t}$ implies that (take $\delta = 1$), on **A**,

$$\|\hat{g}_{n,t} - g_0\| \leq 4(\frac{\lambda_n}{n} + \tau^2(g^*|g_0)), \qquad (94)$$

which is less than $1/2$ for $n \geq N$. So then $\|\hat{g}_{n,t}-g^*\| \leq \|\hat{g}_{n,t}-g_0\|+\|g^*-g_0\| \leq \frac{1}{2}+\|g^*-g_0\|$. But then,

$$\|\hat{g}_n - g^*\| \leq (1 + \|\hat{g}_n - g^*\|/\bar{R})(\frac{1}{2} + \|g^* - g_0\|)$$
$$\leq \frac{1}{2} + \frac{3}{4}\|\hat{g}_n - g^*\| + \|g^* - g_0\|, \qquad (95)$$

by our choice $\bar{R} = 2 + 4\|g^* - g_0\| \geq 1 + 4\|g^* - g_0\|$. And now we found that

$$\|\hat{g}_n - g^*\| \leq 2 + 4\|g^* - g_0\| = \bar{R}. \qquad (96)$$

Thus, our starting point is true on the set **A**. This completes the proof. □

**Proof of Theorem 2.** We go back to Lemma 3, and show it holds uniformly in $\beta$. To this end, let us define

$$\bar{A}_{M,R} = \{(\alpha, \bar{\alpha}) : I(\alpha - \bar{\alpha}) \leq M, \|g_\alpha - g_{\bar{\alpha}}\| \leq R\}. \tag{97}$$

Let

$$w_{n,\beta}(\alpha) = \Gamma_{n,\beta}(g_\alpha) - \Gamma_\beta(g_\alpha), \quad \alpha \in \mathbf{R}^m. \tag{98}$$

It is clear that

$$w_{n,\beta}(\alpha) = \beta v_{1,n}(\alpha) + (1 - \beta) v_{2,n}(\alpha), \tag{99}$$

where

$$v_{1,n}(\alpha) = l_{1,n}(\alpha) - l_1(\alpha), \tag{100}$$

$$l_{1,n}(\alpha) = \frac{1}{n} \sum_{i=1}^n |Y_i - g_\alpha(x_i)| \mathbf{1}\{Y_i - g_\alpha(x_i) < 0\}, \tag{101}$$

and

$$l_1(\alpha) = \mathbf{E} l_{1,n}(\alpha). \tag{102}$$

Moreover,

$$v_{2,n}(\alpha) = l_{2,n}(\alpha) - l_2(\alpha), \tag{103}$$

$$l_{2,n}(\alpha) = \frac{1}{n} \sum_{i=1}^n |Y_i - g_\alpha(x_i)| \mathbf{1}\{Y_i - g_\alpha(x_i) \geq 0\}, \tag{104}$$

and

$$l_2(\alpha) = \mathbf{E} l_{2,n}(\alpha). \tag{105}$$

Now, define

$$Z_1 = \sup_{(\alpha,\bar{\alpha}) \in \bar{A}_{M,R}} |v_{1,n}(\alpha) - v_{1,n}(\bar{\alpha})|, \tag{106}$$

and

$$Z_2 = \sup_{(\alpha,\bar{\alpha}) \in \bar{A}_{M,R}} |v_{2,n}(\alpha) - v_{2,n}(\bar{\alpha})|. \tag{107}$$

Also, let

$$Z = \sup_{\beta \in \mathcal{B}} \sup_{(\alpha,\bar{\alpha}) \in \bar{A}_{M,R}} |w_{n,\beta}(\alpha) - w_{n,\beta}(\bar{\alpha})|. \tag{108}$$

As in the proof of Lemma 3, one can show that $\mathbf{E}Z_1 \leq 12M\sqrt{\log n/n}$ as well as $\mathbf{E}Z_2 \leq 12M\sqrt{\log n/n}$. Thus also

$$\mathbf{E}Z \leq 12M\sqrt{\log n/n}. \tag{109}$$

Arguing as in the proof of Lemma 3 and Lemma 4, we now arrive at a uniform in $\beta$ version of Lemma 4:

$$\mathbf{P}\left(\sup_{\beta \in \mathcal{B}} \sup_{\|g_\alpha - g_{\bar{\alpha}}\| \leq R} \frac{|v_{n,\beta}(\alpha) - v_{n,\beta}(\bar{\alpha})|}{I(\alpha - \bar{\alpha}) + \frac{1}{n}} > (12+u)2\sqrt{\frac{\log n}{n}}\right) \leq 4R^2 \exp[-\frac{u^2 \log n}{64 R^2}], \quad (110)$$

for all $R \geq 1$ and $u \geq 6$.

Now, replace in the proof of Theorem 1, the set **A** by

$$\{|v_{n,\beta}(\alpha) - v_{n,\beta}(\alpha_\beta^*)| \leq \lambda_n I(\alpha - \alpha_\beta^*) + \lambda/n, \text{ for all } \|g_\alpha - g_\beta^*\| \leq \bar{R} \text{ and all } \beta \in \mathcal{B}\}, \quad (111)$$

and proceed as there. □

## REFERENCES

1. Akaike, H. Information theory and an extension of the maximum likelihood principle. Proceedings 2nd International Symposium on Information Theory, P.N. Petrov and F. Csaki (eds.), Akademia Kiado, Budapest (1973) 267-281.
2. Donoho, D.L., and Johnstone, I.M. Ideal spatial adaptation via wavelet shrinkage. Biometrika No. 81 (1994) 425-455.
3. Donoho, D.L. Denoising via soft-thresholding. IEEE Transactions in Information Theory No. 41 (1995) 613-627.
4. Edmunds, E., and Triebel, H. Entropy numbers and approximation numbers in function spaces. II. Proceedings of the London Mathematical Society (3) No. 64 (1992) 153-169.
5. Hastie, T., Tibshirani, R., and Friedman, J. The Elements of Statistical Learning. Data Mining, Inference and Prediction. Springer, New York, 2001.
6. Koenker, R., and Bassett Jr. G. Regression quantiles. Econometrica No. 46 (1978) 33-50.
7. Koenker, R., Ng, P.T., and Portnoy, S.L. Nonparametric estimation of conditional quantile functions. $L_1$ Statistical Analysis and Related Methods, Y. Dodge (ed.), Elsevier, Amsterdam (1992) 217-229.
8. Koenker, R., Ng, P.T., and Portnoy, S.L. Quantile smoothing splines. Biometrika No. 81 (1994) 673-680.
9. Massart, P. Some applications of concentration inequalities to statistics. Ann. Fac. Sci. Toulouse No. 9 (2000) 245-303.
10. Ledoux, M., and Talagrand, M. Probability in Banach Spaces, Isoperimetry and Processes. Springer, Berlin, 1991.
11. Loubes, J.-M., and van de Geer S. Adaptive estimation, using soft thresholding type penalties. Statistica Neerlandica No. 56 (2002) 453-478.
12. Pinkus, A. $n$-widths in Approxiamation Theory. Springer, New York, 1985.
13. Portnoy, S. Local asymptotics for quantile smoothing splines. Ann. Statist. No. 25 (1997) 414-434.
14. Portnoy, S., and Koenker, R. The Guassian hare and the Laplacian tortoise: computability of squared error versus absolute-error estimators, with discussion. Stat. Science No. 12 (1997) 279-300.
15. Schwarz, G. Estimating the dimension of a model. Ann. Statist. No. 6 (1978) 461-464.

16. Tibshirani, R. Regression analysis and selection via the LASSO. Journal Royal Statist. Soc. B No. 58 (1996) 267-288.
17. van de Geer, S. Empirical Processes in M-Estimation. Cambridge University Press, 2000.
18. van de Geer, S. M-estimation using penalties or sieves. J. Statist. Planning Inf. No. 108 (2002) 55-69.
19. van der Vaart, A.W., and Wellner, J.A. Weak Convergence and Empirical Processes, with Applications to Statistics. Springer, New York, 1996.

Recent Advances and Trends in Nonparametric Statistics
Michael G. Akritas and Dimitris N. Politis (Editors)
© 2003 Elsevier Science B.V. All rights reserved.

# Set estimation: an overview and some recent developments

Antonio Cuevas[a]* and Alberto Rodríguez-Casal[b]

[a]Departamento de Matemáticas, Facultad de Ciencias
Universidad Autónoma de Madrid, 28049-Madrid, Spain

[b]Departamento de Estadística e Investigación Operativa
Fac. de Ciencias Económicas y Empresariales, Universidad de Vigo
36200 Vigo, Spain

We consider the problem of estimating a compact set $S \subset \mathbb{R}^d$ from random observations whose distribution is related to $S$. Typically, the target set is the support of the distribution which generates the data, or a level set of the underlying density, or the boundary of any of these sets. This paper provides an overview of the available literature on this topic. In the last section we offer an advance of our recent unpublished work on the subject.

## 1. INTRODUCTION

Set estimation deals with the problem of estimating a compact set $S \subset \mathbb{R}^d$ from a random sample of points whose distribution is, in some sense, related to $S$. For example, the data could be a random sample $X_1, \ldots, X_n$ drawn from a density $f$ on $\mathbb{R}^d$ and the target set $S$ could be the support of $f$ or a level set of type $\{f \geq c\}$. The discussion below will mainly fit in this setup. Most theoretical results (e.g., [16,14,54]) concern the asymptotic behavior, as $n \to \infty$, of suitable estimators $\hat{S}_n$ of $S$. Another useful sampling model, e.g. [47], assumes that the data come from a realization of a Poisson process whose intensity function is $\nu f$, where $f$ is an unknown function with support $S$ and $\nu$ is a constant. In this case the aim is also estimating $S$ but the size of the available data set is random and the asymptotic results are established assuming that $\nu \to \infty$, which amounts to let the expected number of observations grow to infinity.

A different framework leading to a set estimation problem arises in image analysis. A binary (black and white) image can be identified with the set $S \subset [0,1]^2$ of activated pixels on a screen. An usual aim is recovering the image $S$ from observations which are perturbed by a random noise. A simple model is established in regression terms by assuming that the data are $(X_i, Y_i)$, $i = 1, \ldots, n$, where $Y_i = \mathbb{I}_{\{X_i \in S\}} + \epsilon_i$, $\mathbb{I}_A$ is the indicator function of the set $A$, $X_i \in [0,1]^2$ are the design variables (possibly random) and $\epsilon_i$ the error variables. See, e.g., [37] for more details.

*This work has been partially supported by grants BFM2001-0169 (A. Cuevas) and BFM2002-03213 (A. Rodríguez-Casal), from the Spanish Ministry of Science and Technology. The work of the second author has also been partially supported from the University of Vigo and the Xunta de Galicia under grant PGIDIT02PXIA30003PR.

As this example suggests, set estimation could be seen as a particular case of functional estimation (since a set can be identified with its indicator function). However, there is at least an important difference: the geometrical considerations, often expressed in terms of intuitive "visual" conditions, play an important role here. In particular, the theoretical results are usually established in terms of distances between sets, which are reminiscent of functional distances but have an independent meaning in geometrical terms. The main distances so far considered in the theory are the **Hausdorff metric**, $d_H$, and the **distance in measure**, $d_\mu$. See [1] for other distances and additional references.

Given two compact non-empty sets $S$, $T$ in $\mathbb{R}^d$, we define

$$d_H(S,T) = \max\left\{\sup_{x\in T} d(x,S), \sup_{y\in S} d(y,T)\right\} = \inf\{\epsilon > 0 : T \subset B(S,\epsilon), S \subset B(T,\epsilon)\}$$

$$d_\mu(S,T) = \mu(S\Delta T),$$

where $d(x,S) = \inf\{\|x-y\| : y \in S\}$, $B(S,\epsilon) = \{x \in M : d(x,S) \leq \epsilon\}$, $\mu$ is a measure on $\mathbb{R}^d$ (often the Lebesgue measure $\mu_L$) and $S\Delta T = (S\setminus T) \cup (T\setminus S)$.

It is not difficult to see that

$$d_H(S,T) = \sup_{x\in M} |d(x,S) - d(x,T)| \text{ and } d_\mu(S,T) = \int |\mathbb{1}_S - \mathbb{1}_T| d\mu,$$

so that the distances $d_H$ and $d_\mu$ are conceptually related to the supremum ($L_\infty$) and the $L_1$ functional distances, respectively.

The Hausdorff metric is well-suited to quantify the notion of visual proximity between two sets. It has been used in different problems of correspondence theory, fractals, random sets and image analysis; two well-known references are [20] and [39].

The distance in measure is more appropriate for those problems where we are concerned with the set contents rather than with physical proximity between sets. An interesting example arises in the framework of supervised learning (also called pattern recognition, see [17]). As an important particular case, consider the following **binary classification problem**: We want to predict the value of a random response $Y \in \{0,1\}$ from the observation of a random $d$-valued covariate $X$ related to $Y$. Assume that we have available a "training sample" $D_n = \{(X_i, Y_i), i = 1, \ldots, n\}$ of pairs covariate-response. This situation is very common in practice, in connection with different problems of quality control, automatic diagnostic, etc.; see [32]. For each $T \subset \mathbb{R}^d$ we can define a classification rule (i.e., a predictor of $Y$) by $\hat{Y} = \mathbb{1}_T$. It can be seen that the classification error, $R(T) = P\{Y \neq \hat{Y}\}$, is minimized by the rule $\hat{Y} = \mathbb{1}_S$, where $S = \{x : \eta(x) \geq 1/2\}$ and $\eta(x) = P(Y = 1|X = x)$. Then, the classification problem is equivalent to define, from the data $D_n$, a suitable estimator $S_n$ of $S$. The efficiency of the corresponding rule can be measured by

$$R(\hat{S}_n) - R(S) = \int_{S\Delta \hat{S}_n} |2\eta(x) - 1| P_X(dx) = \mu\left(S\Delta \hat{S}_n\right) = d_\mu\left(S, \hat{S}_n\right),$$

where $\mu(dx) = |2\eta(x) - 1| P_X(dx)$; see [53].

This work is mainly devoted to summarize the current state of the set estimation theory. Our point of view is inevitably biased by our own experience in this subject. So, we make no claim of completeness. For convenience, our survey is organized in three sections corresponding, respectively, to the problems of estimating the support of a multivariate density, their level sets and the support boundary. In each case, we will briefly comment the available literature and outline some applications. Finally, in the last section we will summarize our recent unpublished work in this field.

## 2. SUPPORT ESTIMATION

Perhaps the most direct problem in set estimation is approximating a compact set $S \subset \mathbb{R}^d$ from a random sample of points $X_1, \ldots, X_n$ taken into $S$. In more formal terms, the aim is estimating the support of an absolutely continuous probability measure $P_X$ from $n$ independent observations with distribution $P_X$.

### 2.1. The convex case

The case where $S$ is assumed to be convex deserves a separate treatment. In fact, it has been first considered in the literature probably due to the mathematical appeal of the convexity assumption and to the fact that in this case there is a quite natural estimator: the **convex hull of the sample** $H_n = conv(X_1, \ldots, X_n)$. Two early references in the study of $H_n$ (in the bidimensional case $d = 2$) are [45] and [46] where the asymptotic behavior of the expected number of sides of $H_n$, and its expected perimeter and area are analyzed in detail. Further results in this line, with generalizations to higher dimensions, are given in [50]. Maximum likelihood properties for $H_n$ are given in [37] and [47].

Convergence rates for $d_H(H_n, S)$ have been obtained by Dümbgen and Walther in [19]. These authors show that, in general,

$$d_H(H_n, S) = O\left(\frac{\log n}{n}\right)^{1/d}, \tag{1}$$

which in fact turns out to be a typical convergence rate in set estimation.

Under some smoothness conditions on the boundary $\partial S$ the convergence rate is faster as the exponent $1/d$ in (1) can be replaced by $2/(d+1)$.

An interesting survey can be found in [57].

### 2.2. The general case

If $S$ is not convex, the use of $H_n$ makes no sense. A more flexible estimator is needed. Let us note that the sample $\aleph_n = \{X_1, \ldots, X_n\}$ itself is a $d_H$-consistent estimator of the support $S$. This follows from a straightforward probability reasoning which requires no assumption on the shape of $S$. It is also clear, however, that the estimator $\aleph_n$ is of no use if we are interested in the measure metric $d_{\mu_L}$. These simple observations lead to consider an smoothed version of the sample (very much in the nonparametrics spirit) such as the **Devroye-Wise estimator** (see [10,16,23]), defined by

$$\widehat{S}_n = \bigcup_{i=1}^{n} B(X_i, \epsilon_n), \tag{2}$$

where $\epsilon_n$ is a sequence of smoothing parameters which must tend to zero "not too fast" in order to get consistency in measure. In fact, the Devroye and Wise [16] showed that the assumptions $\epsilon_n \to 0$ and $n\epsilon_n^d \to \infty$ (identical to those imposed on the smoothing parameter in nonparametric density estimator to ensure consistency in the integral norms) imply the consistency in probability $d_\mu\left(S, \widehat{S}_n\right) \xrightarrow{P} 0$ for any measure $\mu$ whose restriction to $S$ is absolutely continuous with respect to the distribution $P_X$ of the $X_i$.

Convergence rates with respect to $d_H$ for $\widehat{S}_n$ are studied in [11] and [15].

An application of the estimator (2) in **statistical quality control** (SQC) has been also proposed by Devroye and Wise in [16]: suppose that we sequentially receive independent observations $X_1, X_2,...$ from a system or mechanism which initially (for a large enough number of stages $i = 1, 2 ..., n_0$) is "under control", in the sense that the $X_i$ come from a common, unknown, underlying distribution $P_X$ which corresponds to the "normal behavior" of the system. Suppose further that this distribution changes at some unknown stage $n_0$ in such a way that the distribution of $X_i$ for $i > n_0$ is no longer $P_X$ (in the SQC terminology, "the system goes out of control"). The problem is to estimate, as soon as possible, the changepoint $n_0$ from the observation "on line" of the $X_i$. The detection method proposed by Devroye and Wise is based on a simple idea: we might estimate $n_0$ by the first value $N$ such that $X_{N+1} \notin \widehat{S}_N$.

Notice that this method could be seen as a nonparametric multivariate version of the classical Sewhart control charts. Note also that the total error probability

$$P\left(\{X_{N+1} \notin \widehat{S}_N,\ X_{N+1} \in S\} \cup \{X_{N+1} \in \widehat{S}_N,\ X_{N+1} \notin S\}\right),$$

is given by

$$d_\mu\left(S, \widehat{S}_N\right) = \mu\left(S \triangle \widehat{S}_N\right),$$

where $\mu$ is the distribution of the new observation $X_{N+1}$. So this is a further example of a practical situation where the distance between the target set and its estimator should be evaluated in terms of the measure metric.

This application of the estimator (2) to quality control is also analyzed in [4] where convergence rates for the probability of false alarm $P_X(S \setminus \widehat{S}_n)$ are obtained. A shape condition on the set $S$, called **standardness** (which excludes the presence of sharp peaks in the set) plays an important role here and arises also frequently in set estimation (see, e.g., [11], [14]). A Borel set $S \subset \mathbb{R}^d$ is said to be **standard** with respect to a Borel measure $\mu$ if there exist $\lambda > 0$ and $\delta > 0$ such that

$$\mu(B(x,\epsilon) \cap S) \geq \delta \mu_L(B(x,\epsilon)), \quad \forall x \in S, \quad 0 < \epsilon \leq \lambda, \tag{3}$$

where $\mu_L$ denotes the Lebesgue measure on $\mathbb{R}^d$.

The simplicity of the Devroye-Wise estimator (2) is an important advantage in many respects. For example, its connected components are easy to identify, which has an interesting application in cluster analysis (see [12,13]). Also, this estimator allows us, in some cases, to incorporate shape restrictions to the estimation. For example, if we know in advance that the target set $S$ is connected, we can incorporate this restriction to the estimator $\widehat{S}_n$ by simply taking the smoothing parameter $\epsilon_n$ as the minimum value of $\epsilon$ such that $\widehat{S}_n = \widehat{S}_n(\epsilon)$ is connected. A similar idea has been used in [2] to tackle the problem of estimating a star-shaped set ($S$ is said to be star-shaped if there exists $a \in S$ such that, for all $x \in S$ the segment $[a, x]$ joining $a$ and $x$ is included in $S$).

In other cases, shape restrictions on $S$ arise as technical conditions that make easier the theoretical analysis. Thus in [37] minimax convergence rates are obtained by assuming that the boundary of $S$ fulfills some piecewise Lipschitz conditions.

In spite of its evident advantages, the estimator (2) suffers from some inherent problems which, to some extent, are analogous to those of the histogram density estimator when compared with other more sophisticated alternatives. In particular, the boundary of $\hat{S}_n$ is non-differenciable. A more flexible alternative estimator has been considered by Cuevas and Fraiman in [14] as a by-product of nonparametric density estimation. These authors define $S_n = \{f_n > \alpha_n\}$, where $f_n$ is a sequence of nonparametric density estimators (usually of kernel type, see e.g. [56]), based on the original sample and $\alpha_n \downarrow 0$ is a sequence of tuning parameters. Observe that $\partial S_n$ can be smooth if $f_n$ is suitably chosen. Very general consistency results, as well as $d_{\mu_L}$ and $d_H$-convergence rates, are obtained for these estimators. Again, the standardness condition (3), together with a "blow-up" assumption of type $\mu_L(B(S,\epsilon)) = \mu_L(S) + O(\epsilon)$ arise here in a very natural way. Also, some results by Janson [33] on multivariate spacings (the $k$-spacing is the volume of the $k$-th greatest ball $B$ such that $B \subset S$ and $X_i \notin S$ for all $i$) have been very useful to show a partial optimality condition for the obtained $d_H$ rates (of type $O((\log n/n)^{1/d})$).

Chaudhury et al. [9] have obtained a $d_{\mu_L}$-consistency result and convergence rates for an estimator, which they call **s-shape**, based on a different principle: in the bidimensional case $d = 2$, let us cover the set $S$ by a lattice of square grids of appropriate length $s$. The estimator is then defined as the union of the grids containing sample points (or a smoothed version of this union).

### 3. LEVEL SET ESTIMATION

Given an absolutely continuous probability distribution $P_X$ on $\mathbb{R}^d$ with density $f$, we define for $c > 0$ the $c$-level set $L(c) = \{f \geq c\}$. For $c = 0$ we define the level set $L(0)$ as the topological closure of $\{f > 0\}$. It is not difficult to see that, under very general conditions, the support $S$ of $P_X$ coincides with $L(0)$. So it is natural to see level sets as direct extensions of the concept of support.

There are several practical motivations for the use of level sets. For example, we might define, for $\alpha \in (0,1)$, the $1-\alpha$ significant support of $P_X$ as the level set $\{f \geq c\}$, where $c$ is chosen in such a way that $P_X\{f \geq c\} = 1-\alpha$. This suggests an application, considered in [3], in the spirit of the **nonparametric detection method** commented in the previous section: we might decide that the system has gone "out of control" the first time that we observe $X_{n+1} \notin L_n(c_n)$ where $L_n(c_n) = \{f_n \geq c_n\}$ is a plug-in estimator of the level set $L(c)$, $f_n$ is a kernel estimator of $f$ and $c_n$ is the solution, in $c$, of the equation

$$\int_{\{x:f_n(x)\geq c_n\}} f_n(x)dx = 1 - \alpha.$$

Observe that, in this setup, the values $\alpha$ and

$$\alpha_n = \int_{\{x:f_n(x)<c_n\}} f(x)dx$$

are, respectively, the theoretical and the true probability of false alarm. Convergence rates for the consistency $\alpha_n \to \alpha$ are obtained in [3] for the univariate case. In the most favorable situation (with compact-supported $f$) these rates are of type $n^{-1/3}$ up to factor $(\log n)^{-1/3}$.

Another interesting motivation for using level sets is **cluster analysis**. The techniques for grouping data are sometimes based on heuristic considerations leading to a data-based procedure with no clear population target in mind. However, if we adopt Hartigan's [30] concept of a **population cluster** as a connected component of the $c$-level set $\{f \geq c\}$ ($c$ being a given constant), it is quite obvious that level set estimation could be a useful auxiliary tool in the definition of a clustering algorithm. This has been carried out in [13]. The authors define the **empirical clusters** as the connected components of a plug-in kernel-based estimator $\{f_n \geq c\}$ of the $c$-level set. Then the basic idea of the algorithm is just to classify the observations $X_i$ in **data clusters** according to the empirical cluster they belong. An interesting practical problem arises to effectively identify the connected components of the empirical level set $\{f_n \geq c\}$. In general, this is not an easy task. The procedure suggested in [13] relies on the idea of approximating, via bootstrap, the set $\{f_n \geq c\}$ using a Devroye-Wise estimator of type (2) based on the bootstrap observations. This points out again the practical usefulness of this simple estimator.

A similar idea has been considered in [12] for the problem of estimating the number of clusters.

Given a level set $L(c) = \{f \geq c\}$, a plug-in estimator of type $L_n(c) = \{f_n \geq c\}$ can be a good choice if there is no information available on the shape of $L(c)$. Although, in general terms, $L_n(c)$ has good "universal" properties as an estimator of $L(c)$ (see, e.g., [40]) the properties of $L_n(c)$ depend heavily on the smoothness properties of $f$ rather than on the shape of $L(c)$ itself. However, in some cases is necessary to incorporate some restrictions on $L(c)$ as, for example, convexity [31], elliptic shape [42], or some regularity conditions on the boundary [43], which are not easy to handle with a plug-in approach.

Hartigan [31] and Müller and Sawitzki [41] proposed a methodology to estimate level sets incorporating shape restrictions. This methodology is based on the concept of **excess mass**. Given a Borel set $A$, the $c$-excess mass is defined by

$$M_c(A) = \int_A f(x)dx - c\mu_L(A),$$

It is clear that $M_c(L(c)) \geq M_c(A)$, for any $A$. Let's now define the empirical excess mass by

$$M_{n,c}(A) = \frac{1}{n}\sum_{i=1}^{n} \mathbb{I}_{\{X_i \in A\}} - c\mu_L(A).$$

If we assume that the level sets belong to a certain class $\mathcal{G}$, a natural estimator of $L(c)$ could be obtained by maximizing on $\mathcal{G}$ the empirical excess mass:

$$L_{n,\mathcal{G}}(c) = \mathrm{argmax}_{A \in \mathcal{G}} M_{n,c}(A).$$

For example one could take $\mathcal{G} = \ell_m$, the class of the sets in the real line which are union of at most $m$ closed intervals. Note that if $f : \mathbb{R} \longrightarrow \mathbb{R}$ is continuous and bounded with $m$ modes, all the corresponding level sets belong to $\ell_m$. This property can be used in the construction a **multimodality test** to compare, for a given number of modes $m > 1$, the empirical excess mass $M_{n,c}(L_{n,\ell_m}(c))$ of the level set $L_{n,\ell_m}(c)$ with $M_{n,c}(L_{n,\ell_1}(c))$ for different values of $c$. If $f$ is unimodal, both empirical excess masses should be similar for all values of $c$ since they are evaluated on consistent estimators of the same level set $L(c) \in \ell_1$. See [41] for a detailed account of this test. More general multimodality tests can be found in [43].

Level set estimators based on the excess mass approach have been considered, for different classes $\mathcal{G}$ of admissible level sets, in [31], [41], [42], [43] and [44]. In particular, these papers provide consistency results and convergence orders with respect to the distance in measure.

The problem of analyzing the optimality of the excess mass estimators, in the minimax sense, is considered in [52]. More precisely, given a class of sets $\mathcal{G}$ and a family $\mathcal{F}$ of density functions such that the level set $L(c)$ belongs to $\mathcal{G}$, the optimum rate of convergence is defined as a sequence $\{\psi_n\}$ for which there exists a sequence of estimators $L_n^*$ satisfying

$$\limsup_{n \to \infty} \sup_{f \in \mathcal{F}} \mathbb{E}_f \left( \omega \left( \frac{d(L_n^*, L(c))}{\psi_n} \right) \right) < \infty$$

and

$$\liminf_{n \to \infty} \inf_{L_n} \sup_{f \in \mathcal{F}} \mathbb{E}_f \left( \omega \left( \frac{d(L_n, L(c))}{\psi_n} \right) \right) > 0,$$

where $\mathbb{E}_f$ denotes expectation with respect to the density $f$, $\inf_{L_n}$ denotes the infimum over all possible estimators of $L(c)$, $d$ is a distance between sets (usually $d_H$ or $d_{\mu_L}$) and $\omega$ is a loss function whose growth rate is at most polynomial. Tsybakov [52] has shown that the optimal estimator $L_n^*$ for wide classes $\mathcal{G}$ of level sets and $\mathcal{F}$ of densities, arise as a result of maximizing a local version of the empirical excess mass. This entails that, in some cases, the estimators based on the global maximization of the empirical excess mass do not attain the optimal convergence rates. The estimators proposed in [52] have a piecewise polynomial boundary. The methodology used in this paper is similar to that of [37] minimax rates are obtained for other problems of image analysis or support estimation.

The excess mass approach is a very useful tool to obtain optimal convergence rates which allow us to assess the performance of different procedures. However, generally speaking, the excess mass estimators (either in local or in global version) are not easy to implement in practice. For example, in [31] an algorithm to obtain $L_{n,\mathcal{C}^2}(c)$ is given, $\mathcal{C}^2$ being the class of closed and convex sets on $\mathbb{R}^2$, although such algorithm has a complexity of order $O(n^3)$. For other problems in higher dimensions with broader classes of sets and moderate or large sample sizes the computational burden can be beyond any practical use.

An interesting computationally efficient estimator of level sets has been proposed by Walther [54]. This estimator is similar to (2) in that both are defined as a union of balls, with a common radius, centered at sample points. However, in the case of Walther's estimator not every sample point is selected to compose this union. As we will indicate below, the selection process of the sample points is based on an auxiliary density estimator, which entails some computational cost, but given the simple structure of the final estimator, the whole procedure is computationally efficient even in higher dimensions.

On the other hand, the estimator proposed in [54] is designed to work under a particular shape restriction (of a relevant interest in the mathematical theory of image analysis) which we will comment next in some detail before defining the estimator.

In mathematical morphology (see [51], p. 144) is often considered a class $\mathcal{G}$ of sets, usually called Serra's regular model, which consists of all compact sets $S \subset \mathbb{R}^d$ such that, for some $\lambda > 0$, fulfill

$$S = (S \oplus \lambda B) \ominus \lambda B = (S \ominus \lambda B) \oplus \lambda B, \tag{4}$$

where $B$ denotes the closed ball centered at zero with unit radius, $\lambda B = \{\lambda b : b \in B\} = B(0, \lambda)$ and the symbols $\oplus$, $\ominus$ denote the Minkowski sum and difference, defined respectively by $A \oplus C := \{a + c : a \in A, c \in C\}$, $A \ominus C := \{x : \{x\} \oplus C \subset A\}$ for any $A, C \subset \mathbb{R}^d$. It can be seen that $A \ominus C = (A^c \oplus (-1)C)^c$ (where the superscript $c$ denotes complement). A deep study of Serra's regular model, which in particular provides several useful characterizations for (4), is given in [55].

The problem of level set estimation when the level sets are assumed to satisfy a condition of type (4) has been considered by Walther [54]. More specifically, this author defines the class $\mathcal{G}(r)$ of sets satisfying $S = (S \ominus \lambda B) \oplus \lambda B$, for $0 \leq \lambda \leq r$, and $S = (S \oplus \lambda B) \ominus \lambda B$, for $0 \leq \lambda < r$, $r$ being a given constant. Note that if $s < r$ then $\mathcal{G}(r) \subset \mathcal{G}(s)$; also $\mathcal{G} = \cup_{r>0} \mathcal{G}(r)$. In [54] rather general conditions are given to ensure that the level sets belong to the class $\mathcal{G}(r)$.

The basic idea behind Walther's estimator is that if $L(c) \in \mathcal{G}(r)$ then $L(c)$ should fulfill

$$L(c) = (L(c) \ominus \lambda B) \oplus \lambda B = (L(c)^c \oplus \lambda B)^c \oplus \lambda B, \quad 0 \leq \lambda \leq r,$$

where in the last equality we have used the duality relation $A \ominus C = (A^c \oplus (-1)C)^c$. Then, given a random sample $X_1, \ldots, X_n$ from $f$, a sort of empirical version, see [54], of this morphological operation would lead to the estimator

$$L_n(c) = \left(\left(\aleph_n^-(c) \oplus r_n B\right)^c \cap \aleph_n^+(c)\right) \oplus r_n B,$$

where $r_n$ is a smoothing parameter with $r_n < r$,

$$\aleph_n^+(c) = \{X_i : f_n(X_i) \geq c\}, \quad \aleph_n^-(c) = \{X_i : f_n(X_i) < c\},$$

and $f_n$ is an auxiliary nonparametric density estimator (e.g., of kernel type).

The sample points $X_i$'s used in the estimator are those in $\aleph_n^+(c)$ whose distance to $\aleph_n^-(c)$ is larger than $r_n$. The computational cost of identifying these points is of order $O(dn^2)$. Thus the computational complexity increases only linearly in $d$. Moreover, in [54] it is shown that if $f$ is smooth enough and $L(c) \in \mathcal{G}(r)$ the above estimator can outperform $L_{n,\mathcal{G}(r)}(c)$. Also, if $f$ is discontinuous and $d > 2$ it is possible, with a slight modification of the indicated smoothing procedure, to get an estimator attaining the minimax $d_{\mu_L}$-rates obtained by Mammen and Tsybakov [38] with an estimator based on a local version of the excess mass approach.

## 4. BOUNDARY ESTIMATION

A popular practical motivation for the problem of boundary estimation comes from the field of productivity analysis where an interesting issue is to assess the efficiency of a company which transforms some inputs $x$ (capital investments, human resources,...) into an output $y$ (capital gains).

If we summarize the performance of a company by the pair $(x,y)$, we might assume that, for every company in a given production sector, $(x,y)$ must belong to a **hypograph set** such as

$$S = \{(x,y) : x \in \mathcal{I},\, y \leq g(x)\},$$

where $\mathcal{I}$ is the set of all possible inputs and $g(x)$ represents the maximum output attainable from the input $x$. The function $g$ defines the upper boundary of $S$ which, in view of the above motivation, can be seen as an "efficient frontier". Indeed note that, for a given company with performance $(x_0, y_0)$, $g(x_0) - y_0$ is a measure of economic efficiency.

Of course, in practice the set $S$ and its upper boundary $g$ are unknown. The problem is estimating the efficient frontier from a sample $(x_1, y_1), (x_2, y_2), \ldots$. This is again an example of a set estimation problem where the target set incorporates in a natural way a shape restriction: in this case $S$ must be the hypograph of a function. Additional restrictions can be incorporated through conditions imposed on $g$. For example, if $x$ is univariate and $g$ is assumed to be concave and increasing, the hypograph $S$ turns out to be convex. This situation arises in the setup of the so-called Capital Assets Pricing Models, where the aim is analyzing the behavior of different portfolios $(x,y)$, $x$ being a risk measure (volatility, variance) of an investment strategy and $y$ the corresponding average return. The standard assumptions on this type of models entail that $g$ is increasing and concave and hence the set $S$ of feasible portfolios is convex; see [22] and references therein. A classical reference for this case is [21] where a natural estimator, called **DEA** (*data envelopment analysis*) is proposed for the efficient frontier. This estimator is defined as the lowest increasing concave function $\hat{g}$ whose hypograph contains the sample points. The linear programming techniques (see [8]) are useful in the practical implementation of this estimator. The statistical properties of $\hat{g}$ have been studied more recently: some results of consistency, convergence rates and optimality can be found in [5], [34] and [36]. The asymptotic distribution of $\hat{g}(x_0)$ is obtained in [22]. Other works have considered the estimation of the efficiency boundary assuming other conditions on $g$ (different from convexity) as monotonicity [35] or smoothness [24,26,29].

If we consider the efficient frontier problem within the general framework of set estimation, we may see that the real target here is not properly to estimate a set but a function. As a consequence, the distances between sets are not so relevant here, as we are mainly interested in the vertical difference $g(x) - \hat{g}(x)$. However, the particular structure of the data, which are drawn from a set whose upper border is defined by $g$, makes still reasonable to see this problem from the point of view of set estimation. On the other hand, as mentioned in [37] the estimation of hypographs has some independent interest as these of sets can be seen as "building blocks" to construct other more complicate structures.

The problem of boundary estimation arises also in other contexts motivated by image analysis problems in fields such as agricultural or environmental studies based on satellite images, medical diagnosis, etc. For example, the problem of approximating a black-and-white image is essentially equivalent to that of estimating its boundary. Of course, image analysis is a broad subject. The statistical aspects represent only a part of the interesting issues in this field. Some works, with a statistical orientation, dealing with problems of boundary approximation from the point of view of image analysis are [6,18,25,27,28,37,49]. A common feature of some of these works is the modelization of an image in terms of a regression function whose value represents the color intensity at every point.

In [2] the problem of boundary estimation is considered in the general context of set estimation as outlined at the beginning of this paper. We first have a reasonable estimator $S_n$ of a set $S$ which, typically, will require at least the $d_H$-consistency $d_H(S_n, S) \to 0$. Then, additional results on boundary estimation of type $d_H(\partial S_n, \partial S) \to 0$ are studied. The idea is to exclude the "irregular behavior" $d_H(S_n, S) \to 0$ and $d_H(\partial S_n, \partial S) \not\to 0$, shown, for instance, by the "raw sample" $S_n = \aleph_n$. Sometimes the $d_H$-consistency of $\partial S_n$ towards $\partial S$ is obtained by imposing a shape assumption on $S$ which is incorporated to $S_n$. Thus in [2] is assumed that $S$ is star-shaped and this condition is incorporated to the estimator (2) through a suitable choice of the smoothing parameter $\epsilon$.

Let us finally mention the paper by Carlstein and Krishnamoorthy [7], devoted to the problem of estimating the border between two regions. The data come from random observations made on the nodes of a grid with different distributions in both regions.

## 5. SOME FORTHCOMING DEVELOPMENTS: A PERSONAL ACCOUNT

In this final section we briefly summarize our recent unpublished work in this subject as it was presented in the *International Conference on Current Advances and Trends in Nonparametric Statistics* held on July 15-19, 2002 in Crete (Greece). So our attempts to provide a survey, even partial, stop here. This section is just a personal account of recent results.

Our interest has been mainly focused on boundary estimation. In particular, we have considered the following problem (placed itself in the border between nonparametric statistics and stochastic geometry): under what conditions a smoothed version of the sample as defined by the Devroye-Wise estimator (2) is able to estimate the support boundary $\partial S$? We have proved [15] that, for every compact support $S$,

$$\epsilon_n \to 0 \text{ almost surely (a.s.), together with } S \subset \hat{S}_n \text{ imply } d_H(\partial \hat{S}_n, \partial S) \to 0, \text{ a.s.} \tag{5}$$

Hence the main point is that when the smoothing parameter $\epsilon_n$ is taken large enough to provide an overestimation of $S$, the consistent estimation of $S$ entails that of its boundary. This situation bears some analogy with that of density estimation, where a moderate oversmoothing can be beneficial to estimate qualitative features (inflexion points, bumps,...) related to higher order derivatives.

Our second interest has been to obtain convergence rates for $d_H(\partial \hat{S}_n, \partial S)$. This requires some shape restrictions on $S$. Besides the standardness condition defined in (3), we need to assume an additional condition, which we denominate **partial expandability**. It is defined as follows. A bounded Borel set $S \subset \mathbb{R}^d$ is said to be partially expandable if there exist constants $r > 0$ and $C(S) \geq 1$ such that

$$d_H(\partial S, \partial(S \oplus \epsilon B)) \leq C(S)\epsilon, \quad 0 \leq \epsilon < r. \tag{6}$$

This condition prevents the set $S$ from having sharp inlets. For example, the hypograph $\{(x, y) \in \mathbb{R}^2 : -1 \leq x \leq 1 \text{ and } 0 \leq y \leq 1 + x^{1/3}\}$ is not partially expandable because the sharp inlet in $(0, 0)$.

We have shown [15] that if $S$ is standard and partially expandible and

$$\epsilon_n = C \left(\frac{\log n}{n}\right)^{\frac{1}{d}}, \text{ for some } C > \left(\frac{2}{\delta \omega_d}\right)^{\frac{1}{d}} \tag{7}$$

then, with probability one there exists $n_0$ such that

$$d_H(\partial \hat{S}_n, \partial S) \leq C(S)\epsilon_n, \quad n \geq n_0, \text{ and } d_H\left(\hat{S}_n, S\right) \leq \epsilon_n, \quad n \geq n_0.$$

Condition (7) ensures eventually the crucial condition $S \subset \hat{S}_n$. The unique unknown element in (7) is the standardness constant $\delta$, defined in (3) which, in a way, measures "how spiky" $S$ is. Thus (7) provides a practical guide to select $\epsilon_n$ through a conservative guess about the regularity of $S$, expressed in terms of a value of $\delta$.

The asymptotic behavior of $d_H(\{f_n \geq c\}, \{f \geq c\})$ has been considered by Molchanov [40] in the case that $f_n, f$ are functions (not necessarily densities) defined on a metric space (not necessarily $\mathbb{R}^d$). Such an abstraction level is not a mere formalism. In fact it has some direct practical motivations. For example, if we have spherical data (coming, e.g., from coordinates of double-stars projected onto the unit sphere; see [48]), we should consider an underlying density $f$ and density estimators $f_n$ defined on a variety different from the euclidean space $\mathbb{R}^d$. On the other hand, the interest of studying level sets $\{f \geq c\}$ is not limited to the case where $f$ is a density. For instance, a standard model in image analysis (see e.g. [37]) is

$$Y_i = f(X_i) + \epsilon_i, \quad i = 1\ldots n,$$

where the design points $X_i$ are pixels randomly chosen on $[0,1]^2$, the $\epsilon_i$ are noise iid variables with mean zero and $f$ is the function which expresses the image in grey scale. Hence in this case $f$ is a regression function and the level set $\{f \geq c\}$ corresponds to the image areas where the intensity exceeds a certain value.

In [48], Chapter 3, we have analyzed, in the indicated general setting, the sequence $d_H(\partial\{f_n \geq c\}, \partial\{f \geq c\})$ and have proved that, under very broad conditions, the assumption $d_H(\{f = t\}, \{f = c\}) = O(|t - c|)$ implies

$$d_H(\partial \{f \geq c\}, \partial \{f_n \geq c\}) = O(\|f_n - f\|_\infty).$$

So basically the uniform consistency of $f_n$ to $f$ is a sufficient condition (if we exclude the presence of a plateau in $\{f = c\}$) for $d_H(\partial \{f \geq c\}, \partial \{f_n \geq c\}) \to 0$ and the corresponding convergence rates coincide.

Coming back to support estimation, we have considered in the last chapter of [48], the problem of estimating a support $S$ under the assumption that it exists $r > 0$ such that $S$ belongs to the class $\mathcal{G}(r)$ defined in Section 3 above. Walther (see [55], Th. 1) has proved that this implies the $r$-convexity of $S$, that is, $S$ can be expressed as the intersection of the complementary sets of all closed balls of radius $r$ which are disjoint with $S$. This can be seen a an extension of the concept of convexity; indeed, if we let $r$ tend to infinity we get the convex case in the limit, as a convex set can be expressed as the intersection of the halfspaces which contain the set. Nevertheless, $r$-convexity is a much broader shape condition which allow the set to have inlets (provided they are not too narrow or deep).

Let us also point out that $S \in \mathcal{G}(r)$ is in fact a smoothness condition more restrictive than $r$-convexity; see again [55], Th. 1 for details. In any case, if $S$ fulfils (4) a natural estimator of $S$ from a random sample would be, the $r$-**convex hull**, $C_r(\aleph_n) = (\aleph_n \oplus rB) \ominus rB$, where $\aleph_n$ denotes the sample, and $B$ the unit ball. We have proved ([48], chapter 4)

$$d_H(S, C_r(\aleph_n)) = O\left(\left(\frac{\log n}{n}\right)^{\frac{2}{d+1}}\right), \quad a.s.,$$

$$d_H(\partial S, \partial C_r(\aleph_n)) = O\left(\left(\frac{\log n}{n}\right)^{\frac{2}{d+1}}\right), \quad a.s.,$$

$$d_{\mu_L}(S, C_r(\aleph_n)) = O\left(\left(\frac{\log n}{n}\right)^{\frac{2}{d+1}}\right), \quad a.s.$$

thus obtaining (under an identical smoothness condition) the same convergence rate for $d_H(S, H_n)$ proved in [19] for the convex hull $H_n$ if $S$ is assumed to be convex.

## REFERENCES

1. A.J. Baddeley, Errors in binary images and an $L^p$ version of the Hausdorff metric, Nieuw Arch. Wisk. 4 (1992) 157-183.
2. A. Baíllo and A. Cuevas, On the estimation of a star-shaped set, Adv. Appl. Probab. 33 (2001) 1-10.
3. A. Baíllo, J.A. Cuesta-Albertos and A. Cuevas, Convergence rates in nonparametric estimation of level sets, Statist. Probab. Lett. 53 (2001) 27-35.
4. A. Baíllo, A. Cuevas and A. Justel, Set estimation and nonparametric detection, Canad. J. Statist. 28 (2000) 765-782.
5. R.D. Banker, Maximum likelihood, consistency and data envelopment analysis: a statistical foundation, Management Science, 39 (1993) 1265-1273.
6. E.J. Candès and D.L. Donoho, New tight frames of curvelets and optimal representations of objects with smooth singularities, preprint, Stanford University, 2002.
7. E. Carlstein and C. Krishnamoorty, Boundary estimation, J. Amer. Statist. Assoc. 87 (1992) 430-438.
8. A. Charnes, W. Cooper and E. Rhodes, Mesasuring the inefficiency of decision making units, European J. Oper. Res. 6 (1978) 429-444.
9. A.R. Chaudhury, A. Basu, S.K. Bhandari and B.B. Chaudhury, An efficient approach to consistent set estimation, Sankhya Ser. B 61 (1999) 496-513.
10. J. Chevalier, Estimation du support et du contour de support d'une loi de probabilité, Ann. Inst. H. Poincaré, sec. B 12 (1976) 339-364.
11. A. Cuevas, On pattern analysis in the nonconvex case, Kybernetes 19 (1990) 26-33.
12. A. Cuevas, M. Febrero and R. Fraiman, Estimating the number of clusters, Canad. J. Statist. 28 (2000) 367-382.
13. A. Cuevas, M. Febrero and R. Fraiman, Cluster analysis: a further approach based on density estimation, Comp. Stat. Data Analysis 36 (2001) 441-459.
14. A. Cuevas, A. and R. Fraiman, A plug-in approach to support estimation, Ann. Statist. 25 (1997) 2300-2312.

15. A. Cuevas and A. Rodríguez-Casal, On boundary estimation, preprint, Univ. Autónoma de Madrid and Univ. de Vigo.
16. L. Devroye and G. L. Wise, Detection of abnormal behavior via nonparametric estimation of the support, SIAM J. Appl. Math. 38 (1980) 480-488.
17. L. Devroye, L. Györfi and G. Lugosi, A Probabilisty Theory of Pattern Recognition, Springer-Verlag, New York, 1996.
18. D.L. Donoho, Wedgelets: nearly minimax estimation of edges, Ann. Statist. 27 (1999) 859-897.
19. L. Dümbgen and G. Walther, Rates of convergence for random approximations of convex sets, Adv. Appl. Prob. 28 (1996) 384-393.
20. G.A. Edgar, Measure, Topology and Fractal Geometry, Springer-Verlag, New York, 1990.
21. M.J. Farrell, The measurement of productive efficiency, J. Roy. Statist. Soc. Ser. A 120 (1957) 253-281.
22. I. Gijbels, E. Mammen, B.U. Park, and L. Simar, On estimation of monotones and concave frontier funtions, J. Amer. Statist. Assoc. 94 (1999) 220-228.
23. U. Grenander, Abstract Inference, Wiley, New York, 1980.
24. P. Hall, M. Nussbaum, and S.E. Stern, On the estimation of a support curve of indeterminate sharness, J. Multivariate Anal. 62 (1997) 204-232.
25. P. Hall, L. Peng and C. Rau, Local likelihood tracking of fault lines and boundaries, J. R. Stat. Soc. Ser. B 63 (2001) 569-582.
26. P. Hall, B. U. Park and S.E. Stern, On polynomial estimators of frontiers and boundaries, J. Multivariate Anal. 66 (1998) 71-98.
27. P. Hall and C. Rau, Tracking a smooth fault line in a response surface, Ann. Statist. 28 (2000) 713-733.
28. P. Hall and C. Rau, Likelihood-based confidence bands for fault lines in response surfaces, Probab. Theory Related Fields 124 (2002) 26-49.
29. W. Härdle, B.U. Park, and A.B. Tsybakov, Estimation of non-sharp support boundaries, J. Multivariate Anal. 55 (1995) 205-218.
30. J.A. Hartigan, Clustering Algorithms, Wiley, New York, 1975.
31. J.A. Hartigan, Estimation of a convex density contour in two dimensions, J. Amer. Statist. Assoc. 82 (1987) 267-270.
32. T. Hastie, T., R. Tibshirani, R. and J. Friedman, The Elements of Statistical Learning, Springer-Verlag, New York, 2001.
33. Janson, S., Maximal spacings in several dimensions, Ann. Probab. 15 (1987) 274-280.
34. A. Kneip, B.U. Park and L. Simar, A note on the convergence of nonparametric DEA efficiency measures, Econometric Theory 14 (1998) 783-793.
35. A. Korostelev, L. Simar and A.B. Tsybakov, Efficient estimation of monotone boundaries, Ann. Statist. 23 (1995) 476-489.
36. A. Korostelev, L. Simar and A.B. Tsybakov, On estimation of monotone and convex boundaries, Publ. Inst. Statist. Univ. Paris 39 (1995) 3-18.
37. A.P. Korostelev and A.B. Tsybakov, Minimax Theory of Image Reconstruction, Lecture Notes in Statistics 82, Springer-Verlag, New York, 1993.
38. E. Mammen and A.B. Tsybakov, Asymptotical minimax recovery of sets with smooth boundaries, Ann. Statist. 23 (1995) 502-524.

39. G. Matheron, Random Sets and Integral Geometry, Wiley, New York, 1975.
40. I.S. Molchanov, A limit theorem for solutions of inequalities, Scand. J. Statist. 25 (1998) 235-242.
41. D.W. Müller and G. Sawitzki, Excess mass estimates and tests of multimodality, J. Amer. Statist. Assoc. 86 (1991) 738-746.
42. D. Nolan, The excess-mass ellipsoid, J. Multivariate Anal. 39 (1991) 348-371.
43. W. Polonik, Measuring mass concentration and estimating density contour clusters-an excess mass approach, Ann. Statist. 23 (1995) 855-881.
44. W. Polonik, Concentration and goodness-of-fit in higher dimensions: (Asymptotically) distribution-free methods, Ann. Statist. 27 (1999) 1210-1229.
45. A. Rényi, A. and R. Sulanke, Über die konvexe Hülle von $n$ zufällig gewählten Punkten, Z. Wahrs. und Verw. Gebiete 2 (1963) 75-84.
46. A. Rényi, A. and R. Sulanke, Über die konvexe Hülle von $n$ zufällig gewählten Punkten (II), Z. Wahrs. und Verw. Gebiete 2 (1964) 138-147.
47. B.D. Ripley and J.P. Rasson, Finding the edge of a Poisson forest, J. Appl. Prob. 14 (1977).
48. A. Rodríguez-Casal, Estimación de conjuntos y sus fronteras. Un enfoque geométrico. Ph. D. Thesis, Univ. Santiago de Compostela.
49. M. Rudemo and H. Stryhn, Approximating the distribution of maximum likelihood contour estimators in two-region images, Scand. J. Statist. 21 (1994) 41-55.
50. R. Schneider, Random approximation of convex sets, J. Microscopy 151 (1988) 211-227.
51. J. Serra, Image Analysis and Mathematical Morphology, Academic Press, London, 1982.
52. A.B. Tsybakov, On nonparametric estimation of density level sets, Ann. Statist., 25 (1997) 948-969.
53. A.B. Tsybakov, Optimal aggregation of classifiers in statistical learning, preprint, Université Paris VI, 2001.
54. G. Walther, Granulometric smoothing, Ann. of Statist. 25 (1997) 2273-2299.
55. G. Walther, On a generalization of Blaschke's Rolling Theorem and the smoothing of surfaces, Math. Meth. Appl. Sci., 22 (1999) 301-316.
56. M.P. Wand and M.C. Jones, Kernel Smoothing, Chapman and Hall, London, 1995.
57. W. Weil and J.A. Wieacker, Stochastic Geometry. Handbook of convex geometry, Vol. A, B, 1391-1438, North-Holland, Amsterdam, 1993.

Recent Advances and Trends in Nonparametric Statistics
Michael G. Akritas and Dimitris N. Politis (Editors)
© 2003 Elsevier Science B.V. All rights reserved.

# Nonparametric methods for heavy tailed vector data: A survey with applications from finance and hydrology

Mark M. Meerschaert[a]* and Hans-Peter Scheffler[b]

[a]Department of Mathematics, University of Nevada, Reno NV 89557 USA

[b]Fachbereich Mathematik, University of Dortmund, 44221 Dortmund, Germany

Many real problems in finance and hydrology involve data sets with power law tails. For multivariable data, the power law tail index usually varies with the coordinate, and the coordinates may be dependent. This paper surveys nonparametric methods for modeling such data sets. These models are based on a generalized central limit theorem. The limit laws in the generalized central limit theorem are operator stable, a class that contains the multivariate Gaussian as well as marginally stable random vectors with different tail behavior in each coordinate. Modeling this kind of data requires choosing the right coordinates, estimating the tail index for those coordinates, and characterizing dependence between the coordinates. We illustrate the practical application of these methods with several example data sets from finance and hydrology.

## 1. Introduction

Heavy tailed random variables with power law tails $P(|X| > x) \approx Cx^{-\alpha}$ are observed in many real world applications. Estimation of the tail parameter $\alpha$ is important, because it determines which moments exist. If $\alpha < 2$ then the variance is infinite, and if $\alpha < 1$ the mean is also undefined. For a heavy tailed random vector $\boldsymbol{X} = (X_1, \ldots, X_d)'$ the tail index $\alpha_i$ for the $i$th component may vary with $i$. Choosing the wrong coordinates can mask variations in tail index, since the heaviest tail will dominate.

Modeling dependence is more complicated when $\alpha_i < 2$ since the covariance matrix is undefined. In order to model the tail behavior and dependence structure of heavy tailed vector data, a generalized central limit theorem [10,11,27] can be used. A nonparametric characterization of the heavy tailed vector data identifies the operator stable limit. Since the data distribution belongs to the generalized domain of attraction of that operator stable limit, it inherits the tail behavior, natural coordinate system, and tail dependence structure of that limit.

This paper surveys some nonparametric methods for modeling heavy tailed vector data. These methods begin by estimating the appropriate coordinate system, to unmask variations in tail behavior. Then the tail index is estimated for each coordinate, and finally the dependence structure can be characterized. In the parlance of operator stable laws, the first two steps estimate the exponent of the operator stable law, and the last step estimates

---
*Partially supported by NSF grants DES-9980484 and DMS-0139927.

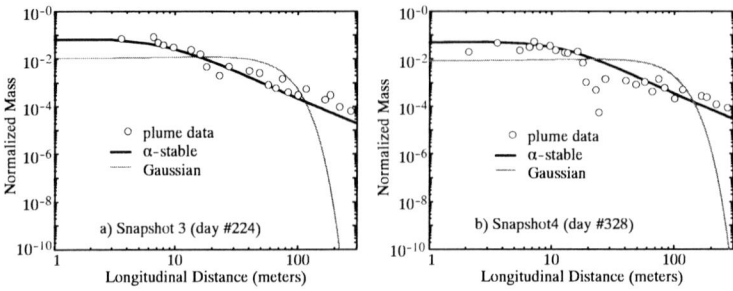

Figure 1. Heavy tailed $\alpha$-stable densities provide a superior fit to relative concentration of a tracer released in an underground aquifer (from [6]).

its spectral measure. All of the methods presented here apply to data whose distribution is attracted to an operator stable limit, not just to operator stable data. Therefore these methods are extremely robust.

## 2. Examples of data sets with heavy tails

Applications of heavy tailed random variables and random vectors occur in many areas, including hydrology and finance. Anderson and Meerschaert [4] find heavy tails in a river flow with $\alpha \approx 3$, so that the variance is finite but the fourth moment is infinite. Tessier, et al. [39] find heavy tails with $2 < \alpha < 4$ for a variety of river flows and rainfall accumulations. Hosking and Wallis [14] find evidence of heavy tails with $\alpha \approx 5$ for annual flood levels of a river in England. Benson, et al. [5,6] model concentration profiles for tracer plumes in groundwater using stochastic models whose heavy tails have $1 < \alpha < 2$, so that the mean is finite but the variance is infinite. Figure 1 shows the best-fitting Gaussian and $\alpha$-stable densities plotted against relative concentration of a passive tracer. The straight lines on these log-log graphs indicate power law tails for the $\alpha$-stable densities. Heavy tail distributions with $1 < \alpha < 2$ are used in physics to model anomalous diffusion, where a cloud of particles spreads faster than classical Brownian motion predicts [7,18,38]. More applications to physics with $0 < \alpha < 2$ are cataloged in Uchaikin and Zolotarev [40]. Resnick and Stărică [34] examine the quiet periods between transmissions for a networked computer terminal, and find heavy tails with $0 < \alpha < 1$, so that the mean and variance are both infinite. Several additional applications to computer science, finance, and signal processing appear in Adler, Feldman, and Taqqu [3]. More applications to signal processing can be found in Nikias and Shao [30].

Mandelbrot [22] and Fama [9] pioneered the use of heavy tail distributions in finance. Mandelbrot [22] presents graphical evidence that historical daily price changes in cotton have heavy tails with $\alpha \approx 1.7$, so that the mean exists but the variance is infinite. Jansen and de Vries [15] argue that daily returns for many stocks and stock indices have heavy tails with $3 < \alpha < 5$, and discuss the possibility that the October 1987 stock market plunge might be just a heavy tailed random fluctuation. Loretan and Phillips [21] use similar methods to estimate heavy tails with $2 < \alpha < 4$ for returns from numerous stock market indices and exchange rates. This indicates that the variance is finite but the fourth moment is infinite. Both daily and monthly returns show heavy tails with similar values of $\alpha$ in this study. Rachev and Mittnik [33] use different methods to find heavy tails with $1 < \alpha < 2$ for a variety of stocks, stock indices, and exchange rates. McCulloch [23] uses similar methods to re-analyze the data in [15,21], and obtains estimates of $1.5 < \alpha < 2$. This is important because the variance of price returns is finite if $\alpha > 2$ and infinite if $\alpha < 2$. While there is disagreement about the true value of $\alpha$, depending on which model is employed, all of these studies agree that financial data is typically heavy tailed, and that the tail parameter $\alpha$ varies between different assets.

## 3. Generalized central limit theorem

For heavy tailed random vectors, a generalized central limit theorem applies. If $\boldsymbol{X}, \boldsymbol{X}_1, \boldsymbol{X}_2, \boldsymbol{X}_3, \ldots$ are IID random vectors on $\mathbb{R}^d$ we say that $\boldsymbol{X}$ belongs to the *generalized domain of attraction* of some full dimensional random vector $\boldsymbol{Y}$ on $\mathbb{R}^d$, and we write $\boldsymbol{X} \in \text{GDOA}(\boldsymbol{Y})$, if

$$A_n(\boldsymbol{X}_1 + \cdots + \boldsymbol{X}_n - \boldsymbol{b}_n) \Rightarrow \boldsymbol{Y} \tag{3.1}$$

for some $d \times d$ matrices $A_n$ and vectors $\boldsymbol{b}_n \in \mathbb{R}^d$. The limits in (3.1) are called *operator stable* [17,27,37]. If $E(\|\boldsymbol{X}\|^2)$ exists then the classical central limit theorem shows that $\boldsymbol{Y}$ is multivariable normal, a special case of operator stable. In this case, we can take $A_n = n^{-1/2}I$ and $\boldsymbol{b}_n = nE(\boldsymbol{X})$. If (3.1) holds with $A_n = n^{-1/\alpha}I$ for any $\alpha \in (0, 2]$, then $\boldsymbol{Y}$ is multivariable stable with index $\alpha$. In this case we say that $\boldsymbol{X}$ belongs to the generalized domain of normal[2] attraction of $\boldsymbol{Y}$. See [27,35] for more information.

Matrix powers provide a natural extension of the stable index $\alpha$, allowing the tail index to vary with the coordinate. Let $\exp(A) = I + A + A^2/2! + A^3/3! + \cdots$ be the usual exponential operator for $d \times d$ matrices, and let $E = \text{diag}(1/\alpha_1, \ldots, 1/\alpha_d)$ be a diagonal matrix. If (3.1) holds with $A_n = n^{-E} = \exp(-E \ln n)$, then $A_n$ is diagonal with entries $n^{-1/\alpha_i}$ for $i = 1, \ldots, d$. Write $\boldsymbol{X}_t = (X_1(t), \ldots, X_d(t))'$, $\boldsymbol{b}_t = (b_1(t), \ldots, b_d(t))'$, $\boldsymbol{Y} = (Y_1, \ldots, Y_d)'$, and project (3.1) onto its $i$th coordinate to see that

$$\frac{X_i(1) + \cdots + X_i(n) - b_i(n)}{n^{1/\alpha_i}} \Rightarrow Y_i \quad \text{for each } i = 1, \ldots, d. \tag{3.2}$$

Each coordinate $X_i(t)$ is in the domain of attraction of a stable law $Y_i$ with index $\alpha_i$, and the matrix $E$ specifies every tail index. The limit $\boldsymbol{Y}$ is called *marginally stable*, a special case of operator stable. The matrix $E$, called an *exponent* of the operator stable random vector $\boldsymbol{Y}$, plays the role of the stable index $\alpha$ for stable laws. The matrix $E$ need not be diagonal. Diagonalizable exponents involve a change of coordinates, degenerate eigenvalues thicken probability tails by a logarithmic factor, and complex eigenvalues introduce rotational scaling, see Meerschaert [24].

A proof of the generalized central limit theorem for matrix scaling can be found in Meerschaert and Scheffler [27]. Since $\boldsymbol{Y}$ is infinitely divisible, the Lévy representation (Theorem

---
[2]This terminology refers to the special form of the norming, not the limit!

3.1.11 in [27]) shows that the characteristic function $E[e^{i\mathbf{k}\cdot\mathbf{Y}}]$ is of the form $e^{\psi(\mathbf{k})}$ where

$$\psi(\mathbf{k}) = i\mathbf{b}\cdot\mathbf{k} - \frac{1}{2}\mathbf{k}\cdot\Sigma\mathbf{k} + \int_{\mathbf{x}\neq 0}\left(e^{i\mathbf{k}\cdot\mathbf{x}} - 1 - \frac{i\mathbf{k}\cdot\mathbf{x}}{1+\|\mathbf{x}\|^2}\right)\phi(d\mathbf{x}) \tag{3.3}$$

for some $\mathbf{b}\in\mathbb{R}^d$, some nonnegative definite symmetric $d\times d$ matrix $\Sigma$ and some Lévy measure $\phi$. The Lévy measure satisfies $\phi\{\mathbf{x}: \|\mathbf{x}\|>1\}<\infty$ and

$$\int_{0<\|\mathbf{x}\|<1}\|\mathbf{x}\|^2\phi(d\mathbf{x}) < \infty.$$

For a multivariable stable law,

$$\phi\{\mathbf{x}: \|\mathbf{x}\|>r, \frac{\mathbf{x}}{\|\mathbf{x}\|}\in B\} = Cr^{-\alpha}M(B)$$

where $M$ is a probability measure on the unit sphere that is not supported on any $d-1$ dimensional subspace of $\mathbb{R}^d$. We call $M$ the *spectral measure*[3]. If $\phi=0$ then $\mathbf{Y}$ is normal with mean $\mathbf{b}$ and covariance matrix $\Sigma$. If $\Sigma=0$ then a necessary and sufficient condition for (3.1) to hold is that

$$nP(A_n\mathbf{X}\in B) \to \phi(B) \quad \text{as } n\to\infty \tag{3.4}$$

for Borel subsets $B$ of $\mathbb{R}^d\setminus\{0\}$ whose boundary have $\phi$-measure zero, where $\phi$ is the Lévy measure of the limit $\mathbf{Y}$. Proposition 6.1.10 in [27] shows that the convergence (3.4) is equivalent to *regular variation* of the probability distribution $\mu(B) = P(\mathbf{X}\in B)$, an analytic condition that extends the idea of power law tails. If (3.4) holds then Proposition 6.1.2 in [27] shows that the Lévy measure satisfies

$$t\phi(d\mathbf{x}) = \phi(t^{-E}d\mathbf{x}) \quad \text{for all } t>0 \tag{3.5}$$

for some $d\times d$ matrix $E$. Then it follows from the characteristic function formula that $\mathbf{Y}$ is operator stable with exponent $E$, and that for $\mathbf{Y}_n$ IID with $\mathbf{Y}$ we have

$$n^{-E}(\mathbf{Y}_1 + \cdots + \mathbf{Y}_n - \mathbf{b}_n) \stackrel{d}{=} \mathbf{Y} \tag{3.6}$$

for some $\mathbf{b}_n$, see Theorem 7.2.1 in [27]. Hence operator stable laws belong to their own GDOA, so that the probability distribution of $\mathbf{Y}$ also varies regularly, and sums of IID operator stable random vectors are again operator stable with the same exponent $E$. If $E=aI$ then $\mathbf{Y}$ is multivariable stable with index $\alpha=1/a$, and (3.4) is equivalent to the balanced tails condition

$$\frac{P(\|\mathbf{X}\|>r, \frac{\mathbf{X}}{\|\mathbf{X}\|}\in B)}{P(\|\mathbf{X}\|>r)} \to M(B) \quad \text{as } r\to\infty \tag{3.7}$$

for all Borel subsets $B$ of the unit sphere $S = \{\boldsymbol{\theta}\in\mathbb{R}^d: \|\boldsymbol{\theta}\|=1\}$ whose boundary has $M$-measure zero.

---
[3] Some authors call $CM$ the spectral measure.

## 4. Spectral decomposition theorem

For any $d \times d$ matrix $E$ there is a unique *spectral decomposition* based on the real parts of the eigenvalues, see for example Theorem 2.1.14 in [27]. Write the minimal polynomial of $E$ as $f_1(x) \cdots f_p(x)$ where every root of $f_j$ has real part $a_j$ and $a_1 < \cdots < a_p$. Define $V_j = \ker f_j(E)$ and let $d_j = \dim V_j$. Then we may write $\mathbb{R}^d = \mathbb{V}_1 \oplus \cdots \oplus V_p$, and in any basis that respects this direct sum decomposition we have

$$E = \begin{pmatrix} E_1 & 0 & \cdots & 0 \\ 0 & E_2 & & 0 \\ \vdots & & \ddots & \vdots \\ 0 & 0 & \cdots & E_p \end{pmatrix} \quad (4.1)$$

where $E_i$ is a $d_i \times d_i$ matrix, every eigenvalue of $E_i$ has real part equal to $a_i$, and $d_1 + \cdots + d_p = d$. This is called the spectral decomposition with respect to $E$. Given a nonzero vector $\boldsymbol{\theta} \in \mathbb{R}^d$, write $\boldsymbol{\theta} = \boldsymbol{\theta}_1 + \cdots + \boldsymbol{\theta}_p$ with $\boldsymbol{\theta}_i \in V_i$ for each $i = 1, \ldots, p$ and define

$$\alpha(\boldsymbol{\theta}) = \min\{1/a_i : \boldsymbol{\theta}_i \neq 0\}. \quad (4.2)$$

If $\boldsymbol{Y}$ is operator stable with exponent $E$, then the probability distribution of $\boldsymbol{Y}$ varies regularly with exponent $E$, and Theorem 6.4.15 in [27] shows that for any small $\delta > 0$ we have

$$r^{-\alpha(\boldsymbol{\theta})-\delta} < P(|\boldsymbol{Y} \cdot \boldsymbol{\theta}| > r) < r^{-\alpha(\boldsymbol{\theta})+\delta}$$

for all $r > 0$ sufficiently large. In other words, the tail behavior of $\boldsymbol{Y}$ is dominated by the component with the heaviest tail. This also means that $E(|\boldsymbol{Y} \cdot \boldsymbol{\theta}|^\beta)$ exists for $0 < \beta < \alpha(\boldsymbol{\theta})$ and diverges for $\beta > \alpha(\boldsymbol{\theta})$. Theorem 7.2.1 in [27] shows that every $a_i \geq 1/2$, so that $0 < \alpha(\boldsymbol{\theta}) \leq 2$. If we write $\boldsymbol{Y} = \boldsymbol{Y}_1 + \cdots + \boldsymbol{Y}_p$ with $\boldsymbol{Y}_i \in V_i$ for each $i = 1, \ldots, p$, then projecting (3.6) onto $V_i$ shows that $\boldsymbol{Y}_i$ is an operator stable random vector on $V_i$ with some exponent $E_i$. We call this the spectral decomposition of $\boldsymbol{Y}$ with respect to $E$. Since every eigenvalue of $E_i$ has the same real part $a_i$ we say that $\boldsymbol{Y}_i$ is spectrally simple, with index $\alpha_i = 1/a_i$. Although $\boldsymbol{Y}_i$ might not be multivariable stable, it has similar tail behavior. For any small $\delta > 0$ we have

$$r^{-\alpha_i-\delta} < P(\|\boldsymbol{Y}_i\| > r) < r^{-\alpha_i+\delta}$$

for all $r > 0$ sufficiently large, so $E(\|\boldsymbol{Y}_i\|^\beta)$ exists for $0 < \beta < \alpha_i$ and diverges for $\beta > \alpha_i$.

If $\boldsymbol{X} \in \text{GDOA}(\boldsymbol{Y})$ then Theorem 8.3.24 in [27] shows that the limit $\boldsymbol{Y}$ and norming matrices $A_n$ in (3.1) can be chosen so that every $V_i$ in the spectral decomposition of $\mathbb{R}^d$ with respect to the exponent $E$ of $\boldsymbol{Y}$ is $A_n$-invariant for every $n$, and $V_1, \ldots, V_p$ are mutually perpendicular. Then the probability distribution of $\boldsymbol{X}$ is regularly varying with exponent $E$ and $\boldsymbol{X}$ has the same tail behavior as $\boldsymbol{Y}$. In particular, for any small $\delta > 0$ we have

$$r^{-\alpha(\boldsymbol{\theta})-\delta} < P(|\boldsymbol{X} \cdot \boldsymbol{\theta}| > r) < r^{-\alpha(\boldsymbol{\theta})+\delta} \quad (4.3)$$

for all $r > 0$ sufficiently large. In this case, we say that $\boldsymbol{Y}$ is spectrally compatible with $\boldsymbol{X}$, and we write $\boldsymbol{X} \in \text{GDOA}_c(\boldsymbol{Y})$.

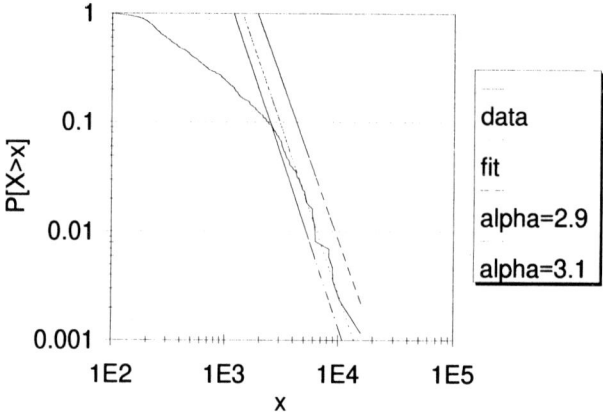

Figure 2. Monthly average river flows (cubic feet per second) for the Salt river near Roosevelt AZ exhibit heavy tails with $\alpha \approx 3$ (from [4]).

## 5. Nonparametric methods for tail estimation

Mandelbrot [22] pioneered a graphical estimation method for tail estimation. If $y = P(X > r) \approx Cr^{-\alpha}$ then $\log y \approx \log C - \alpha \log r$. Ordering the data so that $X_{(1)} \geq X_{(2)} \geq \cdots \geq X_{(n)}$ we should have approximately that $r = X_{(i)}$ when $y = i/n$. Then a plot of $\log X_{(i)}$ versus $\log(i/n)$ should be approximately linear with slope $-\alpha$. If $P(X > r) \approx Cr^{-\alpha}$ for $r$ large, then the upper tail should be approximately linear. We call this a *Mandelbrot plot*. Figure 2 shows a Mandelbrot plot used to estimate the tail index for a river flow time series.

The most popular numerical estimator for $\alpha$ is due to Hill [13], see also Hall [12]. Assuming that $P(X > r) = Cr^{-\alpha}$ for large values of $r > 0$, the maximum likelihood estimates for $\alpha$ and $C$ based on the $m+1$ largest observations are

$$\hat{\alpha} = \left[ \frac{1}{m} \sum_{i=1}^{m} \left( \ln X_{(i)} - \ln X_{(m+1)} \right) \right]^{-1}$$

$$\hat{C} = \frac{m}{n} X_{(m+1)}^{\hat{\alpha}}$$

(5.1)

where $m$ is to be taken as large as possible, but small enough so that the tail condition $P(X > r) = Cr^{-\alpha}$ remains a useful approximation. Finding the best value of $m$ is a practical challenge, and creates a certain amount of controversy [8]. Jansen and de Vries [15] use Hill's estimator with a fixed value of $m = 100$ for several different assets. Loretan and Phillips [21] tabulate several different values of $m$ for each asset. Hill's estimator $\hat{\alpha}$ is consistent and asymptotically normal with variance $\alpha^2/m$, so confidence intervals are easy to construct. These intervals clearly demonstrate that the tail parameters in Jansen and de Vries [15] and Loretan and Phillips [21] vary depending on the asset. Painter, Cvetkovic, and Selroos [32] apply Hill's estimator to data on fluid flow in fractures, to estimate two parameters of interest. Their $\alpha_i$ estimates also show a significant difference between the two parameters. In all of these studies, an appropriate model for vector data must allow $\alpha_i$ to vary with $i$.

Aban and Meerschaert [1] develop a more general Hill's estimator to account for a possible shift in the data. If $P(X > r) = C(r - s)^{-\alpha}$ for $r$ large, the maximum likelihood estimates for $\alpha$ and $C$ based on the $m + 1$ largest observations are

$$\hat{\alpha} = \left[\frac{1}{m}\sum_{i=1}^{m}\left(\ln(X_{(i)} - \hat{s}) - \ln(X_{(m+1)} - \hat{s})\right)\right]^{-1} \quad (5.2)$$

$$\hat{C} = \frac{m}{n}(X_{(m+1)} - \hat{s})^{\hat{\alpha}}$$

where $\hat{s}$ is obtained by numerically solving the equation

$$\hat{\alpha}(X_{(m+1)} - \hat{s})^{-1} = (\hat{\alpha} + 1)\frac{1}{m}\sum_{i=1}^{m}(X_{(i)} - \hat{s})^{-1} \quad (5.3)$$

over $\hat{s} < X_{(m+1)}$. Once the optimal shift is computed, $\hat{\alpha}$ comes from Hill's estimator applied to the shifted data. One practical implication is that, since the Pareto model is not shift-invariant, it is a good idea to try shifting the data to get a linear Mandelbrot plot.

Meerschaert and Scheffler [25] propose a robust estimator

$$\hat{\alpha} = \frac{2\ln n}{\ln n + \ln \hat{\sigma}^2} \quad (5.4)$$

based on the sample variance $\hat{\sigma}^2 = n^{-1}\sum_{t=1}^{n}(X_t - \bar{X}_n)^2$, where as usual $\bar{X}_n = n^{-1}(X_1 + \cdots + X_n)$ is the sample mean. This tail estimator is consistent for IID data in the domain of attraction of a stable law with index $\alpha < 2$. Like Hill's estimator [34], it is also consistent for moving averages. If $X$ is attracted to a normal limit, then $\hat{\alpha} \to 2$. It is interesting, and even somewhat ironic, that the sample variance can be used to estimate tail behavior, and hence tells us something about the spread of typical values, even in this case $0 < \alpha < 2$ where the variance is undefined.

## 6. The right coordinate system

Equation 4.3 shows that tail behavior is dominated by the component with the heaviest tail. In order to unmask variations in tail behavior, one has to find a coordinate system $\boldsymbol{\theta}_1, \ldots, \boldsymbol{\theta}_d$ where $\alpha_i = \alpha(\boldsymbol{\theta}_i)$ varies with $i$. This is equivalent to estimating the spectral decomposition. A useful estimator is based on the (uncentered) sample covariance matrix

$$M_n = \frac{1}{n}\sum_{t=1}^{n}\boldsymbol{X}_t\boldsymbol{X}_t'. \quad (6.1)$$

If $X_t$ are IID with $X$, then $X_t X_t'$ are IID random elements of the vector space $\mathcal{M}_s^d$ of symmetric $d \times d$ matrices, and the extended central limit theorem applies (see Section 10.2 in [27] for complete proofs). If the probability distribution of $X$ is regularly varying with exponent $E$ and (3.4) holds with $t\phi\{d\boldsymbol{x}\} = \phi\{t^{-E} d\boldsymbol{x}\}$ for all $t > 0$, then the distribution of $XX'$ is also regularly varying with

$$nP(A_n XX' A_n' \in B) \to \Phi(B) \quad \text{as } n \to \infty \tag{6.2}$$

for Borel subsets $B$ of $\mathcal{M}_s^d$ that are bounded away from zero and whose boundary has $\Phi$-measure zero. The exponent $\xi$ of the limit measure $\Phi\{d(\boldsymbol{xx'})\} = \phi\{d\boldsymbol{x}\}$ is defined by $\xi M = EM + ME'$ for $M \in \mathcal{M}_s^d$. If every eigenvalue of $E$ has real part $a_i > 1/2$, then

$$n A_n M_n A_n' \Rightarrow W \tag{6.3}$$

holds with $W$ operator stable. The centered sample covariance matrix is defined by

$$\Gamma_n = \frac{1}{n} \sum_{i=1}^n (\boldsymbol{X}_i - \bar{\boldsymbol{X}}_n)(\boldsymbol{X}_i - \bar{\boldsymbol{X}}_n)'$$

where $\bar{\boldsymbol{X}}_n = n^{-1}(\boldsymbol{X}_1 + \cdots + \boldsymbol{X}_n)$ is the sample mean. For heavy tailed data, Theorem 10.6.15 in [27] shows that $\Gamma_n$ and $M_n$ have the same asymptotics. In practice, it is common to mean-center the data, so it does not matter which form we choose.

Since $M_n$ is symmetric and nonnegative definite, there exists an orthonormal basis of eigenvectors for $M_n$ with nonnegative eigenvalues. Sort the eigenvalues

$$\lambda_1 \leq \cdots \leq \lambda_d$$

and the associated unit eigenvectors

$$\boldsymbol{\theta}_1, \ldots, \boldsymbol{\theta}_d$$

so that $M_n \boldsymbol{\theta}_j = \lambda_j \boldsymbol{\theta}_j$ for each $j = 1, \ldots, d$. In the spectral decomposition (4.1) each block $E_i$ is a $d_i \times d_i$ matrix, and every eigenvalue of $E_i$ has the same real part $a_i$ for some $1/2 \leq a_1 < \cdots < a_p$. Let $D_0 = 0$ and $D_i = d_1 + \cdots + d_i$ for $1 \leq i \leq p$. Now Theorem 10.4.5 in [27] shows that

$$\frac{2 \log n}{\log n + \log \lambda_j} \xrightarrow{P} \alpha_i \quad \text{as } n \to \infty$$

for any $D_{i-1} < j \leq D_i$, where $\alpha_i = 1/a_i$ is the tail index. This is a multivariable version of the one variable tail estimator (5.4). Furthermore, Theorem 10.4.8 in [27] shows that the eigenvectors $\boldsymbol{\theta}_j$ converge in probability to $V_1$ when $j \leq D_1$, and to $V_p$ when $j > D_{p-1}$. This shows that the eigenvectors estimate the coordinate vectors in the spectral decomposition, at least for the lightest and heaviest tails. Meerschaert and Scheffler [29] show that the same results hold for moving averages. If $p \leq 3$, this gives a practical method for determining the right coordinate system for modeling heavy tail data: Simply use the eigenvalues of the sample covariance matrix as the coordinate vectors.

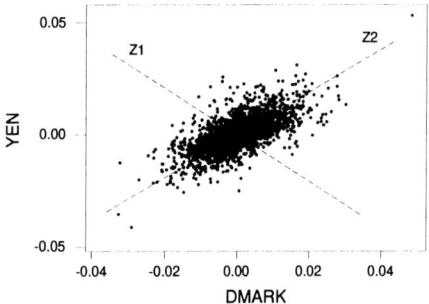

Figure 3. Exchange rates against the US dollar. The new coordinates uncover variations in the tail parameter $\alpha$.

**Example 6.1.** Meerschaert and Scheffler [26] consider $n = 2853$ daily exchange rate log-returns $X_1(t)$ for the German Deutsch Mark and $X_2(t)$ for the Japanese Yen, both taken against the US Dollar. Divide each entry by .004, which is the approximate median for both $|X_1(t)|$ and $|X_2(t)|$. This has no effect on the eigenvectors but helps to obtain good estimates of the tail thickness. Then compute

$$M_n = \frac{1}{n} \sum_{t=1}^{n} \begin{pmatrix} X_1(t)^2 & X_1(t)X_2(t) \\ X_1(t)X_2(t) & X_2(t)^2 \end{pmatrix} = \begin{pmatrix} 3.204 & 2.100 \\ 2.100 & 3.011 \end{pmatrix}$$

which has eigenvalues $\lambda_1 = 1.006$, $\lambda_2 = 5.209$ and associated unit eigenvectors $\boldsymbol{\theta}_1 = (0.69, -0.72)'$, $\boldsymbol{\theta}_2 = (0.72, 0.69)'$. Next compute

$$\begin{aligned} \hat{\alpha}_1 &= \frac{2 \ln 2853}{\ln 2853 + \ln 1.006} = 1.998 \\ \hat{\alpha}_2 &= \frac{2 \ln 2853}{\ln 2853 + \ln 5.209} = 1.656 \end{aligned} \quad (6.4)$$

indicating that one component fits a finite variance model but the other fits a heavy tailed model with $\alpha = 1.656$.

Now the eigenvectors can be used to find a new coordinate system that unmasks the variations in tail behavior. Let

$$P = \begin{pmatrix} 0.69 & -0.72 \\ 0.72 & 0.69 \end{pmatrix}$$

be the change of coordinates matrix whose $i$th row is the eigenvector $\boldsymbol{\theta}_i$, and let $\boldsymbol{Z}_t = P\boldsymbol{X}_t$ be the same data in the new, rotated coordinate system. The random vectors $\boldsymbol{Z}_t \in \text{GDOA}_c(\boldsymbol{Y})$ where $\boldsymbol{Y} = (Y_1, Y_2)'$ is operator stable with exponent

$$E = \begin{pmatrix} 0.50 & 0 \\ 0 & 0.60 \end{pmatrix}$$

since $0.50 = 1/1.998$ and $0.60 = 1/1.656$. Hence the random variables $Z_1(t)$ can be modeled as belonging to the domain of attraction of a normal limit $Y_1$, while the random variables $Z_2(t)$ are attracted to a stable limit $Y_2$ with index $\alpha = 1.656$. Inverting $\boldsymbol{Z}_t = P\boldsymbol{X}_t$ we obtain

$$\begin{aligned} X_1(t) &= 0.69 Z_1(t) + 0.72 Z_2(t) \\ X_2(t) &= -0.72 Z_1(t) + 0.69 Z_2(t). \end{aligned} \tag{6.5}$$

Both exchange rates have a common heavy-tailed term $Z_2(t)$, so both have heavy tails with the same tail index $\alpha = 1.656$. It is tempting to interpret $Z_2(t)$ as the common influence of fluctuations in the US dollar, and the remaining light-tailed factor $Z_1(t)$ as the accumulation of other price shocks independent of the US dollar.

Once the new coordinate system is identified, any one variable tail estimator can be used to approximate the $\alpha_i$. Applying Hill's estimator to the $Z_i(t)$ data with $m \approx 500$ yields estimates similar to those obtained here, providing another justification for a model with $\alpha_1 \neq \alpha_2$. This exchange rate data was also analyzed by Nolan, Panorska and McCulloch [31] using a multivariable stable model with the same tail index $\alpha \approx 1.6$ for both exchange rates. Rotating to a new coordinate system unmasks variations in the tail index that are not apparent in the original coordinates. Kozubowski, et al. [19] compare the fit of both stable and geometric stable laws to the rotated data $Z_2(t)$, see Figure 4. Since operator geometric stable laws (where the number $n$ of summands in (3.1) is replaced by a geometric random variable) have the same tail behavior as operator stable laws, and even the same domains of attraction (see Theorem 3.1 in [20]), the spectral decomposition and its eigenvector estimator are the same for both models.

## 7. Modeling dependence

For heavy tailed random vector data with tail index $\alpha_i < 2$, the covariance matrix is undefined. In this case, dependence can be modeled using the spectral measure. Suppose that $\boldsymbol{X}_t$ are IID in the generalized domain of attraction of an operator stable law $\boldsymbol{Y}$ with no normal component, so that $\Sigma = 0$ in (3.3). In this case, the log-characteristic function can be written in the form

$$\psi(\boldsymbol{k}) = i\boldsymbol{b} \cdot \boldsymbol{k} + C \int_0^{2\pi} \int_0^\infty \left( e^{i\boldsymbol{k} \cdot r^E \boldsymbol{\theta}} - 1 - \frac{i\boldsymbol{k} \cdot r^E \boldsymbol{\theta}}{1+r^2} \right) \frac{dr}{r^2} M(d\boldsymbol{\theta}) \tag{7.1}$$

where $C > 0$, and $M$ is a probability measure on the unit circle $S = \{x \in \mathbb{R}^d : \|x\| = 1\}$ called the *spectral measure* of $\boldsymbol{X}$. Equation (7.1) comes from applying a disintegration formula (Theorem 7.2.5 in [27]) to the Lévy measure in (3.3). The spectral measure $M$ determines the dependence between the components of $\boldsymbol{X}$. For example (cf. [35] for the multivariate

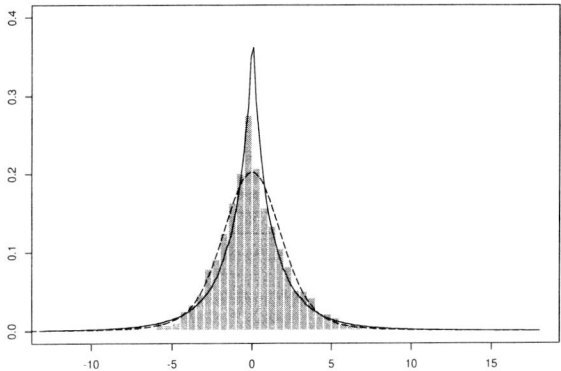

Figure 4. Two models for the rotated exchange rate data $Z_2$: stable (dotted line) and geometric stable (solid line).

stable case and Meerschaert and Scheffler [28] for the general case), the components of $\boldsymbol{X}$ are independent if and only if $M$ is supported on the coordinate axes.

The following method of Scheffler [36] can be used to estimate $C, M$: Any $\boldsymbol{x} \in \mathbb{R}^d \setminus \{0\}$ can be written uniquely in the form $\boldsymbol{x} = \tau(\boldsymbol{x})^E \boldsymbol{\theta}(\boldsymbol{x})$ for some radius $\tau(\boldsymbol{x}) > 0$ and some direction $\boldsymbol{\theta}(\boldsymbol{x}) \in S$. These are called the Jurek coordinates [16]. Define the order statistics $\boldsymbol{X}_{(i)}$ such that $\tau(\boldsymbol{X}_{(1)}) \geq \cdots \geq \tau(\boldsymbol{X}_{(n)})$ where ties are broken arbitrarily. The estimate $\hat{M}_m$ of the mixing measure based on the $m$ largest order statistics is just the empirical measure based on the points $\boldsymbol{\theta}(\boldsymbol{X}_{(i)})$ on the unit sphere $S$. In other words, the probability we assign to any sector $F$ of the unit sphere $S$ is equal to the fraction of the points $\{\boldsymbol{\theta}(\boldsymbol{X}_{(i)}) : 1 \leq i \leq m\}$ falling in this sector. The estimator of $C$ is $\hat{C} = (m/n)\tau(\boldsymbol{X}_{(m)})$, which reduces to Hill's estimator of $C$ in the one variable case. This estimator applies when the data belong to the normal[4] generalized domain of attraction of an operator stable law with no normal component.

**Example 7.1.** Painter, Cvetkovic, and Selroos [32] performed a detailed simulation of fracture flow networks, and extracted data on fracture aperture $q$ (millimeters) and fluid velocity $v$ (meters/year). Mandelbrot plots of an unpublished data set related to that study indicate that both $X_t(1) = 1/v$ and $X_t(2) = 1/qv$ have heavy tails with $1 < \alpha < 2$. The Hill estimator (5.1) yields $\hat{\alpha}_1 = 1.4$ and $\hat{C}_1 = 0.0065$ for the $1/v$ data, $\hat{\alpha}_2 = 1.05$ and $\hat{C}_2 = 0.028$ for the $1/qv$ data. The $\hat{\alpha}$ estimates were plotted as a function of $m$ to see where they stabilized. Linear regression estimates of the parameters based on a log-log plot are consistent with the

---

[4] Here normal means that we can take $A_n = n^{-E}$ in (3.1).

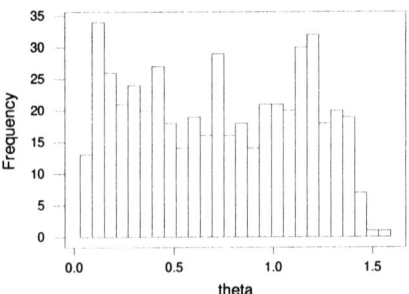

Figure 5. The empirical mixing measure distribution for the fracture flow data is nearly uniform on the interval $[0, \pi/2]$.

results of Hill's estimator. See Aban and Meerschaert [2] for a discussion of linear regression estimates and their relation to Hill's estimator. Since $\alpha_1 \neq \alpha_2$ the original coordinates are appropriate. Also, the eigenvalues of the sample covariance matrix are near the coordinate axes.

We rescale $Y_t(i) = X_t(i)/\hat{C}_i^{a_i}$ so that approximately $P(Y_t(i) > r) = r^{-a_i}$, and then we estimate $C, M$ for the data $\boldsymbol{Y}_1, \ldots, \boldsymbol{Y}_n$. This rescaling gives a clearer picture of the mixing measure. A histogram of $\{\boldsymbol{\theta}(\boldsymbol{Y}_{(i)}) : 1 \leq i \leq m\}$ for $m = 500$, shown in Figure 5, indicates that these unit vectors are approximately uniformly distributed over the first quadrant of the unit circle. Other values of $m$ in the range $100 < m < 1000$ show similar behavior, and we conclude that the mixing measure $M$ is approximately uniform on the first quadrant of the unit circle, indicating strong dependence between $1/v$ and $1/qv$. The estimator $\hat{C}$ based on the $m$ largest order statistics stabilizes at a value near $\pi/2$ for $100 < m < 1000$, which coincides with the arclength of the first quadrant, so we estimate $CM(d\boldsymbol{\theta}) = d\boldsymbol{\theta}$. Painter, Cvetkovic, and Selroos [32] argue that the $1/v$ and $1/qv$ data are well modeled by stable distributions. In that case, it is reasonable to model the rescaled data $\boldsymbol{Y}$ as operator stable with characteristic function

$$F(\boldsymbol{k}) = \exp\left\{i\boldsymbol{b} \cdot \boldsymbol{k} + \int_0^{2\pi}\!\!\int_0^\infty \left(e^{i\boldsymbol{k}\cdot r^E \boldsymbol{\theta}} - 1 - i\boldsymbol{k} \cdot r^E \boldsymbol{\theta}\right) \frac{dr}{r^2} d\boldsymbol{\theta}\right\}$$

where $\boldsymbol{b}$ is the sample mean (this uses an alternative form of the log-characteristic function, see [27] Theorem 3.1.14 and Remark 3.1.15), and $\boldsymbol{k} \cdot r^E \boldsymbol{\theta} = k_1 r^{1/1.4} \cos\boldsymbol{\theta} + k_2 r^{1/1.05} \sin\boldsymbol{\theta}$. The density $f(\boldsymbol{y})$ of this operator stable random vector can then be obtained from the Fourier inversion formula

$$f(\boldsymbol{y}) = (2\pi)^{-1} \int_{\boldsymbol{k} \in \mathbb{R}^2} e^{-i\boldsymbol{k}\cdot\boldsymbol{y}} F(\boldsymbol{k}) d\boldsymbol{k}$$

or perhaps more efficiently via inverse fast Fourier transforms.

## 8. Summary

Vector data sets with heavy tails can be usefully modeled as belonging to the generalized domain of attraction of an operator stable law. This robust model characterizes the tail behavior, which can vary with coordinate, and also the dependence between coordinates. Choosing the right coordinate system is crucial, since variations in tail behavior can otherwise go undetected. A useful coordinate system in this regard is the set of eigenvectors of the sample covariance matrix. Once the right coordinates are chosen, any one variable tail estimator can be used. Then a nonparametric estimator of the spectral measure provides a way to model the dependence between coordinates. These methods have proven useful in a variety of applications to data analysis problems in hydrology and finance.

## REFERENCES

1. I. Aban and M. Meerschaert (2001) Shifted Hill's estimator for heavy tails. *Comm. Statist. Simulation Comput.* **30**, 949–962.
2. Aban, I. and M. Meerschaert (2002) Generalized least squares estimators for the thickness of heavy tails. *J. Statis. Plann. Inf.*, to appear.
3. Adler, R., R. Feldman, and M. Taqqu (1998) *A Practical Guide to Heavy Tails.* Birkhäuser, Boston.
4. Anderson, P. and M. Meerschaert (1998) Modeling river flows with heavy tails. *Water Resour. Res.* **34**, 2271–2280.
5. Benson, D., S. Wheatcraft, and M. Meerschaert (2000) Application of a fractional advection-dispersion equation. *Water Resour. Res.* **36**, 1403–1412.
6. Benson, D., R. Schumer, M. Meerschaert, and S. Wheatcraft (2001) Fractional dispersion, Lévy motions, and the MADE tracer tests. *Transport Porous Med.* **42**, 211–240.
7. Blumen, A., G. Zumofen and J. Klafter (1989) Transport aspects in anomalous diffusion: Lévy walks. *Phys Rev. A* **40** 3964–3973.
8. Drees, H, L. de Haan, and S. Resnick (2000) How to make a Hill plot. *Ann. Statist.* **28**, 254–274.
9. Fama, E. (1965) The behavior of stock market prices. *J. Business* **38**, 34–105.
10. Feller, W. (1971) *An Introduction to Probability Theory and Its Applications.* Vol. II, 2nd Ed., Wiley, New York.
11. Gnedenko, B. and A. Kolmogorov (1968) *Limit Distributions for Sums of Independent Random Variables.* Addison-Wesley, Reading, Mass.
12. Hall, P. (1982) On some simple estimates of an exponent of regular variation. *J. Royal Statist. Soc. B* **44** 37–42.
13. Hill, B. (1975) A simple general approach to inference about the tail of a distribution. *Ann. Statist.* **3** 1163–1173.

14. Hosking, J. and J. Wallis (1987) Parameter and quantile estimation for the generalized Pareto distribution. *Technometrics* **29**, 339–349.
15. Jansen, D. and C. de Vries (1991) On the frequency of large stock market returns: Putting booms and busts into perspective. *Rev. Econ. Statist.* **23**, 18–24.
16. Jurek, Z.J. (1984) Polar Coordinates in Banach Spaces, *Bull. Ac. Pol. Math.* 32, 61-66.
17. Jurek, Z. and J. D. Mason (1993) *Operator-Limit Distributions in Probability Theory.* Wiley, New York.
18. Klafter, J., A. Blumen and M. Shlesinger (1987) Stochastic pathways to anomalous diffusion. *Phys. Rev. A* **35** 3081–3085.
19. Kozubowski, T. J., M.M. Meerschaert, A. K. Panorska and H. P. Scheffler (2003) Operator geometric stable laws, preprint.
20. Kozubowski, T. J., M.M. Meerschaert and H. P. Scheffler (2003) The operator $\nu$-stable laws, preprint.
21. Loretan, M. and P. Phillips (1994) Testing the covariance stationarity of heavy tailed time series. *J. Empirical Finance* **1**, 211–248.
22. Mandelbrot, B. (1963) The variation of certain speculative prices. *J. Business* **36**, 394–419.
23. McCulloch, J. (1997) Measuring tail thickness to estimate the stable index $\alpha$: A critique. *J. Business Econ. Statist.* **15**, 74–81.
24. Meerschaert, M. (1990) Moments of random vectors which belong to some domain of normal attraction. *Ann. Probab.* **18**, 870–876.
25. Meerschaert, M. and H. P. Scheffler (1998) A simple robust estimator for the thickness of heavy tails. *J. Statist. Plann. Inf.* **71**, 19–34.
26. Meerschaert, M. and H. P. Scheffler (1999) Moment estimator for random vectors with heavy tails. *J. Multivariate Anal.* **71**, 145–159.
27. Meerschaert, M. and H. P. Scheffler (2001) *Limit Theorems for Sums of Independent Random Vectors: Heavy Tails in Theory and Practice.* Wiley, New York.
28. Meerschaert, M. and H. P. Scheffler (2001) Sample cross-correlations for moving averages with regularly varying tails. *J. Time Series Anal.* **22**, no. 4, 481–492.
29. M.M. Meerschaert and H.P. Scheffler (2003) Portfolio modeling with heavy tailed random vectors. *Handbook of Heavy Tailed Distributions in Finance*, S.T.Rachev (Ed.), Elsevier Science.
30. Nikias, C. and M. Shao (1995) *Signal Processing with Alpha Stable Distributions and Applications.* Wiley, New York.
31. Nolan, J., Panorska A. and J. H. McCulloch (2001) Estimation of stable spectral measures. *Math. Comput. Modelling*, **34**, no. 9-11, 1113–1122.
32. Painter, S., V. Cvetkovic and J. Selroos (2002) Power-law velocity distributions in fracture networks: Numerical evidence and implications for tracer transport. *Geophysical Research Letters* **29**, no. 14.
33. Rachev, S. and S. Mittnik (2000) *Stable Paretian Models in Finance*, Wiley, Chichester.
34. Resnick, S. and C. Stărică (1995) Consistency of Hill's estimator for dependent data. *J. Appl. Probab.* **32** 139–167.
35. Samorodnitsky, G. and M. Taqqu (1994) *Stable non-Gaussian Random Processes.* Chapman & Hall, New York.
36. Scheffler, H. P. (1999) On estimation of the spectral measure of certain nonnormal op-

erator stable laws. *Statist. Probab. Lett.* **43**, 385–392.
37. Sharpe, M. (1969) Operator-stable probability distributions on vector groups. *Trans. Amer. Math. Soc.* **136**, 51–65.
38. Shlesinger, M., G. Zaslavsky, and U. Frisch (1994) Lévy flights and related topics in physics. *Lecture Notes in Physics* **450**, Springer-Verlag, Berlin.
39. Tessier, Y., S. Lovejoy, P. Hubert and D. Schertzer (1996) Multifractal analysis and modeling of rainfall and river flows and scaling, causal transfer functions. *J. Geophys. Res.*, **101**, 427–440.
40. Uchaikin, V. and V. Zolotarev (1999) *Chance and Stability: Stable Distributions and Their Applications*. VSP, Utrecht.

# Nonparametrics in Finance

# Nonparametric Methods in Continuous-Time Finance: A Selective Review *

Zongwu Cai [a] and Yongmiao Hong [b]

[a]Department of Mathematics, University of North Carolina, Charlotte, NC 28223, E-mail: zcai@uncc.edu

[b]Department of Economics and Department of Statistical Science, Cornell University, Ithaca, NY 14853, E-mail: yh20@cornell.edu

This paper gives a selective review on the recent development of nonparametric methods in continuous-time finance, particularly in the areas of nonparametric estimation of diffusion processes, nonparametric testing of parametric diffusion models, and nonparametric pricing of derivatives. In each financial context, we discuss suitable statistical concepts, models, procedures, as well as some of their applications to financial data. Much theoretical and empirical research is needed in this area, and we point out several aspects that deserve further investigation.

**Key Words**: Continuous time model, derivative pricing, Hermite polynomial, jump process, kernel smoothing, neural network, nonparametric test, options, risk neural density, time-dependent model.

## 1. INTRODUCTION

Nonparametric analysis has become a core area in statistics (see [44,54]) in the last two decades and has been widely used to various fields such as economics and finance (see [77,78]) due to its advantage of requiring little prior information on the data generating process. In finance, nonparametric methods have been proved to be an attractive way to conduct research and gain economic intuition in such core areas as asset and derivative pricing, term structure theory, portfolio choice, and risk management. The purpose of this survey is to review some recent developments of nonparametric methods in continuous-time finance, particularly in the areas of nonparametric estimation and testing of diffusion models, and nonparametric derivative pricing.

Finance is characterized by time and uncertainty. Continuous-time modelling has been a basic analytic tool in modern finance since the seminar papers by [24] and [76]. The rationale is that most of time, news arrives at financial markets in a continuous manner.

---

*We thank Haitao Li for his valuable and helpful comments. Cai's research was supported, in part, by the National Science Foundation grant DMS-0072400 and funds provided by the University of North Carolina at Charlotte and Sonderforschungsbereich 373, Berlin, Germany. Hong's research was supported, in part, by the National Science Foundation grant SES-0111769.

More importantly, derivative pricing in theoretical finance is generally much more convenient and elegant in a continuous-time framework than through binomial or other discrete approximations. However, statistical inference based on continuous-time financial models has just emerged as a field in less than a decade. This is apparently due to the difficulty of estimating and testing continuous-time models using discretely observed data. Financial time series data have distinct important stylized facts, such as persistent volatility clustering, heavy tails, strong serial dependence, and occasionally sudden but large jumps. In addition, financial modelling is closely embedded in a financial theoretical framework. These features imply that standard statistical theory may not be readily applicable to financial time series. Effort is required to accommondate these important features in statistical analysis of financial time series data.

Section 2 introduces various continuous-time diffusion processes and nonparametric estimation methods for diffusion processes. Section 3 reviews the estimation and testing of a parametric diffusion model using nonparametric methods. Section 4 discusses nonparametric derivative pricing, particularly the estimation of risk neutral density functions. Section 5 concludes.

## 2. DIFFUSIONS AND NONPARAMETRIC ESTIMATION

### 2.1. Models

Modeling the dynamics of interest rates, stock prices, foreign exchange rates, and macroeconomic factors, *inter alia*, is one of the most important topics in asset pricing studies. The instantaneous risk-free interest rate or the so-called short rate is, for example, the state variable that determines the evolution of the yield curve in an important class of term structure models, which is of fundamental importance for pricing fixed-income securities. Many models have been developed to describe the short rate movement.

In the theoretical term structure literature, the short rate $\{X_t, t \geq 0\}$ is often modelled as a time-homogeneous diffusion process as

$$dX_t = \mu(X_t)\,dt + \sigma(X_t)\,dB_t, \tag{1}$$

where $\{B_t, t \geq 0\}$ is a standard Brownian motion. The functions $\mu(\cdot)$ and $\sigma^2(\cdot)$ are respectively the drift (instantaneous mean) and the diffusion (instantaneous variance) of $X_t$, which determine the dynamics of the process.

There are two approaches to modeling $\mu(\cdot)$ and $\sigma(\cdot)$. The first is parametric, which assumes some parametric forms $\mu(\cdot, \theta)$ and $\sigma(\cdot, \theta)$, and then estimates the unknown parameters $\theta$. Most existing models in the literature assume that the interest rate exhibits mean-reversion and that the drift $\mu(\cdot)$ is linear in the interest rate level. It is also often assumed that the diffusion $\sigma(\cdot)$ takes the form of $\sigma |X_t|^\gamma$, where $\gamma$ measures the sensitivity of interest rate volatility to the interest rate level. With $\gamma = 0$ and 0.5, model (1) reduces to the well-known CIR ([36]) and Vasicek ([84]) models respectively. The forms of $\mu(\cdot, \theta)$ and $\sigma(\cdot, \theta)$ are typically chosen for theoretical convenience. They may not be consistent with the data generating process.

The second approach is nonparametric, which does not assume any restrictive functional form for $\mu(\cdot)$ and $\sigma(\cdot)$ beyond regularity conditions. In the last few years, great progress has been made in estimating and testing continuous-time models using nonparametric methods. However, empirical analysis on the forms of $\mu(\cdot)$ and $\sigma(\cdot)$ is still not conclusive. For example, recent studies by [4] and [83] using nonparametric methods, overwhelmingly reject all linear drift models for the short rate, indicating that the drift of the short rate is nonlinear in the interest rate level. They show that for the lower and middle ranges of the interest rate, the drift is almost zero but there is strong mean-reversion when the interest rate is high. These findings lead to the development of some nonlinear term structure models such as those of [2].

However, the evidence of nonlinear drift has been challenged by [30] and [79], who find that the nonparametric methods of [4] and [83] have severe finite sample problems near the extreme observations. This casts doubt on the evidence of nonlinear drift, although the findings in [4] and [83] that the drift is nearly flat for the middle range of the interest rate are not much affected. [30] point out that this is a puzzling fact, since "there are strong theoretical reasons to believe that short rate cannot exhibit the asymptotically explosive behavior implied by a random walk model." They conclude that "time series methods alone are not capable of producing evidence of nonlinearity in the drift."

Controversies also exist on the form of the diffusion $\sigma(\cdot)$. [29] show that in a single factor model of the short rate, $\gamma$ roughly equals to 1.5 and all the models with $\gamma \leq 1$ are rejected. [4] finds that $\gamma$ is close to 1, [83] finds that $\gamma$ is about 1.5, and [35] show that $\gamma$ is between 1.5 and 2. However, [25] argue that the result that $\gamma$ equals 1.5 depends on whether the data between October 1979 to September 1982 are included.

## 2.2. Nonparametric estimation

Under some regularity conditions (see [17,62]), the diffusion process in (1) is a one-dimensional, regular, strong Morkov process with continuous sample paths and time-invariant stationary transition density. The drift and diffusion are the first two moments of the infinitesimal conditional distribution of $X_t$ (see [67]):

$$\mu(X_t) = \lim_{\Delta \to 0} E\left[\frac{X_{t+\Delta} - X_t}{\Delta} \bigg| X_t\right] \quad \text{and} \quad \sigma^2(X_t) = \lim_{\Delta \to 0} E\left[\frac{(X_{t+\Delta} - X_t)^2}{\Delta} \bigg| X_t\right]. \quad (2)$$

The drift describes the movement of $X_t$ due to time changes, whereas the diffusion measures the magnitude of random fluctuations around the drift.

Using the Dynkin operator (see [67]), [83] derives the first order approximation

$$\mu(X_t)^{(1)} = \frac{1}{\Delta} E[X_{t+\Delta} - X_t | X_t] + O(\Delta),$$

the second order approximation

$$\mu(X_t)^{(2)} = \frac{1}{2\Delta} \{4 E[X_{t+\Delta} - X_t | X_t] - E[X_{t+2\Delta} - X_t | X_t]\} + O(\Delta^2),$$

and the third; and [46] obtain the higher-order approximations. Also, [83] gives the similar approximation formulas for the diffusion. [17] argue that approximations to the drift and diffusion of any order display the same rate of convergence and limiting variance, so that asymptotic argument in conjunction with computational issues suggest simply using the first order approximations in practice. For more discussions, see [16,17,46].

Now suppose $\{X_{\tau\Delta}\}_{\tau=1}^{n}$ is a discretely observed sample. Then, from (2), the first order approximation to $\mu(\cdot)$ and $\sigma(\cdot)$ leads to

$$\mu(X_{\tau\Delta}) \approx \frac{1}{\Delta} E[X_{(\tau+1)\Delta} - X_{\tau\Delta} | X_{\tau\Delta}] \quad \text{and} \quad \sigma^2(X_{\tau\Delta}) \approx \frac{1}{\Delta} E[(X_{(\tau+1)\Delta} - X_{\tau\Delta})^2 | X_{\tau\Delta}]$$

for $1 \leq \tau \leq n-1$. This becomes a nonparametric regression problem. In [3,4,30,62,83], the Nadaraya-Watson (NW) kernel estimators are used to estimate $\mu(x)$ and $\sigma^2(x)$:

$$\begin{aligned}\widehat{\mu}(x) &= \frac{1}{\Delta} \frac{\sum_{\tau=1}^{n-1}(X_{(\tau+1)\Delta} - X_{\tau\Delta}) K_h(x - X_{\tau\Delta})}{\sum_{\tau=1}^{n-1} K_h(x - X_{\tau\Delta})}, \text{ and} \\ \widehat{\sigma}^2(x) &= \frac{1}{\Delta} \frac{\sum_{\tau=1}^{n-1}(X_{(\tau+1)\Delta} - X_{\tau\Delta})^2 K_h(x - X_{\tau\Delta})}{\sum_{\tau=1}^{n-1} K_h(x - X_{\tau\Delta})},\end{aligned} \quad (3)$$

where $K_h(u) = K(u/h)/h$, $h = h_n \to 0$ is a bandwidth, and $K(\cdot)$ is a standard kernel. [62] suggest first estimating $\sigma^2(\cdot)$ by (3) and then estimating $\mu(\cdot)$ by $\widehat{\mu}(x) = [2\widehat{\pi}(x)]^{-1} \partial[\widehat{\sigma}^2(x) \widehat{\pi}(x)]/\partial x$, where $\widehat{\pi}(\cdot)$ is a consistent estimator of $\pi(\cdot)$ the stationary density of $\{X_t\}$. In (1) the drift is of order $dt$ and the diffusion is of order $\sqrt{dt}$, as $(dB_t)^2 = dt + O((dt)^2)$. Thus, $\sigma(\cdot)$ has lower order than $\mu(\cdot)$ for infinitesimal changes in time, and the local-time dynamics of the sampling path reflects more of $\sigma(\cdot)$ than that of $\mu(\cdot)$. Therefore, when $\Delta$ is very small, identification becomes much easier for $\sigma(\cdot)$ than $\mu(\cdot)$.

It is well known that the NW estimator suffers from some disadvantages such as larger bias, boundary effects, and inferior minimax efficiency (see [44]). To overcome these drawbacks, [46] suggest using the local linear technique, for $k = 1$ and $2$,

$$\sum_{\tau=1}^{n-1} \left[ \Delta^{-1}(X_{(\tau+1)\Delta} - X_{\tau\Delta})^k - \beta_0 - \beta_1(x - X_{\tau\Delta}) \right]^2 K_h(x - X_{\tau\Delta}), \quad (4)$$

which gives the local linear estimator of $\mu(x)$ if $k = 1$, and $\sigma^2(x)$ if $k = 2$. However, the local linear estimator of $\sigma(\cdot)$ cannot be always nonnegative in finite samples. To attenuate this disadvantage of the local linear method, a weighted NW estimator due to [28] can be used to estimate $\sigma(\cdot)$ although the method needs further verification.

The asymptotic theory can be found in [17] and [62] for the NW estimator and in [46] for the local linear estimator. To implement kernel estimates, the bandwidth(s) must be chosen. In the iid setting, there are theoretically optimal bandwidth selections. There are no such results available for diffusion processes.

One key assumption in the above development is the stationarity of $\{X_t\}$. However, it might not hold for real financial time series data. For a nonstationary process $\{X_t\}$, [17] and [19] use the following estimators, termed as *chronological local time estimation*, to estimate $\mu(x)$ and $\sigma^2(x)$:

$$\widehat{\mu}(x) = \frac{\sum_{\tau=1}^{n} K_h(x - X_{\tau\Delta}) \widetilde{\mu}(X_{\tau\Delta})}{\sum_{\tau=1}^{n} K_h(x - X_{\tau\Delta})}, \quad \text{and} \quad \widehat{\sigma}^2(x) = \frac{\sum_{\tau=1}^{n} K_h(x - X_{\tau\Delta}) \widetilde{\sigma}^2(X_{\tau\Delta})}{\sum_{\tau=1}^{n} K_h(x - X_{\tau\Delta})},$$

where $\widetilde{\mu}(x) = \Delta^{-1} \sum_{\tau=1}^{n-1} w_{i,n}(x) (X_{(\tau+1)\Delta} - X_{\tau\Delta})$ and $\widetilde{\sigma}^2(x) = \Delta^{-1} \sum_{\tau=1}^{n-1} w_{i,n}(x) (X_{(\tau+1)\Delta} - X_{\tau\Delta})^2$ with $w_{i,n}(x) = I(|X_{\tau\Delta} - x| \leq b)/\sum_{\tau=1}^{n} I(|X_{\tau\Delta} - x| \leq b)$. Here $b = b_n$ is a bandwidth-like smoothing parameter depending on the time span and on the sample size, which is called the spatial bandwidth in [19]. This approach can deal well with the nonstationary situation. This estimator can be viewed as a two-step smoothing method: The first step defines straight sample analogs to the values that the drift and diffusion take at the sampled points. Indeed, this step uses the smoothing technique (a linear estimator with same weights) to obtain the raw estimates of the two functions $\tilde{\mu}(x)$ and $\tilde{\sigma}^2(x)$, respectively. To implement this method, an empirical and theoretical study on the selection of two bandwidths $b$ and $h$ is needed.

One limitation of the time-homogeneous diffusion model in (1) is that it cannot capture the time effect. To overcome this disadvantage, a variety of time-dependent diffusion models have been proposed in the literature. A time-dependent diffusion process is

$$dX_t = \mu(X_t, t)\, dt + \sigma(X_t, t)\, dB_t, \tag{5}$$

which includes the following well known models in literature: [55] with $\mu(X_t) = \mu(t)$ and $\sigma(X_t) = \sigma(t)$, [58] with $\mu(X_t) = \alpha_0 + \alpha_1(t) X_t$ and $\sigma(X_t) = \sigma(t) X_t^k$ for $k = 0$ or $1/2$, [22] with $\mu(X_t) = \alpha_0(t) X_t + \alpha_1(t) X_t \log(X_t)$ and $\sigma(X_t) = \sigma(t) X_t$, and [23] with $\mu(X_t) = \alpha_0(t) X_t + \alpha_1(t) X_t \log(X_t)$ and $\sigma(X_t) = \sigma(t) X_t$ where $\alpha_1(t) = \sigma'(t)/\sigma(t)$. Similar to (2), one has

$$\mu(X_t, t) = \lim_{\Delta \to 0} \frac{E[X_{t+\Delta} - X_t \mid X_t]}{\Delta} \quad \text{and} \quad \sigma^2(X_t, t) = \lim_{\Delta \to 0} \frac{E[(X_{t+\Delta} - X_t)^2 \mid X_t]}{\Delta},$$

which provide a regression form for estimating $\mu(\cdot, t)$ and $\sigma^2(\cdot, t)$. Recently, [45] consider the following time-varying coefficient single factor model

$$dX_t = [\alpha_0(t) + \alpha_1(t) X_t]\, dt + \beta_0(t) X_t^{\beta_1(t)}\, dB_t, \tag{6}$$

and use the local linear technique in (4) to estimate the coefficient functions $\{\alpha_j(\cdot)\}$ and $\{\beta_j(\cdot)\}$. Since the coefficients depend on time, $\{X_t\}$ might not be stationary. The asymptotic properties of the resulting estimators are still unknown.

All aforementioned models are a special case of the following more general time-varying coefficient multi-factor diffusion models

$$dX_t = \mu(X_t, t)\, dt + \sigma(X_t, t)\, dB_t, \tag{7}$$

where $\mu(X_t, t) = \alpha_0(t) + \alpha_1(t) g(X_t)$, $[\sigma(X_t, t)\sigma(X_t, t)']_{ij} = \beta_{0,ij}(t) + \beta_{1,ij}(t)' h_{ij}(X_t)$, and $g(\cdot)$ and $\{h_{ij}(\cdot)\}$ are known functions. This is the time-dependent version of the multi-factor affine models studied in [39]. It allows time-varying coefficients in a multi-factor affine model. A further theoretical and empirical study of model (7) is warranted.

## 2.3. Jump diffusion models

There has been a vast literature on the study of diffusion models with jumps; see, for example, [21,26,32,38,39,43,65,70,82]. The main purpose of adding jumps into diffusion models or stochastic volatility diffusion models is to accommodate impact of sudden and large shocks to financial markets. For more discussions on why it is necessary to add jumps into diffusion models; see, for example, [26,65,70,72].

For simplicity, we only consider a single factor diffusion model with jump:

$$dX_t = \mu(X_t)\, dt + \sigma(X_t)\, dB_t + dJ_t, \tag{8}$$

where $J_t$ is a compensated jump process (zero conditional mean) with arrival rate $\lambda_t = \lambda(X_t) \geq 0$, which is an instantaneous intensity, and the jump size, $\xi$, has a time-invariant distribution $\Pi(\cdot)$ with mean zero. For example, $J_t = \xi P_t$, where $P_t$ is a Poisson process with an intensity $\lambda(X_t)$ or a binomial distribution with probability $\lambda(X_t)$, and $\Pi(\cdot)$ can be either normal or uniform. If $\lambda_t(\cdot) = 0$ or $E(\xi^2) = 0$, the jump-diffusion model in (8) becomes the diffusion model in (1). More generally, [32] consider a Lévy process for $\{J_t\}$. In practice, $\lambda(\cdot)$ might be assumed to have a particular form. For example, [32] consider three different types of special forms, each having the appealing feature of yielding analytic option pricing formula for European type contracts written on the stock price index. There are some open issues for the jump-diffusion model: (i) jumps are not observed and it is not possible to say surely if they exist; (ii) if they exist, a natural question arises how to estimate a jump time $\tau$, which is defined to be the discontinuous time at which $X_{\tau+} \neq X_{\tau-}$, and the jump size $\xi$, which is $\xi = X_{\tau+} - X_{\tau-}$.

Similar to (2), the first two conditional moments are given by

$$\mu_1(X_t) = \lim_{\Delta \to 0} \Delta^{-1} E[\{X_{t+\Delta} - X_t\} | X_t] = \mu(X_t) + \lambda(X_t) E(\xi),$$

and

$$\mu_2(X_t) = \lim_{\Delta \to 0} \Delta^{-1} E\left[\{X_{t+\Delta} - X_t\}^2 | X_t\right] = \sigma^2(X_t) + \lambda(X_t) E(\xi^2).$$

This implies that the first two moments are the same as those for a diffusion model by using a new drift $\tilde{\mu}(X_t) = \mu(X_t) + \lambda(X_t) E(\xi)$ and a new diffusion $\tilde{\sigma}^2(x) = \sigma^2(x) + \lambda(x) E(\xi^2)$. However, the fundamental difference between a diffusion model and a jump diffusion model arises in higher order moments. Using the infinitesimal generator of $X_t$, we can compute, $j > 2$,

$$\mu_j(X_t) = \lim_{\Delta \to 0} \Delta^{-1} E[(X_{t+\Delta} - X_t)^j | X_t] = \lambda(X_t) E(\xi^j).$$

Obviously, jumps provide a simple and intuitive mechanism for capturing the heavy tail behavior of $X_t$. In particular, the conditional skewness and kurtosis are given by $s(X_t) \equiv \lambda(X_t) E(\xi^3)/\tilde{\sigma}^3(x)$ and $k(X_t) \equiv \lambda(X_t) E(\xi^4)/\tilde{\sigma}^4(x)$. Note that $s(X_t) = 0$ if $\xi$ is symmetric. By assuming $\xi \sim N(0, \sigma_\xi^2)$, [65] uses the conditional kurtosis to measure the departures for the treasury bill data from normality, finding that interest rate changes are extremely non-normal.

The NW estimation of $\mu_j(\cdot)$ is considered by [18] and [65]. Moreover, [18] provide a general asymptotic theory for the resulting estimators. Further, by specifying a particular form of $\Pi(\lambda) = \Pi_0(\lambda, \theta)$, say, $\xi \sim N(0, \sigma_\xi^2)$, [18] propose consistent estimators of $\lambda(\cdot)$, $\sigma_\xi^2$, and $\sigma^2(\cdot)$ and derive their asymptotic properties.

A natural question arises on how to measure the departures from a pure diffusion model statistically. That is to test model (8) vs. model (1). It is equivalent to checking whether $\lambda(\cdot) \equiv 0$ or $\xi = 0$. Instead of using the conditional skewness or kurtosis, a test statistic can be constructed based on higher order conditional moments. This is still open.

To capture the time effect, [39] consider the time-varying coefficient intensity $\lambda(X_t, t) = \lambda_0(t) + \lambda_1(t) X_t$, and [32] consider a more general stochastic volatility model with the stochastic intensity $\lambda(\xi_0, X_t, t) = \lambda_0(\xi_0, t) + \lambda_1(\xi_0, t) X_t$, where $\xi_0$ is the size of the previous jump. This specification yields a class of jump Lévy measures which combine the features of jump intensities depending on, say volatility, as well as the size of the previous jump. [66] also propose a class of jump diffusion processes with a jump intensity depending on the past jump time and the absolute return. Moreover, as pointed out by [32], another potentially useful specification of the intensity would include the past duration, i.e., the time since the last jump, say $\tau(t)$, which is the time that has elapsed between the last jump and $t$ where $\tau(t)$ is a continuous function of $t$, such as

$$\lambda(\xi_0, X_t, \tau, t) = [\lambda_0(t) + \lambda_1(t) X_t] \lambda[\tau(t)] \exp[G(\xi_0)]. \tag{9}$$

This can accommodate increasing, decreasing or hump-shaped hazard functions of the size of the previous jump, and the duration dependence of jump intensities. However, there has been no attempt in the literature to discuss the estimation and test of the intensity $\lambda(X_t)$ or $\lambda(X_t, \theta)$ nonparametrically in the above settings.

A natural question arises is how to generalize model (8) to a more general time-dependent jump diffusion model with the time-dependent intensity $\lambda(\xi_0, X_t, \tau, t)$ unspecified or having some nonparametric structure, say, like (9). Clearly, they include the aforementioned models as a special case, studied by [32,39,66]. This is still an open problem.

## 3. NONPARAMETRIC ESTIMATION AND TESTING OF PARAMETRIC DIFFUSIONS

### 3.1. Nonparametric estimation of parametric diffusion models

As is well-known, derivative pricing in mathematical finance is conveniently tractable in continuous-time. In empirical study, however, it is an usual practice to abandon continuous-time modeling when estimating derivative pricing models. This is mainly because the transition density for most continuous-time models with discrete observations has no closed form and so the maximum likelihood estimation (MLE) is infeasible.

One major focus of the continuous-time literature is on developing methods to estimate continuous-time models using discretely-sampled data. This is motivated by the fact that using the discrete version of a continuous-time model can result in inconsistent parameter estimates ([71]). Available estimation procedures include the MLE method of [71], the simulated methods of moments of [40] and [50], the generalized method of moments (GMM) of [53], the efficient method of moments (EMM) of [48], the Markov chain Monte Carlo (MCMC) of [61], and the methods based on the empirical characteristic functions of [63,82].

Below we focus on some nonparametric estimation methods of a parametric model

$$dX_t = \mu(X_t, \theta) dt + \sigma(X_t, \theta) dB_t, \tag{10}$$

where $\mu(\cdot, \cdot)$ and $\sigma(\cdot, \cdot)$ are of known form, and $\theta$ is an unknown parameter vector in a parameter space $\Theta$. [4] proposes a minimum distance estimator

$$\widehat{\theta} = \arg\min_{\theta \in \Theta} n^{-1} \sum_{\tau=1}^{n} [\widehat{\pi}_0(X_{\tau\Delta}) - \pi(X_{\tau\Delta}, \theta)]^2, \tag{11}$$

where $\widehat{\pi}_0(x) = n^{-1}\sum_{\tau=1}^n K_h(x - X_{\tau\Delta})$ is a kernel estimator for the marginal density of $X_t$, and

$$\pi(x, \theta) = \frac{c(\theta)}{\sigma^2(x, \theta)} \exp\left[\int_{x_0^*}^x \frac{2\mu(u, \theta)}{\sigma^2(u, \theta)} du\right], \tag{12}$$

is the marginal density estimator implied by model (10). The factor $c(\theta)$ ensures that $\pi(\cdot, \theta)$ integrates to 1 for every $\theta \in \Theta$, and $x_0^*$ is the lower bound of the support of $X_t$. Because the marginal density cannot capture the full dynamics of $\{X_t\}$, $\widehat{\theta}$ is not be asymptotically most efficient, although it is root-$n$ consistent for $\theta_0$.

Recently, [6] proposes an approximated likelihood approach. Let $p_x(\Delta, x \mid x_0, \theta)$ be the conditional density of $X_{\tau\Delta} = x$ given $X_{(\tau-1)\Delta} = x_0$ induced by model (10). The log-likelihood function of the model for the sample $\{X_{\tau\Delta}\}_{\tau=1}^n$ is

$$l_n(\theta) = \sum_{\tau=1}^n \ln p_x(\Delta, X_{\tau\Delta} \mid X_{(\tau-1)\Delta}, \theta).$$

The MLE estimator that maximizes $l_n(\theta)$ is asymptotically most efficient. Unfortunately, except for some simple models, $p_x(\Delta, x\mid x_0, \theta)$ has no closed form. [6] uses a Hermite polynomial approximation $p_x^{(J)}(\Delta, x\mid x_0, \theta)$ for $p_x(\Delta, x\mid x_0, \theta)$, and then obtains an estimator $\widehat{\theta}_n^{(J)}$ that maximizes the approximated model likelihood. This estimator enjoys the same asymptotic efficiency as the (infeasible) MLE as $J = J_n \to \infty$. More specifically, [6] considers a transform

$$Z_t = \Delta^{-1/2}(Y_t - y_0), \quad \text{where} \quad Y_t \equiv \gamma(X_t, \theta) = \int_{-\infty}^{X_t} \sigma^{-1}(u, \theta) du,$$

and then approximates the transition density of $Z_t$ by the Hermite polynomials:

$$p_z^{(J)}(\Delta, z \mid z_0, \theta) = \phi(z)\sum_{j=0}^J \eta_z^{(j)}(z_0, \theta) H_j(z),$$

where $\phi(\cdot)$ is the $N(0, 1)$ density, and $\{H_j(z)\}$ is the Hermite polynomial series. The coefficients $\{\eta_z^{(j)}(z_0, \theta)\}$ are specific conditional moments of process $Z_t$, and can be calculated using the Monte Carlo method or a higher Taylor series expansion in $\Delta$. The approximated transition density of $X_t$ is then given as follows:

$$p_x(\Delta, x \mid x_0, \theta) = \Delta^{-1/2} p_z\{\Delta^{-1/2}[\gamma(x, \theta) - \gamma(x_0, \theta)] \mid \gamma(x_0, \theta), \theta\}.$$

Under suitable conditions, the estimator $\widehat{\theta}_n^{(J)} = \arg\min_{\theta \in \Theta} \sum_{t=1}^n \ln p_x^{(J)}(\Delta, X_{\tau\Delta}|X_{(\tau-1)\Delta}, \theta)$ is asymptotically equivalent to the infeasible MLE. [5] applies this method to estimate a variety of diffusion models for spot interest rates, and finds that $J = 2$ or 3 gives accurate approximations for most financial diffusion models. [42] extend this approach to stationary time-inhomogeneous diffusion models, [7] to general multivariate diffusion models and [8] to affine multi-factor term structure models.

In a rather general continuous-time setup which allows for stationary multi-factor diffusion models with partially observable state variables (e.g., stochastic volatility model), [48] propose an EMM estimator that also enjoys the asymptotic efficiency as the MLE. The basic idea of EMM is to first use a Hermite-polynomial based semi-nonparametric (SNP) density estimator to approximate the transition density of the observed state variables. This is called the auxiliary SNP model and its score is called the score generator, which has expectation zero under the model-implied distribution when the parametric model is correctly specified. Then, given a parameter setting for the multi-factor model, one may use simulation to evaluate the expectation of the score under the stationary density of the model and compute a chi-square criterion function. A nonlinear optimizer is used to find the parameter values that minimize the proposed criterion.

Specifically, suppose $\{X_t\}$ is a stationary possibly vector-valued process with the conditional density $p_0(\Delta, X_{\tau\Delta} | X_{s\Delta}, s \leq \tau - 1) = p_0(\Delta, X_{\tau\Delta} | Y_{\tau\Delta})$, where $Y_{\tau\Delta} = (X_{(\tau-1)\Delta}, \ldots, X_{(\tau-d)\Delta})'$ for some fixed integer $d \geq 0$. This is a Markovian process of order $d$. To estimate parameters in model (10) or its multivariate extension, [48] propose to check whether the following moment condition holds:

$$M(\beta_n, \theta) \equiv \int \frac{\partial \log f(\Delta, x, y; \beta_n)}{\partial \beta_n} p(\Delta, x, y; \theta) dx dy = 0, \quad \text{if} \quad \theta = \theta_0 \in \Theta, \tag{13}$$

where $p(\Delta, x, y; \theta)$ is the model-implied joint density for $(X_{\tau\Delta}, Y'_{\tau\Delta})'$, $\theta_0$ is the unknown true parameter value, and $f(\Delta, x, y; \beta_n)$ is an auxiliary SNP model for the joint density of $(X_{\tau\Delta}, Y'_{\tau\Delta})'$. Note that $\beta_n$ is the parameter vector in $f(\Delta, x, y; \beta_n)$ and may not nest parameter $\theta$. By allowing the dimension of $\beta_n$ to grow with the sample size $n$, the SNP density $f(\Delta, x, y; \beta_n)$ will eventually span the true density $p_0(\Delta, x, y)$ of $(X_{\tau\Delta}, Y'_{\tau\Delta})'$, and thus is free of misspecification asymptotically. [48] use a Hermite polynomial approximation for $f(\Delta, x, y; \beta_n)$, with the dimension of $\beta_n$ determined by such model selection criteria as BIC. Because $p(\Delta, x, y; \theta)$ usually has no closed form, the integration in (13) can be computed by simulating a large number of realizations under model (10).

The EMM estimator is defined as follows:

$$\widehat{\theta} = \arg\min_{\theta \in \Theta} M(\widehat{\beta}_n, \theta)' \widehat{I}^{-1}(\theta) M(\widehat{\beta}_n, \theta),$$

where $\widehat{\beta}$ is the quasi-MLE for $\beta_n$, the coefficients in the SNP density model $f(x, y, \beta_n)$ and the matrix $\widehat{I}(\theta)$ is an estimate of the asymptotic variance of $\sqrt{n} \partial M_n(\widehat{\beta}_n, \theta)/\partial \theta$ (see [49]). This estimator $\widehat{\theta}$ is asymptotically as efficient as the (infeasible) MLE.

The EMM has been applied widely in financial applications. See, for example, [1,15,37] for interest rate applications, [13,32,69] for estimating stochastic volatility models for stock prices with such complications as long memory and jumps, [33] for estimating and testing target zero models of exchange rates, and [64] for price option pricing. It would be interesting to compare the EMM method and approximate MLE of [6] in finite samples.

### 3.2. Nonparametric testing of diffusion models

In financial applications, most continuous-time models are parametric. It is important to test whether a parametric diffusion model adequately captures the dynamics of $\{X_t\}$. Model misspecification generally renders inconsistent model parameter estimators and their variance-covariance matrix estimators, leading to misleading conclusions in inference and hypothesis testing. More importantly, a misspecified model can yield large errors in hedging, pricing and risk management.

Unlike the vast literature of estimation of parametric diffusion models, there are relatively few specification tests for parametric diffusion models using discrete observations.

Suppose $\{X_t\}$ follows a continuous-time possibly time-inhomogenous diffusion process in (5). Often it is assumed that the drift and diffusion $\mu(\cdot,t)$ and $\sigma(\cdot,t)$ have some parametric forms $\mu(\cdot,t,\theta)$ and $\sigma(\cdot,t,\theta)$, where $\theta \in \Theta$. We say that models $\mu(\cdot,t,\theta)$ and $\sigma(\cdot,t,\theta)$ are correctly specified for $\mu(\cdot,t)$ and $\sigma(\cdot,t)$ respectively if

$$H_0: P\left[\mu(X_t,t,\theta_0) = \mu(X_t,t), \sigma(X_t,t,\theta_0) = \sigma(X_t,t)\right] = 1 \text{ for some } \theta_0 \in \Theta. \tag{14}$$

Available estimation methods for model (5) take (14) as given. However, these methods generally cannot deliver consistent parameter estimates if $\mu(\cdot,t,\theta)$ or $\sigma(\cdot,t,\theta)$ is misspecified in the sense that

$$H_A: P\left[\mu(X_t,t,\theta) = \mu(X_t,t), \sigma(X_t,t,\theta) = \sigma(X_t,t)\right] < 1 \text{ for all } \theta \in \Theta. \tag{15}$$

Under $H_A$ of (15), there exists no parameter value $\theta \in \Theta$ such that the drift model $\mu(\cdot,t,\theta)$ and the diffusion model $\sigma(\cdot,t,\theta)$ coincide with $\mu(\cdot,t)$ and $\sigma(\cdot,t)$ respectively.

There is a growing interest in testing whether a continuous-time model is correctly specified using a discrete sample $\{X_{\tau\Delta}\}_{\tau=1}^n$. [4] observes that for a stationary time-homogeneous diffusion process in (10), a pair of drift and diffusion models $\mu(\cdot,\theta)$ and $\sigma(\cdot,\theta)$ uniquely determines the stationary density $\pi(\cdot,\theta)$ given in (12). [4] compares a parametric estimator $\pi(\cdot,\widehat{\theta})$ with a nonparametric estimator $\widehat{\pi}_0(\cdot)$ via the quadratic form

$$M \equiv \int_{x_0^*}^{x_1^*} \left[\widehat{\pi}_0(x) - \pi(x,\widehat{\theta})\right]^2 \widehat{\pi}_0(x)dx, \tag{16}$$

where $x_1^*$ is the upper bound for $X_t$, $\widehat{\theta}$ is the minimum distance estimator given by (11). The $M$ statistic, after demeaning and scaling, is asymptotically normal under $H_0$.

The $M$ test makes no restrictive assumptions on the data generating process and can detect a wide range of alternatives. This appealing power property is not shared by parametric approaches such as generalized method of moment tests (see [35]). The latter has optimal power against certain alternatives but may be completely silent against other alternatives. In an application to Euro-dollar interest rates, [4] rejects all existing one-factor linear drift models and finds that "the principal source of rejection of existing models is the strong nonlinearity of the drift."

However, several limitations of this test may hinder its empirical applicability. First, as [4] has pointed out, the marginal density cannot capture the full dynamics of $\{X_t\}$. Second, subject to some regularity conditions, the asymptotic distribution of a normalized version of $M$ in (16) remains the same whether the sample $\{X_{\tau\Delta}\}_{\tau=1}^n$ is iid or persistently dependent ([4]). This convenient asymptotic property unfortunately results in a substantial discrepancy between the asymptotic and finite sample distributions, particularly when data display persistent dependence ([79]). This casts some doubt on the applicability of first order asymptotic theory of nonparametric methods in finance, since persistent serial dependence is a stylized fact for interest rates and many other high frequency financial data. Third, a kernel density estimator produces finite sample biased estimates near the boundaries of the data, which may generate spurious nonlinear drifts (see [31]).

Recently, [57] have developed a nonparametric test for model (5) using the transition density. Let $p_{x,0}(x,t \mid x_0,s)$ be the transition density of the diffusion process $X_t$; that is, the conditional density of $X_t = x$ given $X_s = x_0$, $s < t$. For a given pair of drift and diffusion models $\mu(\cdot, t, \theta)$ and $\sigma(\cdot, t, \theta)$, a certain family of transition densities $\{p_x(x,t\,|\,x_0,s,\theta)\}$ is characterized. When (and only when) $H_0$ in (14) holds, there exists some $\theta_0 \in \Theta$ such that $p_x(x,t\,|\,x_0,s,\theta_0) = p_{x,0}(x,t\,|\,x_0,s)$ almost everywhere for all $t > s$. Hence, the hypotheses $H_0$ in (14) versus $H_A$ in (15) can be equivalently written as follows:

$$H_0 : p_x(x,t\,|\,y,s,\theta_0) = p_{x,0}(x,t\,|\,y,s) \text{ almost everywhere for some } \theta_0 \in \Theta \qquad (17)$$

versus the alternative hypothesis

$$H_A : p_x(x,t\,|\,y,s,\theta) \neq p_{x,0}(x,t\,|\,y,s) \text{ for some } t > s \text{ and for all } \theta \in \Theta. \qquad (18)$$

A natural approach to testing $H_0$ in (17) versus $H_A$ in (18) would be to compare a model transition density estimator $p_x(x,t\,|\,x_0,s,\widehat{\theta})$ with a nonparametric transition density estimator, say $\widehat{p}_{x,0}(x,t\,|\,x_0,s)$. Instead of comparing $p_x(x,t\,|\,x_0,s,\widehat{\theta})$ and $\widehat{p}_{x,0}(x,t\,|\,x_0,s)$ directly, [57] first consider the transformation

$$Z_\tau(\theta) \equiv \int_{-\infty}^{X_{\tau\Delta}} p_x[x, \tau\Delta | X_{(\tau-1)\Delta}, (\tau-1)\Delta, \theta] dx, \quad \tau = 1, \ldots, n. \qquad (19)$$

Under (and only under) $H_0$ in (17), there exists some $\theta_0 \in \Theta$ such that for all $\Delta > 0$, $p_x[x, \tau\Delta | X_{(\tau-1)\Delta}, (\tau-1)\Delta, \theta_0] = p_{x,0}[x, \tau\Delta | X_{(\tau-1)\Delta}, (\tau-1)\Delta]$ almost surely. Consequently, the transformed series $\{Z_\tau \equiv Z_\tau(\theta_0)\}_{\tau=1}^n$ is iid $U[0,1]$ under $H_0$ in (17).

To test $H_0$ in (17), [57] check whether $\{Z_\tau\}_{\tau=1}^n$ is both iid and $U[0,1]$. They compare a kernel estimator $\widehat{g}_j(z_1, z_2)$ for the joint density of $\{Z_\tau, Z_{\tau-j}\}$ with unity, the product of two $U[0,1]$ densities. The estimator $\widehat{g}_j(z_1, z_2)$ is defined as

$$\widehat{g}_j(z_1, z_2) \equiv (n-j)^{-1} \sum_{\tau=j+1}^n K_h(z_1, \widehat{Z}_\tau) K_h(z_2, \widehat{Z}_{\tau-j}), \qquad j > 0,$$

where $\widehat{Z}_\tau = Z_\tau(\widehat{\theta})$, $\widehat{\theta}$ is any $\sqrt{n}$-consistent estimator for $\theta_0$, and $K_h(x,y)$ is the kernel with boundary correction (see [57]). The test statistic proposed by [57] is

$$\widehat{Q}(j) \equiv \left[(n-j)h \int_0^1 \int_0^1 [\widehat{g}_j(z_1, z_2) - 1]^2 dz_1 dz_2 - A_h^0\right] / V_0^{1/2},$$

where $A_h^0$ and $V_0$ are non-stochastic centering and scale factors that depends on $h$ and $K(\cdot, \cdot)$. Since there is no serial dependence in $\{Z_\tau\}$ under $H_0$ in (17), nonparametric estimators are expected to perform much better in finite samples. In particular, the finite sample distribution of $\widehat{Q}(j)$ is expected to be robust to persistent dependence in data. Also, there is no asymptotic bias for $\widehat{g}_j(z_1, z_2)$ under $H_0$ in (17). Finally, even if $\{X_t\}$ is time-inhomogeneous or nonstationary, $\{Z_\tau\}$ is always iid $U[0,1]$ under $H_0$.

In a simulation mimicking the dynamics of U.S. interest rates via the Vasicek model, [57] find that $\widehat{Q}(j)$ has rather reasonable and robust sizes for $n = 500$. Moreover, $\widehat{Q}(j)$ has better power than the marginal density test. [57] find extremely strong evidence against a variety of existing one-factor diffusion models for the spot interest rate and affine models for interest rate term structures. They also propose a class of separate inference procedures to gauge possible sources of model misspecification. [41] have recently extended [57] to evaluate out-of-sample of density forecasts of a multivariate diffusion model possibly with jumps and partially unobservable state variables.

[48] also propose an EMM-based minimum chi-square specification test for stationary continuous-time models. They examine the moment condition in (13), which holds under correct model specification. This approach is applicable to a wide range of stationary continuous-time processes, including both one-factor and multi-factor diffusion processes with partially observable state variables. In addition to the minimum chi-square test for generic model mis-specifications, the EMM approach also provides a class of individual $t$-statistics that are informative in revealing sources of model misspecification. This is perhaps the most appealing strength of the EMM approach, and is similar in spirit to the separate inference procedures of [57].

Another feature of the EMM tests is that they avoid estimating long-run variance-covariances, thus resulting in reasonable finite sample size performance ([14]). In practice, however, it may not be easy to find an adequate SNP density model for financial time series, as is illustrated in [56].

There has been also an interest in separately testing the drift model and the diffusion model in (10). For example, it has been controversial whether the drift of interest rates is linear. To test the linearity of the drift term, [46] apply the generalized likelihood ratio (GLR) test developed by [47]. They find that the null hypothesis of linear drift is not rejected for the short-term interest rates. Also, they use the GLR test to test whether the diffusion function has a particular parametric form. It is noted that the GLR test is developed for the iid samples. It is still unknown whether it is valid for a time series context. On the other hand, [31] consider an empirical likelihood goodness-of-ft test for a time series regression model, and they apply the test to test a discrete version of the drift model of a diffusion process.

There has been also interest in testing the diffusion model $\sigma(\cdot, \theta)$. The motivation comes from the fact that derivative pricing with an underlying equity process only depends on the diffusion $\sigma(\cdot)$. [68] recently proposes a nonparametric test for $\sigma(\cdot, \theta)$. More specifically, [68] compares a nonparametric diffusion estimator $\hat{\sigma}^2(\cdot)$ with a parametric diffusion estimator $\sigma^2(\cdot, \theta)$ via an asymptotically $\chi^2$ test statistic $\hat{T}_k = \Sigma_{l=1}^k [\hat{T}(x_l)]^2$, where $\hat{T}(x) = [nh\hat{\pi}(x)]^{1/2} [\hat{\sigma}^2(x)/\tilde{\sigma}^2(x,\hat{\theta}) - 1]$, $\hat{\theta}$ is an $\sqrt{n}$-consistent estimator for $\theta_0$ and $\tilde{\sigma}^2(x, \theta) = (n h \hat{\pi}(x))^{-1} \Sigma_{\tau=1}^n \sigma^2(x, \hat{\theta}) K_h[(x - X_\tau)/h]$ is a smooth version of $\sigma^2(x, \theta)$. The use of $\tilde{\sigma}^2(x, \hat{\theta})$ instead of $\sigma^2(x, \hat{\theta})$ directly reduces kernel estimation bias in $\hat{T}(x)$, thus allowing the use of the optimal bandwidth $h$ for $\hat{\sigma}^2(x)$. [68] finds that the empirical level of $\hat{T}_k$ is too large relative to the significance level in finite samples and then proposes a modified test statistic using an empirical likelihood approach, which endogenously studentizes conditional heteroscedasticity. As expected, the empirical level of the modified test improves in finite samples, though the power of the test may not.

Finally, using the GLR test of [47], [45] test whether the coefficients in the time-varying coefficient one-factor diffusion model of (6) are indeed time-varying.

## 4. DERIVATIVE PRICING AND RISK NEURAL DENSITY ESTIMATION

### 4.1. Risk neutral density

The pricing of contingent claims is important in modern finance, given the phenomenal growth in turnover and volume of financial derivatives over the past decades. Derivative pricing formulas are highly nonlinear even when they are available in closed form. Nonparametric techniques are expected to be very useful in this area. In a dynamic setting, the equilibrium price of a security at date $t$ with a single liquidating payoff $Y(C_T)$ at date $T$ that is a function of aggregate consumption $C_T$ is given by

$$P_t = E_t\left[Y(C_T)M_{t,T}\right], \tag{20}$$

where the conditional expectation is taken with respect to the information set available to the representative economic agent at time $t$, $M_{t,T} = \delta^{T-1} U'(C_T)/U'(C_t)$, the so-called stochastic discount factor, is the marginal rate of substitution between dates $t$ and $T$, $\delta$ is the rate of time preference, and $U(\cdot)$ is the utility function of the economic agent. This is the stochastic Euler equation, or the first order condition of the intertemporal utility maximization of the economic agent with suitable budget constraints ([34]). It holds for all securities, including assets and various derivatives. All capital asset pricing models (CAPM) and derivative pricing models can be embedded in (20) — each model can be viewed as a specific specification of $M_{t,T}$. See [34] for an excellent discussion.

There have been some parametric tests for CAMP models ([52]). To our knowledge, there seems to be only one nonparametric test for CAMP models based on the kernel method ([85]). Also, all the tests for CAMP models are formulated in terms of discrete time frameworks. Below, we focus on nonparametric derivative pricing.

Assuming that the conditional distribution of future consumption $C_T$ has a density representation $p_t(\cdot)$, then the conditional expectation can be expressed as

$$E_t\left[Y(C_T)M_{t,T}\right] = e^{-r\tau} \int Y(C_T) p_t^*(C_T) dC_T = e^{-r\tau} E_t^*\left[Y(C_t)\right],$$

where $r_t$ is the risk free interest rate, $\tau = T - t$, and $p_t^*(c) = M_{t,T} p_t(c) / \int M_{t,T} p_t(c) dc$ is called the risk neutral density (RND) function, or the risk-neural pricing probability, or the equivalent martingale measure, or the state-price density. It contains rich information on the pricing and hedging of risky assets, and can be used to price other assets, or to recover the information about market preferences and asset price dynamics. Obviously, the RND function differs from $p_t(\cdot)$, the physical density function of $C_T$ conditional on the information available at time $t$.

### 4.2. Nonparametric pricing

To calculate an option price from (20), one has to make some assumption on the data generating process of the underlying asset, say $\{P_t\}$. For example, [24] assume that the underlying asset follows a geometric Brwonian motion: $dP_t = \mu P_t dt + \sigma P_t dB_t$, where $\mu$ and $\sigma$ are two constants. Applying Ito's lemma, one can show that $P_\tau$ follows a lognormal distribution with parameter $(\mu - \frac{1}{2}\sigma^2)\tau$ and $\sigma\sqrt{\tau}$. Using a no-arbitrage argument, [24] show that options can be priced if investors are risk neutral by setting the expected rate of return in the underlying asset, $\mu$, equal to the risk-free interest rate, $r$. Specifically, the European call option price is

$$\pi(K_t, P_t, r, \tau) = P_t \Phi(d_t) - e^{-r_t \tau} K_t \Phi(d_t - \sigma\sqrt{\tau}), \tag{21}$$

where $K_t$ is the strike price, $\Phi(\cdot)$ is the N(0,1) cumulative distribution function and $d_t = [\ln(P_t/K_t) + (r + \frac{1}{2}\sigma^2)\tau]/(\sigma\sqrt{\tau})$. In (21), the only parameter that is not observable a time $t$ is $\sigma$. This parameter, when multiplied with $\sqrt{\tau}$, is the underlying asset return volatility over the remaining life of the option. A knowledge of $\sigma$ can be inferred from the prices of options traded in the markets: given an observed option price, one can solve an appropriate option pricing model for $\sigma$ via (21), which gives a market estimate of the future volatility of the underlying asset return. This estimate of $\sigma$ is known as "implied volatility".

The most important implication of Black-Scholes option pricing is that when the option is correctly priced, the implied volatility $\sigma^2$ should be the same across all exercise prices of options on the same underlying asset and with the same maturity date. However, the implied volatility observed in the market is usually convex in the exercise price, which is often referred to as the "volatility smile". This implies that market participants make more complicated assumptions than the geometric Bownian motion for the dynamics of the underlying asset. In particular, the convexity of "volatility smile" indicates the degree to which the market RND function has a heavier tail than a lognormal density. A great deal of effort has been made to use alternative models for the underlying asset to smooth out the volatility smile and achieve higher accuracy in pricing and hedging.

A more general approach to derivative pricing is to estimate the RND function directly from the observed option prices and then use it to price derivatives or extract market information. To obtain better estimation of the RND function, several econometric techniques have been introduced. They are all based on a fundamental relation between option prices and RNDs, which first shown by [27] in a time-state preference framework: Suppose $G_t = G(K_t, P_t, r_t, \tau)$ is the option pricing formula. Then there is a close relation between the second derivative of $G_t$ with respect to the strike price $K_t$ and the RND function:

$$\frac{\partial^2 G_t}{\partial K_t^2} = e^{-\tau r_t} p_t^*(P_T). \tag{22}$$

Most commonly used estimation methods for RNDs are parametric. One of them is to assume that the underlying asset follows a parametric diffusion process, from which one can obtain the option pricing formula by a no-arbitrage argument, and then obtain the RND function from (22) (see [11,20,21]). Another approach is to directly impose some form for the RND function and then estimate unknown parameters by minimizing the distance between the observed option prices and those generated by the assumed RND function (see [60,75,80]). A third approach is to assume and estimate a parametric form for the call pricing function or the implied volatility smile curve and then apply (22) to get the RND function (see [20,73,74,81]).

All aforementioned parametric approaches impose certain restrictive assumptions, directly or indirectly, on the data generating process as well as the stochastic discount factor in some cases. The obtained RND function is not robust to the violation of these restrictions. To avoid this drawback, [9] use a nonparametric method to extract the RND function from option prices. Given observed call option prices $\{G_t, K_t, \tau\}$, the price of the underlying asset $\{P_t\}$, and the risk free rate of interest $\{r_t\}$, [9] construct a kernel-estimator for $E(G_t|P_t, K_t, \tau, r_t)$. Under regularity conditions, [9] show that the RND estimator is consistent and asymptotically normal and they provide explicit expressions for the asymptotic variance of the estimator. [10] use a nonparametric RND estimator to compute the economic value at risk, that is, the value at risk of the RND function.

The artificial neural network (ANN) has received much attention in economics and finance over the last decade. [12,51,59] have successfully applied the ANN models to estimate pricing formulas of financial derivatives. In particular, [59] use the ANN to examine whether ANN can "learn" the Black-Scholes formula, if option prices are indeed determined by the Black-Scholes formula. They conduct simulation experiments in which various ANNs are trained on artificially generated Black-Scholes formula and then compare the estimated price formula to the Black-Scholes formula in an out-of-sample hedging setup. By simulating a two-year sample of daily stock prices, and creating a across-section of options each day according to the rules used by the Chicago Broad Options Exchange with prices given by the Black-Scholes formula, they find that even with training sets of only six months of daily data, ANN pricing formulas can approximate the Black-Scholes formula reasonably well. ANN models yield estimates option prices very similar to the true Black-Scholes prices.

There are several directions of further research on nonparametric analysis of RNDs for derivative pricing. First, how to evaluate the quality of a RND function estimated from option prices? In other words, how to judge how well an estimated RND function reflects the market expected uncertainty of the underlying asset? Because the RND function differs from the physical probability density of the underlying process, the valuation of the RND function is rather challenging. The method developed by [57] cannot, for example, be applied directly. One possible way to evaluate the RND function is to assume a certain family of utility functions for the representative investor, as in [80] and [11]. One can obtain the stochastic discount factor and the physical probability density, to which the test of ([57]) can be applied. However, the utility function of the economic agent is not observable. When the test delivers a rejection, it may be due to either misspecification of the utility function or misspecification of the data generating process, or both. More fundamentally, it is not clear whether the economy can be simply proxied by an representative agent.

Most estimation methods for the RND function are restricted to European options, while many of the more liquid exchange-traded options are American. Rather complex extensions of the existing methods, including the nonparametric ones, are required to estimate the RND functions from the prices of American options. This is an interesting and important direction for further research.

## 5. CONCLUSION

Over the last several years, nonparametric continuous-time methods have become an integral part of research in finance. The literature is already vast and continues to grow swiftly, involving a full spread of participants for both financial economists and statisticians and engaging a wide sweep of academic journals. The field has left indelible mark on almost all core areas in finance such as asset pricing theory, consumption-portfolio selection, derivatives and risk analysis. The popularity of this field is also witnessed by the fact that the graduate students at both Master and doctoral levels in economics, finance, mathematics and statistics are expected to take courses in this discipline or alike and review important articles in this area to search for their own research interests, particularly dissertation topics for doctoral students. On the other hand, this area also has made an impact in the financial industry as the sophisticated nonparametric techniques can be of practical assistance in the industry. We hope that this selective review has provided the reader a perspective on this important field in finance and statistics and some open research problems.

## REFERENCES

1. Ahn, D.H., R.F. Dittmar and A.R. Gallant (2002), Quadratic term structure models: Theory and evidence, *Review of Financial Studies* **15**, 243-288.
2. Ahn, D.H. and B. Gao (1999), A parametric nonlinear model of term structure dynamics, *Review of Financial Studies* **12**, 721-762.
3. Ait-Sahalia, Y. (1996a), Nonparametric pricing of interest rate derivative securities, *Econometrica* **64**, 527-560.
4. Ait-Sahalia, Y. (1996b), Testing continuous-time models of the spot interest rate, *Review of Financial Studies* **9**, 385-426.
5. Ait-Sahalia, Y. (1999), Transition densities for interest rate and other nonlinear diffusions, *Journal of Finance* **54**, 1361-1395.
6. Ait-Sahalia, Y. (2002a), Maximum likelihood estimation of discretely sampled diffusions: A closed-form approach, *Econometrica* **70**, 223-262.
7. Ait-Sahalia, Y. (2002b), Closed-form likelihood expansions for multivariate diffusion, *Working Paper*, Department of Economics, Princeton University.
8. Ait-Sahalia, Y. and Kimmel, R. (2002), Estimating affine multifactor term structure models using closed-form likelihood expansions, *Working Paper*, Department of Economics, Princeton University.
9. Ait-Sahalia, Y. and Lo, A.W. (1998), Nonparametric estimation of state-price densities implicit in financial asset prices, *Journal of Fiance* **53**, 499-547.
10. Ait-Sahalia, Y. and Lo, A.W. (2000), Nonparametric risk management and implied risk aversion, *Journal of Econometric* **94**, 9-51.
11. Anagnou, I., M. Bedendo, S. Hodges and R. Tompkins (2001), The relation between implied and realized probability density functions, *Working Paper*, University of Technology, Vienna.
12. Anders, U., O. Korn and C. Schmitt (1998), Improving the pricing of options: A neural network approach, *Journal of Forecasting* **17**, 369-388.
13. Andersen, T.G., L. Benzoni and J. Lund (2001), Towards an empirical foundation for continuous-time equity return models, *Working Paper*, Graduate School of Management, Northwestern University.
14. Andersen, T.G., H.-J. Chung and B.E. Sorensen (1999), Efficient method of moments estimation of a stochastic volatility model: A Monte Carlo study, *Journal of Econometrics* **91**, 61-87.
15. Andersen, T.G. and J. Lund (1997), Estimating continuous-time stochastic volatility models of the short-term interest rate, *Journal of Econometrics* **77**, 343-377.
16. Bandi, F. (2000), Nonparametric fixed income pricing: theoretical issues, *Working Paper*, Graduate School of Business, The University of Chicago.
17. Bandi, F. and T.H. Nguyen (2000a), Fully nonparametric estimators for diffusions: a small sample analysis, *Working Paper*, Graduate School of Business, The University

of Chicago.
18. Bandi, F. and T.H. Nguyen (2000b), On the functional estimation of jump-diffusion models, *Working Paper*, Graduate School of Business, The University of Chicago.
19. Bandi, F. and P.C.B. Phillips (2003), Fully nonparametric estimation of scalar diffusion models, *Econometrica* **71**, 241-283.
20. Bates, D.S. (1991), The Crash of '87: Was it expected? The evidence from options markets, *Journal of Finance* **46**, 1009-44.
21. Bates, D.S. (2000), Post-'87 crash fears in the S&P 500 futures option market, *Journal of Econometrics* **94**, 181-238.
22. Black, F., E. Derman and W. Toy (1990), A one-factor model of interest rates and its application to treasury bond options, *Financial Analysts Journal* **46**, 33-39.
23. Black, F. and P. Karasinski (1991), Bond and option pricing when short rates are log-normal, *Financial Analysts Journal* **47**, 52-59.
24. Black, F. and M. Scholes (1973), The pricing of options and corporate liabilities, *Journal of Political Economy* **71**, 637-654.
25. Bliss, R.R. and D. Smith (1998), The elasticity of interest rate volatility: Chan, Karolyi, Longstaff, and Sanders revisited, *Journal of Risk* **1**, 21-46.
26. Bollerslev, T. and H. Zhou (1999), Estimating stochastic volatility diffusion using conditional moments of integrated volatility, *Working Paper*, Department of Economics, Duke University.
27. Breeden, D.T. and R.H. Litzenberger (1978), Prices of state-contingent claims implicit in option prices, *Journal of Business* **51**, 621-51.
28. Cai, Z. (2001), Weighted Nadaraya-Watson regression estimation, *Statistics and Probability Letters* **51**, 307-318.
29. Chan, K.C., G.A. Karolyi, F.A. Longstaff and A.B. Sanders (1992), An empirical comparison of alternative models of the short-term interest rate, *Journal of Finance* **47**, 1209-1227.
30. Chapman, D. and N. Pearson (2000), Is the short rate drift actually nonlinear? *Journal of Finance* **55**, 355-388.
31. Chen, S.X., W. Härdle and T. Kleinow (2001), An empirical likelihood goodness-of-fit test for time series, *Working paper*, Institute of Statistics and Economics, Humboldt University, Germany.
32. Chernov, M., A.R. Gallant, E. Ghysels and G. Tauchen (2002), A new class of stochastic volatility models with jumps: Theory and estimation, *Working Paper*, Department of Economics, University of North Carolina at Chapel Hill.
33. Chung, C.C. and G. Tauchen (2001), Testing target zone models using efficient method of moments, *Journal of Business and Economic Statistics* **19**, 255-277.
34. Cochrane, J.H. (2001), *Asset Pricing*, New Jersey: Princeton University Press.
35. Conley, T.G., L.P. Hansen, E.G.J. Luttmer and J.A. Scheinkman (1997), Short-term interest rates as subordinated diffusions, *Review of Financial Studies* **10**, 525-577.
36. Cox, J.C., J.E. Ingersoll and S.A. Ross (1985), A theory of the term structure of interest rates, *Econometrica* **53**, 385-407.
37. Dai, Q. and K.J. Singleton (2000), Specification analysis of affine term structure models, *Journal of Finance* **55**, 1943-1978.
38. Duffie, D. and J. Pan (2001), Analytical value-at-risk with jumps and credit risk,

*Finance and Stochastics* **5**, 15-180.
39. Duffie, D., J. Pan and K.J. Singleton (2000), Transform analysis and asset pricing for affine jump-diffusions, *Econometrica* **68**, 1343-1376.
40. Duffie, D. and K.J. Singleton (1993), Simulated moments estimation of Markov models of asset prices, *Econometrica* **61**, 929-952.
41. Egorov, A., Y. Hong and H. Li (2003), Evaluation of out-of-sample density forecasts of multifactor diffusion models with application to affine models of term structures, *Working Paper*, Department of Economics, Cornell University.
42. Egorov, A., H. Li and Y. Xu (2003), Maximum likelihood estimation of time-inhomogeneous diffusions, *Journal of Econometrics*, in press.
43. Eraker, B., M.S. Johannes and N.G. Polson (1999), The impact of jumps in volatility and returns, *Working Paper*, Graduate School of Business, Columbia University.
44. Fan, J. and I. Gijbels (1996), *Local Polynomial Modeling and its Applications*, London: Chapman and Hall.
45. Fan, J., J. Jiang, C. Zhang and Z. Zhou (2001), Time-dependent diffusion models for term structure dynamics and the stock price volatility, *Working Paper*, Department of Statistics, University of North Carolina at Chapel Hill.
46. Fan, J. and C. Zhang (2001), A re-examination of diffusion estimators with applications to financial model validation, *Journal of the American Statistical Association*, in press.
47. Fan, J., C. Zhang and J. Zhang (2001), Generalized likelihood ratio statistics and Wilks phenomenon, *The Annals of Statistics* **29**, 153-193.
48. Gallant, A.R. and G. Tauchen (1996), Which moments to match? *Econometric Theory* **12**, 657-681.
49. Gallant, A.R. and G. Tauchen (2001), Efficient method of moments, *Working Paper*, Department of Economics, University of North Carolina and Department of Economics, Duke University.
50. Gourieroux, C., A. Monfort and E. Renault (1993), Indirect inference, *Journal of Applied Econometrics* **8**, 85-118.
51. Hanke, M. (1999), Neural networks versus Black-Scholes: An empirical comparison of the pricing accuracy of two fundamentally different option pricing methods, *Journal of Computational Finance* **5**, 26-34.
52. Hansen, L.P. and R. Janaganan (1997), Assessing specification errors in stochastic discount factor models, *Journal of Finance* **52**, 557-590.
53. Hansen, L.P. and J.A. Scheinkman (1995), Back to the future: Generating moment implications for continuous time Markov processes, *Econometrica* **63**, 767-804.
54. Härdle, W. (1990), *Applied Nonparametric Regression*, New York: Cambridge University Press.
55. Ho, T.S.Y. and S.B. Lee (1986), Term structure movements and pricing interest rate contingent claims, *Journal of Finance* **41**, 1011-1029.
56. Hong, Y. and T.H. Lee (2003), Diagnostic checking for nonlinear time series models, *Econometric Theory*, in press.
57. Hong, Y. and H. Li (2002), Nonparametric specification testing for continuous-time models with applications to interest rate term structures. Submitted to *Review of Financial Studies*.

58. Hull, J. and H. White (1990), Pricing interest-rate derivative securities, *Review of Financial Studies* **3**, 573-592.
59. Hutchinson, J., A.W. Lo and T. Poggio (1994), A nonparametric approach to pricing and hedging derivative securities via learning networks, *Journal of Finance* **49**, 851-889.
60. Jackwerth J.C. and M. Rubinstein (1996), Recovering probability distributions from contemporary security prices, *Journal of Finance* **51**, 1611-1631.
61. Jacquier, E., N.G. Polson and P. Rossi (1994), Bayesian analysis of stochastic volatility models, *Journal of Business and Economic Statistics* **12**, 371-389.
62. Jiang, G.J. and J.L. Knight (1997), A nonparametric approach to the estimation of diffusion processes, with an application to a short-term interest rate model, *Econometric Theory* **13**, 615-645.
63. Jiang, G.J. and J.L. Knight (2002), Estimation of continuous-time processes via the empirical characteristic function, *Journal of Business and Economic Statistics* **20**, 198-212.
64. Jiang, G.J. and P.J. van der Sluis (2000), Option pricing with the efficient method of moments, in *Computational Finance* (Y.S. Abu-Mostafa, B. LeBaron, A.W. Lo, and A.S. Weigend, eds.), Cambridge: MIT Press.
65. Johannes, M.S. (2000), A nonparametric view of the role of jumps to interest rates, *Working Paper*, Graduate school od Business, Columbia University.
66. Johannes, M.S., R. Kumar and N.G. Polson (1999), State dependent jump models: How do US equity indices jump? *Working Paper*, Graduate School of Business, University of Chicago.
67. Karatzas, I. and S.E. Shreve (1988), *Brownian Motion and Stochastic Calculus*, second edition, New York: Spring-Verlag.
68. Kleinow, T. (2002), Testing the diffusion coefficients, *Working paper*, Institute of Statistics and Economics, Humboldt University, Germany.
69. Liu, M. (2000), Modeling long memory in stock market volatility, *Journal of Econometrics* **99**, 139-171.
70. Liu, J., F.A. Longstaff and J. Pan (2002), Dynamic asset allocation with event risk, *Journal of Finance* **58**, 231-259.
71. Lo, A.W. (1988), Maximum likelihood estimation of generalized Ito processes with discretely sampled data, *Econometric Theory* **4**, 231-247.
72. Lobo, B.J. (1999), Jump risk in the U.S. stock market: Evidence using political information, *Review of Financial Economics* **8**, 149-163.
73. Longstaff, F.A. (1992), Multiple equilibria and term structure models, *Journal of Financial Economics* **32**, 333-344.
74. Longstaff, F.A. (1995), Option pricing and the martingale restriction, *Review of Financial Studies* **8**, 1091-1124.
75. Melick, W.R. and C.P. Thomas (1997), Recovering an asset's implied PDF from option prices: An application to crude oil during the Gulf crisis, *Journal of Financial and Quantitative Analysis* **32**, 91-115.
76. Merton, R.C. (1973), Theory of rational option pricing, *Bell Journal of Economics and Management Science* **4**, 141-183.
77. Mittelhammer, R.C., G.G. Judge and D.J. Miller (2000), *Econometrics Foundation*,

New York: Cambridge University Press.
78. Pagan, A. and A. Ullah (1999), *Nonparametric Econometrics*, New York: Cambridge University Press.
79. Pritsker, M. (1998), Nonparametric density estimation and tests of continuous time interest rate models, *Review of Financial Studies* **11**, 449-487.
80. Rubinstein M. (1994), Implied binomial trees, *Journal of Finance* **49**, 771-818.
81. Shimko, D. (1993), Bounds of probability, *Risk* **6**, 33-37.
82. Singleton, K.J. (2001), Estimation of affine asset pricing models using the empirical characteristic function, *Journal of Econometrics* **102**, 111-141.
83. Stanton, R. (1997), A nonparametric model of term structure dynamics and the market price of interest rate risk, *Journal of Finance* **52**, 1973-2002.
84. Vasicek, O. (1977), An equilibrium characterization of the term structure, *Journal of Financial Economics* **5**, 177-188.
85. Wang, K. (2002), Asset pricing with conditioning information: A new test, *Journal of Finance* **58**, 161-196.

Recent Advances and Trends in Nonparametric Statistics
Michael G. Akritas and Dimitris N. Politis (Editors)
© 2003 Elsevier Science B.V. All rights reserved.

# Nonparametric Estimation in a Stochastic Volatility Model

Jürgen Franke[a] and Wolfgang Härdle[b] and Jens-Peter Kreiss[c]

[a]Universität Kaiserslautern,
Erwin-Schrödinger-Straße, D-67663 Kaiserslautern, Germany

[b]Humboldt-Universität zu Berlin,
Spandauer Straße 1, D-10178 Berlin, Germany

[c]Technische Universität Braunschweig,
Pockelsstraße 14, D-38106 Braunschweig, Germany

In this paper we derive nonparametric stochastic volatility models in discrete time. These models generalize parametric autoregressive random variance models, which have been applied quite successfully to financial time series. For the proposed models we investigate nonparametric kernel smoothers. It is seen that so-called nonparametric deconvolution estimators could be applied in this situation and that consistency results known for nonparametric errors-in-variables models carry over to the situation considered herein.

## 1. Introduction

Many methods of financial engineering like option pricing or portfolio management crucially depend on the stochastic model of the underlying asset. If $S(t)$ denotes the stock price at time $t$, then, e.g., the Black-Scholes approach to option pricing is based on modelling $\log S(t)$ as a Wiener process with drift $\mu$ and diffusion coefficient or volatility $\sigma$ :

$$d(\log S(t)) = \alpha \, dt + \sigma dW(t)$$

where $W(t)$ is a standard Wiener process. This particular model is known to be inappropriate in various circumstances. For instance, $\sigma$ can no longer be assumed to be constant if the time up to exercising the option is rather short. Replacing the constant $\sigma$ by a positive stochastic process $\sigma(t)$ we arrive at the following equation for the asset price:

$$d(\log S(t)) = \alpha \, dt + \sigma(t) dW(t). \qquad (1.1)$$

In the literature, several specific parametric models for the stochastic volatility $\sigma(t)$ have been proposed and used for option pricing. Here, we restrict ourselves to models which characterize $\sigma(t)$ as the solution of a stochastic differential equation for $\log \sigma(t)$ known up to a few parameters. An example is the equation

$$d(\log \sigma(t)) = \lambda(\kappa - \log \sigma(t))dt + \gamma dW^*(t) \qquad (1.2)$$

considered by Scott (1987, 1991), Wiggins (1987) and Chesney and Scott (1989). Here, $W^*(t)$ is another standard Wiener process correlated with $W(t)$ of (1.1)

$$dW(t)\, dW^*(t) = \rho\, dt,$$

and $\alpha$ of (1.1), $\lambda, \kappa, \gamma$ and $\rho$ are the unknown model parameters. Other models of a similar structure have been proposed in the literature.

To help to answer the question which stochastic volatility model is appropriate for a particular data set we consider a rather general type of model avoiding the assumption of a particular parametric form of the equation defining $\sigma(t)$. At the beginning, we discretize time, as is also frequently done for parametric models for the purpose of estimating the model parameters. The log-volatility will then satisfy a general nonlinear stochastic difference equation or nonlinear autoregressive scheme. As $\sigma(t)$ is not directly observable, the now quite familiar kernel estimates for the autoregression function are not applicable. We use instead nonparametric deconvolution estimators similar to those discussed in regression analysis by Fan and Truong (1993). These estimators are consistent and provide a convenient tool for exploratory data analysis helping in the decision which particular parametric model to choose for further analysis of the data.

## 2. A nonparametric stochastic volatility model

We consider some asset with price $S(t)$ at time $t$ and, following Taylor (1994), define the return from an integer time $t-1$ to time $t$ as

$$R_t = \log \frac{S(t)}{S(t-1)}.$$

To estimate a stochastic volatility model like (1.1) and (1.2), discretized versions of these equations are considered. Wiggins (1987) and Chesney and Scott (1989) use the Euler approximation

$$R_t = \mu + \sigma_{t-1} W_t \tag{2.3}$$

$$\log \sigma_t = \alpha + \phi\{\log \sigma_{t-1} - \alpha\} + \vartheta W_t^* \tag{2.4}$$

$(W_t, W_t^*)$ denote i.i.d. bivariate standard normal random variables with zero mean and correlation $\rho$. In (2.3), the lagged quantity $\sigma_{t-1}$ appears as the stochastic volatility for period $t$. This is rather advantageous for statistical purposes, as we will clearly see later on.

As another simplification of (1.1), Taylor (1994) considers

$$R_t = \mu + \sigma_t W_t, \tag{2.5}$$

and he called (2.3), (2.4) a lagged autoregressive random variance (LARV) model, as $\log \sigma_t$ follows a linear autoregressive scheme. Analogously, (2.5), (2.4), together, is called a contemporaneous autoregressive random variance (CARV) model.

In this paper, we consider nonparametric generalizations of these models. We start with the lagged case and study it in detail, whereas we give a short discussion of the contemporaneous case at the end of Section 3.
We replace (2.4) by a nonlinear nonparametric model for $\xi_t = \log \sigma_t$:

$$\xi_t = m(\xi_{t-1}) + \eta_t, \tag{2.6}$$

where $\eta_t$ denote i.i.d. zero-mean normal random variables with variance $\sigma_\eta^2$, and $m$ is an arbitrary autoregression function for which we only require certain smoothness assumptions.
In order to ensure that the Markov chain $(\xi_t)$ possesses nice probabilistic properties - e.g. geometric ergodicity and $\beta$-mixing (absolute regularity) or $\alpha$-mixing (strongly mixing) with geometrically decaying mixing coefficients - it suffices (because of the assumption of normally distributed innovations $\eta_t$) to assume an appropriate drift condition on $m$, e.g.

$$\limsup_{|x| \to \infty} \left| \frac{m(x)}{x} \right| < 1, \tag{A1}$$

cf. Doukhan (1994), Proposition 6 (page 107). Then, in particular, $\xi_t$ has a unique stationary distribution with density $p_\xi$.
We want to estimate $m$ using kernel-type estimates. The usual Nadaraya-Watson estimates are, however, not applicable as we cannot observe the volatility $\sigma_t$ or its logarithm $\xi_t$ directly. The available data are the asset prices $S_t$ or the returns $R_t$ which are related to $\sigma_t$ by (2.3). Taking logarithms and using the abbreviations

$$X_t = \frac{1}{2} \log(R_t - \mu)^2 - \mu_\varepsilon, \quad \varepsilon_t = \frac{1}{2} \log W_t^2 - \mu_\varepsilon$$

with $\mu_\varepsilon = \mathcal{E}\left(\frac{1}{2} \log W_t^2\right) = -0.63518...$ (Scott (1987)), we get

$$X_t = \xi_{t-1} + \varepsilon_t, \tag{2.5}$$

where the $\varepsilon_t$ are i.i.d. zero-mean random variables distributed as $\frac{1}{2}$ times the logarithm of a $\chi_1^2$-random variable centered around 0. The correlation between the standard normal random variable $W_t$, appearing in the definition of $\varepsilon_t$, and $\eta_t$ of (2.6) is $\rho$. (2.6), (2.5), together, form a nonparametric autoregressive model with errors-in-variables as $\xi_t$ cannot be observed directly but is known only through its convolution with the i.i.d. random variables $\varepsilon_t$. Plugging (2.5) into (2.6) we obtain the following equation for $X_t$ alone

$$X_t = m(X_{t-1} - \varepsilon_{t-1}) + \eta_{t-1} + \varepsilon_t. \tag{2.6}$$

**Remark** Assumption (A.1) also implies geometric ergodicity including geometrically $\beta$- and strong mixing for the process $(X_t)$.

## 3. Kernel estimates for the autoregressive volatility function

Fan and Truong (1993) have studied a nonparametric regression model with errors-in-variables similar to the nonparametric autoregressive model (2.6), (2.5). Following their approach, we construct nonparametric estimates for $m$ based on a sample $X_1, \ldots, X_T$. Let us assume that the parameter $\mu$, which is the expectation of the returns $R_t$, is known such that the $X_t$ are observable. From applications it can be justified that this expectation is close to zero. In case $\mu \neq 0$, the returns have to be centered before the procedure described below should be applied.

If we could observe $\xi_1, \ldots, \xi_T$ then we could estimate their stationary density, $p_\xi(x)$ by the kernel estimate

$$\hat{p}_\xi(x,h) = \frac{1}{Th} \sum_{t=1}^{T} K(\frac{x-\xi_t}{h}),$$

where $K$ denotes a probability density and $h > 0$ denotes the bandwidth. The strongly mixing property of $(\xi_t)$, which is ensured by (A1), immediately implies consistency via a covariance inequality.

As we only observe $X_1, \ldots, X_T$, whose stationary density is the convolution of $p_\xi$ with the known density of the i.i.d. random variables $\varepsilon_t$, we have to use a deconvolution density estimate instead:

$$\hat{p}(x,h) = \frac{1}{2\pi} \int_{-\infty}^{\infty} e^{-iwx} \phi_K(wh) \frac{\hat{\phi}_x(w)}{\phi_\varepsilon(w)} dw \tag{3.7}$$

with

$\phi_\varepsilon(w) = \mathcal{E} \, e^{iw\varepsilon_1}$, the characteristic function of $\varepsilon_t$,
$\phi_K(w) = \int_{-\infty}^{\infty} e^{iwx} K(x) dx$, the Fourier transform of the kernel $K$,
$\hat{\phi}_x(w) = \frac{1}{T} \sum_{t=1}^{T} e^{iwX_t}$, the sample characteristic function of $X_1, \ldots, X_T$.

The bandwidth $h$, depending on the sample size $T$, acts as a smoothing parameter as usual. For i.i.d. observations $\xi_1, \ldots, \xi_T$, the estimate $\hat{p}(x,h)$ for $p_\xi(x)$ has been investigated in detail by Stefanski and Carroll (1990), Carroll and Hall (1988), Fan (1991a,b) and Liu and Taylor (1989). Note that (3.7) can be written as a kernel estimator similar to $\hat{p}_\xi(x,h)$, namely

$$\hat{p}(x,h) = \frac{1}{Th} \sum_{t=1}^{T} K_h(\frac{x-X_t}{h})$$

with a kernel $K_h$ depending on $h$ and on the known distribution of the $\varepsilon_t$

$$K_h(x) = \frac{1}{2\pi} \int_{-\infty}^{\infty} e^{-iwx} \frac{\phi_K(w)}{\phi_\varepsilon(w/h)} dw. \tag{3.8}$$

**Remark.** It should be noted, that without knowing anything of the distribution of the $\varepsilon_t$ it is completely impossible to recover the stationary density $p_\xi$.

Now, the nonparametric estimate for $m(x)$ is defined as a Nadaraya-Watson estimate with kernel $K_h$ and with $X_t$ replacing $\xi_t$, more exactly

$$\hat{m}(x, h) = \frac{1}{Th} \cdot \sum_{t=1}^{T} K_h(\frac{x - X_t}{h}) X_{t+1}/\hat{p}(x, h). \quad (3.9)$$

In order to apply this estimator it is necessary to evaluate the characteristic function $\phi_\varepsilon$ of $\varepsilon_t$ and to make use of a kernel $K$ for which the Fourier transform $\phi_K$ takes a convenient form. Concerning the explicit form and the asymptotic behaviour of $\phi_\varepsilon$ we have the following result.

**Lemma 3.1:** Assume $W \sim \mathcal{N}(0, 1)$, and let the density of the standard normal distribution be $\varphi$. The distribution of the centered random variable $\varepsilon = \frac{1}{2} \log W^2 - \mu_\varepsilon$ possesses the following density

$$p_\varepsilon(x) = 2\, \varphi(e^{x+\mu_\varepsilon})\, e^{x+\mu_\varepsilon}, \quad x \in \mathbb{R}.$$

Here $\mu_\varepsilon = (\kappa + \log 2)/2 \approx 0.63518$ ($\kappa$ denotes Eulers constant).
Let us denote by $\Gamma$ the Gamma function. We have

$$\phi_\varepsilon(w) = \frac{e^{(\frac{\log 2}{2} - \mu_\varepsilon)iw}}{\sqrt{\pi}} \Gamma(\frac{1+iw}{2}), \quad w \in \mathbb{R}.$$

Concerning the tail behaviour of $\phi_\varepsilon$ we have for all $d_0, d_1$ with $0 < d_0 < \sqrt{2} < d_1 < \infty$:

$$d_0\, e^{-|w|\pi/4} \leq |\phi_\varepsilon(w)| \leq d_1 e^{-|w|\pi/4} \text{ as } |w| \longrightarrow \infty. \quad (3.10)$$

**Proof:** The explicit expressions for $p_\varepsilon$ and $\phi_\varepsilon$ can be obtained by direct computation, while (3.10) is an immediate consequence of the tail-behaviour of $\Gamma$, which can be found for example in Gradstein and Ryshik (1981) (No. 8.328, page 331). ∎

Now, let us investigate the asymptotic behaviour of the kernel estimator $\hat{m}(\cdot, h)$, cf. (3.9). We have

$$\hat{m}(x, h) - m(x) = \frac{\frac{1}{Th} \sum_t K_h(\frac{x-X_t}{h})(X_{t+1} - m(x))}{\frac{1}{Th} \sum_t K_h(\frac{x-X_t}{h})}. \quad (3.11)$$

The following lemmas imply the consistency of $\hat{m}(\cdot, h)$.

**Lemma 3.2:** Assume that $m$ is twice continuously differentiable and that $p_\xi$ is continuously differentiable. Assume that $\phi_K$ has a bounded support, $[-M_0, M_0]$ say, and that $h = h(T) = c/\log T$ where $c > M_0\pi/2$.

(i) $\quad E \frac{1}{Th} \sum_t K_h(\frac{x-X_t}{h})(X_{t+1} - m(x)) \;=\; \int_{-\infty}^{\infty} \{m(u) - m(x)\} \frac{1}{h} K(\frac{x-u}{h}) p_\xi(u) du$
$\qquad\qquad\qquad\qquad\qquad\qquad\qquad\qquad\quad\;= O(h^2)$

(ii) $\quad \mathrm{Var}\left(\frac{1}{Th} \sum_t K_h(\frac{x-X_t}{h})(X_{t+1} - m(x))\right) = o(1).$

**Lemma 3.3:** *Assume that $p_\xi$ is twice times continuously differentiable. Assume that $\phi_K$ has a bounded support, $[-M_0, M_0]$, say, and that $h = h(T) \sim c/\log T$ where $c > M_0\pi/2$. Then*

(i) $\quad E \frac{1}{Th} \sum_t K_h(\frac{x-X_t}{h}) = \int_{-\infty}^{\infty} p_\xi(u) \frac{1}{h} K(\frac{x-u}{h}) du$
$\qquad\qquad\qquad\qquad\quad = p_\xi(x) + O(h^2)$

(ii) $\quad Var\left(\frac{1}{Th} \sum_t K_h(\frac{x-X_t}{h})\right) = o(1).$

As an immediate consequence of Lemma 3.2 and 3.3 we obtain

**Theorem 3.4:** *Under the assumptions of Lemma 3.2 and 3.3 we obtain for all $x \in \mathbb{R}$*

$$(\log T)^2(\hat{m}(x,h) - m(x)) = O_p(1).$$

The nonparametric generalization of the contemporaneous autoregressive random variance model, where

$$X_t = \xi_t + \varepsilon_t \qquad (3.12)$$

holds instead of (2.5), while the structure of $(\xi_t)$ stated in (2.6) remains valid, is much more complicated to deal with. The problems arise from the fact that $\xi_t$ and $\varepsilon_t$ are not independent (as $\xi_{t-1}$ and $\varepsilon_t$ were before). To see this recall that $\xi_t$ depends on $\eta_t$ which itself is correlated to $W_t$ (correlation $\rho$) appearing in the definition of $\varepsilon_t$. Thus, the stationary density of our observations $X_t$ is for the contemporaneous case not the convolution of $p_\xi$ (which we are interested in) with the known density of the i.i.d. random variables $\varepsilon_t$. To overcome the difficulties one could assume that $\rho = 0$ which together with the assumption of normality for the distribution of $(\eta, W)$ implies independence even of $\varepsilon_t$ and $\xi_t$. Under this assumption $\rho = 0$ all above results remain valid as can be easily seen.

In case we want to stay with the assumption $\rho \neq 0$ one has to look for another possibility to estimate $p_\xi$. One proposal may be as follows. Since

$$X_t = \xi_t + \varepsilon_t = m(\xi_{t-1}) + (\eta_t + \varepsilon_t)$$

we could estimate the characteristic function of $\mathcal{L}(m(\xi_o))$ by $\hat{\phi}_x(w)/\phi_{\eta+\varepsilon}(w)$. Here $\phi_{\eta+\varepsilon}$ denotes the characteristic function of the known distribution of $\eta_1 + \varepsilon_1$. Now $\xi_1 = m(\xi_0) + \eta_1$ which suggests the following deconvolution estimator for $p_\xi$

$$\tilde{p}(x,h) = \frac{1}{2\pi} \int_{-\infty}^{\infty} e^{-iwx} \phi_K(wh) \frac{\hat{\phi}_x(w)}{\phi_{\eta+\varepsilon}(w)} \phi_\eta(w) dw$$
$$= \frac{1}{Th} \sum_{t=1}^{T} \tilde{K}_h\left(\frac{x-X_t}{h}\right)$$

where $\tilde{K}_h(n) = \dfrac{1}{2\pi} \displaystyle\int_{-\infty}^{\infty} e^{-iwn} \phi_K(w) \dfrac{\phi_\eta(w/h)}{\phi_{\eta+\varepsilon}(w/h)} dw.$

Finally, as a nonparametric estimator for $m$ we propose

$$\tilde{m}(x,h) = \frac{1}{Th} \sum_{t=1}^{T} \tilde{K}_h\!\left(\frac{x-X_t}{h}\right) X_{t+1}/\tilde{p}(x,h).$$

We have, as before

**Lemma 3.5:** *Under suitable assumptions we have*

$$E\,\tilde{p}(x,h) = \int_{\mathbb{R}} p_\xi(x-hu) K(u)\,du = p_\xi(x) + O(h^2).$$

In order to obtain consistency of $\hat{p}(x,h)$ we computed above the variance and obtained that it converges to zero. For the proof (cf. proof of Lemma 3.3) it was rather essential to know the asymptotic behaviour of the characteristic function $\phi_\varepsilon$ appearing in the denominator of $K_h$. Similarily, we need for a consistency result for $\tilde{p}(x,h)$ some information on the asymptotic behaviour of $\phi_{\varepsilon+\eta}$, which seems to be a rather delicate problem. A direct computation of $\phi_{\varepsilon+\eta}(w)$ leads to explicit expressions containing functions related to the so-called parabolic-cylinder functions $D_\nu(x)$. The argument $w$ appears in the argument and in the parameter of $D$, and we were not able to quantify the asymptotic behaviour of such functions as $|w| \longrightarrow \infty$.

The same problems arise when dealing with the numerator of $\tilde{m}(x,h)$, For the numerator even the computation of its expectations does not lead to such nice expressions as in the lagged case.

**Proofs.**

**Proof of Lemma 3.2:**

(i) The expectation is equal to

$$\frac{1}{h} E\, K_h\!\left(\frac{x-X_1}{h}\right)(X_2 - m(x))$$
$$= \frac{1}{h} E\, K_h\!\left(\frac{x-\xi_0-\varepsilon_1}{h}\right)(m(\xi_0) + \eta_1 + \varepsilon_2 - m(x))$$
$$= \frac{1}{h} E\, K_h\!\left(\frac{x-\xi_0-\varepsilon_1}{h}\right)(m(\xi_0) - m(x)) + \frac{1}{h} E\, K_h\!\left(\frac{x-\xi_0-\varepsilon_1}{h}\right)\eta_1.$$

Recall that $E\,\varepsilon_2 = 0$ and that $\varepsilon_2$ is independent of $\xi_0$ and $\varepsilon_1$. Unfortunately $\eta_1$ and $\varepsilon_1$ are not independent. But, because of the independence of $\xi_0$ and $(\varepsilon_1, \eta_1) = (\frac{1}{2}\log W_1^2 - \mu_\varepsilon, \eta_1)$ and $W_1 \sim \mathcal{N}(0,1)$

$$E\, K_h\!\left(\frac{x-\xi_0-\varepsilon_1}{h}\right)\eta_1$$

$$= \int_{\mathbb{R}^{\nu}} K_h(\frac{x - u - \frac{1}{2}\log w^2 + \mu_\varepsilon}{h}) \; v \; p_\xi(u) \; p_{\eta|W=w}(v)\varphi(w) du \; dv \; dw$$

$$= \int_{\mathbb{R}^{\nu}} K_h(\frac{x - u - \frac{1}{2}\log w^2 + \mu_\varepsilon}{h}) \; \rho \sigma_\eta w \; p_\xi(u) \; \varphi(w) \; du \; dw$$

since the conditional distribution of $\eta$ given $W = w$ is $\mathcal{N}(\rho\sigma_\eta w, \sigma_\eta^2(1-\rho^2))$ by our assumptions. The latter integral is equal to zero by symmetry arguments (recall that the normal density $\varphi$ is a symmetric function). Thus, the expectation under investigations equals

$$\frac{1}{h} E \; K_h(\frac{x - \xi_0 - \varepsilon_1}{h})(m(\xi_0) - m(x))$$

$$= \frac{1}{h} \int_{\mathbb{R}^{\nu}} K_h(\frac{x - u - v}{h})(m(u) - m(x)) \; p_\xi(u) \; p_\varepsilon(v) \; du \; dv$$

$$= \frac{1}{2\pi h} \int_{\mathbb{R}^{\nu}} e^{-i\frac{w}{h}(x-u-v)} \frac{\phi_K(w)}{\phi_\varepsilon(\frac{w}{h})}(m(u) - m(x)) \; p_\xi(u) \; p_\varepsilon(v) \; du \; dv \; dw$$

$$= \frac{1}{2\pi h} \int_{\mathbb{R}^{\nu}} e^{-i\frac{w}{h}(x-u)} \phi_K(w)(m(u) - m(x)) \; p_\xi(u) \; du \; dw$$

$$= \frac{1}{h} \int_{\mathbb{R}} \{ \int_{\mathbb{R}} \frac{1}{2\pi} e^{-iw(x-u)/h} \phi_K(w) \; dw \} \; (m(u) - m(x)) \; p_\xi(u) \; du.$$

The expression in curly bracket is by Fourier inversion equal to $K((x-u)/h)$, thus we have proved the first part of (i).
A Taylor-expansion of $m$ and $p_\xi$ up to second (first) order yields because of $\int_\mathbb{R} v \; K(v) \; dv = 0$ :

$$\int_\mathbb{R} \{m(u) - m(x)\} \frac{1}{h} \; K(\frac{x-u}{h}) \; p_\xi(u) \; du$$

$$= \int_\mathbb{R} K(v)\{-h \; v \; m'(x) + \frac{1}{2}h^2 v^2 m''(\hat{x}_1)\}\{p_\xi(x) - h \; v \; p'_\xi(\hat{x}_2)\} \; dv$$

$$= O(h^2).$$

$\hat{x}_1$ and $\hat{x}_2$ denote suitable values between $x - h \; v$ and $x$, possibly depending on $v$.

(ii) Concerning the variance we obtain

$$\text{var}\left(\frac{1}{Th} \sum_t K_h(\frac{x - X_t}{h})(X_{t+1} - m(x))\right)$$

$$= \frac{1}{Th^2} \cdot \text{var}\left(K_h(\frac{x - X_1}{h})(X_2 - m(x))\right) +$$

$$+ \frac{2}{T^2 h^2} \cdot \sum_{s<t} \text{cov}\left(K_h(\frac{x - X_s}{h})(X_{s+1} - m(x)), \; K_h(\frac{x - X_t}{h})(X_{t+1} - m(x))\right).$$

Using a covariance-inequality for strongly mixing sequences with geometrically decaying mixing coefficient (cf. Bosq (1996), Corollary 1.1 (page 19) we obtain the following bound of the above expression

$$\frac{1}{Th^2} \sup_{u \in \mathbb{R}} |K_h(u)|^2 \, E(X_2 - m(x))^2 + \frac{O(1)}{Th^2} \left( E|K_h(\frac{x - X_1}{h})(X_2 - m(x))| \right)^{\frac{2}{2+\delta}}$$

for $\delta > 0$ arbitrarily small. Since

$$\left( E|K_h(\frac{x-X_1}{h})(X_2 - m(x))|^{2+\delta} \right)^{\frac{2}{2+\delta}} \leq \sup_{u \in \mathbb{R}} |K_h(u)|^2 \cdot \left( E|X_2 - m(x)|^{2+\delta} \right)^{\frac{2}{2+\delta}},$$

and since, from Fan and Truong (1993), (7.8), we have for $\chi = \frac{1}{4} M_0 \pi > 0$

$$\sup_{u \in \mathbb{R}} |K_h(u)| = O(h) + O(\frac{\exp(\chi/h)}{h}),$$

we can bound the variance through $O(\frac{\exp(2\chi/h)}{Th^4})$. This expression converges to zero for $h = c/\log T$ and $c > 2\chi$.

**Proof of Lemma 3.3:**

(i) We have by independence of $\xi_0$ and $\varepsilon_1$

$$E \frac{1}{h} K_h(\frac{x - X_1}{h})$$
$$= \frac{1}{h} E K_h(\frac{x - \xi_0 - \varepsilon_1}{h})$$
$$= \frac{1}{h} \int_{\mathbb{R}^2} K_h(\frac{x - u - v}{h}) \, p_\xi(u) \, p_\varepsilon(v) \, du \, dv$$
$$= \frac{1}{2\pi h} \int_{\mathbb{R}^2} e^{-i\frac{w}{h}(x-u)} \phi_K(w) \, p_\xi(u) \, du \, dw$$
$$= \frac{1}{h} \int_{\mathbb{R}} K(\frac{x - u}{h}) \, p_\xi(u) \, du$$
$$= \int_{\mathbb{R}} p_\xi(x - h v) \, K(v) \, dv = p_\xi(x) + O(h^2).$$

The last equality is based on a second order Taylor-approximation of $p_\xi$.

(ii) Along the same lines as in the proof of Lemma 3.2 we obtain the wanted assertion.

**Proof of Lemma 3.5:**

$$E \frac{1}{h} \tilde{K}_h(\frac{x - X_1}{h}) = \frac{1}{h} E \tilde{K}_h(\frac{x - m(\xi_0) - \varepsilon_1 - \eta_1}{h})$$

$$= \frac{1}{2\pi h} \int_{\mathbb{R}^k} e^{-iw\frac{x-r-s-t}{h}} \phi_K(w) \frac{\phi_\eta(w/h)}{\phi_{\eta+\varepsilon}(w/h)} dP^{m(\xi_0)}(r) dP^{(\varepsilon,\eta)}(s,t) \, dw$$

$$= \frac{1}{2\pi h} \int_{\mathbb{R}^k} e^{-iw\frac{x-r}{h}} \int_{\mathbb{R}^k} e^{iw\frac{s+t}{h}} dP^{(\varepsilon,\eta)}(s,t) \, dP^{m(\xi_0)}(r) \, \phi_K(w) \frac{\phi_\eta^{(w/h)}}{\phi_{\eta+\varepsilon}(w/h)} \, dw$$

$$= \frac{1}{2\pi h} \int_{\mathbb{R}^k} e^{-iw\frac{x-r}{h}} \phi_K(w) \, \phi_\eta(w/h) \, dw \, dP^{m(\xi_0)}(r)$$

$$= \frac{1}{h} \int_{\mathbb{R}^k} \frac{h}{\sigma_\eta} \varphi\left(\frac{\frac{x-r}{h} - u}{\sigma_\eta/h}\right) K(u) \, du \, dP^{m(\xi_0)}(r)$$

because $\phi_K(w) \cdot \phi_\eta(w/h)$ is the characteristic function of $K * \mathcal{N}(0, \sigma_\eta^2/h^2)$ with density $\frac{h}{\sigma_\eta} \int_{\mathbb{R}} \varphi(\frac{\cdot - u}{\sigma_\eta/h}) K(u) \, du$. ($\varphi$ denotes the density of the standard normal distribution) and the Fourier inversion formula.

$$= \int_{\mathbb{R}} \left\{ \frac{1}{\sigma_\eta} \int_{\mathbb{R}} \varphi(\frac{x - hu - r}{\sigma_\eta}) \, dP^{m(\xi_0)}(r) \right\} K(u) \, du$$

The term in curly brackets is the density of $\mathcal{L}(m(\xi_0) + \eta_1)$ which is $p_\xi$

$$= \int_{\mathbb{R}} p_\xi(x - hu) \, K(u) \, du = p_\xi(x) + O(h^2)$$

using the usual arguments and $\int_{\mathbb{R}} u \, K(u) \, du = 0$.

## REFERENCES

1. Bosq, D. (1996). Nonparametric Statistics for Stochastic Processes. Lecture Notes in Statistics **110**, Springer-Verlag, New York.
2. Carroll, R.J. and Hall, P. (1988). Optimal rates of convergence for deconvoluting a density. *J. Amer. Statist. Assoc.* **83**, 1184-1186.
3. Chesney, M. and Scott, L.O. (1989). Pricing European Currency Options: A Comparison of the Modified Black-Scholes Model and a Random Variance Model. *J. Financial Quant. Anal.* **24**, 267-284.
4. Doukhan, P. (1994). Mixing: Properties on Examples. Lecture Notes in Statistics **85**, Springer-Verlag, New York.
5. Fan, J. (1991a). On the optimal rates of convergence for nonparametric deconvolution problems. *Ann. Statist.* **19**, 1257-1272.
6. Fan, J (1991b). Asymptotic normality for deconvolution kernel density estimators. *Sankhyā Ser. A* **53**, 97-110.
7. Fan, J. and Truong, Y.K. (1993). Nonparametric regression with errors in variables. *Ann. Statist.* **21**, 1900-1925.
8. Gradstein, I. and Ryshik, I. (1981). Tables. Verlag MiR, Moskau and Verlag Harri Deutsch, Thun.
9. Liu, M.C. and Taylor, R.L. (1989). A consistent nonparametric density estimator for the deconvolution problem. *Canad. J. Statist.* **17**, 427-438.
10. Scott, L.O. (1987). Option Pricing When the Variance Changes Randomly: Theory, Estimation and an Application. *J. Financial Quant. Anal.* **22**, 419-438.
11. Scott, L.O. (1991). Random Variance Option Pricing: Empirical Tests of the Model and Delta-Sigma Hedging. *Adv. Futures Options Res.* **5**, 113-135.

12. Stefanski, L.A. and Carroll, R.J. (1990). Deconvoluting kernel density estimators. *Statistics* **21**, 169-184.
13. Taylor, S.J. (1994). Modelling Stochastic Volatility: A Review and Comparative Study. *Mathematical Finance* **4**, 183-204.
14. Wiggins, J.B. (1987). Option Values Under Stochastic Volatility: Theory and Empirical Estimates. *J. Financial Econ.* **19**, 351-372.

# Dynamic nonparametric filtering with application to volatility estimation

Ming-Yen Cheng[a] [*] and Jianqing Fan[b] [†] and Vladimir Spokoiny[c]

[a]Department of Mathematics, National Taiwan University
Taipei 106, Taiwan

[b]Department of Operation Research and Financial Engineering, Princeton University
Princeton, NJ 08544

[c]Weierstrass-Institute
Mohrenstr. 39, 10117 Berlin, Germany

Problems of nonparametric filtering arises frequently in engineering and financial economics. Nonparametric filters often involve some filtering parameters to choose. These parameters can be chosen to optimize the performance locally at each time point or globally over a time interval. In this article, the filtering parameters are chosen via minimizing the prediction error for a large class of filters. Under a general martingale setting, with mild conditions on the time series structure and virtually no assumption on filters, we show that the adaptive filter with filtering parameter chosen by historical data performs nearly as well as the one with the ideal filter in the class, in terms of filtering errors. The theoretical result is also verified via intensive simulations. Our approach is also useful for choosing the orders of parametric models such as AR or GARCH processes. It can also be applied to volatility estimation in financial economics. We illustrate the proposed methods by estimating the volatility of the returns of the S&P500 index and the yields of the three-month Treasury bills.

## 1. Introduction

Problems of nonparametric filtering arises frequently in engineering, financial economics, and many other scientific disciplines. Given a time series $\{Y_t\}$, the nonparametric filtering problem is to dynamically predict $Y_t$ based on the observations preceding $t$. This is a specific problem of the time domain smoothing (see §6.2 of [10]), but allows to use only historical data at each time. A traditional class of nonparametric filters is the moving average filtering which is the average of last $m$ time periods (see Example 1 below). Other classes include exponential smoothing (Example 2 below), kernel smoothing (Example 2), autoregressive filtering (Example 5) and ARCH and GARCH filtering (see §4.2). All of these filters depend on certain parameters, called filtering parameters

---
[*]Partially supported by NSC grant 91-2118-M-002-001
[†]Partially supported by NSF grant DMS-0204329 and a direct allocation RGC grant of the Chinese University of Hong Kong.

in this paper. An interesting and challenging issue is how to choose these parameters so that they are adaptive automatically to the data.

There are basically two versions of filtering parameters, local and global versions. The local version is that at each time point $t$, we choose the filtering parameters $\widehat{\lambda}_t$, say, to optimize the performance near $t$. The global version is to set an in-sample period, $[1,T]$, say, then to choose filtering parameters $\widehat{\lambda}$ to optimize the performance in the time interval $[1,T]$, and finally to predict the data in the out-sample period $[T+1, T+n]$, say, using the filtering parameters $\widehat{\lambda}$. The local choice of ideal filtering parameters is more powerful than the global one. However, owing to stochastic errors, data-driven choices of local filtering parameters, while are more flexible, do not necessarily outperform the global choice. However, they are very useful in the situation where there are structural changes of an underlying series over time. The situation is very similar to the local bandwidth and global bandwidth selection in the nonparametric smoothing literature (see e.g. [5] and [7]).

A natural criterion for choosing filtering parameters is the prediction error, since only the historical data have been used in the construction of filters. Because of this, semi-martingale structures remain valid in the computation of the prediction error. This enables us to show, with mild conditions on the time series structure and virtually no assumption on filters, that the resulting adaptive filter performs nearly as well as the ideal choice of filtering parameters. This property is also verified via intensive numerical computation. The nice property encourages us to apply the techniques to volatility estimation in financial econometrics.

The concept of volatility is associated with the notation of risks. It is very critical for portfolio optimization, option pricing and management of financial risks. As shown in Section 4.1, the problem of dynamic prediction of volatility is strongly associated with a filtering problem. In fact, a family of power transform can even be accommodated to estimate volatility (see [15]), with our filtering techniques. This yields a family of volatility estimators: some aim at robustness, while others at efficiency. The family of nonparametric methods compares favorably with GARCH techniques in volatility estimation, from our numerical experiments.

The paper is organized as follows. Section 2 outlines various filtering techniques. Their filtering parameters are selected in Section 3 where the properties of the adaptive filters are investigated both theoretically and empirically. Problems of volatility estimation and their associations with nonparametric filtering are investigated in Section 4.

## 2. Problems of dynamic filtering

Consider a time series $Y_1, \ldots, Y_T$, which is progressively measurable with respect to a filtration $\mathbb{F} = (\mathcal{F}_t)$ and allows a semi-martingale representation:

$$Y_t = f_t + v_t \varepsilon_t, \tag{1}$$

where $f_t$ and $v_t$ are predictable and the innovations $\varepsilon_t$ form a standardized martingale difference. They satisfy

$$E[\varepsilon_t \mid \mathcal{F}_{t-1}] = 0, \qquad E[\varepsilon_t^2 \mid \mathcal{F}_{t-1}] = 1.$$

The function $f_t$ is a trend or a drift series and $v_t$ is the conditional standard deviation or the diffusion of the series. In statistical forecasting, one wishes to estimate the conditional mean $f_t$ based on the past data $Y_1, \ldots, Y_t$. Such a problem is also referred to as filtering or one-step forecasting of the process $\{Y_t\}$.

Depending on the background of applications, as detailed below, filters or predictors depend on some tuning parameters. Suppose that we are given a family of different filters (predictors) $\widehat{f}_{t,\lambda}$ indexed by some parameter $\lambda$. Each predictor $\widehat{f}_{t,\lambda}$ estimates the unknown value $f_t$ from the 'past' observations $Y_1, \ldots, Y_{t-1}$. Our goal is to construct one predictor which does the job nearly as well as the best filter in the family $\{\widehat{f}_{t,\lambda}, \lambda \in \Lambda\}$. We use a few examples to illustrate the versatility of the scope of our study.

**Example 1** (*Moving average filtering*) A traditional approach to estimate the trend of a time series is the moving average estimator. For every integer $m$, one defines

$$\widehat{f}_{t,m} = \frac{1}{m} \sum_{s=t-m}^{t-1} Y_s.$$

Here the parameter $\lambda$ coincides with the window size $m$. One may consider a family of such predictors for different window sizes and the problem of adaptive estimation is to choose a window size from data.

**Example 2** (*Exponential smoothing*) An improvement of the moving average (MA) filtering is the exponential smoothing (ES) which weighs down the observed data from the past. The family of exponential smoothing is defined by

$$\widehat{f}_{t,\lambda} = \frac{1}{1 - e^{-\lambda}} \sum_{s<t} e^{-(t-s)\lambda} Y_s, \qquad (2)$$

for a positive parameter $\lambda$. Formally this estimate $\widehat{f}_{t,\lambda}$ depends on all the past observations, but this dependence decreases exponentially. Our problem becomes to choose the parameter $\lambda$ from data such that the resulting adaptive estimator performs nearly as well as the ideal exponential filtering among this family.

The MA and ES filtering are a member of the kernel estimator. See, for example, [8] and §6.2 of [10]. In fact, the ES corresponds to the kernel regression estimator in time domain

$$\widehat{f}_{t,h} = \sum_{s<t} K_h(t-s) Y_s \Big/ \sum_{s<t} K_h(t-s), \qquad K_h(x) = h^{-1} K(x/h)$$

with the one-sided kernel $K(x) = \exp(-x) I(x > 0)$ and $\lambda = 1/h$. Further discussion on this subject, including bandwidth selection and asymptotic theory, can be found in [11].

**Example 3** (*J.P. Morgan's RiskMetrics*) An important measure to gauge the risk of a portfolio is the Value-at-Risk, which is the worst loss to be expected with certain confidence for a given time horizon. See [13]. An important contribution to the calculation of VaR is the RiskMetrics of J.P. Morgan[16]. The method is to first estimate the volatility for holding a portfolio for one day before converting this into the volatility for multiple days and to then compute the quantile of standardized return processes through the assumption that the processes follow a standard normal distribution. Let $S_t$ be the price of a portfolio at time $t$ and $R_t = \log(S_t/S_{t-1})$ be the observed return at time $t$. The J.P. Morgan estimate of volatility $\widehat{\sigma}_t^2$ for one-period return is

$$\widehat{\sigma}_t^2 = (1 - \lambda_0) R_{t-1}^2 + \lambda_0 \widehat{\sigma}_{t-1}^2.$$

By iterating the above formula, it can easily be seen that

$$\widehat{\sigma}_t^2 = (1 - \lambda_0)\{R_{t-1}^2 + \lambda_0 R_{t-2}^2 + \lambda_0^2 R_{t-3}^2 + \cdots\}.$$

This is an alternative form of the ES (2) with $\lambda_0 = \exp(-\lambda)$. Our adaptive dynamic filtering is to choose $\lambda_0$ from data to ameliorate the performance. Such an approach has been introduced by [8]. Our current study gives further theoretical endorsement of their approach.

The above estimator is basically a discretized approach to estimate the diffusion function $\theta(t)$ in the following geometric Brownian motion $d\log(S_u) = \theta(u) dW_u$ via a local constant approximation. See [8] for derivations and connections.

**Example 4** (*Adaptive estimation of volatility*) The problem of estimating $v_t$ in (1) can also be regarded as adaptive filtering problem. Let $\widehat{R}_t = Y_t - \widehat{f}_t$ be the residual from model fitting (1). Then, define a family of the filters for square residuals as

$$\widehat{v}_{t,h}^2 = \sum_{s<t} K_h(t-s) \widehat{R}_s^2 / \sum_{s<t} K_h(t-s).$$

As shown in [9] and [17], the errors in estimating $\widehat{f}_t$ are usually negligible in estimating $\widehat{v}_{t,h}^2$. Hence, our methodology and theory continue to apply.

**Example 5** (*Autoregression*) Suppose that the process $\{Y_t\}$ is to be approximated by an autoregressive (AR) equation

$$Y_t = \alpha_1 Y_{t-1} + \ldots + \alpha_p Y_{t-p} + v_t \varepsilon_t$$

where $\varepsilon_t$ are conditionally independent innovations, $v_t$ is an unknown predictable process, and $\alpha_1, \ldots, \alpha_p$ are unknown autoregression coefficients. Denote by

$$X_{t,p} = (Y_{t-1}, \ldots, Y_{t-p})^\mathsf{T} \quad \text{and} \quad \alpha = (\alpha_1, \ldots, \alpha_p)^\mathsf{T}.$$

Then, the above autoregressive equation can be written in the form $Y_t = X_{t,p}^\mathsf{T} \alpha + v_t \varepsilon_t$. The least-squares estimate of the parameter $\alpha$ from the observations $Y_s$ for $t_0 \leq s < t$ reads as follows:

$$\widehat{\alpha}_{t,p} = \left( \sum_{s=t_0}^{t-1} X_{s,p} X_{s,p}^\mathsf{T} \right)^{-1} \sum_{t=t_0}^{t-1} X_{s,p} Y_s$$

and the corresponding filter $\widehat{f}_{t,p}$ of $Y_t$ is defined as $\widehat{f}_{t,p} = X_{t,p}^\mathsf{T} \widehat{\alpha}_{t,p}$. The problem of adaptive dynamic filtering is to choose a $p$ using the available data before time $t$ such that it performs nearly as well as the ideal choice of $p$.

Similar idea can be applied to choose the order of GARCH models (see §4.2 of [10] and [12])

## 3. Choice of filtering parameters

There are two possible choices of filtering parameters. A local choice aims at choosing a $\lambda$, which depends on $t$, such that it optimizes the performance of the filter at each time point $t$. In other words, we use $\widehat{f}_{t,\widehat{\lambda}_t}$ to estimate $f_t$, where $\widehat{\lambda}_t$ is selected based the data collected up to $t-1$. A global choice aims at choosing a $\lambda$, which is independent of $t$, such that it optimizes the performance of the filter over an interval, $[t_0, T]$. The local choice of the filtering parameters is more flexible than the global one and the resulting filter is more capable of adapting to the dynamic change of the underlying time series. On the other hand, the local choice of filtering parameters is harder and more variable, since only the local data are involved in choosing the filtering parameters. The problem is very analogous to the local and global bandwidths in nonparametric smoothing, studied, for example, by [5] and [7].

### 3.1. Global adaptation via minimal prediction error

The performance of any filter can be measured by the sum of squared filtering errors:

$$R(\lambda) = \sum_{t=t_0}^{T} \left(f_t - \widehat{f}_{t,\lambda}\right)^2,$$

where $T$ is the length of a time series and $t_0$ is a predefined time point large enough to avoid the boundary effect. An ideal choice of the parameter $\lambda$ can be defined as the one that minimizes the global loss:

$$\lambda_{II} = \arginf_{\lambda \in \Lambda} \sum_{t=t_0}^{T} \left(f_t - \widehat{f}_{t,\lambda}\right)^2. \tag{3}$$

This choice is ideal since it relies on the unknown target function.

An empirical analog of the filtering error is the prediction error defined as:

$$\rho(\lambda) = \sum_{t=t_0}^{T} \left(Y_t - \widehat{f}_{t,\lambda}\right)^2.$$

This criterion leads to the following data-driven selection rule:

$$\widehat{\lambda} = \arginf_{\lambda \in \Lambda} \rho(\lambda) = \arginf_{\lambda \in \Lambda} \sum_{t=t_0}^{T} \left(Y_t - \widehat{f}_{t,\lambda}\right)^2. \tag{4}$$

The resulting adaptive filter is given by $\widehat{f}_t = \widehat{f}_{t,\widehat{\lambda}}$. The filtering error of this estimator is given by

$$R(\widehat{\lambda}) = \sum_{t=t_0}^{T} \left(f_t - \widehat{f}_t\right)^2 = \sum_{t=t_0}^{T} \left(f_t - \widehat{f}_{t,\widehat{\lambda}}\right)^2.$$

An interesting question is how the the quality (filtering error) of the data-driven selector from (4) is compared with the "ideal" selector from (3). We attempt to answer this question in the next section.

## 3.2. Properties of the adaptive selector

For every $\lambda \in \Lambda$, it holds that

$$\rho(\lambda) = \sum_{t=t_0}^{T} \left(Y_t - \widehat{f}_{t,\lambda}\right)^2 = R(\lambda) + 2\sum_{t=t_0}^{T} \left(f_t - \widehat{f}_{t,\lambda}\right) v_t \varepsilon_t + \sum_{t=t_0}^{T} v_t^2 \varepsilon_t^2.$$

The last sum in the above decomposition does not depend on $\lambda$ and hence does not affect the minimization in (4). Hence, minimizing the prediction error $\rho(\lambda)$ corresponds to minimizing the filtering error $R(\lambda)$ plus the cross term

$$S_{cross,\lambda} = 2\sum_{t=t_0}^{T} \left(f_t - \widehat{f}_{t,\lambda}\right) v_t \varepsilon_t.$$

If one could show that this cross term is relatively small, then these two minimization procedures would be nearly equivalent.

To bound the cross term $S_{cross,\lambda}$, we apply a result for martingales from [14]. Define

$$M_{t,\lambda} = \sum_{s=t_0}^{t} \left(f_s - \widehat{f}_{s,\lambda}\right) v_s \varepsilon_s, \quad \text{and} \quad V_{t,\lambda}^2 = \sum_{s=t_0}^{t} \left(f_s - \widehat{f}_{s,\lambda}\right)^2 v_s^2.$$

Note that for homoscedastic error $v_s \equiv v$, $V_{T,\lambda}^2 = v^2 R^2(\lambda)$. In general, $V_{T,\lambda}$ is of the same order as $R^2(\lambda)$ as long as $v_s$ is bounded from below and above. Since both the drift $f_t$ and the estimator $\widehat{f}_{t,\lambda}$ are predictable processes with respect to the filtration $\mathcal{F}_{t-1}$, $M_{t,\lambda}$ is a square integrable martingale with the quadratic variation $V_{t,\lambda}^2$.

**Lemma 1** *Let the innovations $\varepsilon_t$ fulfill $\boldsymbol{E}e^{u\varepsilon_t} \leq \exp\{u^2/(2a)\}$ for some positive $a$ and all $u \geq 0$. Then, for all $\gamma \geq 1$*

$$\boldsymbol{P}\left(M_{T,\lambda} > \gamma V_{T,\lambda}, \mathcal{A}_\lambda\right) \leq \alpha_\lambda(\gamma)$$

*where $\mathcal{A}_\lambda = \{\vartheta \leq V_{T,\lambda}^2 \leq \vartheta B\}$ with some deterministic values $\vartheta, B$ and $\alpha_\lambda(\gamma) = 4\sqrt{e}(1 + \log B)\gamma e^{-\gamma^2/(2a)}$.*

Note that the constants $\vartheta, B$ may also depend on $\lambda$. We suppress this dependence to facilitate the notation. As a corollary of Lemma 1:

$$\sum_{\lambda \in \Lambda} \boldsymbol{P}\left(S_{cross,\lambda} > 2\gamma V_{T,\lambda}, \mathcal{A}_\lambda\right) \leq \sum_{\lambda \in \Lambda} \alpha_\lambda(\gamma). \tag{5}$$

As noted above, $V_{T,\lambda}^2 \leq v^2 R(\lambda)$ when $v_t \leq v$. It follows that $S_{cross,\lambda} \leq 2\gamma v \sqrt{R(\lambda)}, \forall \lambda \in \Lambda$ with a probability at least $1 - \sum_{\alpha \in \Lambda} \alpha(\gamma)$. This yields the following results.

**Theorem 1** *It holds for every $\gamma \geq 0$ and every $v > 0$*

$$\boldsymbol{P}\left(\sqrt{R(\widehat{\lambda})} \geq \sqrt{R(\lambda_{\#})} + 3v\gamma, \mathcal{A}\right) \leq \sum_{\lambda \in \Lambda} \alpha_\lambda(\gamma)$$

*with $\mathcal{A} = \bigcap_{\lambda \in \Lambda} \mathcal{A}_\lambda \cap \{V_{T,\lambda}^2 \leq v^2 R(\lambda)\}$.*

**Proof.** In view of (5) it suffices to show that the inequalities

$$S_{cross,\lambda} \leq 2\gamma V_{T,\lambda} \leq 2\gamma v \sqrt{R(\lambda)} \qquad \forall \lambda \in \Lambda \tag{6}$$

imply that

$$\sqrt{R(\widehat{\lambda})} \leq \sqrt{R(\lambda_{I\!I})} + 3v\gamma/2.$$

To this end, define $\rho_\lambda^* = R(\lambda) + S_{cross,\lambda}$ and denote by $R_\lambda = R(\lambda)$. Then $\widehat{\lambda}$ is the minimizer of $\rho_\lambda^*$, while $\lambda_{I\!I}$ is the minimizer of $R_\lambda$. The condition $\rho_\lambda^* \leq R_\lambda + 2\gamma v \sqrt{R_\lambda}$ implies $\sqrt{\rho_\lambda^*} \leq \sqrt{R_\lambda} + \gamma v$. Similarly, (6) implies $\rho_\lambda^* \geq R_\lambda - 2\gamma v \sqrt{R_\lambda}$. Since $\sqrt{x-y} \geq \sqrt{x} - y/\sqrt{x}$ for every positive numbers $x, y$, it follows that for each $\lambda \in \Lambda$:

$$\sqrt{R_\lambda} - 2\gamma v \leq \sqrt{\rho_\lambda} \leq \sqrt{R_\lambda} + \gamma v.$$

Since $\widehat{\lambda}$ is the minimizer of $\rho_\lambda$, $\rho_{\widehat{\lambda}} \leq \rho_{\lambda_{I\!I}}$. Therefore,

$$\sqrt{R_{\widehat{\lambda}}} \leq \sqrt{\rho_{\widehat{\lambda}}} + 2\gamma v \leq \sqrt{\rho_{I\!I}} + 2\gamma v \leq \sqrt{R_{I\!I}} + 3\gamma v \tag{7}$$

as required. ∎

Note that $\alpha_\lambda(\gamma) = o(T^{-b})$, when $\gamma > (2ab \log T)^{1/2}$ for a given positive $b$. Thus, when the number of elements in $\Lambda$ is of order $O(T^b)$, $\sum_{\lambda \in \Lambda} \alpha_\lambda(\gamma) \to 0$. Thus, the extra term $3v\gamma$ required for adaptation in Theorem 1 is not excessive and is only of order $O\{(\log T)^{1/2}\}$. For example, for a parametric model such as an $AR(p)$ model, the filtering error $|f_t - \widehat{f}_{t,\lambda}|^2$ is typically of order $t^{-1}$ so that $R_{\lambda,T}$ is of order $\sum_{t \leq T} t^{-1} \approx \log T$. The extra term $3v\gamma = O\{(\log T)^{1/2}\}$ is negligible. For nonparametric filters like moving average with window size $m$, the filtering errors are of order $O(T^{-2/5})$ for $m = T^{1/5}$, which are much larger than those of parameter models. Hence, the extra term of order $O\{(\log T)^{1/2}\}$ is also negligible, comparing with $R(\lambda_{I\!I})$. In summary, with probability tending to one, the data-driven filters perform as well as filters with an ideal choice of filtering parameter.

### 3.3. Local choice

The aforementioned global procedure chooses one filter parameter to fit the whole observed path. Such a method can be efficient in many situations where there are virtually no structural breaks in the observed time series. However, it has a serious drawback of being slow in reacting to spontaneous changes in the structure of the observed process. We illustrate this issue using the moving average filter with window size $m$. See also Example 6. For a large $m$, the accuracy of estimation is very good provided that the underlying process $f_t$ is nearly constant within this window. However, if the value $f_t$ changes abruptly at some time point, then the filter with a large $m$ will react to this change with a long delay of order $m$. On the other hand, a filter with a small $m$ allows for a fast reaction to the sudden changes in structure, but is not as precise and stable as a filter with larger $m$ over stationary regions.

To enhance the flexibility of the family of filters $\{\widehat{f}_{t,\lambda}\}$ to adapt to possible structural changes over time, the parameter $\lambda$ should be allowed to vary over time. For each given time $t$, $\lambda_t$ should be chosen so that it optimizes the performance near the time point $t$. Following (4), we choose

$$\widehat{\lambda}_t = \underset{\lambda \in \Lambda}{\arg\inf} \sum_{t=t-M+1}^{t} \left(Y_t - \widehat{f}_{t,\lambda}\right)^2, \qquad (8)$$

where $M$ is the size of the neighborhood preceding the time point $t$, over which we wish to optimize the performance. One can also regard $M$ as another filtering parameter and wishes to choose $\lambda_t$ and $M$ simultaneously. But, simultaneous choices of $M$ and $\lambda$ face the challenges of instability and computational cost.

In the local bandwidth selection setting, [7] employed a similar idea. However, the resulting parameters $\{\widehat{\lambda}_t\}$ are smoothed further to enhance the smoothness of the resulting estimate $\widehat{f}_{t,\widehat{\lambda}_t}$. In our time domain smoothing, such a step can be avoided, since the smoothness of $\widehat{f}_{t,\widehat{\lambda}_t}$ in time domain is not a critical visual requirement.

Applying Theorem 1 to the local choice of the filter parameters, we can obtain a similar result on the bound of the filtering errors around the time $t$. Again, as long as the number of elements in $\Lambda$ is not excessively large, the performance of the data-driven choice of local filtering parameters is nearly as good as their ideal choice.

### 3.4. Numerical Results

We illustrate the performance of the global and local choices of filtering parameters via two different classes of underlying processes: piecewise constant processes and autoregressive processes. For the first class, an application of moving average or exponential smoothers is quite reasonable, while the second class is oriented towards autoregressive filtering in Example 5. The effectiveness of each filter can be assessed by the Mean Absolute Filtering Error (MAFE) or the Mean Squared Filtering Error (MSFE):

$$\text{MAFE} = \frac{1}{n} \sum_{t=T+1}^{T+n} |f_t - \widehat{f}_t|, \qquad \text{MSFE} = \frac{1}{n} \sum_{t=T+1}^{T+n} |f_t - \widehat{f}_t|^2,$$

for a post sample of size $n$. The in-sample period is taken to be $t = 1, \cdots, T$. Since the results are similar by using MAFE and MSFE, we only report the MAFE.

**Example 6** Let the process $f_t$ take only two values $\{-1, 1\}$, with transitions between these two states at random stopping times $\tau_1 < \tau_2 < \ldots < \tau_m < \ldots$. These stopping times were generated from a Poisson process with rate $1/\mu$, namely, the intervals $\tau_k - \tau_{k-1}$ were generated from the exponential with mean $\mu = 150$. The observed process is

$$Y_t = f_t + \sigma \varepsilon_t, \quad \varepsilon_t \sim N(0, 1).$$

Figure 1(a) depicts a simulated series of length 500.

To estimate the function $f_t$, we apply the moving average (MA) and exponential smoothing (ES) methods to estimate the time trend. We first apply the global method to choose the filtering parameters, the window size $m$ in MA and the decay parameter $\lambda$ in ES. The initial value $t_0 = 101$ is taken. The filtering parameters are chosen to minimize (4) among 15 geometric grids. Figure 1(b) shows the resulting estimates for the realization

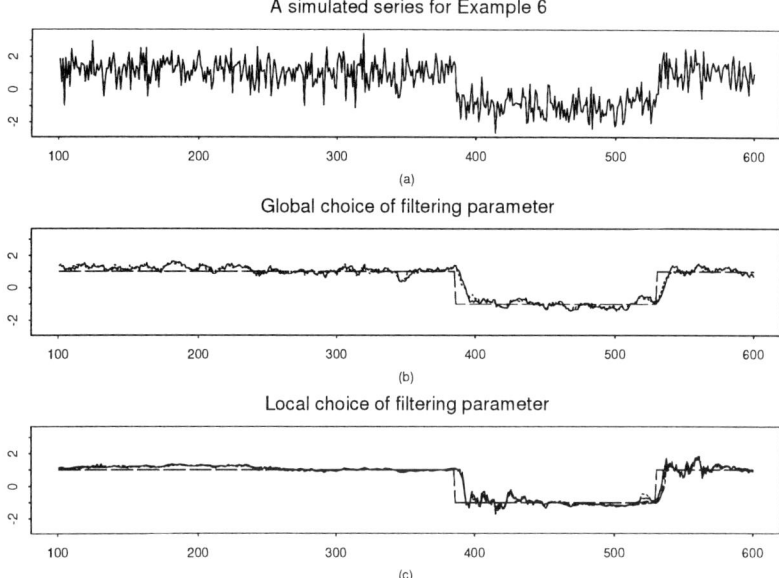

Figure 1. (a) A simulated series from Example 6. (b) Filtered series with global choices of filtering parameters. True $f_t$ — long-dashed curve; MA — solid curve; ES — dotted curve. (c) Filtered series with local choices of filtering parameters. MA — dotted curve; MA ideal — dot-and-dash curve; ES — solid; ES ideal — dashed.

in Figure 1(a). Both the adaptive MA and ES methods recover reasonably well the mean process $f_t$ and detect the jumps in $f_t$. The jumps in the process $f_t$ force the methods to choose small values of $m$ and $\lambda$. As a result, the estimates are somewhat undersmoothed and have rough appearance in the constant regions.

For the signal function $f_t$ in this example, it is reasonable to expect that a large smoothing parameter is used in the first part of the data and a smaller one is applied to the last piece of series. To achieve such a scheme adaptively, we appeal to (8) with $M = 40$. The resulting estimates by using the MA and ES with local choice of filtering parameters are shown in Figure 1(c). To compare with their performance with the ideal choices of the local filtering parameters, which minimize the corresponding local version of (3), Figure 1(c) also depicts the MA and ES estimates using the ideal local filtering parameters. The four estimates are hard to differentiate, which in turn endorses the performance of our adaptive local version of selecting filtering parameters.

Comparing the estimates with the local choices of filtering parameters to those with the global ones, the local version tends to choose larger smoothing parameters for the first part of the series. At the point of the structure break, smaller smoothing parameters are chosen so that "leakages" (biases around the change point) have been reduced by the

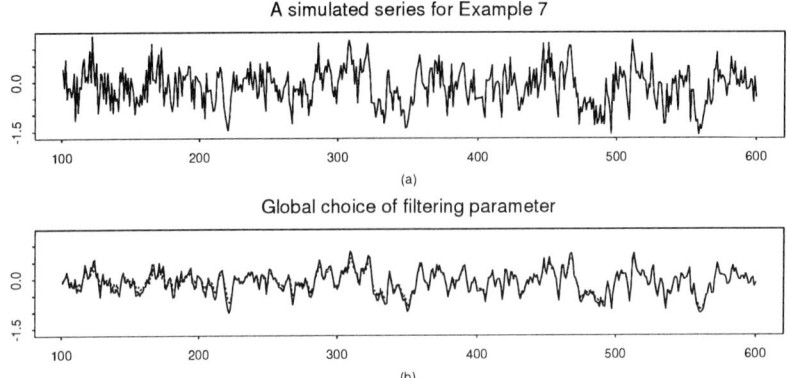

Figure 2. (a) A simulated series from Example 7. (b) Filtered series with global choices of filtering parameters. True $f_t$ — solid curve; filtered series — dotted curve.

local methods, but at the expenses of increasing variability.

The simulation results in terms of MAFE are reported in Table 1. The post-sample size is $n = 500$ and the in-sample period is [1,1000].

Next, we consider an application of the proposed methods to selecting the order of autoregressive processes.

**Example 7** We generate a series from the following AR(2) model:

$$Y_t = 0.4\,Y_{t-1} + 0.32\,Y_{t-2} + \varepsilon_t, \qquad \varepsilon_t \sim N(0, 0.5^2).$$

Figure 2(a) depicted a realization of length 600. As in Example 5, the aim is to choose an order $p$ to best predict the series.

Recall the filter $\widehat{f}_{t,p}$ defined in Example 5 is based on an autoregressive model of order $p$. For the global choice, we choose $p$ to minimize $\sum_{t=101}^{600} \left(Y_t - \widehat{f}_{t,p}\right)^2$, among the set $\mathcal{P} = \{1, 2, 4, 8\}$. For this realization, the above order selection rule yields $\widehat{p} = 2$, which is the same as the true value of $p$. The resulting filter is depicted in Figure 2(b). The estimate is very well in accordance with the mean process $f_t$.

The simulation results in terms of MAFE are reported in Table 1.

The next example deals with the situation where the dynamic of an underlying process changes over time. This reflects some extent in the real world, where stochastic dynamics such as stock markets can change over time.

**Example 8** We simulated a process from the AR(2) process $Y_t = 0.3\,Y_{t-1} + 0.4\,Y_{t-2} + 0.3\varepsilon_t$ till the time point $t = 450$ and then from the AR(1)-process $Y_t = 0.7\,Y_{t-1} + 0.3\,\varepsilon_t$ after $t = 450$. Here, $\varepsilon_t$ is a standard Gaussian noise. Figure 3(a) depicts a realization from the

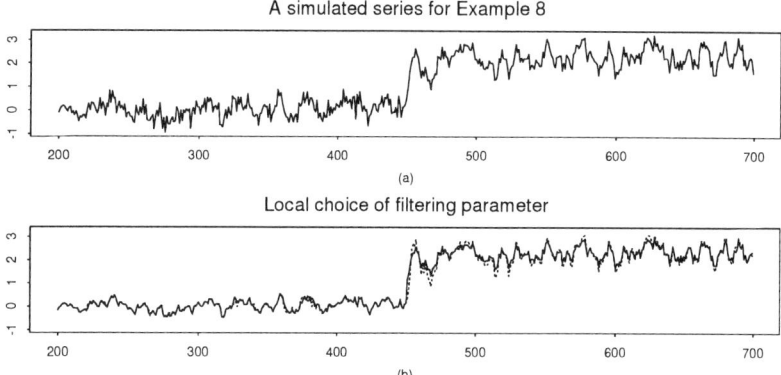

Figure 3. (a) A simulated series from Example 7. (b) Filtered series with global choices of filtering parameters. True $f_t$ — solid curve; filtered series — dotted curve.

model. This model is similar to the thresholding autoregressive model (see [10] and [18]), but the structure change occurs in the time domain rather than the state domain.

To accommodate the possible structural break over time, it is natural to use only a local stretch of data around the time $t$. Similarly to Example 5, we consider the family of filters

$$\widehat{f}_{t,p,m} = X_{t,p}^\top \left( \sum_{s=t-m}^{t-1} X_{s,p} X_{s,p}^\top \right)^{-1} \sum_{s=t-m}^{t-1} X_{s,p} Y_s.$$

At each time point, a local selection procedure is applied to choose both $p$ and $m$ to minimize $\sum_{s=t-M+1}^{t} \left( Y_s - \widehat{f}_{s,p,m} \right)^2$. We searched $p$ in $\mathcal{P} = \{1, 2, 4, 8\}$ and $m$ in $\{20, 40, 80, 160\}$ and took $M = 20$. The result filter is plotted in Figure 3(b). The result illustrates how this procedure works in the stationary region before the change and immediately after it. In particular, the delay between the change and the first moment when the procedure starts to selects a small $m$ can be interpreted as the sensitivity to changes.

We now briefly summarize the simulation results using MAFE. The relative performance of a filter to another one is measured by the ratio of the MAFE of the former filter to the latter. This ratio is independent of the scale of a simulated data. Table 1 summarizes the distributions of these ratios over 500 simulations by computing their mean, SD, the first, second and third quartiles. From the right block of Table 1, one can see easily that the relative performance between the filters with their parameters chosen by data and those using the ideal ones is nearly the same. This is consistent with the theoretical result given by Theorem 1. For each realization, we computed the relative MAFE of the filters with filtering parameters chosen by data to that with ideal filtering parameters. The in-sample period is set to be $[1, 1000]$ and the post-sample period is $[1001, 1500]$.

Table 1
Relative MAFE performance. Empirical mean (first row), sample standard deviation (second row), first quartile (third row), median (fourth row), and third quartile (fifth row) of MAFE ratios.

|  | Relative to $\widehat{f}_{t,\widehat{p}}$ | | | Relative to ideal counterparts | | | |
| --- | --- | --- | --- | --- | --- | --- | --- |
|  | ES global | ES local | AR local | ES global | ES local | AR global | AR local |
|  | 0.931 | 0.743 | 1.228 | 1.000 | 1.076 | 0.100 | 1.104 |
|  | 0.065 | 0.138 | 0.069 | 0.034 | 0.057 | 0.0197 | 0.041 |
| Ex. 6 | 0.901 | 0.655 | 1.177 | 0.996 | 1.036 | 1.000 | 1.075 |
|  | 0.936 | 0.745 | 1.226 | 1.000 | 1.065 | 1.000 | 1.101 |
|  | 0.967 | 0.845 | 1.268 | 1.000 | 1.111 | 1.000 | 1.131 |
|  | 10.789 | 14.254 | 8.200 | 1.003 | 1.290 | 1.089 | 2.559 |
|  | 6.534 | 8.619 | 4.940 | 0.010 | 0.103 | 0.608 | 0.449 |
| Ex. 7 | 6.191 | 8.450 | 4.785 | 1.000 | 1.217 | 1.000 | 2.252 |
|  | 8.988 | 11.790 | 6.912 | 1.000 | 1.280 | 1.000 | 2.499 |
|  | 13.453 | 17.695 | 10.039 | 1.004 | 1.352 | 1.000 | 2.810 |
|  | 1.001 | 0.952 | 0.759 | 1.014 | 1.149 | 1.001 | 1.320 |
|  | 0.086 | 0.083 | 0.081 | 0.049 | 0.068 | 0.010 | 0.093 |
| Ex. 8 | 0.941 | 0.899 | 0.705 | 1.000 | 1.097 | 1.000 | 1.251 |
|  | 0.993 | 0.945 | 0.753 | 1.000 | 1.138 | 1.000 | 1.312 |
|  | 1.057 | 1.009 | 0.809 | 1.000 | 1.194 | 1.000 | 1.381 |

The results in the left block of Table 1 summarize the relative performance among 4 different filters: ES global (using $\widehat{\lambda}$), ES local (using $\widehat{\lambda}_t$), AR global (using $\widehat{p}$) and AR local (using $\widehat{p}_t$ and $\widehat{m}_t$). All filters are compared with the AR global filter. This avoids the scale problems, which vary from one simulation to another. For Example 6, the best procedure among 4 competitors is ES local, followed by ES global and AR global. This is consistent with our intuition, since the data were not generated from an AR model, but a piecewise AR model. For Example 7, since the data were generated from an AR(2) model, AR global performs the best, followed by AR local. The performance of the AR local filter can be much better than what we presented here, if we allow the upper bound of $m$ to take a larger value. ES global outperforms the ES local, since the data are stationary. The AR local performs the best for Example 8, since the model is a piecewise AR model. The ES local performs outstandingly, thanks to its flexibility.

## 4. Applications to volatility estimation

Let $S_1, \ldots, S_T$ be the prices of an asset or yields of a bond. The return of such an asset or bond process is usually described via the conditional heteroscedastic model:

$$R_t = \sigma_t \varepsilon_t \qquad (9)$$

where $R_t = \log(S_t/S_{t-1})$, $\sigma_t$ is the predictable volatility process and $\varepsilon_t'$s are the standardized innovations.

### 4.1. Connections with filtering problems

The volatility is associated with the notion of risks. For many purposes in financial practice, such as portfolio optimization, option pricing and prediction of Value-at-Risk, one would be interested in predicting the volatility $\sigma_t$ based on the past observations $S_1, \ldots, S_{t-1}$ of the asset process. The distribution of the innovation process can be skewed or have heavy tails. To produce a robust procedure, following [15], we consider the power transformation $Y_t = |R_t|^\gamma$ for some $\gamma$. Then, the model (9) can be written in the following semi-martingale form:

$$Y_t = C_\gamma \sigma_t^\gamma + D_\gamma \sigma_t^\gamma \xi_t \equiv f_t + v_t \xi_t \qquad (10)$$

with $C_\gamma = \boldsymbol{E}|\varepsilon_t|^\gamma$, $D_\gamma^2 = \text{Var}\,|\varepsilon_t|^\gamma$ and $\xi_t = D_\gamma^{-1}(|\varepsilon_t|^\gamma - C_\gamma)$. Mercurio and Spokoiny [15] argued that the choice $\gamma = 1/2$ leads to a nearly Gaussian distribution of the 'innovations' $\xi_t$, when $\varepsilon_t \sim N(0,1)$. In particular, $\boldsymbol{E}e^{u\xi_t} \leq e^{u^2/(2a)}$ with $a \approx 1.005$, a condition in Lemma 1.

Now the original problem is clearly equivalent to estimating the drift coefficient $f_t = C_\gamma \sigma_t^\gamma$ from the 'observations' $Y_s = |R_s|^\gamma$, $s = 1, \ldots, t-1$. The semi-martingale representation (10) is a specific case of the model (1) with $v_t = D_\gamma \sigma_t^\gamma$. Hence, the techniques introduced in Section 3 are still applicable.

There is a large literature on the estimation of volatility. In addition to the famous parametric models such as ARCH and GARCH (see [10] and [12]), stochastic volatility models have also received a lot of attention (see, for example, [1], [2] and [4] and references therein). We here consider only the ARCH and GARCH models in addition to the nonparametric methods (MA and ES) in Section 3.

### 4.2. Choice of orders of ARCH and GARCH

Commonly used parametric techniques for modeling volatility are ARCH [6] and GARCH [3] models. See [10] and [12] for an overview of the field. In the current context, ARCH model assumes the following autoregressive structure:

$$\boldsymbol{E}\left[Y_t \mid \mathcal{F}_{t-1}\right] = \theta_1 Y_{t-1} + \ldots + \theta_p Y_{t-p}$$

The coefficients $\theta = (\theta_1, \ldots, \theta_p)^\top$ can be estimated by using the least-squares approach:

$$\widehat{\theta}_p = \left(\sum_{s=t_0}^{t-1} X_{s,p} X_{s,p}^\top\right)^{-1} \sum_{s=t_0}^{t-1} X_{s,p} Y_s$$

with $X_{s,p} = (Y_{s-1}, \ldots, Y_{s-p})^\top$. The estimate $\widehat{f}_{t,p}$ is then defined by $\widehat{f}_{t,p} = X_{t,p}^\top \widehat{\theta}_p$. As in Section 3, the order $p$ can be chosen by minimizing the prediction error:

$$\widehat{p} = \underset{p \leq p^*}{\arg\inf} \sum_{s=t_0}^{t} (Y_s - \widehat{f}_{s,p})^2 \qquad (11)$$

The upper bound $p^*$ should be sufficiently large to reduce possible approximation errors. To facilitate computation, $t$ in (11) can be replaced by $T$, the length of the time series in the in-sample period. The approach is a global choice of the order of an ARCH model.

The volatility process $\sigma_t$ in GARCH($p,q$) is modeled as

$$\sigma_t^2 = c_0 + \alpha_1 \sigma_{t-1}^2 + \ldots + \alpha_p \sigma_{t-p}^2 + \beta_1 R_{t-1}^2 + \ldots + \beta_q R_{t-q}^2.$$

The coefficients $\alpha_j, \beta_k$ can be estimated by using the maximum likelihood method. See for example Fan and Yao [10]. The estimates $\widehat{\alpha}_j$ and $\widehat{\beta}_k$ are then used to construct the filter

$$\widehat{f}_{t,p,q} = C_\gamma \left( \sum_{j=1}^p \widehat{\alpha}_j \sigma_{t-j}^2 + \sum_{k=1}^q \widehat{\beta}_k R_{t-k}^2 \right)^{\gamma/2}.$$

The order $(p,q)$ can be chosen to minimize a quantity that is similar to (11).

GARCH(1,1) is one of most frequently used models in volatility estimation in financial time series. It has been observed to fit well many financial time series. To simplify the computation efforts, we mainly focus on the GARCH(1,1) rather than general GARCH($p,q$) in our simulation studies.

### 4.3. Simulated financial time series

We simulated time series from the volatility model

GARCH(1,1): $\quad \sigma_t^2 = 0.00005 + 0.85\sigma_{t-1}^2 + 0.1 R_{t-1}^2$

GARCH(1,3): $\quad \sigma_t^2 = 0.00002 + 0.8\sigma_{t-1}^2 + 0.02 R_{t-1}^2 + 0.05 R_{t-2}^2 + 0.11 R_{t-3}^2.$

ARCH(2): $\quad \sigma_t^2 = 0.00085 + 0.1 R_{t-1}^2 + 0.05 R_{t-2}^2.$

As shown in (10), the problem of volatility estimation is closely related to the filtering problems in Section 3. Therefore, the measure of effectiveness of each method can be gauged by MAFE and MSFE in Section 3.4. Tables 2 and 3 summarize the result for $\gamma = 0.5$ and $\gamma = 2$ in a similar format to Table 1. Table 4 summarizes the results using "rank" as a measure. For example, for the GARCH(1,3) model (second block), using untransformed data transformation (right block), in terms of MAFE, among 500 simulations, the GARCH(1,1), ES and AR methods ranked respectively, 334, 162 and 4 times in the first place, 159, 309 and 32 times in the second place and 7, 29 and 464 times in the third place.

First of all, from Tables 2 and 3, the ES and AR with their parameters chosen from data perform nearly as well as their corresponding estimators using the ideal filtering parameters. This is consistent with our theoretical result, which is the theme of our study. The GARCH(1,1) and ES estimation methods are quite robust. When the true model is GARCH(1,1), the GARCH(1,1) method performs the best, as expected, followed by ES global and then AR global. When the true model is the GARCH(1,3), which can still reasonably be well approximated by a GARCH(1,1) model, the performance of the GARCH(1,1) method and the ES method is nearly the same, though the GARCH(1,1) method performs somewhat better. It is clear that the relative performance of the GARCH(1,1) method gets deteriorated from the GARCH(1,1) model to the GARCH(1,3) model. When the series comes from the ARCH(2) model, the AR filter performs the best, as expected.

Table 2
Relative MAFE performance. ES and AR filtering of $Y_t = |R_t|^{1/2}$. Empirical mean (first row), sample standard deviation (second row), first quartile (third row), median (fourth row), and third quartile (fifth row) of MAFE ratios.

| Model | Relative to GARCH(1,1) | | Relative to ideal counterparts | |
|---|---|---|---|---|
| | ES global | AR global | ES global | AR global |
| GARCH(1,1) | 2.898 | 3.078 | 1.026 | 1.095 |
| | 1.747 | 2.045 | 0.060 | 0.164 |
| | 1.816 | 1.900 | 0.998 | 1.000 |
| | 2.464 | 2.544 | 1.006 | 1.060 |
| | 3.381 | 3.564 | 1.050 | 1.187 |
| GARCH(1,3) | 1.485 | 1.610 | 1.034 | 1.063 |
| | 0.246 | 0.304 | 0.070 | 0.122 |
| | 1.314 | 1.401 | 1.000 | 1.000 |
| | 1.482 | 1.571 | 1.000 | 1.034 |
| | 1.639 | 1.794 | 1.051 | 1.120 |
| ARCH(2) | 2.914 | 1.330 | 1.000 | 1.061 |
| | 1.448 | 0.731 | 0.000 | 0.131 |
| | 1.899 | 0.797 | 1.000 | 1.000 |
| | 2.575 | 1.139 | 1.000 | 1.000 |
| | 3.473 | 1.626 | 1.000 | 1.111 |

Table 3
Relative MAFE performance. ES and AR filtering of $Y_t = R_t^2$. Empirical mean (first row), sample standard deviation (second row), first quartile (third row), median (fourth row), and third quartile (fifth row) of MAFE ratios.

| Model | Relative to GARCH(1,1) | | Relative to ideal counterparts | |
|---|---|---|---|---|
| | ES global | AR global | ES global | AR global |
| GARCH(1,1) | 2.119 | 2.789 | 1.115 | 1.171 |
| | 1.413 | 1.823 | 0.152 | 0.249 |
| | 1.283 | 1.655 | 1.010 | 1.000 |
| | 1.815 | 2.340 | 1.055 | 1.101 |
| | 2.449 | 3.318 | 1.165 | 1.308 |
| GARCH(1,3) | 1.111 | 2.147 | 1.132 | 1.179 |
| | 0.222 | 2.381 | 0.181 | 0.291 |
| | 0.971 | 1.448 | 1.000 | 1.000 |
| | 1.092 | 1.778 | 1.070 | 1.108 |
| | 1.220 | 1.171 | 1.181 | 1.325 |
| ARCH(2) | 2.484 | 0.964 | 1.002 | 1.152 |
| | 1.229 | 0.562 | 0.032 | 0.353 |
| | 1.632 | 0.565 | 1.000 | 1.000 |
| | 2.166 | 0.816 | 1.000 | 1.000 |
| | 2.939 | 1.237 | 1.000 | 1.213 |

Table 4
Rank performance of GARCH(1,1), ES global, and AR global.

| Model | Filtering $Y_t = |R_t|^{1/2}$ | | | Filtering $Y_t = R_t^2$ | | |
|---|---|---|---|---|---|---|
| | GARCH(1,1) | ES | AR | GARCH(1,1) | ES | AR |
| GARCH(1,1) | 487 | 9 | 4 | 451 | 42 | 7 |
| | 11 | 286 | 203 | 40 | 383 | 77 |
| | 2 | 205 | 293 | 9 | 75 | 416 |
| GARCH(1,3) | 491 | 7 | 2 | 334 | 162 | 4 |
| | 8 | 347 | 145 | 159 | 309 | 32 |
| | 1 | 146 | 353 | 7 | 29 | 464 |
| ARCH(2) | 299 | 0 | 201 | 183 | 0 | 317 |
| | 197 | 4 | 299 | 310 | 8 | 182 |
| | 4 | 496 | 0 | 7 | 492 | 1 |

### 4.4. Applications

We apply the GARCH(1,1) approach $\widehat{f}_{t,1,1}$, the adaptive global ES smoothing $\widehat{f}_{t,\widehat{\lambda}}$, and the global AR smoothing $\widehat{f}_{t,\widehat{p}}$ to estimate the volatility of the log-returns of the S&P500 index and the three-month Treasury Bills. For the ES and AR approaches, we consider the square root transformation $Y_t = |R_t|^{1/2}$, which yields more stable estimates than the square transformation $Y_t = R_t^2$. The order of the AR filtering was searched among the candidate set $\mathcal{P} = \{1, \cdots, 15\}$ and the collection of grids of ES smoothing parameters was taken to be $\Lambda = \{[5 \times 1.2^k], k = 0, 1, \cdots, 15\}$. For the real data, we don't know the true volatility. Hence, we use the Average of Prediction Errors (APE) as a measure of effectiveness:

$$APE1 = \frac{1}{T - t_0 + 1} \sum_{t=t_0}^{T} (|R_t| - C_1 \widehat{\sigma}_t)^2 \quad \text{and} \quad APE2 = \frac{1}{T - t_0 + 1} \sum_{t=t_0}^{T} |R_t^2 - \widehat{\sigma}_t^2|.$$

As noted in [8], the prediction errors consist of stochastic errors and estimation (filtering) errors. The former is independent of estimation methods and dominates the latter. Therefore, a small percentage of improvement in prediction errors implies a large improvement in the filtering error.

Table 5
Relative prediction performance for yields of three-month Treasury Bills over four different periods

| Time period | ES $\widehat{f}_{t,\widehat{\lambda}}$ relative to GARCH(1,1) | | AR $\widehat{f}_{t,\widehat{p}}$ relative to GARCH(1,1) | |
|---|---|---|---|---|
| | $APE1$ | $APE2$ | $APE1$ | $APE2$ |
| 12/09/55–07/02/65 | 1.012 | 1.038 | 1.051 | 0.979 |
| 06/09/67–12/31/76 | 0.956 | 0.889 | 0.983 | 0.858 |
| 12/08/78–07/01/88 | 0.772 | 0.696 | 0.840 | 0.724 |
| 06/08/90–12/31/99 | 1.004 | 0.879 | 0.989 | 0.948 |

The three-month Treasury bills data consist of weekly observations (Fridays' closing) of interest rates of the three-month Treasury bills, from August 1954 to December 1999. The rates are based on quotes at the official close of the U.S. government securities market on a discount basis. To attenuate the time effects, we divided the entire series into four sub-series. The gaps between the time periods are the length $t_0$ used for the subsequent series. The volatility is computed based on the difference of the yields series. The relative performance of global ES and global AR smoothing and GARCH(1,1) is given in Table 5. The values are smaller than one most of the time and are sometimes as small as 0.696. This implies that with the adaptive choice of filtering parameters, the exponential smoothing and the autoregressive model outperform the GARCH(1,1) model for the periods studied. Figure 4 depicts first one hundred lags of the autocorrelation of the absolute returns and the absolute returns divided by the standard deviations estimated by the three methods. The horizontal lines indicate the 95% confidence limits. All of the three estimation methods explain well the volatility: the standardized returns rarely exhibit significant correlations.

Table 6
Relative prediction performance for the Standard and Poor 500 index over two time periods.

| Time period | ES $\widehat{f}_{t,\widehat{\lambda}}$ relative to GARCH(1,1) | | AR $\widehat{f}_{t,\widehat{p}}$ relative to GARCH(1,1) | |
|---|---|---|---|---|
| | $APE1$ | $APE2$ | $APE1$ | $APE2$ |
| 03/08/90–18/07/94 | 0.950 | 0.883 | 1.002 | 0.983 |
| 08/12/94–20/11/98 | 0.993 | 0.952 | 1.031 | 0.898 |

The S&P500 data consist of the daily closing of the Standard and Poor 500 index. The volatility estimation methods are applied to the data in the time periods 03/08/90–18/07/94 and 08/12/94–20/11/98. Again the AR and ES methods with our adaptive choice of filtering parameters provide satisfactory estimate of the underlying volatility. The ACF plots of the standardized log-returns (not shown here, similar to Figure 4) indicate success of the three methods. The relative performance against GARCH(1,1) is shown in Table 6. Again, the ES and AR filters with filtering parameters chosen by data outperform the GARCH(1,1).

The adaptive local ES filter and local AR filter were also applied to the above two data sets. We do not report the details here to save space. They both perform reasonably well. However, the local ES method does not perform as well as global one. The local AR filter performs quite well and is often better than the global AR filter, for the two financial series data that we examined.

## REFERENCES

1. Barndoff-Neilsen, O.E. and Shephard, N. (2001). Non-Gaussian Ornstein-Uhlenbeck-based models and some of their uses in financial economics (with discussion). *Jour. Roy. Statist. Soc. B*, **63**, 167-241.

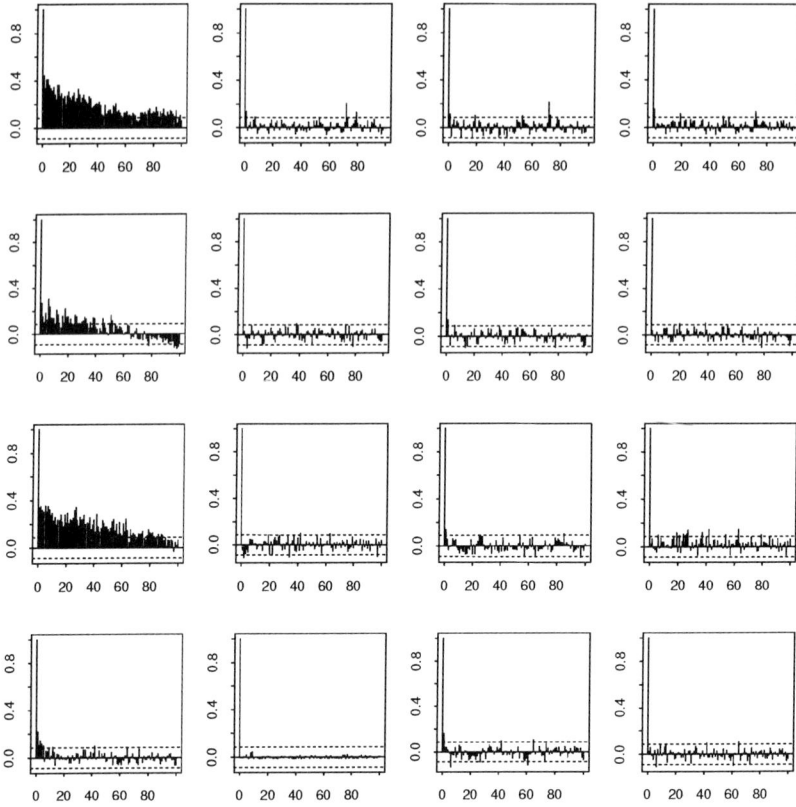

Figure 4. First one hundred lags of the autocorrelation function (ACF). Left to right: ACF of the absolute log-returns, ACF of the absolute log-returns divided by volatility estimated by GARCH(1,1) model, global ES, and global AR. From top to bottom: time periods 12/09/55–07/02/65, 06/09/67–12/31/76, 12/08/78–07/01/88, and 06/08/90–12/31/99.

2. Barndoff-Neilsen, O.E. and Shephard, N. (2002). Econometric analysis of realized volatility and its use in estimating stochastic volatility models. *Jour. Roy. Statist. Soc. B*, **64**, 253-280.
3. Bollerslev, T. (1986). Generalized autoregressive conditional heteroscedasticity. *Journal of Econometrics*, **31**, 307-327.
4. Bollerslev, T. and Zhou, H. (2002). Estimating stochastic volatility diffusion using conditional moments of integrated volatility. *Jour. Econometrics*, **109**, 33-65.
5. Brockmann, M., Gasser, T. and Herrmann, E. (1993). Locally adaptive bandwidth choice for kernel regression estimators. *Jour. Amer. Statist. Assoc.*, **88**, 1302–1309.

6. Engle, R.F. (1982). Autoregressive conditional heteroscedasticy with estimates of the variance of U.K. inflation. *Econometrica*, **50**, 987-1008.
7. Fan, J. and Gijbels, I. (1995). Data-driven bandwidth selection in local polynomial fitting: variable bandwidth and spatial adaptation. *J. Royal Statist. Soc. B*, **57**, 371–394.
8. Fan, J. and Gu, J. (2003). Data-analytic approaches to the estimation of value-at-risk. *2003 International Conference on Computational Intelligence for Financial Engineering*, to appear.
9. Fan, J. and Yao, Q. (1998). Efficient estimation of conditional variance functions in stochastic regression. *Biometrika*, **85**, 645-660.
10. Fan, J. and Yao, Q. (2003). *Nonlinear Time Series: Nonparametric and Parametric Methods*. Springer-verlag, New York.
11. Gijbels, I., Pope, A., Wand, M.P. (1999). Understanding exponential smoothing via kernel regression. *Journal of the Royal Statistical Society, Series B*, **61**, 39–50.
12. Gouriéroux, C. (1997) ARCH models and financial applications. Springer-Verlag, New York.
13. Jorion, P. (2000). *Value at Risk: The new benchmark for managing financial risk* (2nd ed.). McGraw-Hill, New York.
14. Liptser, R. and Spokoiny, V. (2000). Deviation probability bound for martingales with applications to statistical estimation. *Statist. Probab. Lett.*, **46**, 347–357.
15. Mercurio, D. and Spokoiny, V. (2003). Statistical inference for time-inhomogeneous volatility models. *Ann. Statist.*, to appear.
16. Morgan, J. P. (1996). *RiskMetrics Technical Document*. Fourth edition, New York.
17. Ruppert, D., Wand, M.P., Holst, U. and Hössjer, O. (1997). Local polynomial variance function estimation. *Technometrics*, **39**, 262-273.
18. Tong, H. (1990). *Non-Linear Time Series: A Dynamical System Approach*. Oxford University Press, Oxford.

Recent Advances and Trends in Nonparametric Statistics
Michael G. Akritas and Dimitris N. Politis (Editors)
© 2003 Elsevier Science B.V. All rights reserved.

# A normalizing and variance-stabilizing transformation for financial time series

Dimitris N. Politis[a]*

[a]Department of Mathematics, University of California—San Diego,
La Jolla, CA 92093-0112, USA; email: dpolitis@ucsd.edu

The well-known ARCH/GARCH models with normal errors can account only partly for the degree of heavy tails empirically found in the distribution of financial returns series. Instead of resorting to an arbitrary nonnormal distribution for the ARCH/GARCH residuals we propose a different viewpoint and introduce a nonparametric normalizing and variance-stabilizing transformation that can be seen as an alternative to parametric modelling. The properties of this transformation are discussed, and algorithms for optimizing it are given. For illustration, the proposed transformation is implemented in connection with some real data.

## 1. INTRODUCTION

Consider data $X_1, \ldots, X_n$ arising as an observed stretch from a financial *returns* time series $\{X_t, t = 0, \pm 1, \pm 2, \ldots\}$ such as the percentage returns of a stock price, stock index or foreign exchange rate; the returns may be daily, weekly, or calculated at different (discrete) intervals. The returns series $\{X_t\}$ will be assumed (strictly) stationary with mean zero which—from a practical point of view—implies that trends and other nonstationarities have been successfully removed.

Bachelier's (1900) pioneering work suggested the Gaussian random walk model for (the logarithm of) stock market prices. Because of the approximate equivalence of percentage returns to differences in the (logarithm of the) price series, the implication of Bachelier's thesis was that the returns series $\{X_t\}$ can be modelled as independent, identically distributed (i.i.d.) random variables with Gaussian $N(0, \sigma^2)$ distribution.

The assumption of Gaussianity was challenged in the 1960s when it was noticed that the distribution of returns seemed to have fatter tails than the normal; see e.g. Fama (1965). The adoption of some non-normal, heavy-tailed distribution for the returns seemed—at the time—to be the solution.

Nevertheless, in the early paper of Mandelbrot (1963) the phenomenon of 'volatility clustering' was pointed out, i.e., the fact that high volatility days are clustered together and the same is true for low volatility days; this is effectively negating the assumption of independence of the returns in the implication that the absolute values (or squares) of the returns are positively correlated.

---
*Research partially supported by NSF grant DMS-01-04059.

The celebrated ARCH (Auto-Regressive Conditional Heteroscedasticity) models of Engle (1982) were designed to capture the phenomenon of volatility clustering by postulating a particular structure of dependence for the time series of squared returns $\{X_t^2\}$. A typical ARCH($p$) model is described by the following equation:

$$X_t = Z_t \sqrt{a + \sum_{i=1}^{p} a_i X_{t-i}^2} \tag{1}$$

where the series $\{Z_t\}$ is assumed to be i.i.d. $N(0,1)$ and $p$ is an integer indicating the order of the model.

Volatility clustering as captured by model (1) does indeed imply a marginal distribution for the $\{X_t\}$ returns that has heavier tails than the normal. However, model (1) can account only partly for the degree of heavy tails empirically found in the distribution of returns, and the same is true for the ARCH spin-off models (GARCH, EGARCH, etc.); see Bollerslev et al. (1992) for a review. For example, the market crash of October 1987 is still an outlier 6-7 standard deviations away even after the best ARCH model is employed; see e.g. Nelson (1991).

Consequently, researchers and practitioners have been resorting to ARCH models with heavy-tailed errors. A popular assumption for the distribution of the $\{Z_t\}$ is the $t$-distribution with degrees of freedom empirically chosen to match the apparent degree of heavy tails as measured by higher-order moments such as the kyrtosis; see e.g. Shephard (1996) and the references therein.

Nevertheless, this situation is not very satisfactory since the choice of a $t$-distribution seems quite arbitrary. In a certain sense, it seems that we have come full-circle back to the 60s in trying to model the excess kyrtosis by an arbitrarily chosen heavy-tailed distribution.

Perhaps the real issue is that a simple and neat parametric model such as (1) could not be expected to perfectly capture the behavior of a complicated real-world phenomenon such as the evolution of financial returns that—almost by definition of market 'efficiency'—ranks at the top in terms of difficulty of modelling/prediction.

As a more realistic alternative, one may resort to an exploratory, nonparametric approach in trying to understand this type of data; such an approach is outlined in the paper at hand. In the next section, a normalizing and variance–stabilizing transformation for financial returns series is proposed, and its properties are discussed. An application to real data is given in Section 3, while a more general form of the NoVaS transformation is put forth in Section 4—including a comparison to the ARCH model (1).

## 2. NORMALIZING AND VARIANCE–STABILIZING TRANSFORMATION

### 2.1. Definition of the NoVaS transformation

Recall that, under model (1), the quantity

$$\frac{X_t}{\sqrt{a + \sum_{i=1}^{p} a_i X_{t-i}^2}}$$

is thought of as perfectly normalized and variance–stabilized as it is assumed to be i.i.d. $N(0,1)$. From an applied statistics point of view, the above ratio can be interpreted as an attempt to 'studentize' the return $X_t$ by dividing with a (time-localized) measure of the standard deviation of $X_t$.

Nevertheless, there seems to be no reason—other than coming up with a neat model— to exclude the value of $X_t$ from an empirical (causal) estimate of the standard deviation of $X_t$. Hence, we now define the new 'studentized' quantity

$$W_{t,p} := \frac{X_t}{\sqrt{a_0 X_t^2 + \sum_{i=1}^{p} a_i X_{t-i}^2}}. \tag{2}$$

Equation (2) describes our proposed normalizing and variance–stabilizing (NoVaS, for short) transformation under which the series $\{X_t\}$ is mapped to the new series $\{W_{t,p}\}$. The order $p(\geq 0)$ and the parameters $a_0, \ldots, a_p$ are chosen by the practitioner with the twin goals of normalization/variance-stabilization in mind that will be made more precise shortly.

We can re-arrange the NoVaS equation (2) to make it look more like model (1):

$$X_t = W_{t,p} \sqrt{a_0 X_t^2 + \sum_{i=1}^{p} a_i X_{t-i}^2}. \tag{3}$$

The only real difference is the presence of the term $X_t^2$ paired with the coefficient $a_0$. Equation (3) is very useful but should not be interpreted as a "model" for the $\{X_t\}$ series; rather, the focus should remain on equation (2) and the effort to render the transformed series $\{W_{t,p}, t = 1, 2, \cdots\}$ close (in some sense) to behaving like the standard normal ideal. For instance, it is immediate that the series $\{W_{t,p}\}$ can never be thought to be *exactly* distributed as $N(0,1)$. To see this, note that

$$\frac{1}{W_{t,p}^2} = \frac{a_0 X_t^2 + \sum_{i=1}^{p} a_i X_{t-i}^2}{X_t^2} \geq a_0$$

if all the $a_i$s are chosen to be nonnegative. Thus, $|W_{t,p}| \leq 1/\sqrt{a_0}$ almost surely, indicating that, if $a_0 \neq 0$, then the series $\{W_{t,p}\}$ comprises of bounded random variables. However, with $a_0$ chosen small enough, the boundedness of the $\{W_{t,p}\}$ series is effectively not noticeable while yielding the following side-benefit: regardless of the degree of heavy tails present in the $\{X_t\}$ series, the $\{W_{t,p}\}$ series will never exhibit heavy tails, i.e., there will never be any 'excess kyrtosis' to account for. Consequently, great advantages in practical data modelling and/or prediction may be derived; an illustration is given in Section 3.

### 2.2. Optimizing the NoVaS transformation

In choosing the order $p(\geq 0)$ and the parameters $a_0, \ldots, a_p$ the twin goals of normalization and variance–stabilization of the transformed series $\{W_{t,p}\}$ are taken into account. The target of variance-stabilization is easier and—given the assumed structure of the return series—amounts to constructing a local estimator of scale for studentization purposes; for this reason we require

$$a_0 > 0, \quad a_i \geq 0 \quad \text{for all } i, \quad \text{and} \quad \sum_{i=0}^{p} a_i = 1. \tag{4}$$

Equation (4) has the interesting implication that the $\{W_{t,p}\}$ series can be assumed to have an (unconditional) variance that is (approximately) unity. Nevertheless, note that $p$ and $a_0, \ldots, a_p$ must be carefully chosen to achieve a degree of conditional homoscedasticity as well; to do this, one must necessarily take $p$ not too large so that a local (as opposed to global) estimator of scale is obtained. An additional intuitive—but not obligatory—constraint involves monotonicity:

$$a_i \geq a_j \quad \text{if} \quad 1 \leq i < j \leq p. \tag{5}$$

It is practically advisable that a simple structure for the $a_i$ coefficients is employed satisfying (4) and (5). The simplest such example is to let $a_i = 1/(p+1)$ for all $0 \leq i \leq p$; this specification will be called the 'simple' NoVaS transformation, and involves only one parameter, namely the order $p$, to be chosen by the practitioner. Another example is given by the exponential decay NoVaS, i.e., $a_i = c'e^{-ci}$ for all $0 \leq i \leq p$; this involves choosing two parameters: $p$ and $c > 0$ since $c'$ is determined by (4).

Subject to the variance stabilization condition (4)—together with (5) if desirable—one then proceeds to choose (the parameters needed to identify) $p$ and $a_0, a_1, \ldots, a_p$ with the optimization goal of making the $\{W_{t,p}\}$ transformed series as close to normal as possible. To quantify this target it is suggested that one matches the empirical kyrtosis (and possibly some higher order even moments) of $W_{t,p}$ to those of a standard normal random variable. In order to render joint distributions of the $\{W_{t,p}\}$ series more normal, one may also apply the previous moment matching idea to a few specific linear combinations of $W_{t,p}$ random variables.

However, in view of the bound $|W_{t,p}| \leq 1/\sqrt{a_0}$, one must be careful to ensure that the $\{W_{t,p}\}$ random variables have a range large enough so that the boundedness is not seen as spoiling the normality. Thus, we also require

$$1/\sqrt{a_0} \geq C \quad \text{i.e.,} \quad a_0 \leq 1/C^2 \tag{6}$$

for some appropriate $C$ of the practitioner's choice. Recalling that 99.7% of the mass of the $N(0,1)$ distribution is found in the range $\pm 3$, the simple choice $C = 3$ can be suggested; this choice seems to work reasonably well—at least for the samples sizes typically encountered in practice. Alternatively, one may let $C$ depend on the sample size $n$; taking into account that the maximum of $n$ i.i.d. $N(0,1)$ random variables is of the order of $\sqrt{2 \ln n}$, one may let $C$ be equal (or proportional) to $\sqrt{2 \ln n}$.

### 2.3. Simple NoVaS algorithm

We now give specific algorithms for optimizing the NoVaS transformation in the two previously mentioned examples, the simple and exponential NoVaS transformation. For this, let $KYRT(Y_t)$ denote the empirical kyrtosis of data $\{Y_t, t = 1, \ldots, n\}$, i.e.,

$$KYRT(Y_t) = \frac{n^{-1}\sum_{t=1}^{n}(Y_t - \bar{Y})^4}{(n^{-1}\sum_{t=1}^{n}(Y_t - \bar{Y})^2)^2}$$

where $\bar{Y} = n^{-1}\sum_{t=1}^{n} Y_t$ is the sample mean. Although even moments of order higher than four may also be used, in what follows we focus on the fourth moment for concreteness.

ALGORITHM FOR SIMPLE NoVaS:

- Let $a_i = 1/(p+1)$ for all $0 \leq i \leq p$.
- Pick $p$ such that $|KYRT(W_{t,p}) - 3|$ is minimized.

The last step of the above algorithm was described as an optimization problem for mathematical concreteness. Nevertheless, it could be better understood as a moment matching, i.e.,

- Pick $p$ such that $KYRT(W_{t,p}) \simeq 3$,

where of course the value 3 for kyrtosis corresponds to the Gaussian distribution.

To see that the moment matching goal is a feasible one, note first that for $p = 0$ we have $a_0 = 1$, $W_{t,0} = sign(X_t)$, and $KYRT(W_{t,0}) = 1$. It is also to be expected that for large $p$, $KYRT(W_{t,p})$ will be bigger than 3; as a matter of fact, if $X_t$ happens to have a finite 2nd moment, then the law of large numbers would imply that $KYRT(W_{t,p}) \simeq KYRT(X_t)$ for large $p$. Therefore, viewing $KYRT(W_{t,p})$ as a (hopefully smooth) function of $p$, one would expect that for an intermediate value of $p$ the level 3 would be (approximately) attained; this is actually what happens in practice.

Thus, to actually carry out the search for the optimal $p$ in the Simple NoVaS Algorithm, one sequentially computes $KYRT(W_{t,p})$ for $p = 1, 2, \cdots$, stopping when $KYRT(W_{t,p})$ first hits or just passes the value 3. Interestingly, $KYRT(W_{t,p})$ is typically an increasing function of $p$ which makes this scheme very intuitive.

The above simple algorithm seems to work remarkably well; see Section 3 for illustration. A caveat, however, is that the range condition (6) might not be satisfied. If this is the case, the following 'range-adjustement' step can be added to algorithm.

- If $p$ (and $a_0$) as found above are such that (6) is not satisfied, then increase $p$ accordingly; in other words, redefine $p$ to be the smallest integer such that $1/(p+1) \leq 1/C^2$, and let $a_i = 1/(p+1)$ for all $0 \leq i \leq p$.

It goes without saying that this range-adjustment should be used with care, that is, the choice of $C$ in (6) should be reasonable; both concrete suggestions, i.e., $C = 3$ or $C \simeq \sqrt{2 \ln n}$, seem to work well in practice.

In the above algorithm, the target was 4th moment matching of $W_{t,p}$ to the corresponding Gaussian moment, i.e., $KYRT(W_{t,p}) \simeq 3$; this procedure has the goal of (approximately) normalizing the marginal distribution of $W_{t,p}$. Interestingly, this simple procedure seems to be somehow effective in normalizing joint distributions as well, e.g. the joint distribution of $W_{t,p}$ and its lagged version $W_{t-1,p}$, which is a highly desirable objective; see the illustrative example in Section 3.

Nevertheless, if one wants to *ensure* that some joint distributions are normalized—at least as far as 4th moments are concerned—then the moment matching criterion of the algorithm can be modified. To fix ideas, consider the target of normalizing the joint distribution of $W_{t,p}$ and $W_{t-1,p}$. The Cramer-Wold device suggests to simultaneously consider some linear combinations of the type:

$$\tilde{W}_{t,p,i} = W_{t,p} + \lambda_i W_{t-1,p} \quad \text{for} \quad i = 1, \ldots, K,$$

where the $\lambda_i$'s are some chosen constants; a concrete suggestion is to take $K = 3$ and $(\lambda_1, \lambda_2, \lambda_3) = (0, 1, -1)$.

The NoVaS algorithm is then altered to focus on the $\tilde{W}_{t,p,i}$s instead of $W_{t,p}$; to elaborate:

- Pick $p$ such that $\max_i |KYRT(\tilde{W}_{t,p,i}) - 3|$ is minimized.

### 2.4. Exponential NoVaS algorithm

In the Exponential NoVaS, to specify all the $a_i$s, one just needs to specify the two parameters $p$ and $c > 0$, in view of (4). However, because of the exponential decay, the parameter $p$ is of secondary significance as the following algorithm suggests.

ALGORITHM FOR EXPONENTIAL NOVAS:

- Let $a_i = c'e^{-ci}$ for all $0 \le i \le p$.

- Let $p$ take a very high starting value, e.g., let $p \simeq n/4$ or $n/5$.

- Pick $c$ in such a way that $|KYRT(W_{t,p}) - 3|$ is minimized.

It is apparent that the above search will be conducted over a discrete grid of $c$–values; let $c_0$ denote the resulting minimizer. Consequently, the following range-adjustement safeguard may be added.

- If $c_0$ (and $a_0$) as found above are such that (6) is not satisfied, then decrease $c$ stepwise (starting from $c_0$) over the discrete grid until (6) is satisfied.

Finally, the value of $p$ must be trimmed for efficiency of usage of the available sample. The "tolerance level" $\epsilon$ below is the practitioner's choice; a value of 0.01 is reasonable in connection with the $a_i$ which—it should be stressed—are normalized to sum to one.

- Trim the value of $p$ by a criterion of the type: if $a_i < \epsilon$, then $a_i = 0$. Thus, if $a_i < \epsilon$, for all $i \ge i_0$, then let $p = i_0$, and renormalize the $a_i$s so that their sum (for $i = 0, 1, \ldots, i_0$) equals one.

As in the simple NoVaS algorithm, here too we could focus on moment matching for the linear combinations $\tilde{W}_{t,p,i}$ instead of $W_{t,p}$. In addition, the Exponential NoVaS algorithm could be extended to include a sum of many exponentials, i.e., a situation where $a_i = b'e^{-bi} + c'e^{-ci} + d'e^{-di} \cdots$. The generalization may well include higher order moment matching and/or looking at linear combinations of higher order lags.

## 3. ILLUSTRATION WITH REAL DATA

Figure 1(a) shows a plot of the daily returns of the S&P500 stock index from January 1, 1928, to August 30, 1991; this unusually long data stretch is available as part of the garch module in Splus. Figure 1(b) shows our target series $\{X_t\}$, namely the S&P500 returns; those were computed by differencing the (logarithm of) the original series of Figure 1(a).

The time series plot in Figure 1(b) prominently shows the previously discussed phenomenon of "volatility clustering". In addition, the $\{X_t\}$ dataset appears quite non-normal

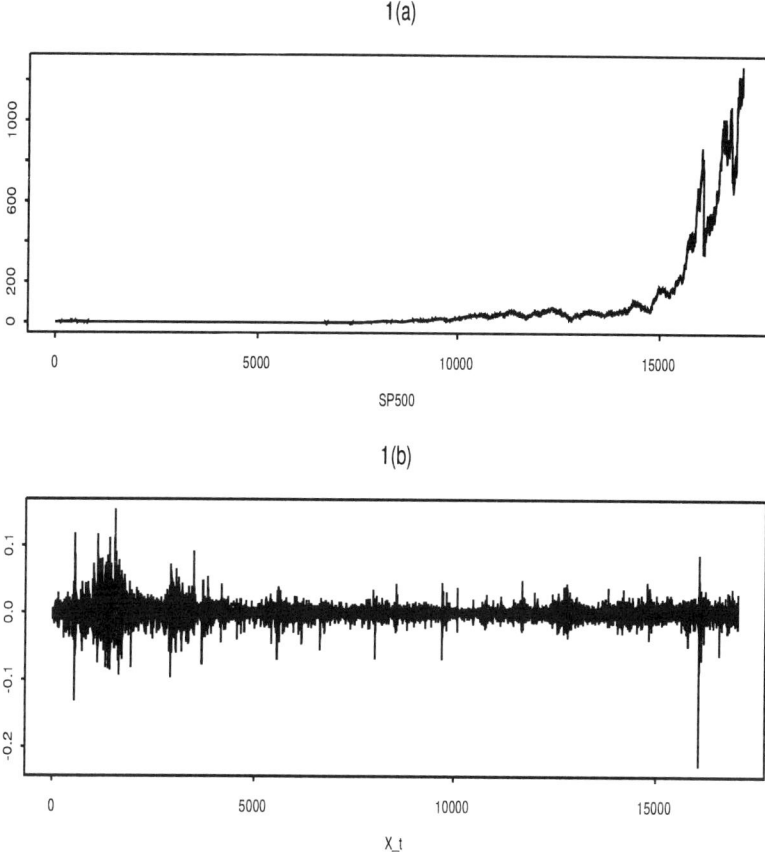

Figure 1. (a) Time series plot of the daily S&P500 stock index from January 1, 1928, to August 30, 1991; (b) Time series plot of the $\{X_t\}$ series, i.e., S&P500 returns.

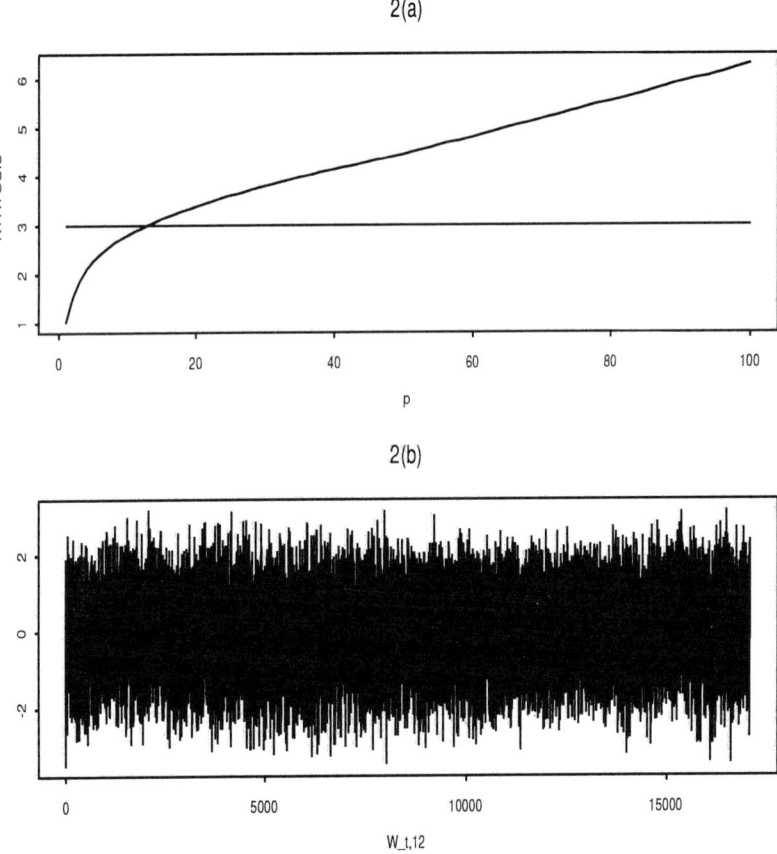

Figure 2. (a) Plot of $KYRT(W_{t,p})$ as a function of $p$; the solid line indicates the Gaussian kyrtosis of 3; (b) Time series plot of the optimal simple NoVaS series $\{W_{t,12}\}$.

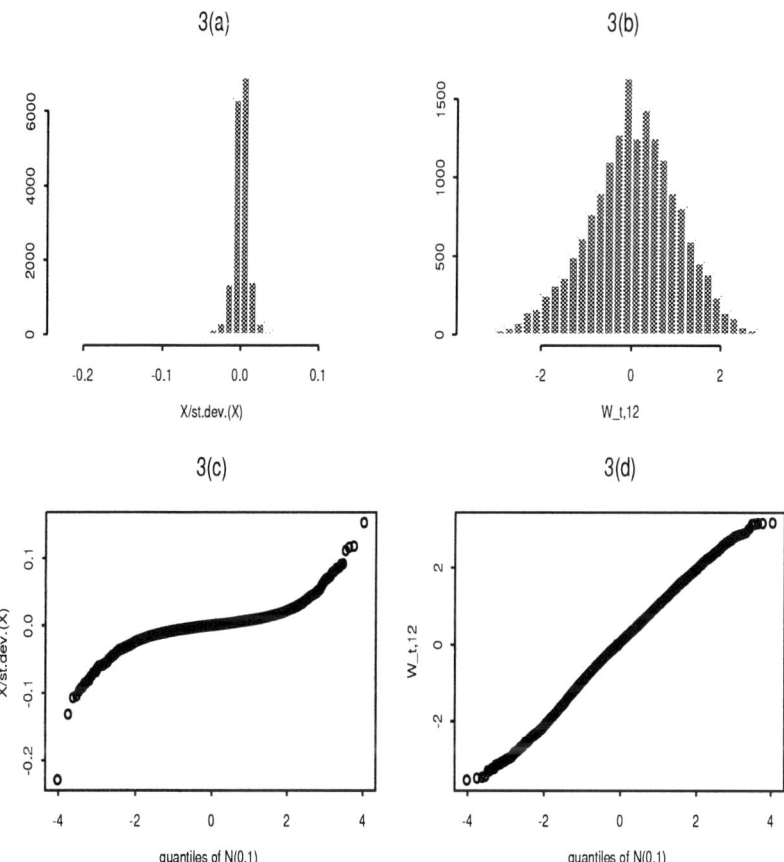

Figure 3. (a) Histogram of the $\{X_t\}$ series divided by its sample standard deviation; (b) Histogram of the simple NoVaS series $\{W_{t,12}\}$; (c) Q-Q plot of the series in part (a); (d) Q-Q plot of the series in part (b).

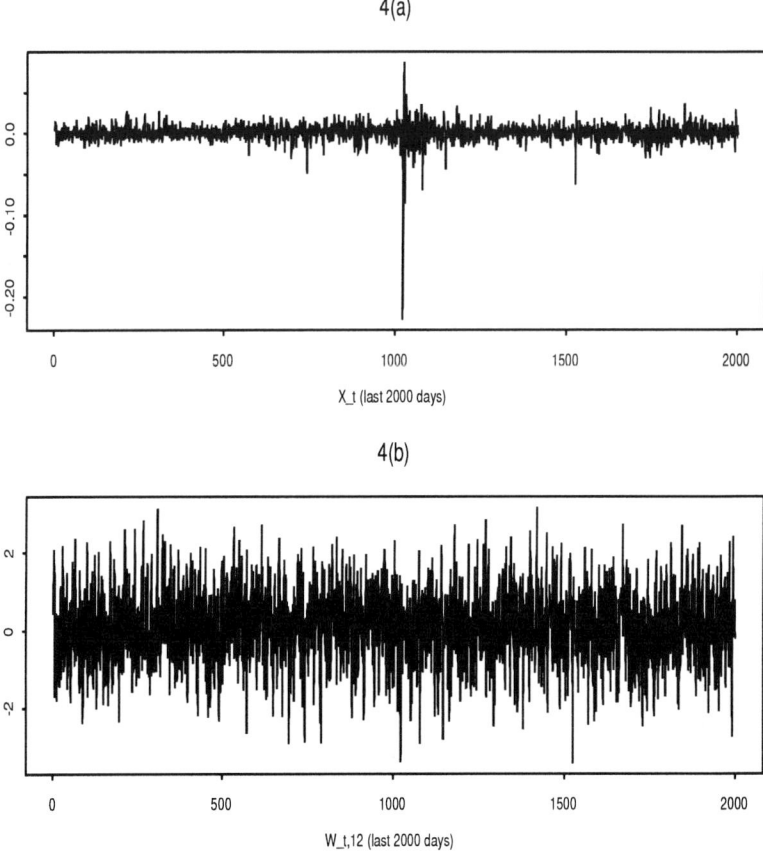

Figure 4. (a) Time series plot of the last 2000 days of the $\{X_t\}$ series; (b) Time series plot of the last 2000 days of the simple NoVaS series $\{W_{t,12}\}$.

and fat-tailed as the histogram in Figure 3(a) shows. Indeed, the (sample) kyrtosis of the $\{X_t\}$ data is: $KYRT(X_t) = 25.4$ which is quite high.

In order to implement the simple NoVaS algorithm, $KYRT(W_{t,p})$ was computed for $p$ ranging from 1 to 100; see the plot in Figure 2(a). The monotonic increase of $KYRT(W_{t,p})$ is apparent, rendering the NoVaS algorithm easy to implement. Notably, $KYRT(W_{t,p})$ is closest to 3 for $p = 12$; actually, $KYRT(W_{t,12}) = 3.01$.

The optimal simple NoVaS transformed series $\{W_{t,12}\}$ is plotted in Figure 2(b). Although $\{W_{t,12}\}$ is related in a simple way to the original data of Figure 1(b), the regions of "volatility clustering" corresponding to the $\{X_t\}$ series are hardly (if at all) discernible in the plot of the NoVaS series $\{W_{t,12}\}$.

In order to capture some more detail, Figure 4 focuses on the last 2000 days of the $\{X_t\}$ and $\{W_{t,12}\}$ series. The crash of 1987 is very prominent around the middle point of Figure 4(a); by contrast, the events of October 1987 are hardly noticeable in the corresponding plot of the NoVaS series $\{W_{t,12}\}$ in Figure 4(b).

The histogram of the full-length series $\{W_{t,12}\}$ is shown in Figure 3(b); it is quite close visually to a normal histogram, at least as compared to the heavy-tailed histogram of the $\{X_t\}$ original series of Figure 3(a). This observation is further confirmed by an examination of the two Q-Q plots in Figures 3(c) and 3(d). It seems that simple NoVaS has achieved its objective of normalizing, as well as variance-stabilizing, the $\{X_t\}$ returns. The Exponential NoVaS algorithm gave qualitatively similar results; the details are omitted due to lack of space.

It should be noted that the effective range of the $\{W_{t,12}\}$ series is about 3.51 which is acceptable in terms of (6) being satisfied with the simple choice $C = 3$. If one opted for the choice $C = \sqrt{2 \ln n}$, then in this case $C$ would be about 4.41 and a range-adjustement step would be required leading to the choice $p = 18$; note that $KYRT(W_{t,18}) = 3.33$ which is still quite close to the target value of 3. As a matter of fact, a Q-Q plot (not shown) of the simple NoVaS series $\{W_{t,18}\}$ is very similar to the one in Figure 3(d).

Finally, it is interesting to observe that—as alluded to in Section 2—it seems that the NoVaS transformation approximately normalizes some joint distributions as well, giving a strong indication that $\{W_{t,12}, t \in \mathbf{N}\}$ is close to a Gaussian *process*. Table 1 gives the sample kyrtosis of the series $\tilde{W}_{t,12,i} = W_{t,12} + \lambda_i W_{t-1,12}$ for different values of $\lambda_i$; all the entries of Table 1 are close to the nominal value of 3 supporting this claim.

| $\lambda_i$ | -4 | -1 | -0.5 | 0 | 0.5 | 1 | 4 |
|---|---|---|---|---|---|---|---|
| $KYRT(\tilde{W}_{t,12,i})$ | 2.99 | 3.11 | 3.14 | 3.01 | 2.97 | 3.01 | 3.04 |

Table 1: Sample kyrtosis of $\tilde{W}_{t,12,i} = W_{t,12} + \lambda_i W_{t-1,12}$ for different values of $\lambda_i$.

## 4. A GENERAL FORM OF THE NoVaS TRANSFORMATION

### 4.1. A general form of NoVaS

Comparing the NoVaS eq. (2) to the ARCH model (1) we notice the absence of a constant term in the denominator of NoVaS. Such a term can indeed be included, paired with a (causal) estimator of the (unconditional) variance of $X_1$. Thus, the general form of the NoVaS transformation is given below:

$$W_{t,p} := \frac{X_t}{\sqrt{\alpha \hat{\sigma}_t^2 + a_0 X_t^2 + \sum_{i=1}^{p} a_i X_{t-i}^2}}. \tag{7}$$

In the above, $\hat{\sigma}_t^2$ is a (causal) consistent estimator of $Var(X_1)$, e.g., the sample variance or sample second moment of $\{X_1, \ldots, X_t\}$, and $\alpha(\geq 0)$ is one more NoVaS parameter to be chosen by the practitioner. Note that if it known (or even suspected) that $X_t$ might not have a finite second moment, then $\alpha$ should be taken to be zero. In general, for proper studentization we now need:

$$\alpha \geq 0, \quad a_0 > 0, \quad a_i \geq 0 \quad \text{for all } i, \quad \text{and} \quad \alpha + \sum_{i=0}^{p} a_i = 1 \tag{8}$$

instead of eq. (4). Eq. (8) above implies that $\alpha$ must be less than one; in practice, $\alpha$ must be chosen much less than one in order for the kyrtosis matching to be successful. A rule-of-thumb is that $\alpha$ should be of the order of magnitude of $a_0$ or $a_1$; doing so will ensure that the denominator in eq. (7) is indeed a local (as opposed to global) estimator of standard deviation. More details on choosing the parameter $\alpha$ will be given elsewhere.

### 4.2. Comparison to ARCH models

We can re-arrange the general NoVaS equation (7) to yield:

$$X_t^2 = \frac{W_{t,p}^2}{1 - a_0 W_{t,p}^2} \left( \alpha \hat{\sigma}_t^2 + \sum_{i=1}^{p} a_i X_{t-i}^2 \right) \tag{9}$$

and

$$X_t = \frac{W_{t,p}}{\sqrt{1 - a_0 W_{t,p}^2}} \sqrt{\alpha \hat{\sigma}_t^2 + \sum_{i=1}^{p} a_i X_{t-i}^2}. \tag{10}$$

Recall that—by construction—$W_{t,p}$ should be approximately equal to a standard normal random variable that is truncated to the range $\pm 1/\sqrt{a_0}$; in other words, $W_{t,p}$ can be thought to (approximately) have the probability density given by

$$\frac{\phi(x) \mathbf{1}\{|x| \leq C_0\}}{\int_{-C_0}^{C_0} \phi(y) dy}, \tag{11}$$

where $\phi$ denotes the standard normal density, and $C_0 = 1/\sqrt{a_0}$.

Equation (10) can now be compared to the ARCH model (1). It is apparent that the general NoVaS transformation as manifested by (10) yields a particular implication for the $Z_t$ innovation term appearing in (1). Thus, it is hardly surprising that this innovation term has been found empirically to be heavy-tailed; eq. (10) suggests that its degree of heavy tails is exactly that associated with the distribution of the random variable $W/\sqrt{1 - a_0 W^2}$, where $W$ has (approximately) the density given in (11).

# REFERENCES

1. Bachelier, L. (1900). *Theory of Speculation*. Reprinted in *The Random Character of Stock Market Prices*, P.H. Cootner (Ed.), Cambridge, Mass.: MIT Press, pp. 17-78, 1964.
2. Bollerslev, T., Chou, R. and Kroner, K. (1992). ARCH modelling in finance: a review of theory and empirical evidence, *J. Econometrics*, 52, 5-60.
3. Engle, R. (1982). Autoregressive conditional heteroscedasticity with estimates of the variance of UK inflation, *Econometrica*, 50, 987-1008.
4. Fama, E.F. (1965). The behaviour of stock market prices, *J. Business*, 38, 34-105.
5. Mandelbrot, B. (1963). The variation of certain speculative prices, *J. Business*, 36, 394-419.
6. Nelson, D. (1991). Conditional heteroscedasticity in asset returns: a new approach, *Econometrica*, 59, 347-370.
7. Shephard, N. (1996). Statistical aspects of ARCH and stochastic volatility. in *Time Series Models in Econometrics, Finance and Other Fields*, D.R. Cox, David V. Hinkley and Ole E. Barndorff-Nielsen (eds.), London: Chapman & Hall, pp. 1-67.

# 9. Bioinformatics and Biostatistics

# Biostochastics and nonparametrics: Oranges and Apples?

Pranab Kumar Sen[a] *

[a] Departments of Biostatistics and Statistics,
University of North Carolina, Chapel Hill, NC 27599-7420, USA

Biostochastics is envisaged as the field dealing with the stochastics (statistical modeling and analysis) in large biological systems, not necessarily having dominant genetic undercurrents; bioinformatics (dealing with genomics, computational biology, and large genetic data models) is an important subfield of biostochastics. In bioinformatics, stochastic evolutionary forces provide ample room for comprehending the basic differences between mathematical exactness and biological diversity. In biostochastics, stochastics may be latent to a greater extent, and usually categorical (qualitative) responses or multivariate counting processes with imprecise dependence patterns crop up. As such, standard (multivariate) statistical modeling and analysis, attuned to the likelihood principle, may not be appropriate. Also, statistical learning approaches exploits mainly computational algorithms and may lack proper methodological support. Alternative (nonparametric and semiparametric) approaches that take into account underlying biological implications to a greater (and parametrics to a lesser) extent are appraised in the light of validity and robustness considerations.

## 1. Introduction

Stochastic modeling and analysis of large biological systems (specially the ones with significant molecular undercurrents) have received considerable attention in the recent past. Such macro-biological models differ considerably from conventional biometric models, studied extensively during the second half of the last century). A greater part of this evolution is due to the rampage of (super-)computers in every walk of life and science. Nonparametrics, at its genesis, emerged as *quick and dirty methods* when nothing but hand calculators were providing the basic service. It is therefore quite conceivable that in the awake of information technology, statistical science would lean more towards computational statistics, and in that way, to bioinformatics. Yet, the picture is not that rosy: in one hand, too much of information (to disseminate), and on the other, the curse of (high-)dimensionality have posed some challenging stochastic problems. There is a basic concern: To what extent standard stochastic modeling and analysis may be appropriate in biostochastics, the evolving field of stochastics for large biological systems?

With the advent of modern information technology and scientific curiosity in human

---
*The present version is mostly adapted from the Third Senior Noether Award Lecture presentation at the Joint Statistical Meetings in New York City, August 14, 2002.

genomics, we are at the dawn of bioinformatics; it is not precisely known what really constitutes the domain of bioinformatics: pharmaceutical researchers, molecular biologists, computer scientists, biomathematicians, and statisticians differ in their definitions and interpretations, though to a much lesser extent than in their basic emphasis on methodological versus computational aspects. Faced with this dilemma, let me quote a few lines from a very recent statistics text on bioinformatics by Ewens and Grant (2001): *We take bioinformatics to mean the emerging field of science growing from the application of mathematics, statistics, and information technology, including computers and the theory surrounding them, to the study and analysis of very large biological and in particular, genetic data sets. The field has been fueled by the increase in the DNA data generation.* In a similar manner, an image of bioinformatics has been drawn earlier by Waterman (1995), with more emphasis on computational biology. Durbin et al. (1998) appealed to *Hidden Markov models* (HMM) in their treatise of computational biology, albeit without much emphasis on the underlying biostochastics.

At the current stage, gene scientists can not scramble fast enough to keep up with the genomics, emerging at a furious rate and in astounding detail. This is particularly significant in view of the nearly completed status of the mapping of human genome sequence (IHGSC 2001, Venter et al. 2001). Faced with this monumental genomic task, it is more important to see how sound methodolgy could be tied to superb computational schemes to resolve the mysteries of the molecular activities in living organisms, including man. But, to formulate molecular biological theory based on sound statistical methodology, we need to have the dusts of (mostly) empirical findings (based mainly on countless statistical packages and algorithms) settled before trying to resolve the basic theoretical foundations, if that can be done at all. In some way, bioinformatics is designed, at least at the present, as the custodian of *data mining* for enormously large molecular biological data sets, and as such, at least at this stage, as a discipline, it does not aim to lay down fundamental mathematical laws that govern biological systems parallel to those laid down in physics or even theoretical statistics. Frankly, given the enormous biodiversity, genetic complexities, and extreme variability in human metabolisim, such laws may not exist in bioinformatics, and even if they do, are a long way from being determined for biological systems by computational devices alone. Nevertheless, at this stage, in bioinformatics, there is mathematical utility in the creation of tools (mostly, in the form of computer graphics and algorithms) that investigators can use to analyse exceedingly large data sets arising in this context, without bypassing the biological undercurrents just for *theory for theory's sake*. Because of underlying stochastic evolutionary forces, such tools may generally involve statistical modeling of (molecular) biological systems, which in turn, requires incorporation of probability theory, statistics, and stochastic processes. This need for statistical rationality could not be any less for the greater field of biostochastics. Although *knowledge discovery and data mining* (KDDM) procedures dominate computational biology, and bioinformatics in general, from the above discussion we may gather that it would be improper to arrive at statistical conclusions based on data analysis alone (even under the umbrella of KDDM). Elements of *statistical learning* (Hastie et al. 2001) have therefore emerged as an endeavor to impart statistical reasoning to data mining; *bagging, boosting, bootstrapping* and *addtive trees* are emerging as useful tools in KDDM explorations. Yet there is a genuine need to grasp the genetic and molecular biologic bases of bioinformatics to sort out the stochastic aspects from purely empirical computational aspects, and to lead to some resolutions that synchronize methodology and algorithms towards a common goal. Primarily driven by this motivation, we use the following terminology (Sen 2001): *Biostochastics to deal primarily with stochastic modeling and analysis (i.e., stochastics) for very large biological (including genetic and genomic) data sets.* In this formulation, biostochastics attempts to cover large biological systems which may not have predominant genetic factors; *neuronal spatio-temporal* models are noteworthy examples. In this scenario, we embark on an excursion of biostochastics from the traditional biometry to modern biostatistics to the evolutionary field of bioinformatics, without being confronted solely in the molecular genetics frontier. Yet identifying that molecular biology and large genetic models are important members, biostochastics is charged with the development of sound methodologic support for valuable computer intensive algorithms that are currently widely used in genomic studies. It is in this sense somewhat complementary to statistical learning that places primary emphasis on the monumental computational aspects.

As we look into the genesis of nonparametrics, we gather that the tremendous impact it has made in statistical science is primiarily due to its *model-flexibility, robustness* and *efficacy* in a broader situation. In view of the lack of robustness of parametrics, and the fact that in biostochastics, as we shall see, likelihoods are often too complex to formulate, it is natural to see how far nonparametrics can be incorporated in biostochastics? This is the main theme of the present study. With this motivation, in Section 2, an introduction to the basic need of statistical reasoning in biostochastics is outlined. Section 3 is devoted to likelihoods and variants such as the quasi-likelihood and pseudolikelihood. The relevance and limitations of nonparametrics are discussed in Section 4. For lack of space, we sacrifice much of the discussion on *semiparametrics*, which in the recent past, has flooded the dissertation baskets of would be biostatisticians. Though it has some mathematical convenience (compared to both parametrics and nonparametrics), it might not be robust for complex biological systems; the impasse is due to some stringent assumptions underlying semiparametrics that are unlikely to be tenable in macrostochastic models. We shall therefore *let the cliff fall where it belongs to.*

## 2. Biostochastics : A Preamble

Perhaps, it might be more convenient to trace the evolution of bioinformatics from biometry (to biostatistics), and relay the transition to biostochastics. The evolution of biometry goes back to more than hundred years ago, though stochastic analysis and statistical inference in standard biometric problems have been extensively pursued during the past sixty years or so. While the modeling aspect varies from simple biometric studies to *bioassays* (and *bioequivalence trials*) to *dosimetry* in general, they share a common statistical methodology foundation. In the transition from biometry to biostatistics, aimed to meet the need of a wider field of applications comprising biomedical, clinical, and other public health disciplines, the 'controlled' laboratory type of study in biometry is often not the case, and hence, the scenario has changed. In randomized clinical trials, survival analysis, environmental health problems, and epidemiologic studies, among others, human beings are used as subjects, and that may completely change the design modeling and analysis aspects of such biostatistics problems. Medical ethics, cost-benefit factors, geo-political undercurrents and other factors may impose retraints on a study plan, and hence, the biometric experience may not be totally relevent. Nevertheless, with some genuine extension and modefication of existing statistical methodology, it has been possible to provide broad statistical reasoning in this interdisciplinary field. Incorporation of statistical methodology in biostatistics has been made judiciously by blending methodology with nontrivial and fruitful applications requiring more and more advanced computational facilities. In course of this, the mathematical abstractions that are characteristically associated with theoretical statistics and probability theory were diffused to a greater extent to facilitate the utility and effectiveness of contemplated applications; yet, there remains much to bridge the gap between sound methodology and superb applications in bio-environmental sciences. In clinical trials, environmental health sciences, and epidemiology, even now there is a dearth of statistical methodology in some genuine applications. This deficiency arises in statistical planning (design) and modeling, and as a result, more profoundly in statistical analysis. No wonder nonparametrics and semiparametrics have marched into this arena knocking out the likelihood based parametrics to a certain extent. This factor is emerging as a basic issue in the evolving field of bioinformatics, aimed primarily for quantitative modeling and analysis of genomics, computational biology, and multi-loci (functional) quantitative genetics.

In bioinformatics, generally, there are very high-dimensional qualitative response variables so that conventional multinormal parametric models are not usually appropriate. Even one tries discrete multivariate analysis there could be too many parameters (relative to the size of the dataset), creating roadblocks to parametric statistical modeling and analysis. Thus, the *curse of dimensionality* could be the prime impasse for statistical modeling and analysis. There are other molecular biological undercurrents that also create considerable difficulties in the adoption of standard parametrics for bioinformatics. For this reason, we outline the very basic of genomics to illustrate this secnario.

*Genome* is the sum of all the genetic material in any organism; the precise sequence of the four component chemicals $\{A, C, G, T\}$ determines who we are as well as how we function. Each human cell has 46 chromosomes, 23 from each parent, and genes are precise sequences of $\{A, C, G, T\}$ arrayed at definite sites or *loci* on chromosomes. $DNA$ is the carrier of genetic information; it is double helix model, made up of the four nucleotides where $A$ pairs with $T$ and $G$ with $C$; $A$ and $G$ are purines while $C$ and $T$ are pyrimidines. Like the $DNA$, $RNA$ and *proteins* are macromolecules of a cell, though differ in their forms and constitution; $RNA$ is single-stranded (with $T$ replaced by $U$), and proteins are complex chains of 20 amino acids that carry out tasks necessary for life, while *enzymes* are proteins that take other proteins apart or put them together. Both $RNA$ and proteins are made from instructions in the $DNA$, and new $DNA$ molecules are made from copying existing $DNA$ molecules (in a process called synthesis). $DNA$ has a nearly constant diameter with regularly spaced and repeated structures, irrespective of the base composition or the order of the four bases. Recent researches in human genome analysis have revealed that each gene produces, on an average, three proteins, and sometimes, as many as five.

The *Central Dogma* states that once (genetic) information passes into protein, it can not get out again. The transfer of information from nucleic acid to nucleic acid, or from nucleic acid to protein is possible, but transfer from protein to protein or protein to nucleic acid is not possible. The loop from $DNA$ to $DNA$ is called replication, from $DNA$ to $RNA$ is called transcription, and the loop from $RNA$ to protein is called translation. $RNA$ that is translated into protein is called messenger $RNA$ or $_mRNA$, and the transfer $RNA$ or $_tRNA$ translates the genetic code into amino acids. The central dogma has been extended in later years; *retroviruses* can copy their $RNA$ genomes into $DNA$ by a mechanism called *reverse transcription*.

If we accept the basic hypothesis that $DNA$ is the blueprint for a living organism then it becomes natural to conclude that *molecular evolution* is directly related to changes in $DNA$; during the course of molecular evolution, *substitutions* occur. Recall that $A - T$ and $G - C$ are pairs formed by hydrogen bonds. As such, the substitution $A \leftrightarrow G$ or $C \leftrightarrow T$ are called transitions, while $A \leftrightarrow C$, $A \leftrightarrow T$, $G \leftrightarrow C$, $G \leftrightarrow T$ are called transversions. Thus, two purins or two pyrimidines are said to differ by a transition while a purine and a pyrimidine are said to differ by a transversion. Also, it may be noted that amino acids are encoded by triplets of nucleotides of $DNA$ called *codons*. Let us define $\mathcal{N}_R = \{A, C, G, U\}$ as the set of nucleic acids, and let

$$\mathcal{C} = \{(x_1, x_2, x_3) : x_j \in \mathcal{N}_R, \ j = 1, 2, 3\} \tag{1}$$

be the codon. Finally, let $\mathcal{A}$ be the set of aminoacids and termination codons. Then the *genetic code* can be defined as a map:

$$g : \mathcal{C} \to \mathcal{A}, \ g \in \mathcal{G}. \tag{2}$$

Thus, $\mathcal{G}$ is the set of all genetic codes.

Stochastic evolutionary forces act on genomes (molecular evolution). Probability models have been advocated recently by a host of researchers; we refer to Ewens and Grant (2001) for a up-to-date account. However, genes are not simple: the very high dimensionality and yet unknown nature of the battery of activities, specially in the evolutionary phase, have created an enormous task for molecular biologists and geneticists in the years to come. With the nearly completed picture of the human genome project, there are other formidable statistical tasks too.

It is also wise to heed to the pharmacologic undercurrents in bioinformatics. Pharmacology is the science of drugs including materia medica, toxicology and therapeutics, dealing with the properties and reactions of drugs, specially with relation to their therapeutic values. Phracokinetics relates to the study of the bodily absorption, distribution, metabolism, and excretion of drugs, and pharmacodynamics deals with reactions between drugs and living structures. Pharmacogenetics deals with the study of interrelation of heredity constitution and response to drugs. In quest of complex disease gene discovery, and drug-development and therapy, study of gene-environment interaction is therefore essential to incorporate vital information; pharmacogenomics is, in that way, an essential component of bioinformatics.

In some related matters which were once thought of as purely EBD and health outcome phenomena, genomics is providing us with more pertinent genetic information that could bring an evolutionary change in clinical practice and medical diagnostics. As a result, HRQoL studies are likely to be highly affected by such striking research findings. As two most notable cases, we refer to the *Malaria* and *Inflamatory Bowel Disease* (IBD); the malaria episode in a genomic version has been reported in the leading magazines *Science* and *Nature*, both in October 2002, while the IBD was reported in in Nature in May 2001. Malaria is the third most infectious disease, ranked only next to HIV/AIDS and tuberculosis. Researchers have mapped the genes of the parasite that causes malaria and the mosquito that spreads it. A mosquito is a keypart of the three-stage life-cycle of the malaria parasite. A female A. gambiae requires blood meal to mature its eggs. An insect infected with the malaria injects the parasite into a human when it sucks up blood. The parasite invades first the lever and then the blood cells. When another mosquito bites parasites transfer into the new insect, which then bites another human, and the cycle begins anew. In nature, researchers reported identity of nearly 5300 genes distributed across the 24 million base pairs of DNA that make up the malaria genome. In the mosquito, the science researchers found about 14,000 genes among the 278 million pairs of DNA. This striking discovery raises a number of queries that merit serious statistical appraisal. For example, whether or not the malaria parasite is injected into human blood everytime a mosquito bites, or the hitting probability depends on the colony-size of the parasites in the mosquito? Secondly, if a person has already acquired the malaria parasite through previous mosquito bites, does does a new bitting change his/her immunity status to such parasite invasion? What can be said about the human immunosystem to defend against such parasite invasion? How external drugs react to strengthen the defense line of the immunosystem against malaria parasite injection? What can be said about the (malaria) gene-environment interaction (in terms of the breeding of mosquitos and the prevalance of the malaria parasite)? It is quite clear that statistical studies related to such complex biostochastics problems require a good deal of biological, clinical as well as genetic information.

The IBD or the Crohn's disease is a chronic inflamation that shreds the lining of the digestive track - an inherited form of the Crohn's disease. It was supposed to be mainly due to *environmental burden of disease* (EBD), food habits, hygenic conditions, familial factors etc.. French and US scientists have reported (in May 2001) in *Nature* that a *mutation* in the gene on *chromosome* 16 increases a person's chances for developing IBD by at least twenty-five percent. This finding involved (i) mapping of disease genes, (ii) mutation, and (iii) susceptibility and chances, which all have significant stochastic undercurrents. Therefore, no matter whether or not genomics suggests suitable bioinformatic clues, they need to be followed up by sound statistical modeling and analysis protocols.

We may also draw our attention to some other biostochastics models where there might not be so much emphasis on gentics under currents. Among such models, special mention may be made of (i) SARI (structure-activity relationship information) in environmental toxicology (Sen 2003) and (ii) Neuronal spatio-temporal spike train models (Sen 2002a). Both of these have significant molecular biological components (and genetic under currents may not be ruled out), though they have somewhat different data models and different resolutions. In (i), slow-progression toxico-chemical contamination (structure) in bodily intake leads to biomolecular activity where toxicodynamics and toxicokinetics (TDTK) need to be properly incorporated in physiologically based pharmacokinetics (PBPK) models. A greater part of the problem is due to inadequate means of measuring such environmental toxicants and pollutants, as well as their bioconcentration factors. Moreover, there could be a multitude of such environmental factors which in a synergic mode can relate to a multitude of disease and disorders, not all of which may be known.

The biomolecular activity process is also obscured due to extreme variability in human metabolism, immunity, and their exposure patterns to such toxicants, and yet the stochastics are overwhelming. In (ii), though it might be appealing to use point processes for modeling and statistical analysis of spike trains, the primary roadblock stems from the enormously large number of neurons (nerve cells), packed densely, with diverse activities, their inhomogeneity and unknown spatial dependence patterns. On top of that the experimental process of observing the response pattern from the neurons might be destructive. Thus, a dimension reduction technique is needed on one hand to record the response pattern from a gridded set and functional data analysis on the other hand to tie-up the gridded data sets with functional models for the entire cortex. Such models are not to be confused with neural *network models* which are in use in *cognitive sciences (artificial* or *machine intelligence)*. Statistical learning or KDDM are also somewhat related though more heavily computationally oriented.

The main characteristics of biostochastics are presented as follows:

(1) Very high-dimensional data models.

(2) Significant spatio-temporal patterns.

(3) Lack of homogeneity and stationarity.

(4) Spatial/temporal topology not always properly defined.

(5) Discrete, count, and often purely categorical data models.

(6) Sans multinormality assumption, standard multivariate analysis and related GLLM may not be that appropriate.

(6) Variogram, serial correlations etc., may not be usable.

(7) Detection and elimination of outliers could be big problem.

(8) Change-point problems are pertinent but their formulation could be very complex.

(9) Likelihoods are to be critically appraised in favor of alternatives among which nonparametrics is highly visible. Though the adoption of nonparametrics might not be routine.

(10) There is a genuine need to assess amenability of of biological factors to adopted statistical resolutions.

Some of these are elaborated in the next section, albeit tailored to some simple examples.

## 3. Likelihoods and their Limitations

Conventional likelihood function based statistical inference procedures, though appropriate, may encounter prohibitively laborious computational complexities as well as lack of robustness prospects in large parameter space models. In genomics we typically have data sets on a large number ($K$) of sites or positions, where in each site, there is a purely qualitative ( i.e., categorical) response with 4 to 20 categories depending on the $DNA$ or the protein sequence. Neither these sites can be taken to be stochastically independent nor their spatial dependence or association pattern may be precisely known. Moreover, as typically $K$ is large, there are roadblocks to implementing simple patterns in this complex setup. On the other hand, the embedded variability in these responses and the nearly identical structures of the $DNA$ molecules suggest that alternative variational studies should be more appropriate from statistical modeling and analysis perspectives. For example, in the context of judging whether or not mutations at different sites take place independently of each other, consider a reduction in modeling based on the count of whether or not there is a mutation in position $j$ at a given time, for $j = 1, \ldots, K$. If we let $Y_j = 1$ or $0$, according as there is a mutation or not in position $j (= 1, \ldots, K)$, and define the stochastic vector $\mathbf{Y} = (Y_1, \ldots, Y_K)'$, then the probability law of $\mathbf{Y}$ is defined on $\Omega = \{(i_1, \ldots, i_K) : i_j = 0, 1; j = 1, \ldots, K\}$ (whose cardinality is $2^K$). Even for moderately large values of $K$, the number of parameters ($2^K - 1$) associated with this probability law is quite large, and for such a large parameter space, conventional likelihood approaches stumble into computational as well as conceptual impasses. Replacement of the classical likelihood function by suitable conditional, partial, profile, pseudo or quasi-likelihood functions may generally lead to more severe nonrobustness properties of associated statistical tests and estimates.

If we let $P(\mathbf{y}) = P\{\mathbf{Y} = \mathbf{y}\}$, $\mathbf{y} \in \Omega$, and define

$$Q(\mathbf{y}) = \log\{P(\mathbf{y})/P(\mathbf{0})\}, \quad \mathbf{y} \in \Omega \tag{1}$$

then, we obtain by routine computation that

$$P(\mathbf{y}) = e^{Q(\mathbf{y})} / \sum_{\mathbf{z} \in \Omega} e^{Q(\mathbf{z})}. \tag{2}$$

One can then use the Bahadur (1961) representation of multivariate dichotomous random variables (see Liang, Zeger and Qaqish (1992) for further extensions), and write

$$\begin{aligned} P(\mathbf{y}) &= \exp\Big\{\sum_{k=1}^{K} u_k y_k + \sum_{1 \le s < k \le K} y_s y_k u_{sk} \\ &\quad + \ldots + y_1 \cdots y_K u_{1\ldots K}\Big\} \end{aligned} \tag{3}$$

where the $u_k$ are the conditional *logits*, $u_{sk}$ are the conditional *log odd-ratio* etc. In this setup, if we assume further that the $u_{i_1 \ldots i_l}$ for $l \geq 3$ are all null, we end-up with a pairwise dependence model:

$$Q(\mathbf{y}) = \sum_{k=1}^{K} \alpha_k y_k + \sum_{1 \le s < k \le K} \gamma_{sk} y_s y_k, \tag{4}$$

wherein the $\alpha_k$ and $\gamma_{sk}$ represent the main effect and first-order interactions.

This pairwise dependence model has been incorporated in a *pseudolikelihood function* approach, albeit with insufficient theoretical justifications without a conditional, partial, or even profile likelihood interpretation (Besag 1974). The specific form of this function (based on $n$ independent sequences $\mathbf{Y}_1, \ldots, \mathbf{Y}_n$) is given by

$$\prod_{i=1}^{n} \prod_{k=1}^{K} \left\{ \frac{e^{Y_{ik}(\alpha_k + \sum_{j=1}^{K} \gamma_{kj} Y_{ij})}}{1 + e^{Y_{ik}(\alpha_k + \sum_{j=1}^{K} \gamma_{kj} Y_{ij})}} \right\}; \tag{5}$$

this model is also termed the *autologistic model*. Although, highly computational incentive Markov chain monte carlo (MCMC) methods can be prescribed for finding the maximum pseudo-likelihood estimator (MPLE) of the associated parameters, their robustness and efficiency properties may not be usually tenable, and moreover, simulated likelihood ratio techniques using the Gibbs sampling or the Metropolis-Hastings algorithm may stumble into computational roadblocks when $K$ is large (compared to $n$), as may be the case in the present context. We refer to Sen (2002) for some discussion.

As mentioned before, in genomic sequence analysis, we have sequences of data sets on a large number of sites, and we may like to know about their interrelations as well as possible lack of homogeneity over different groups of subjects; this is known as *computational sequence analysis* (CSA). Often, we need to test for homogeneity of $G(\geq 2)$ independent groups of sequences, each group having in turn a number of presumably independent sequences. For this external CSA problem, analogous to the classical multivariate analysis of variance (MANOVA) proble, several approaches have been advocated (Sen 2001). First, we may consider the classical likelihood ratio type test. But, in view of the unassessed nature of the total (or conditional, or partial) likelihood function, such a procedure is difficult to formulate. As such, we may consider the autologistic model, as described above, where for the $g$th group, we denote the associated parameters by $\boldsymbol{\theta}_g = (\boldsymbol{\alpha}_g', \boldsymbol{\gamma}_g')'$, for $g = 1, \ldots, G$. Based on the PMLE of the $\boldsymbol{\theta}_g$ one may consider then a Wald-type test. But, it could be quite cumbersome to obtain a good estimator of the (asymptotic) covariance matrices of the MPLE (specially when $K$ is large), and this in turn may require a very large sample size (compared to $K$ and $G$) in order that asymptotics may yield reasonably good approximations. Moreover, no (asymptotic) optiality properties for such tests have been precisely formulated, nor they are likely to be true. For this reason, it might be more attractive to use suitable aligned scores statistics (Sen 2002), albeit in a permutation model to come up with more robust and reasonably simple tests for homogeneity. Empirical Bayes and hierarchical Bayes procedures have also been advocated. These procedures also depend on suitable likelihood formulations, and in addition, on the choice of appropriate priors on the associated parameters. There is still ample room for further developments in this direction.

As a second example, we consider a *quantitative trait loci* (QTL) model involving a (large) number of loci and quantitative phenotypic observable variables. It is typically assumed that the observable random variable (say, $Y$) depends on a number of loci with quantitative traits (say $Q_1, \ldots, Q_m$) and possibly under the surveillance of genetic markers at some of these loci, and in view of the multiplicity of the traits and markers, it is generally taken for granted that $Y$ has a (mixed) normal distribution, given these extraneous variables; since some of these are not observable, there is a need to use suitable conditional normal laws, and on integrating on the unobservable variables one can then arrive at the appropriate likelihood function. No matter how we proceed, there could be a very large number of parameters associated with such a likelihood formulation. Although EM algorithm can be used for computational facilities, there remains the basic concern: For a large parameter space with (moderately large) $n$, number of observations, what could be said about properties of derived maximum likelihood estimators? Such estimates are generally not efficient (even asymptotically), are biased, and in some extreme cases, may even be inconsistent; the Neyman-Scott problem is a glaring example of this type. The same criticism may be labelled against the likelihood ratio (or allied Wald-type or Rao's scores) tests. A greater concern is how robust would be a likelihood based test or estimate for model departures (e.g., contamination by heavy tail distribution to assumed normal ones)? As additional examples, we may consider any other problem that crops up in testing for independence of mutations at multiple sites, genetic mapping of disease genes, gene-environment interaction, and other problems referred to in the preceding section. Although, for most of these genetic models, some (pseudo-)likelihood formulations have been advocated, they are generally far from being ideal from statistical modeling and analysis perspectives. For lack of space, we shall not enter into detailed discussions on individual problems; rather, in the next section, we shall discuss some alternative procedures that attach less emphasis on likelihood formulations and more on suiable nonparametrics.

## 4. Whither Nonparametrics

In order to illustrate the relative merits and demerits of likelihood based approaches and some alternative ones, let us consider the genomic (external) CSA problem, treated in the preceding section. Instead of binary response variables, treated there, we consider here a more general model that arises in genomics. We consider $K$ positions or sites where at each position the response variable is purely qualitative with $C$ possible outcomes, so that we have a full model involving $C^K$ possible response vectors. Thus essentially, we are to test for the homogeneity of $G$ high-dimensional contingency tables. In a single-site model, this reduces to a conventional $C$-category multinomial law, so that we have a (categorical) CATANOVA model, treated nicely by Light and Margolin (1971, 1974), and more elaborately in the genomic context by Pinheiro et al. (2000), where the *Hamming distance* in a simple formulation has been exploited. We consider here the general case, and formulate some alternative procedures.

Let $\mathbf{X}_i = (X_{i1}, \ldots, X_{iK})'$ be a random vector where the coordinate $X_{ik}$ stands for the category outcome $c(=1, \ldots, C)$ for the $i$th sequence at site $k$, $i = 1, \ldots, n$; $k = 1, \ldots, K$. Note that the responses are purely categorical variables, and hence, conventional norms or distances may not be usable here. For a pair $(i, i')$ of sequences with responses $\mathbf{X}_i$ and $\mathbf{X}_{i'}$ respectively, the Hamming distance is defined as

$$D_{ii'} = \frac{1}{K} \sum_{k=1}^{K} I(X_{ik} \neq X_{i'k}). \tag{1}$$

This leads us to the sample measure

$$\bar{D}_n = \binom{n}{2}^{-1} \sum_{1 \leq i < i' \leq n} D_{ii'}, \tag{2}$$

which is a $U$-statistic (Hoeffding, 1948) with a kernel of degree 2. Note that this formulation does not assume that the coordinates of the $\mathbf{X}_i$ are all stochastically independent.

Moreover, $\bar{D}_n$, being a $U$-statistic, unbiasedly estimates

$$\Delta_H = \frac{1}{K}\sum_{k=1}^K P\{X_{ik} \neq X_{i'k}\}$$
$$= \frac{1}{K}\sum_{k=1}^K I_{GS}^{(k)}, \tag{3}$$

where $I_{GS}^{(k)}$ is the Gini-Simpson index of biodiversity, as adapted in the present context (Pinheiro et al. 2000).

Consider now $G$ independent groups, where the $g$th group consists of $n_g$ independent sequences $\mathbf{X}_{gi}$, $i = 1, \ldots, n_g$, for $g = 1, \ldots, G$. We let $n = \sum_{g=1}^G n_g$. For the $g$th group we denote the sample and population Hamming distances as $\bar{D}_{n,g}$ and $\Delta_{H,g}$ respectively, for $g = 1, \ldots, G$. Also, we denote the pooled sample measure by $\bar{D}_n$ and its population counterpart as $\bar{\Delta}_H$. Basically, we are interested in testing the homogeneity of the $\Delta_{H,g}$ based on their sample counterparts $\bar{D}_{n,g}$ which are all $U$-statistics. Pinheiro et al. (2000) followed the conventional ANOVA approach based on $U$-statistics and their estimated variance-covariance, with the main emphasis on the partition of the total sum of squares into within and between groups components. Since in the present context, we do not have a (generalized) linear model, their suggested test procedure encounters some complex distributional problems, even under the hypothesis of homogeneity. Note that for each $k (= 1, \ldots, K)$ and $g (= 1, \ldots, G)$, $X_{gi}^{(k)}$ has a probability law on the $C$-simplex, and on top of that, for different $k$, the $X_{gi}^{(k)}$ are not necessarily independent. This renders some degeneracy in the null hypothesis distribution theory, and instead of anticipated asymptotic chi-squared distribution, we end up with more complex Cramér - von Mises type distributions involving a linear combination of independent chi-squared variables with one degree of freedom. It is to be noted that the pooled sample $\bar{D}_n$ can be decomposed into two nonnegative quantities, representing the between and within group components (see Chatterjee and Sen (2000) for some more general results), so that one may compare the between groups component with the within group component to draw statistical conclusions. We end up with similar Cramér - von Mises type distributions. We refer to Pinheiro et al. (2003) for some details. Basically, in this setup, there is no need to exploit the second order moments of the two components of $\bar{D}_n$ that are themselves nonnegative and represent the 'within group' and 'between group' parts. The asymptotic distribution theory resembles the von Mises- Cramér type laws, referred to above, and there is some scope for permutation tests, similar to the ones to be considered now.

Let us consider here two related tests based on the within group estimates $\bar{G}_{n,g}$, $g = 1, \ldots, G$), under appropriate regularity assumptions. First, we consider here a test for the homogeneity of the $\Delta_{H,g}$, in a little bit more stringent form of homogeneity of the $G$ probability laws defined on the same probability space; in this setup, of course the alternative hypotheses relate to the part of the parameter space where the $\Delta_{H,g}$ are not the same. Recall that under this null hypothesis, all the $G$ probability laws are the same (though unknown), so that conventional permutation tests can be formulated. We could generate all possible $N = \{(n!)/\prod_{g=1}^G (n_g)!\}$ partitioning of the $n$ sequences (in the pooled group) into $G$ subsets of sizes $n_1, \ldots, n_G$ respectively, and under the null hypothesis, all these permutations are equally likely; we denote this conditional (permutational) probability measure by $\mathcal{P}_n$. It is then easy to verify that for each $g(=1,\ldots,G)$,

$$E\{\bar{D}_{n,g}|\mathcal{P}_n\} = \bar{D}_n, \tag{4}$$

which is the same measure based on the pooled set,

$$Var\{\bar{D}_{n,g}|\mathcal{P}_n\} = \frac{4(n_g-2)(n-n_g)}{n_g(n_g-1)(n-3)}\zeta_{1,n} + \binom{n_g}{2}^{-1}[1 - \frac{\binom{n_g-2}{2}}{\binom{n-2}{2}}]\zeta_{2,n}, \tag{5}$$

and for $g \neq h(=1,\ldots,G)$,

$$Cov\{\bar{D}_{n,g}, \bar{D}_{n,h}|\mathcal{P}_n\} = \frac{-4}{n-3}\zeta_{1,n} - \binom{n}{2}^{-1}\zeta_{2,n}, \tag{6}$$

where

$$\zeta_{1,n} = \frac{1}{n(n-1)(n-2)} \sum_{1 \leq i \neq j \neq l \leq n} D_{ij}D_{il} - \bar{D}_n^2,$$

$$\zeta_{2,n} = \binom{n}{2}^{-1} \sum_{1 \leq i < j \leq n} D_{ij}^2 - \bar{D}_n^2, \tag{7}$$

and the $D_{ij}$ refer to the Hamming distance between the $i$th and $j$th observations in the pooled sample. A little more algebraic manipulations lead us to conclude that

$$Cov\{\bar{D}_{n,g}, \bar{D}_{n,h}|\mathcal{P}_n\} = \frac{4(n\delta_{gh} - n_g)}{n_g(n-3)}\zeta_{1,n} + O(n_g^{-2}), \tag{8}$$

for $g, h = 1, \ldots, G$, where $\delta_{gh}$, the Kronecker delta, is equal to 1 or 0 according as $g = h$ or not. All these results follow from the general theory of $U$-statistics under simple random sampling without replacement scheme (from the finite population of size $n$), and a detailed account of this computation and its approximations may be found in Nandi and Sen (1963) This permutation variance can be well approximated by its jackknifed version. Based on the above computation we consider a quadratic form in the $\bar{D}_{n,g}$ with their permutation means and variance-covariance terms, and arrive at the following test statistic:

$$\mathcal{L}_n^* = [\sum_{g=1}^{G}(n_g-1)\{\bar{D}_{n_g} - \bar{D}_n\}^2]/4\zeta_{1,n}. \tag{9}$$

Note that $\zeta_{1,n}$, the pooled sequence estimator, is invariant under any permutation of the $n$ sequences among themselves, and so is $\bar{D}_n$. On the other hand, for each of the $N$ permuted set, we could compute the corresponding values of $\bar{D}_{n,g}, g = 1, \ldots, G$, and hence, generate a version of $\mathcal{L}_n^*$. With all these $N$ realizations, one could order them and generate the excact permutation distribution of $\mathcal{L}_n$. Though this procedure works out well for small values of the $n_g$, it becomes prohibitively laborious as the $n_g$ become large. Fortunately, using the results of Sen (1981, Ch. 3), it is possible to verify that under the permutation law, asymptotically, $\mathcal{L}_n^*$ has the central chi square distribution with $G-1$ degrees of freedom (DF). This suggests that even if we do not use the permutation principle, the test statistic $\mathcal{L}_n^*$ can be used in an asymptotic setup for testing the homogeneity of the $G$ groups with respect to the parameters $\Delta_{H_g}$.

Next, we note that the $\Delta_{H,g}$ may be equal without requiring that the $G$ probability laws are all the same. Hence, it may be more desirable to formulate suitable tests for the homogeneity of the $\Delta_{H,g}$ without assuming the homogeneity of the underlying probability laws. Since the within group measures $\bar{D}_{n,g}$ are all $U$-statistics, we could do it by estimating consistently the variance of each measure, and exploiting the asymptotic normality of the estimators $\bar{D}_{n,g}$. Toward this end, we let $\mathbf{X}_{gi} = (X_{gi}^{(1)}, \ldots, X_{gi}^{(K)})'$, $i = 1, \ldots, n_g; g = 1, \ldots, G$. Let then

$$\pi_g(k,c) = P(X_{gi}^{(k)} = c); \quad \pi_g(k,c;l,d) = P(X_{gi}^{(k)} = c, X_{gi}^{(l)} = d). \tag{10}$$

for $c(\neq d) = 1, \ldots, C$, $k(\neq l) = 1, \ldots, K$, $g = 1, \ldots, G$. Let then

$$\gamma_g^2 = \frac{1}{K^2} \sum_{k=1}^{K} \sum_{l=1}^{K} \sum_{c=1}^{C} \sum_{d=1}^{C} \pi_g(k,c)\pi_g(l,d)\{\pi_g(k,c;l,d) - \pi_g(k,c)\pi_g(l,d)\}. \tag{11}$$

for $g = 1, \ldots, G$. Then using the Hoeffding projection result, it follows (Sen 1981, Ch.3) that as $n_g$ increases,

$$\sqrt{n_g}(\bar{D}_{n,g} - \Delta_{H,g}) \stackrel{\mathcal{D}}{\Rightarrow} \mathcal{N}(0, 4\gamma_g^2), \tag{12}$$

for each $g(=1, \ldots, G)$. Note that even under the null hypothesis, the $\gamma_g^2$ may not be all equal. However, for each $g$, we may estimate $\gamma_g^2$ by using the same formula as in $\zeta_{1,n}$ (in the pooled sample case, treated earlier) but solely using the $n_g$ sequences in the group $g$; we denote these estimators by $\hat{\gamma}_g^2$, $g = 1, \ldots, G$. Let then

$$\hat{D}_n^* = \frac{\sum_{g=1}^{G} n_g \bar{D}_{n,g}/\hat{\gamma}_g^2}{\sum_{g=1}^{G} n_g/\hat{\gamma}_g^2}. \tag{13}$$

Let us then define

$$\mathcal{L}_n^{o*} = \sum_{g=1}^{G} \frac{n_g}{\hat{\gamma}_g^2} \{\bar{D}_{n,g} - \bar{D}_n^*\}^2. \tag{14}$$

It follows from the Cochran theorem (along with the Slutzky theorem) (cf. Sen and Singer 1993, Ch.3 ) that under the null hypothesis of homogeneity of the $\Delta_{H,g}$, $\mathcal{L}_n^{o*}$ has asymptotically chi square distribution with $G - 1$ degrees of freedom (DF). This is then used in the determination of the asymptotic critical level for the test based on the test statistic $\mathcal{L}_n^{o*}$. Note that when the $\gamma_g^2$ are not all the same, the test statistic $\mathcal{L}_n^*$ may not have asymptotically chi-square distribution with $G - 1$ DF (but a Cramér - von Mises type distribution), so that it might not have the correct (asymptotic) significance level, although both the tests will be consistent against any possible heterogeneity of the $\Delta_{H,g}$.

There has been a systematic development in statistical genetics dealing with quantitative traits (Lange 1997). Even if such quantitative trait models are conceived in bioinformatics, there remains clouds over the plausible dimensionality of the traits, as well as, the validity of the conventional multinormality assumption (sans which all statistical models in use at the present time may lose their rationality and validity). It may be quite appealing to adopt suitable location-scale family of distributions, and to examine, to what extent, the robustness of assumed normal models for such alternative situations. Again, the high dimensionality of the model arising in bioinformatics can create a complete impasse for such multinormality based statistical approaches. In this context too, there is a genuine need to explore alternative nonparametric approches that are more viable under the biological and genetic setups. Formulation of such alternative approaches, though is a challenging task, should be given due consideration.

As has been mentioned before pharmacogenetics occupies a very focal point in bioinformatics, the main impetus being the tremendous scope for genomics in drug research and marketing perspectives. The principal difficulty in the implementation of phramacogenomics stems from the fact that experimental evidence to justify clinical conclusions often precludes human subject (due to medical ethics and basic clinical considerations). In the name of KDDM, simulation studies have therefore permeated the drug research arena. These ventures are mostly based on some algorithms, and often, without much statistical insights. In such a case, it might be wistful to conceive of simpler probabilistic models where such conventional algorithms can be validly and efficiently applied with full likelihood appreciation. It would be therefore more appropriate to conceive of alternative statistical approaches that put lesser emphasis on likelihoods and greater emphasis on alternative reasoning that adapts well to the biological and genetic explanations. Nonparametrics seems to have advantages in this setup, and we advocate the use of biologically motivated, genetics based, nonparametrics in bioinformatics. It is a challenge, and with the steady flow of research in nonparametrics in all its horizons, the success should be in the reach of statisticians' theory and methodology basket.

# References

Bahadur, R. R. (1961). A representation of the joint distribution of responses to $n$ dichotomous items. In *Studies in Item Analysis and Prediction* (ed. H. Solomon), Stanford Univ. Press, Calif, pp. 158 - 176.

Besag, J. (1974). Spatial interaction and the statistical analysis of life systems (with discussion). *J. Roy. Statist. Soc. B48*, 192 - 236.

Chatterjee, S. K. and Sen, P. K. (2000). On stochastic ordering and Gini-Simpson type poverty indexes. *Calcutta Statist. Assoc. Bull. 50*, 137 - 156.

Durbin, R., Eddy, S., Krogh, A. and Mitchison, G. (1988). *Biological Sequence Analysis: Probabilistic Models for Proteins and Nucleic Acids.* Cambridge Univ. Press, UK.

Ewens, W. J. and Grant, G. R. (2001). *Statistical Methods in Bioinformatics: An Introduction.* Springer-Verlag, New York.

Geyer, C. J. and Thompson, E. A. (1995). Annealing Markov chain monte carlo with applications to ancestral inference. *J. Amer. Statist. Assoc. 90*, 909-920.

Hastie, T., Tibshirani, R. and Friedman, J. (2001). *The Elements of Statistical Learning: Data Mining, Inference, and Prediction.* Springer- Verlag, New York.

Hoeffding, W. (1948). A class of statistics with asymptotically normal distribution. *Ann. Math. Statist. 19*, 293 - 325.

International Human Genome Sequencing Consortium (2001). Initial sequencing and analysis of the human genome. *Nature 409*, 860 - 921.

Karnoub, M. C., Seillier- Moiseiwitsch, F., and Sen, P. K. (1999). A conditional approach to the detection of correlated mutations. *Inst. Math. Statist. Lect. Notes Monogr. Ser. 33*, 221 - 235.

Lange, K. (1997). *Mathematical and Statistical Methods for Genetic Analysis.* Springer-Verlag, New York.

Liang, K., Zeger, S. L. and Qaqish, B. (1992). Multivariate regression analysis for categorical data (with discussion). *J. Roy. Statist. Soc. B 54*, 3 - 40.

Light, R. H. and Margilin, B. H. (1971). An analysis of variance for categorical data. *J. Amer. Statist. Assoc. 66*, 534 - 544.

Light, R. H. and Margolin, B. H. (1974). An analysis of variance for categorical data, II : Small sample comparisons with chi square and other competitors. *J. Amer. Statist. Assoc. 69*, 755-764.

Nandi, H. K. and Sen, P. K. (1963). On the properties of $U$-statistics when the observations are not independent. Part Two: Unbised estimation of the parameters of a finite population. *Calcutta Statist. Assoc. Bull. 12*, 124-148.

Pinheiro, H., Seillier-Moiseiwitsch, F., Sen, P. K. and Eron, J. (2000). Genomic sequence analysis and quasi-multivariate CATANOVA. In *Handbook of Statistics, Volume 18: Bioenvironmental and Public Health Statistics* (eds. P. K. Sen and C. R. Rao), Elsevier, Amsterdam, pp. 713-746.

Pinheiro, H., Pinheiro, A. and Sen, P. K. (2003). Comparison of genomic sequences using the Hamming distance. *Journal of Statistical Planning and Inference*, to appear.

Sen, P. K. (1981). *Sequential Nonparametrics: Invariance Principles and Statistical Inference.* John Wiley, New York.

Sen, P. K. (2001). *Ecxursions in Biostochastics: Biometry to Biostatics to Bioinformatics*, Lecture Notes, Inst. Statist. Sci. Academia Sinica, Taipei, Taiwan.

Sen, P. K. (2002a). Neuronal spatio-temporal models: High- dimensional implications and statistical perspectives. *Scientie Mathematicae Japonicae 56*, 613-648.

Sen, P. K. (2002b). Computational sequence analysis: Genomics and statistical controversies. In *Recent Advances in Statistical Methods* (ed. Y. P. Chaubey), Impirial College press, London, pp. 274-289.

Sen, P. K. (2003). Structure activity relationship information incorporation in environmental risk assessment. *Environmetrics 14*, 223-234.

Sen, P. K. and Singer, J. M. (1993). *Large Sample Methods in Statistics: An Introduction with Applications.* Chapman-Hall, UK.

Simpson, E. H. (1949). The measurement of diversity. *Nature 163*, 688.

Speed, T. and Waterman, M. S. (eds.) (1996). *Genetic Mapping and DNA Sequencing.* Springer-Verlag, New York.

Venter, J. C. et al. (2001). The sequence of the human genome. *Science 291*, 1304 - 1351.

Thompson, E. A. (1986). *Pedigree Analysis in Human Genetics*, Johns Hopkins Press, Baltimore, MD.

Waterman, M. S. (1995) *Introduction to Computational Biology: Maps, Sequences and Genomes.* Chapman-Hall, UK.

Watson, J. D. and Crick, F. H. C. (1953). Genetical implications of the structure of deoxyribonucleic acid. *Nature 171*, 964-967.

# Some Issues Concerning Length-biased Sampling in Survival Analysis

M. Asgharian [a] and David B. Wolfson

[a]Department of mathematics and statistics, McGill University, Burnside Hall, 805 Sherbrooke Street West, Montreal, Quebec, Canada H3A 2K6

When survival data are left truncated and right censored, special methods are required for their analysis. Much of the literature on the analysis of such data has focussed on conditional methods that are required in the so-called "non-stationary" setting; that is the truncation intervals are allowed to have an arbitrary distribution. In the stationary setting, the truncation intervals are assumed, essentially, to be uniformly distributed on some large interval, and this assumption, when appropriate, is accompanied by increased precision in the statistical inference. We present, here, a brief overview of the analysis of left-truncated and right-censored data in the stationary setting, emphasizing the asymptotic behaviour of both non-parametric and parametric maximum likelihood estimators.

## 1. Introduction

In 1991, as part of the Canadian study of Health and Aging, known as CSHA-1, roughly 10,000 subjects over the age of sixty five, were screened for the presence of dementia (Wolfson et al(2001)). The age of onset for each person diagnosed with dementia was obtained through care giver interviews. Those subjects with dementia were then followed until either death or the end of the study in 1996, CSHA-2 , at which point surviving subjects were deemed to have been right censored. There were comparatively few subjects who were lost-to-follow-up. At the time of diagnosis patients were classified into three main categories of dementia: possible Alzheimer's disease, probable Alzheimer's disease and vascular dementia. Principal goals of this part of the CSHA were to nonparametrically estimate the survival function, from onset, of patients with a general diagnosis of dementia, as well as with more specific subdiagnosis of one of the three aforementioned types of dementia. Moreover, it was desired to assess the combined effect of several covariates on survival from onset with dementia. Now, the cross-sectional ascertainment of prevalent cases of dementia with the subsequent follow-up of theses cases resulted in left truncated and right censored survival times from onset

---

*Supported in part by FCAR and NSERC of Canada

(Asgharian et al (2002b)). An additional feature of these data often accompanying survival analyses from cross-sectional follow-up studies, is that the onset times, including those not observed, because their corresponding survival times are left truncated, may be assumed to occur according to a stationary Poisson process. For the CSHA data this assumption is plausible in the absence of any evidence that the incidence rate of dementia has changed over time. For a disease such as AIDS, however, this assumption would, clearly, not be valid.

The lay-out of this paper is as follows. In section 2 we introduce notation and the topics to be discussed in the sequel. In section 3 we outline several different possible sampling schemes that lead to informative censoring, while resulting in the same likelihood, and propose one of these as the most useful in the analysis of survival data. In section 4 we sketch the approach to nonparametric maximum likelihood estimation of the survival function under the preferred sampling scheme, emphasizing the asymptotics of the nonparametric maximum likelihood estimator (NPMLE). In section 5 we introduce covariates, and discuss the asymptotics of maximum likelihood parameter estimators in fully parametric survival models, when there is length-bias. Finally, in section 7 some further lines of research are proposed.

## 2. Preliminaries and Sampling Schemes

When the left truncation time distribution is not specified the approach to estimating the unbiased survivor function is to condition on the observed truncation times (e.g. Wang et al(1986), Tsai et al(1987) and Andersen et al(1993)). However, when there is good reason to assume that the truncation times are independently and uniformly distriubted on some interval (the so called "stationary" assumption, Asgharian et al.(2002b)), an unconditional approach may be prefered as it results in a more efficient estimator of the unbiased survivor function (see Asgharian et al(2002b)). In this paper we focus on the stationary case and the more efficient unconditional estimator. We shall reserve the use of the terminology "length-biased" for the stationary case.

The general approach to the unconditional NPML estimation of the survival function is to first obtain the NPMLE of the length-biased survival function, (e.g. Vardi(1982, 1985) and Gill et al(1988)) and then the NPMLE of the unbiased survival function is a smooth operator, under some general conditions, of the length-biased estimator. Therefore, the asymptotics of the NPMLE of the unbiased survival function heavily depends on the asymptotics of the NPMLE of the length-biased survival function. It should also be emphasized that the asymptotics depend, in turn, on the underlying sampling scheme used to collect the data. We present three different but related sampling schemes and propose, ulimately, that one of these has greater applicability in cross-sectional follow-up studies than the other two. The three sampling schemes are

(i) The number of subjects indentified at the cross-sectional stage is $N$. All subjects are followed for a fixed time period (the end of the study period). Some of these $N$ subjects are censored before the end of the follow-up period, some at the end of this period and the remainder having failed in this time period. It is assumed that the censoring of the forward recurrence times is random. If $M$ denotes the (random) number of uncensored subjects by the end of the study period, then $N - M$ denotes the number of censored observations.

(ii) The scenario is the same as that of (i) except that $M$ and $N - M$ are fixed at the observed values $M = m$ and $N - M = n$ respectively and analyses are carried out conditionally.

(iii) The number of subjects identified at the cross-sectional stage is $N = m + n$. At this stage, a fixed number $n$, are immediately censored, while the remaining $m$ are followed until failure. It is easily seen that this sampling scheme is equivalent to that of multiplicative censoring (Vardi(1989)).

We present, without derivations, the likelihoods under the three sampling schemes (for further details see Asgharian and Wolfson (2001)). We first introduce the notation used in the sequel.

Let $A_i$ and $B_i$ denote, respectively, the time from onset to recruitment, and the time from recruitment to failure for subject $i$. In renewal process terminology these are known as the backward and forward recurrence times respectively. The data associated with subject $i$ may be denoted by the triple $(A_i, R_i \wedge C_i, \delta_i)$, $i = 1, 2, \cdots, N$, where $C_i$, is a nonnegative random variable independent of $(A_i, R_i)$, and which right censors the forward recurrence time, $R_i$. As usual the censoring indicator

$$\delta_i = \begin{cases} 1 & \text{if the } i\text{th subject is not censored} \\ 0 & \text{otherwise.} \end{cases}$$

The vectors $(A_i, R_i \wedge C_i, \delta_i)$, $i = 1, 2, \cdots, N$ are assumed to be independent. Let $S_u(F_u)$ and $f_u$ denote the unbiased survival (distribution) function and its density respectively, and let $\mu_u$ denote the unbiased mean. Let $S_{LB}(F_{LB})$ and $f_{LB}$ denote the length-biased survival (distribution) and its density function, respectively. Let $G^*(t) = P(A + R \leq t \mid \delta = 1)$, and its density be $g^*(t)$. Let $F^*(t) = P(A + C \leq t \mid \delta = 0)$, and its density be $f^*(t)$. Let $f_{FR}(t) = \frac{S_u(t)}{\mu_u}$, which is, as is well known, the forward recurrence time density. Finally, let $p = P(\delta = 1) = P(R \leq C)$.

The likelihood for $S_{LB}$ under the three sampling schemes (i), (ii) and (iii) above, may be shown (Asgharian and Wolfson(2001)) to have the following very similar forms. As we shall see shortly, by invariance of the $MLE$ the NPMLE of $S_u$, $\hat{S}_u$, may be obtained explicitly from the NPMLE of $S_{LB}$, $\hat{S}_{LB}$. The three likelihood for $S_{LB}$ are as follows:

*Random Censoring:* ($M$ random. Scheme (i)),

$$L_R \propto \prod_{i=1}^{N} \left(dS_{LB}(x_i)\right)^{\delta_i} \left(\int_{y_i \leq z} z^{-1} dS_{LB}(z)\right)^{1-\delta_i}, \qquad (1)$$

*Random Censoring:* (Conditional on $M$. Scheme (ii)),

$$L_C \propto \Big( \prod_{i=1}^{m} dS_{LB}(x_i) \Big) \Big( \prod_{j=1}^{n} \int_{y_i \leq z} z^{-1} dS_{LB}(z) \Big), \qquad (2)$$

*Multiplicative censoring:* (Scheme (iii)),

$$L_{MC} \propto \Big( \prod_{i=1}^{m} dS_{LB}(x_i) \Big) \Big( \prod_{j=1}^{n} \int_{a_i \leq z} z^{-1} dS_{LB}(z) \Big), \qquad (3)$$

where $x_i = a_i + r_i$ and $y_i = a_i + c_i$.

Now, although the maximum likelihood estimator of the length-biased survival function, are the same under the three sampling schemes (see e.g. Vardi(1989)), their asymptotic sampling properties are not. Indeed, as has been pointed out by Vardi(1989), the asymptotics must be derived afresh each time a new sampling scheme is considered. Vardi(1989) showed how one may implement the EM algorithm to obtain the NPMLE under sampling scheme (iii), that of multiplicative censoring, which as we have said also yields the NPMLE under sampling schemes (i) and (ii). The difficulty in deriving the asymptotics in all three cases is that the censoring is informative, so that random censoring methodologies may not be implemented. The censoring is informative since the potential failure and censoring times have a common truncation interval, which renders them correlated. One way to avoid the informative censoring difficulty is to take a conditional approach (Wang (1991)), although this is not discussed further here. Instead, we concentrate on an unconditional approach and on sampling scheme (i), for the following reasons. It is very rare to condition on the number of censored lifetimes in survival analysis as this sharply curtails the scope of applicability. This renders sampling scheme (ii) inappropriate. Next, in the context of the analysis of survival data collected following a cross-sectional survey, multiplicative censoring (scheme (iii)) would require that a priori, the length-biased survival times of $m$ subjects be fully observed while for the remaining $N - m = n$ subjects, only their backward recurrence times be observed (Vardi and Zhang(1992)). It is difficult to justify a scenario that does not allow for censoring due to "loss-to-follow-up" or to "end-of-study". We are, therefore, left with sampling scheme (i) which assumes random censorship of the forward recurrence times $R_i$.

## 3. The NPMLE of $S_u$ and Its Asymptotics

It is well-known (Cox(1969)) that

$$S_u(x) = \frac{\int_x^{\infty} \frac{1}{y} dS_{LB}(y)}{\int_0^{\infty} \frac{1}{y} dS_{LB}(y)}, \qquad (4)$$

and therefore, by invariance of the MLE, the NPMLE of $S_u$ is given by

$$\hat{S}_u(x) = \frac{\int_x^{\infty} \frac{1}{y} d\hat{S}_{LB}(y)}{\int_0^{\infty} \frac{1}{y} d\hat{S}_{LB}(y)}, \qquad (5)$$

where $\hat{S}_u$ maximizes the likelihood (1). The main result concerning the asymptotics of $\hat{S}_u$ is

**Theorem 1** *Suppose that there is some $\gamma > 0$ such that $S_u(y) = 1$ for $y < \gamma$ and that $\mu_u = \int_0^\infty y dF_u(y)$. Let $\hat{S}_u$ be the unique NPMLE for the unbiased survival function $S_u$. Define*

$$L_y(x) = \frac{I_{[y,\infty]}(x) - S_u(y)}{x},$$

*where $I_A(x)$ is the indicator function of the set A. Then under condition (6) of Asgharian et al (2002b), we have*

$$\sup_{0 \leq y < \infty} |\hat{S}_u(y) - S_u(y)| = O\left(\sqrt{\frac{\log \log N}{N}}\right) \quad a.s. \text{ as } N \to \infty ; \qquad (6)$$

*and*

$$\sqrt{N}\left(\hat{S}_u(y) - S_u(y)\right) \xrightarrow{D} \mu_u \int_0^\infty L_y(x) dU(x) \quad \text{as } N \to \infty, \qquad (7)$$

*on $D_0[0, \infty)$ (the space of cadlag functions on $[0, \infty)$), where $U$ is a Gaussian process (defined in Asgharian et al(2002)), and the process on the right of (6) has covariance function*

$$r(y,z) = \mu_u^2 \int_0^\infty \int_0^\infty \psi(s,t) dL_y(t) dL_y(s),$$

*with $\psi(s,t) = cov(U(s), U(t))$.*

In order to establish Theorem 1, it is clear from (5) that one must first establish the asymptotic properties of $\hat{S}_{LB}$ under the sampling scheme **(i)**. Avoiding a long regress of notation and terminology (the details may be found in Asgharian and Wolfson(2001)) we merely mention that the approach is to begin with the score equation and use it to obtain what might be termed the "master equation":

$$\left(\frac{\hat{p}-p}{1-p}\right)\left[U_N(t) - \hat{f}_{FR}(t) \int_{0 < y \leq t} y \int_{y \leq z} \frac{U_N(z)}{Z^2} dz d\frac{1}{\hat{f}_{FR}(y)}\right]$$

$$+(1-p)\hat{f}_{FR}(t) \int_{0 < y \leq t} \frac{yf^*(y)}{f_{FR}(y)} \int_{y \leq z} \frac{U_N(z)}{z^2} dz d\frac{1}{\hat{f}_{FR}(y)}$$

$$+p\left(\frac{1-\hat{p}}{1-p}\right) \int_{0 < y \leq t} \frac{g^*(y)}{f_{LB}(y)} dU_N(y)$$

$$= W_{m,n}(t) + p\sqrt{\frac{p}{1-p}}\left(G^*(t) - F_{LB}(t)\right)\sqrt{N}\frac{(\hat{p}-p)}{\sqrt{p(1-p)}} \qquad (8)$$

where $m = \sum_{i=1}^N \delta_i$, $n = N - m$, $\hat{p} = \frac{m}{n}$, $U_N = \sqrt{N}\left(\hat{S}_{LB} - S_{LB}\right)$, and $W_{m,n}(t)$ is a function of the empirical processes corresponding to the uncensored and censored observations. Under sampling scheme **(i)**, $m$ and $n$ are, of course, random, and as we let $N \to \infty$, the randomness of $m$ and $n$ must be taken into account. This contrasts with the derivation of the asymptotics under sampling scheme **(iii)** by Vardi and Zhang(1992), under which $m$ and $n$ are constants.

## 4. The Introduction of Covariates

In the study of the natural history of a disease there is currently no satisfactory means of using Cox's proportional hazards model, when the observed survival times are left truncated and right censored. In particular, the assumptions that must be made are unreasonable (e.g. Wang et al(1993)) and the asymptotics have yet to be developed. For example, Wang(1989) only considers truncated observations, while Wang(1996) does not allow for censoring. Therefore, our position is that parametric models can be important in survival analysis particularly when the data are length-biased and right censored, and when covariates must be included. Now, since we are forced to resort to parametric models, their use is only justified if the asymptotic properties of the MLE's can be established. We believe that the asymptotics are crucial for the use of these parametric models.

A general approach is to begin by proposing a regression model with covariates for the unbiased model so that the covariate effects, if any, may easily be interpreted. Next, transform this model to accommodate the length-biased data that will actually be observed. The likelihood based on the length-biased data may then be maximized with respect to all of the finitely many original model parameters, including the regression parameters. Of course, in the length-biased model, the parameters will, in general, be difficult to interpret. However, since the parameters "began their lives" in a model in which they are easy to interpret, the apparent difficulty is not real. It should be pointed out that there is no advantage to begin by conveniently modeling the length-biased data; for then the parameters will lose their interpretability in the dual unbiased model, which is the one of ultimate interest. The above discussion uncovers the difficulty with using Cox's semi-parameteric proportional hazards model; if one begins with Cox's model for the unbiased distribution (as one should), proportionality and its accompanying conveniences are lost when the change is made to length-biased form. On the other hand, starting with Cox's model for the length-biased distribution, although convenient for parametric estimation via partial likelihood methods, leads to uninterpretable parameters in the unbiased distribution.

In principle, if one accepts the utility of parametric models used as described above, there are three main steps: (a) propose a model for the unbiased distribution, and transform it to length-biased form, (b) obtain the MLE's of the finite set of parameters, and (c) use these parameter estimates to carry out statistical inference. Step (c) requires that one establishes the asymptotic properties of the MLE's for left truncated right censored data. In the uncoditional approach proposed here the induced informative censoring renders blind invocation of standard maximum likelihood theory inappropriate. We therefore sketch an approach that shows that even in this non-standard setting the MLE's are consistent and asymptotically Normal.

Consider the observed data as $Y_i = (A_i, R_i \wedge C_i, z_i, \delta_i)$, $i = 1, 2, \cdots, N$, where $z_i$ is a $1 \times p$ vector of covariates for subject $i$. This representation for $Y_i$ replaces the conventional $(X_i, z_i, \delta_i)$, where $X_i = A_i + R_i \wedge C_i$. We shall assume, as in pervious sections, that $C_i$ and $(A_i, R_i)$, $i = 1, 2, \cdots, N$ are independent conditional on $z_i$. The $Y_i$'s are assumed to be independent. As is the case with nonparametric likelihood estimation the difficulty caused by the informative censoring is circumvented by considering the forward recurrence times, $R_i$, and their censoring times, $C_i$, in isolation, rather in the combined form of $X_i$.

The likelihood under sampling scheme **(i)** (Asgharian et al (2002b)) then has the parametric form,

$$l(\theta) = \prod_{i=1}^{N} l_i(\theta, u_i, v_i, z_i) \tag{9}$$

where

$$l_i(\theta_i, u_i, v_i, z_i) = \Big(\frac{f_U(u_i; \theta, z_i)}{\mu_U(\theta, z_i)}\Big)^{\delta_i} \Big(\int_{\omega \geq v_i} \frac{f_U(\omega; \theta, z_i)}{\mu_U(\theta, z_i)} d\omega\Big)^{1-\delta_i}, \tag{10}$$

where $\theta$ is a $1 \times q$ ($q > p$) vector of model parameters including the regression parameters of the $p$ covariates $z_{i_1}, z_{i_2}, \cdots, z_{i_p}$, and where $u_i = a_i + r_i$ and $v_i = a_i + c_i$. Under conditions given by Asgharian et al(2002a) we have

**Theorem 2** *Let $\hat{\theta}_N$ be the MLE of $\theta$. Then $\hat{\theta}_N \to \theta$ almost surely as $N \to \infty$.*

The main idea of the proof is as follows. Define the likelihood ratio,

$$\lambda_N(\xi) = \prod_{i=1}^{N} \frac{l_i(\theta + \xi; u_i, v_i, z_i, \delta_i)}{l_i(\theta; u_i, v_i, z_i, \delta_i)}$$

where $\xi$ is located in a small sphere, $\Gamma$, in $\mathcal{U} = \times -\theta$, assuming that $\theta$, the true parameter value, belongs to the parameter space $\Theta$. Show that with probability approaching zero, $\sup_{\xi \in \Gamma} \lambda_N(\xi) \geq 1$. Using a compactness argument show that $\sup_{\xi \in \mathcal{U}} \lambda_N(\xi) \geq 1$ with probability approaching zero as $N \to \infty$. Finally, using an argument similar to that of Ibragimov and Ha'sminskii (1981, page 36) conclude that $\hat{\theta} \to \theta$ with probability 1 as $N \to \infty$.

Next, let $U_i = A_i + R_i$ and $V_i = A_i + T_i$. Then under conditions given by Asgharian et al(2002) we have

**Theorem 3** *Let $\hat{\theta}_N$ be the MLE of $\theta_0 \in \Theta$. Then*

$$\sqrt{N}(\hat{\theta}_N - \theta_0) \xrightarrow{\mathcal{D}} N(0, I_{q \times q}^{-1}),$$

$I = I^1 + I^2$, $I^l = [\lim_{n \to \infty} \frac{1}{n} \sum_{i=1}^{n} I_{rs}^{l,i}]_{rs=1,2,\cdots,q}$ *for $l = 1, 2$ and*

$$I_{rs}^{1,i} = \mathrm{cov}\Big(\delta \frac{\partial}{\partial \theta_r} \log f_{LB}(U_i; z_i, \theta_0), \delta \frac{\partial}{\partial \theta_s} \log f_{LB}(U_i; z_i, \theta_0)\Big) \tag{11}$$

$$= E\Big[\delta\Big(\frac{\partial}{\partial \theta_r} \log f_{LB}(U_i; z_i, \theta_0)\Big)\Big(\frac{\partial}{\partial \theta_s} \log f_{LB}(U_i; z_i, \theta_0)\Big)\Big] \tag{12}$$

and

$$I_{rs}^{2,i} = \text{cov}\Big((1-\delta)\frac{\partial}{\partial \theta_r}\log f_{FR}(V_i;z_i,\theta_0), (1-\delta)\frac{\partial}{\partial \theta_s}\log f_{FR}(V_i;z_i,\theta_0)\Big) \quad (13)$$

$$= E\Big[(1-\delta)\big(\frac{\partial}{\partial \theta_r}\log f_{FR}(V_i;z_i,\theta_0)\big)\big(\frac{\partial}{\partial \theta_s}\log f_{FR}(V_i;z_i,\theta_0)\big)\Big]. \quad (14)$$

$r, s = 1, ..., q$.

The proof of Theorem 3 depends on first order Taylor expansions of the score function of $f_{LB}$ and of $f_{FR}$ respectively. The details may be found in Asgharian et al(2002a).

## 5. Concluding Remarks

**1.** It is interesting to note that the informative censoring induced by sampling scheme (i) allows for consistency of the MLE even if all observations are censored. This observation contrasts with the failure of consistency of the MLE when the censoring is random. Roughly, when there is informative censoring, censored observations provide some direct information on the failure time, even though it is not observed.

**2.** The conditions refered to for Theorem 2 and Theorem 3 must be checked for each parametric family separately. They are certainly satisfied though, for the Weibull family of distributions, the most common parametric family used in survival analysis, when the covariates can take on at most a finite set of values and are introduced as is usually the case, as a regression model through the hazard.

**3.** The unconditional (on the truncation times) approach discussed here has the advantage over the conditional approach (Wang(1991)) of being more efficient when the assumption of stationarity is satisfied (see Asgharian et al (2002a)). The conditional approach, however, enjoys greater flexibility, in that the distribution of the truncation times can be arbitrary.

**4.** In the study of the natural history of a disease there is currently no satisfactory means of using Cox's proportional hazards model, when the observed survival times are left truncated and right censored. In particular, the assumption that must be made are untenable and the asymptotics have yet to be developed. This remains a major challenge in the analysis of length-biased right censored data.

## REFERENCES

1. Andersen, P. K., Borgan, Ø., Gill, R. D. and Keiding, N. (1993). *Statistical models based on counting processes*. Springer Series in Statistics. Springer-Verlag.
2. Asgharian, M. and Wolfson, David B. (2001). Asymptotic behaviour of the $NPMLE$ of the survivor function when the data are length-biased and subject to right censoring. *McGill University Department of Mathematics and Statistics, Tehnical Report No. 2001-01*.

3. Asgharian, M., Wolfson, David B. and Cyr Emile M'Lan (2002a). Inference Based on Cross-Sectional Sampling for Diseases with Stationary Incidence. *McGill University Department of Mathematics and Statistics, Tehnical Report No. 2002-01.*
4. Asgharian, M., Cyr Emile M'Lan and Wolfson, David B. (2002b). Length-Biased Sampling With Right Censoring: An Unconditional Approach. *The Journal of the American Statistical Association*, Vol. 97, No. 457, pp 201-209.
5. Cox, D. R. (1969). Some sampling problems in technology. In *New Developments in Survey Sampling*. Edited by Johnson & Smith. Wiley.
6. Gill, R. D., Vardi, Y. and Wellner, J. A. (1988). Large sample theory of empirical distributions in biased sampling models. *Annals of Statistics* 16: 1069–1112.
7. Ibragimov, I.A. and Has'minskii, R.Z.(1981). *Statistical Estimation: Asymptotic Theory*. Springer-Verlag, New York.
8. Tsai, W.-Y., Jewell, N. P. and Wang, M.-C. (1987). A note on the product-limit estimator under right censoring and left truncation. *Biometrika* 74: 883–886.
9. Vardi, Y. (1982). Nonparametric estimation in the presence of length bias. *The Annals of Statistics* 10: 616–620.
10. Vardi, Y. (1985). Empirical distributions in selection bias models. *The Annals of Statistics* 13: 178–205.
11. Vardi, Y. (1989). Multiplicative censoring, renewal processes, deconvolution and decreasing density. *Biometrika* 76: 751–761.
12. Vardi, Y. and Zhang, C.-H. (1992). Large sample study of empirical distributions in a random-multiplicative censoring model. *The Annals of Statistics* 25: 1022–1039.
13. Wang, M.-C. (1996). Hazards regression analysis for length-biased data. *Biometrika* 83: 343–354.
14. Wang, M.-C. (1991). Nonparametric estimation from cross-sectional survival data. *Journal of the American Statistical Association* 86: 130–143.
15. Wang, M.-C. (1989). A semi-parametric model for randomly truncated data. *Journal of the American Statistical Association* 84: 742–748.
16. Wang, M.-C., Brookmeyer, R. and Jewell, N. P. (1993). Statistical models for prevalent cohort data. *Biometrics* 49: 1–11.
17. Wang, M.-C., Jewell, N. P. and Tsai, W.-Y. (1986). Asymptotic properties of the product limit estimator under random truncation. *The Annals of Statistics* 14: 1597–1605.
18. Wolfson, C., Wolfson, D., Asgharian, M., M'Lan, C-E., Østbye, T., Rockwood, K. and Hogan, M., for the Clinical Progression of Dementia Study Group. (2001). A re-evaluation of the duration of survival after the onset of dementia. *The New England Journal of Medicine* 344-15: 1111–1116.

*Recent Advances and Trends in Nonparametric Statistics*
Michael G. Akritas and Dimitris N. Politis (Editors)
© 2003 Elsevier Science B.V. All rights reserved.

# Covariate Centering and Scaling in Varying-Coefficient Regression with Application to Longitudinal Growth Studies

Colin O. Wu[a] * Kai F. Yu[b] and Vivian W. S. Yuan[c]

[a]Office of Biostatistics Research, DECA,
National Heart Lung and Blood Institute, DHHS, Bethesda, MD 20892, U.S.A.

[b]Division of Epidemiology, Statistics and Prevention Research
National Institute of Child Health and Human Development, DHHS,
Bethesda, MD 20852, U.S.A.

[c]Allied Technologies Group, Inc., Bethesda, MD 20852, U.S.A

We propose a class of time-varying coefficient models with centered or standardized (both centered and scaled) covariates, discuss the practical benefits of covariate transformation, and develop a comprehensive set of nonparametric estimation and confidence procedures based on these models. We demonstrate the practical properties of our methods through a fetal growth study and a simulation. Our results suggest that covariate centering or standardization can lead to superior statistical estimates and better biological interpretations.

## 1. INTRODUCTION

Recently there has been an increased interest in modeling longitudinal data using nonparametric regression. This type of data include repeatedly measured responses over time and either time-dependent or time-invariant covariates of interest. In practice, the observed time points are often irregularly spaced. A typical example is the Alabama Small-for-Gestational-Age Study (ASGA Study) which involves 1475 women who were repeatedly observed over time during their pregnancy. An important objective of this study is to assess the associations between maternal risk factors and fetal growth patterns. Although the visits were scheduled at 17, 25, 31 and 36 weeks of gestation, the actual numbers of visits ranged from 1 to 7 and the time of measurements was scattered throughout 12 to 43 weeks. Figure 1 shows the scatter plot of the observed fetal abdominal circumferences at their corresponding gestational age in weeks. Although a general linear growth pattern can be observed for the fetal abdominal circumference, the effects of other risk factors on the growth of fetal abdominal circumference are not seen from these graphs.

Let $T$ be a real-valued variable of time, and $Y(t)$ and $\mathbf{X}(t) = (X_0(t), X_1(t), \ldots, X_k(t))^T$

---

*Partial support for Colin O. Wu was provided by the National Science Foundation, DMS-0103832, while he was with the Johns Hopkins University.

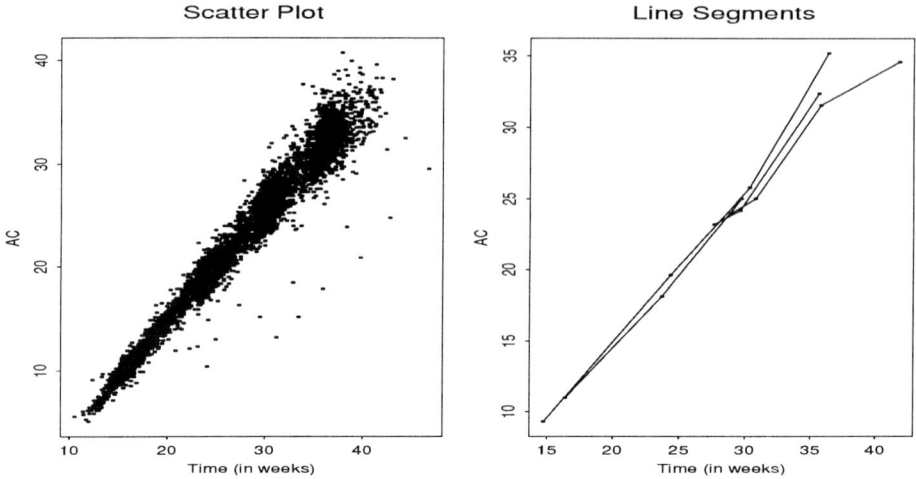

Figure 1. The left panel gives the observed fetal abdominal circumferences (in centimeters). The right panel shows the sequences of the fetal abdominal circumferences for four randomly chosen fetuses.

be a real-valued response and an $R^{k+1}$-valued covariate vector, respectively, at time $t$. The $j$th measurement of $(Y(T), \mathbf{X}(T), T)$ for the $i$th of the $n$ independent subjects is $(Y_{ij}, \mathbf{X}_{ij}, t_{ij})$, where $t_{ij}$ is the measurement time, and $Y_{ij} = Y_i(t_{ij})$ and $\mathbf{X}_{ij} = (X_{0ij}, \ldots, X_{kij})^T$ are the observed response and covariates at time $t_{ij}$. As in most regression models, we set $X_0(t) = X_{0ij} \equiv 1$.

We present in this paper a nonparametric approach for modeling and estimating the time-dependent covariate effects with either the original or some transformed covariates based on the varying-coefficient models. Two popular covariate transformations that are discussed here are covariate centering and standardization: Let $\mu_l(t) = E[X_l(T)|T=t]$ and $\sigma_l^2(t) = \text{var}[X_l(T)|T=t]$ be the mean and variance of $X_l(T)$ given $T=t$. The centered and the standardized transformations of $X_l(t)$ are then $X_{l(c)}(t) = \{X_l(t) - \mu_l(t)\}$ and $X_{l(s)}(t) = \{X_{l(c)}(t)/\sigma_l(t)\}$, respectively. When the original covariates are used, our models are the same as those discussed in [8,5,1], among others. When the transformed covariates are used, our models and procedures may lead to better biological interpretations and superior estimators. In particular, our application to the ASGA Study and simulation show that, when the covariates strongly depend on time, the local linear estimators or kernel estimators with covariate centering are more stable and have smaller biases than the kernel estimators without covariate centering. We present our models in Section 2; describe our estimation methods in Section 3; and propose a set of procedures for constructing pointwise and simultaneous confidence bands in Section 4. We illustrate the application of our procedures in Section 5, and present a simulation study in Section 6.

## 2. THE VARYING-COEFFICIENT MODELS

As a generalization of the linear and partially linear models [13], a time-varying coefficient model with the original covariates $\mathbf{X}(t)$ is

$$Y(t) = \sum_{l=0}^{k} \{X_l(t)\beta_l(t)\} + \epsilon(t), \qquad (1)$$

where $\beta_l(t)$, $l = 0, \ldots, k$, are smooth functions of $t$ and $\epsilon(t)$ is a mean zero stochastic process with $E[\epsilon^2(t)] = \sigma_\epsilon^2(t) < \infty$. Since $X_0(t) \equiv 1$, the baseline curve $\beta_0(t)$ represents the mean of $Y(t)$ when the covariates $X_l(t)$, $l = 1, \ldots, k$, are zero. The coefficient curves $\beta_l(t)$, $l = 1, \ldots, k$, describe the marginal effects of the corresponding covariates. Let $\epsilon_{ij} = \epsilon_i(t_{ij})$ be the $i$th subject's error at time $t_{ij}$. If $\epsilon_{ij}$ is the sum of a random-effect and a measurement error, then (1) reduces to a nonparametric mixed-effects model [10]. Since the marginal effects are of primary interest in this paper, we do not impose further structures on $\epsilon(t)$, except assuming that it is a mean zero stochastic process with finite second moment.

There are two potential disadvantages associated with the model (1). First, the baseline curve $\beta_0(t)$ may not have a meaningful biological interpretation when some of the covariates in $\{X_l(t)\}_{1 \le l \le k}$ can not be zero. Second, as discussed in [12], smoothing estimators based on the popular local least squares, particularly the kernel estimators, may not have desirable theoretical and practical properties when the covariates strongly depend on the time $t$.

If we substitute $(X_{m+1}(t), \ldots, X_k(t))$ for some $0 \le m \le k-1$ with their centered version $(X_{(m+1)(c)}(t), \ldots, X_{k(c)}(t))$, then (1) is equivalent to

$$Y(t) = \beta_{0(c)}(t) + \sum_{l=1}^{m} \{X_l(t)\beta_l(t)\} + \sum_{l^*=m+1}^{k} \{X_{l^*(c)}(t)\beta_{l^*}(t)\} + \epsilon(t), \qquad (2)$$

where the baseline curve $\beta_{0(c)}(t) = \beta_0(t) + \sum_{l^*=m+1}^{k} \{\mu_{l^*}(t)\beta_{l^*}(t)\}$ is the mean of $Y(t)$ when $X_1(t), \ldots, X_m(t)$ are zero while $X_{m+1}(t), \ldots, X_k(t)$ take values at their corresponding local averages $\mu_{m+1}(t), \ldots, \mu_k(t)$.

The coefficient curves for either (1) or (2), however, do not take the local variances of the covariates into account. If, for some $l^*$ with $(m+1) \le l^* \le k$, the effect of $X_{l^*}(t)$ on $Y(t)$ is described through the mean change of $Y(t)$ associated with a unit increment on the standardized value of $X_{l^*}(t)$, we can model the relationship between $Y(t)$ and $(t, X_1(t), \ldots, X_k(t))$ through

$$Y(t) = \beta_{0(s)}(t) + \sum_{l=1}^{m} \{X_l(t)\beta_l(t)\} + \sum_{l^*=m+1}^{k} \{X_{l^*(s)}(t)\beta_{l^*(s)}(t)\} + \delta(t) \qquad (3)$$

with $\delta(t)$ a mean zero stochastic process. Now the baseline curve, $\beta_{0(s)}(t)$, is the mean of $Y(t)$ when the values of $X_l(t)$ and $X_{l^*(s)}(t)$ for all $l = 1, \ldots, m$ and $l^* = m+1, \ldots, k$

are zero. The coefficient curve $\beta_{l^*(s)}(t)$ is the mean change of $Y(t)$ associated with a unit increment on $X_{l^*(s)}(t)$.

The interpretation of (3) is very different from the two previous models. In (1) and (2), a covariate effect is described through the mean change of the response at time $t$ when the covariate gains one unit increment. By associating the mean changes of the response with the increments on the standardized units of $X_{l^*}(t)$, the coefficient curves $\beta_{l^*(s)}(t)$ of (3) have taken the local variations of the corresponding covariates into account. When $X_{l^*}(t)$, $l^* = m+1, \ldots, k$, are heterosdecastic, (3) is a viable and often more realistic alternative to (1) and (2).

## 3. ESTIMATION METHODS

Estimators for the coefficient curves of (1) can be constructed based on a number of smoothing methods, each with its own advantages and disadvantages in practice. Two popular smoothing methods in the literature are the local linear and kernel methods [4]. If, for all $l = 0, \ldots, k$, $\beta_l(t)$ can be approximated by the Taylor expansion

$$\beta_l(t_{ij}) \approx \beta_l(t) + \beta_l'(t)(t_{ij} - t), \qquad (4)$$

the least squares local linear estimators are obtained by minimizing

$$\sum_{i=1}^{n} \sum_{j=1}^{n_i} w_i \left[ Y_{ij} - \sum_{l=0}^{k} (a_l(t) X_{lij}) - \sum_{l=0}^{k} (b_l(t)(t_{ij} - t) X_{lij}) \right]^2 K_h(t_{ij} - t) \qquad (5)$$

with respect to $a_l(\cdot)$ and $b_l(\cdot)$, where $w_i$ are non-negative weight functions, $h$ is a positive bandwidth, $K_h(u) = K(u/h)/h$ and $K(u)$ is a kernel function, usually taken to be a probability density function. Practical choices of $w_i$ usually include $w_i = 1/N$ and $w_i = 1/(nn_i)$. The unique minimizers $\hat{a}_l(t;h)$ and $\hat{b}_l(t;h)$ of (5), when exist, are the local linear estimators of $\beta_l(t)$ and $\beta_l'(t)$, respectively.

To give an explicit expression for the above estimators, we define

$$\mathbf{U}_i(t) = \begin{pmatrix} 1 & (t_{i1}-t) & X_{1i1} & X_{1i1}(t_{i1}-t) & \cdots & X_{ki1} & X_{ki1}(t_{i1}-t) \\ \vdots & \vdots & \vdots & \vdots & & \vdots & \vdots \\ 1 & (t_{in_i}-t) & X_{1in_i} & X_{1in_i}(t_{in_i}-t) & \cdots & X_{kin_i} & X_{kin_i}(t_{in_i}-t) \end{pmatrix},$$

$Y_i = (Y_{i1}, \ldots, Y_{in_i})^T$, $\mathbf{K}_i(t;h)$ to be the $n_i \times n_i$ diagonal matrix whose $j$th diagonal element is $w_i K_h(t - t_{ij})$ and $A(t) = (a_0(t), b_0(t), a_1(t), b_1(t), \ldots, a_k(t), b_k(t))^T$. The optimization problem in (5) is equivalent to minimizing

$$\sum_{i=1}^{n} \left\{ (Y_i - \mathbf{U}_i(t)A(t))^T \mathbf{K}_i(t;h) (Y_i - \mathbf{U}_i(t)A(t)) \right\} \qquad (6)$$

with respect to $A(t)$. If $\sum_{i=1}^{n} [\mathbf{U}_i^T(t)\mathbf{K}_i(t;h)\mathbf{U}_i(t)]$ is invertible, (6) is minimized by

$$\widehat{A}(t;h) = \left\{ \sum_{i=1}^{n} \left[ \mathbf{U}_i^T(t)\mathbf{K}_i(t;h)\mathbf{U}_i(t) \right] \right\}^{-1} \left\{ \sum_{i=1}^{n} \left[ \mathbf{U}_i^T(t)\mathbf{K}_i(t;h)Y_i \right] \right\}. \qquad (7)$$

Let $e_q$ be the $2(k+1)$ column vector with 1 being its $q$th element and zero elsewhere. The local linear estimators of $\beta_l(t)$ and $\beta_l'(t)$ are, respectively,

$$\widehat{a}_l(t;h) = e_{2l+1}^T \widehat{A}(t;h) \quad \text{and} \quad \widehat{b}_l(t;h) = e_{2l+2}^T \widehat{A}(t;h). \tag{8}$$

Since (8) uses the same bandwidth for all $l = 0, \ldots, k$, it may be unable to simultaneously provide adequate smoothings for all the coefficient curves. This motivates the use of possibly different bandwidths $h_l$ and $h_l^*$ in $\widehat{a}_l(t;h_l)$ and $\widehat{b}_l(t;h_l^*)$ for each $l = 0, \ldots, k$.

Let $\mathbf{X}_i$ be the $(k+1) \times n_i$ matrix of covariates whose $j$th row is $(1, X_{1ij}, \ldots, X_{kij})$. If $\sum_{i=1}^n [\mathbf{X}_i^T \mathbf{K}_i(t;h) \mathbf{X}_i]$ is invertible, the kernel estimator of $(\beta_0(t), \ldots, \beta_k(t))^T$, which minimizes $\sum_{i=1}^n \{(Y_i - \mathbf{X}_i \boldsymbol{\beta}(t))^T \mathbf{K}_i(t;h)(Y_i - \mathbf{X}_i \boldsymbol{\beta}(t))\}$, is

$$\widehat{\boldsymbol{\beta}}(t;h) = \left\{ \sum_{i=1}^n \mathbf{X}_i^T \mathbf{K}_i(t;h) \mathbf{X}_i \right\}^{-1} \left\{ \sum_{i=1}^n \mathbf{X}_i^T \mathbf{K}_i(t;h) Y_i \right\}. \tag{9}$$

Similarly, by using a specific bandwidth for each coefficient curve, $\widehat{\beta}_l(t;h_l)$ is a least squares kernel estimator of $\beta_l(t)$.

If we replace $X_{l^*ij(c)}$ and $X_{l^*ij(s)}$ with their estimates, the above smoothing methods can be modified to estimate the coefficient curves in (2) and (3). Using the approximation $\mu_l(t_{ij}) \approx \mu_l(t) + \mu_l'(t)(t_{ij} - t)$, $\mu_l(t)$ can be estimated by the local linear estimator $\widehat{\mu}_{l0}(t;h)$, such that $(\widehat{\mu}_{l0}(t;h), \widehat{\mu}_{l1}(t;h))$ minimizes

$$\sum_{i=1}^n \sum_{j=1}^{n_i} \left\{ w_i \left[ X_{lij} - \mu_{l0}(t) - \mu_{l1}(t)(t_{ij} - t) \right]^2 K_h(t - t_{ij}) \right\}.$$

The local linear estimator of $X_{l^*ij(c)}$ is $\widehat{X}_{l^*ij(c)}^{linear} = X_{l^*ij} - \widehat{\mu}_{l^*0}(t_{ij})$. The local linear estimators $\widehat{a}_{l(c)}(t;h_l)$ and $\widehat{b}_{l(c)}(t;h_l)$ of $\beta_{l(c)}(t)$ and $\beta_{l(c)}'(t)$ are computed by substituting $X_{l^*ij}$ in (8) with $\widehat{X}_{l^*ij(c)}^{linear}$.

To estimate the local variance $\sigma_{l^*}^2(t)$ of $X_{l^*}(t)$, we smooth the squares of the residuals $r_{l^*ij} = X_{l^*ij} - \widehat{\mu}_{l^*0}(t)$. Let $\widehat{V}_{l^*0}(t)$ be the local linear estimator of $\sigma_{l^*}^2(t)$ and $\widehat{\sigma}_{l^*0}(t) = \{\widehat{V}_{l^*0}(t)\}^{1/2}$ be the estimator of $\sigma_{l^*}(t)$. The local linear estimator of $X_{l^*ij(s)}$ is $\widehat{X}_{l^*ij(s)}^{linear} = \widehat{X}_{l^*ij(c)}^{linear}/\widehat{\sigma}_{l^*0}(t_{ij})$. Substituting $X_{l^*ij}$ of (8) with $\widehat{X}_{l^*ij(s)}^{linear}$, we obtain the local linear estimators $\widehat{a}_{l(s)}(t;h_l)$ of $\beta_{l(s)}(t)$. Similarly, we can construct the kernel versions $\widehat{X}_{l^*ij(c)}^{kernel}$ and $\widehat{X}_{l^*ij(s)}^{kernel}$, and obtain the kernel estimators $\widehat{\beta}_{l^*(c)}(t;h_l)$ and $\widehat{\beta}_{l^*(s)}(t;h_l)$ by substituting the $X_{l^*ij}$ in (9) with $\widehat{X}_{l^*ij(c)}^{kernel}$ and $\widehat{X}_{l^*ij(s)}^{kernel}$, respectively.

## 4. INFERENCES

### 4.1. Asymptotic Intervals for Kernel Method

Under some mild conditions and the $w_i = 1/N$ weight, [11] investigated the asymptotic distributions for the kernel estimator defined in (9). Their results give the explicit expressions for the asymptotic bias and variance, $B_l(t)$ and $D_l(t)$, of $\widehat{\beta}_l(t;h_l)$, and show that, as $n \to \infty$,

$$(Nh)^{-1/2} \left[ \widehat{\beta}_l(t;h) - \beta_l(t) \right] \to \mathcal{N}(B_l(t), D_l(t)) \tag{10}$$

in distribution. These authors also proposed some consistent kernel estimators $\widehat{B}_l(t)$ and $\widehat{D}_l(t)$ for $B_l(t)$ and $D_l(t)$, and suggested to use the approximate $(1-\alpha)$ conficence interval

$$\left(\widehat{L}_\alpha(t), \widehat{U}_\alpha(t)\right) = \left[\widehat{\beta}_l(t;h) - (Nh)^{-1/2}\widehat{B}_l(t)\right] \pm Z_{\alpha/2}(Nh)^{-1/2}\left(\widehat{D}_l(t)\right)^{1/2}, \qquad (11)$$

where $Z_{\alpha/2}$ is the $(1-\alpha/2)$th quantile of the standard normal distribution. Since $B_l(t)$ depends on some unknown functions, such as the derivatives of $\beta_l(t)$ and the density of the time points, which are often difficult to estimate, an alternative, perhaps more practical, approach is to make the bias term asymptotically negligible relative to the variance by using a relatively small bandwidth $h_l$, so that an approximate $(1-\alpha)$ confidence interval for $\beta_l(t)$ without the bias adjustment is

$$\left(\widehat{L}_\alpha(t), \widehat{U}_\alpha(t)\right) = \widehat{\beta}_l(t;h) \pm Z_{\alpha/2}(Nh)^{-1/2}\left(\widehat{D}_l(t)\right)^{1/2}. \qquad (12)$$

[11] showed that both (11) and (12) had satisfactory coverage probabilities when the covariates were time-independent. However, extensions to the local linear estimators or the models with centered or standardized covariates have not been established, as their asymptotic distributions and consistent estimators for the unknown biases and variances have not been developed.

### 4.2. Bootstrap Variability Bands

A more general inference approach for longitudinal data is the "resampling-subject" bootstrap [8]. Let $\beta_{l(*)}(t)$, $l = 0, \ldots, k$, be the coefficient curves defined in either (1), (2) or (3) and $\widehat{\beta}_{l(*)}(t)$ be the estimator of $\beta_{l(*)}(t)$ based on any of the smoothing methods described in Section 2. A percentile variability band for $\beta_{l(*)}(t)$ can be constructed by the following bootstrap procedure:

*Step 1.* Resample $n$ subjects randomly with replacement from the original data, and denote the bootstrap sample by $\{(Y_{ij}^*, \mathbf{X}_{ij}^*, t_{ij}^*); \, i = 1, \ldots, n, \, j = 1, \ldots, n_i^*\}$.

*Step 2.* Compute the bootstrap estimator $\widehat{\beta}_{l(*)}^{boot}(t)$ of $\beta_{l(*)}(t)$ based on $\{(Y_{ij}^*, \mathbf{X}_{ij}^*, t_{ij}^*); \, i = 1, \ldots, n, \, j = 1, \ldots, n_i^*\}$.

*Step 3.* Repeat the above steps $B$ times. The approximate $(1-\alpha)$th bootstrap variability band for $\beta_{l(*)}(t)$ is $(L_\alpha^{boot}(t), U_\alpha^{boot}(t))$ with $L_\alpha^{boot}(t)$ and $U_\alpha^{boot}(t)$ being the lower and upper $(\alpha/2)$th percentiles of the $B$ bootstrap estimators.

In general, $(L_\alpha^{boot}(t), U_\alpha^{boot}(t))$ gives an approximate $(1-\alpha)$th confidence interval for the expectation of $\widehat{\beta}_{l(*)}(t)$. When the bias of $\widehat{\beta}_{l(*)}(t)$ is negligible relative to its standard deviation, this variability band gives an adequate confidence interval for $\beta_{l(*)}(t)$.

### 4.3. Simultaneous Bands

Although a number of approaches, such as those based on large sample extreme value Gaussian processes [3], have been used in the literature for constructing nonparametric simultaneous inferences, their extensions to the longitudinal data have been limited by the lack of relevant asymptotic results. A simple and practical idea suggested by [11] is to extend the approaches considered by [9,6,7], among others, to the current longitudinal setting. This approach relies on bridging the gaps between simultaneous pointwise intervals over a set of grid points.

Following [11], we need to assume either (a) $\sup_{t\in[a,b]} |\beta'_{l(*)}(t)| \leq c_1$ or (b) $\sup_{t\in[a,b]} |\beta''_{l(*)}(t)| \leq c_2$ for some known positive constant $c_1$ and $c_2$. A Bonferroni-type confidence band for $\beta_{l(*)}(t)$ over a time interval $t \in [a,b]$ can be constructed as follows:

**Step 1.** Select $M+1$ equally spaced grid points $\{\xi_1, \ldots, \xi_{M+1}\}$ so that $a = \xi_1 < \cdots < \xi_{M+1} = b$ and, for $1 \leq m \leq M$, $\xi_{m+1} - \xi_m = (b-a)/M$.

**Step 2.** Construct a set of $(1-\alpha)$ simultaneous intervals for $\beta_{l(*)}(\xi_m)$, $m = 1, \ldots, M+1$, based on the Bonferroni adjustment

$$\left(\widehat{l}_\alpha(\xi_m), \widehat{u}_\alpha(\xi_m)\right) = \left(L_{\alpha/(M+1)}(\xi_m), U_{\alpha/(M+1)}(\xi_m)\right), \tag{13}$$

where $(L_\alpha(t), U_\alpha(t))$ is a $(1-\alpha)$ confidence interval for $\beta_{l(*)}(t)$.

**Step 3.** Construct a $(1-\alpha)$ confidence band $(\widehat{l}^{(I)}_\alpha(t), \widehat{u}^{(I)}_\alpha(t))$ for the linear interpolation

$$\beta^{(I)}_{l(*)}(t) = \left\{\frac{M(\xi_{m+1}-t)}{b-a}\right\}\beta_{l(*)}(\xi_m) + \left\{\frac{M(t-\xi_m)}{b-a}\right\}\beta_{l(*)}(\xi_{m+1}) \tag{14}$$

for $\xi_m \leq t \leq \xi_{m+1}$ through the linear interpolations of $\widehat{l}_\alpha(t)$ and $\widehat{u}_\alpha(t)$.

**Step 4.** The $(1-\alpha)$ simultaneous bands for $\beta_{l(*)}(t)$ are, if (a) is satisfied,

$$\left(\widehat{l}^{(I)}_\alpha(t) - 2c_1\frac{(\xi_{m+1}-t)(t-\xi_m)}{(b-a)/M}, \widehat{u}^{(I)}_\alpha(t) + 2c_1\frac{(\xi_{m+1}-t)(t-\xi_m)}{(b-a)/M}\right) \tag{15}$$

and, if (b) is satisfied,

$$\left(\widehat{l}^{(I)}_\alpha(t) - (c_2/2)(\xi_{m+1}-t)(t-\xi_m), \widehat{u}^{(I)}_\alpha(t) + (c_2/2)(\xi_{m+1}-t)(t-\xi_m)\right). \tag{16}$$

The above confidence bands are derived directly from (13), (14) and

$$\left|\beta_{l(*)}(t) - \beta^{(I)}_{l(*)}(t)\right| \leq \begin{cases} 2c_1[M/(b-a)](\xi_{m+1}-t)(t-\xi_m), & \text{if (a) is satisfied}; \\ (1/2)c_2(\xi_{m+1}-t)(t-\xi_m), & \text{if (b) is satisfied}. \end{cases}$$

The smoothness conditions and their bounds $c_1$ and $c_2$ may be chosen from the prior information or biological interpretations of the coefficient curves. Different choices of $M$ generally lead to bands with different shapes. Optimal choices of $M$, either in theory or in practice, are still unknown. Although these bands are conservative, they are often useful to give an initial assessment of the coefficient curves.

## 5. APPLICATION TO THE ASGA STUDY

Our primary outcome is the fetal abdominal circumference measured at different gestational age. Two important factors that might be associated with fetal growth are the mother's status on cigarette smoking and alcohol consumption. Because the time-varying patterns of smoking and drinking were not observed, we categorized the mother as a non-smoker if she never smoked a cigarette during her pregnancy, and a smoker if she ever smoked cigarettes. Similarly, we categorize a mother as a non- or light-drinker if she consumed less than one beer or one glass of wine per day on average during her pregnancy, and a drinker if she consumed more alcohol daily on average. This makes smoking and drinking binary time-invariant covariates. Other covariates of interest are the mother's size measured by the pre-pregnancy body mass index and the placental size measured by its thickness.

Let $X_1 = 0$ if the mother is a non-smoker, and 1 if she is a smoker; $X_2 = 0$ if she is a non-drinker/light-drinker, and 1 if a heavy-drinker; $X_3$ be the maternal body mass index; and $X_4(t)$ and $Y(t)$ be the placental thickness and the fetal abdominal circumference, respectively, at $t$ weeks of gestation. We did not transform $X_1$ and $X_2$ as they were binary and could be clearly interpreted in a varying-coefficient model. For the transformation of $X_3$, we first estimated its mean and standard devation by the sample mean $\bar{X}_3$ and the sample standard deviation $s(X_3)$, and then estimated the centered and standardized values for the $i$th mother by $\widehat{X}_{3i(c)} = X_{3i} - \bar{X}_3$ and $\widehat{X}_{3i(s)} = \widehat{X}_{3i(c)}/[s(X_3)]$, respectively. For the transformation of $X_4(t)$, we estimated its centered and standardized values at time $t_{ij}$ using the kernel and the local linear methods of Section 3.2.

When the original covariates $X_1$, $X_2$, $X_3$ and $X_4(t)$ are used, the baseline curve $\beta_0(t)$ does not have a practical interpretation, and the $\beta_l(t)$ for $l = 1, \ldots, 4$ represent the change of the mean fetal abdominal circumference when the corresponding covariate has one unit increment. When $X_3$ and $X_4(t)$ are replaced by their centered transformations $X_{3(c)}$ and $X_{4(c)}(t)$, the baseline curve $\beta_{0(c)}(t)$ now represents the mean fetal abdominal circumference at $t$ weeks of gestation for a non-smoking and non- or light-drinking mother with average pre-pregnancy body mass index and average placental thickness at time $t$. For model (3) with the original $X_1$ and $X_2$ and the standardized $X_{3(s)}$ and $X_{4(s)}(t)$, the baseline curve $\beta_{0(s)}(t)$ has the same interpretation as that of $\beta_{0(c)}(t)$, but the coefficient curves $\beta_{3(s)}(t)$ and $\beta_{4(s)}(t)$ now represent the mean change of the fetal abdominal circumference when $X_3$ and $X_4(t)$ have one unit increase in their corresponding standard deviations. Clearly, model (2) is more desirable than (1), but neither (2) or (3) are uniformly superior to the other.

To compare the practical properties between different estimation procedures, we fitted all three models with the kernel and the local linear procedures using the Epanechnikov kernel, the bandwidth 1.50 and the cross-validated bandwidths based on the "deleting-subject" cross-validation of [8]. For both the kernel and the local linear estimators, the cross-validated bandwidths are approximately 0.45. Figure 2 shows the estimators of $\beta_0(t)$, $\beta_3(t)$ and $\beta_4(t)$ from 15 to 36 weeks of gestation (growth patterns at the early and later stages of pregnancy may be interpreted differently), and their pointwise 95% bootstrap percentile intervals computed based on the procedure of Section 3.2 with 500 bootstrap replications. Since both kernel and local linear fittings showed non-significant effects for cigarette smoking and alcohol consumption, the estimates for $\beta_1(t)$ and $\beta_2(t)$ were omitted from the figure.

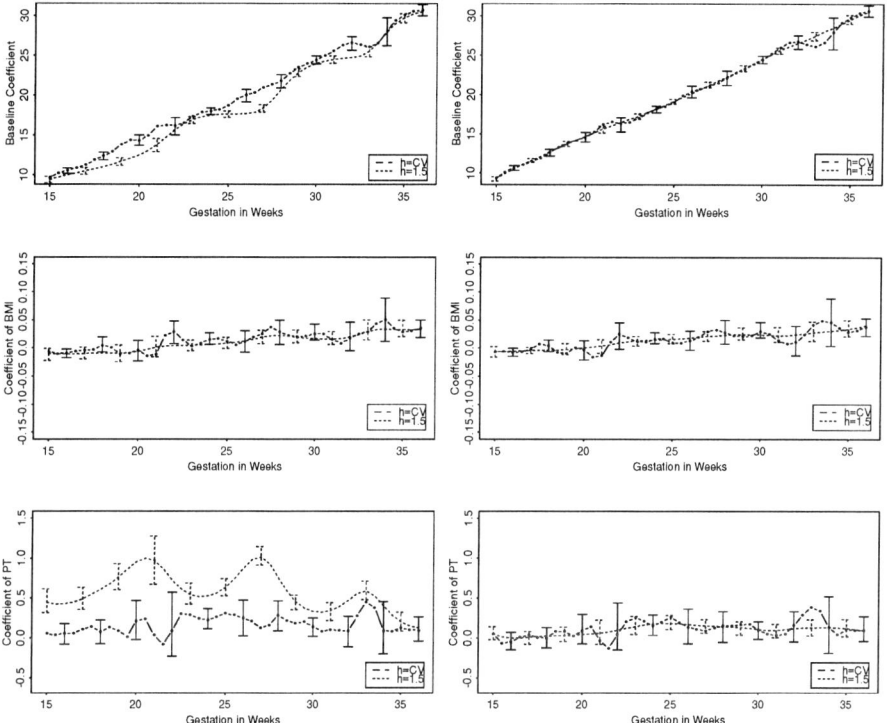

Figure 2. Figure 2: The plots show the kernel (left panels) and local linear (right panels) estimators of $\beta_0(t)$, $\beta_3(t)$ and $\beta_4(t)$ between 15 and 36 weeks of gestation computed based on the model (1), the Epanechnikov kernel, the cross-validated bandwidths and the subjective bandwidth 1.50. The verticle bars represent the 95% bootstrap percentile intervals computed using the procedure of Section 4.2.

The kernel estimates for $\beta_0(t)$ and $\beta_4(t)$ in Figure 2 are very sensitive to the bandwidth choices. The estimate for $\beta_0(t)$ exhibits a significant downward shift when the bandwidth changes from its cross-validated value (approximately 0.45) to 1.50, while the estimate for $\beta_4(t)$ exhibits an upward shift. This is in contrast to the classical kernel regression that, in the interior points of the support, different bandwidths only affect the smoothness of the estimated curves but do not cause upward or downward shifts. The reason for the shifts is that the placental thickness $X_3(t)$ grows quickly with the gestational age $t$. [12] attributes the shifts to the large biases of the estimators when the covariates strongly depend on time. In contrast, the local linear estimators appear to be more desirable, as they exhibit much smaller biases in Figure 2 when the bandwidth increases.

Figure 3 shows the kernel and local linear estimates for (2) based on $X_1$, $X_2$ and the centered covariates $X_{3(c)}$ and $X_{4(c)}(t)$. Similar to Figure 2, the smoking and drinking effects were non-significant and were omitted. Comparing with Figure 2, we note that the kernel estimates for $\beta_{0(c)}(t)$ and $\beta_4(t)$, shown in the left panel of the figure, no longer exhibit upward or downward shifts as the bandwidth changes, suggesting the effects of stability provided by covariate centering.

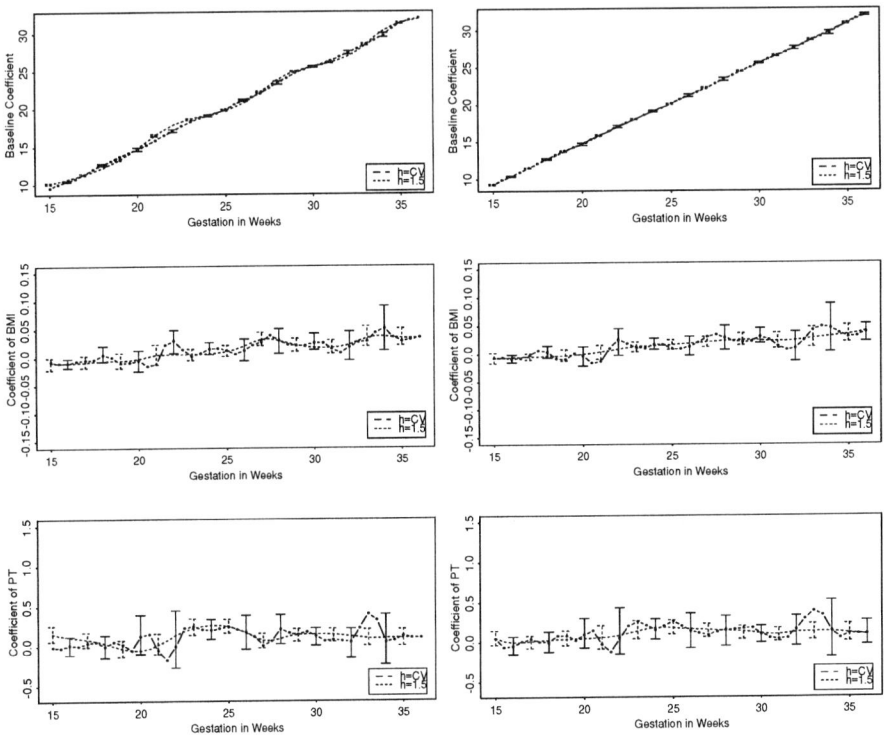

Figure 3. The plots are the kernel (left panels) and local linear (right panels) estimators of $\beta_{0(c)}(t)$, $\beta_3(t)$ and $\beta_4(t)$ between 15 and 36 weeks of gestation. The verticle bars represent the pointwise 95% bootstrap percentile intervals based on 500 bootstrap replications.

Biologically, the findings in Figures 2 and 3 suggest that the placental thickness is positively associated with the fetal abdominal circumference between approximately 22 and 27 weeks of gestation, but this association tappers down at both the beginning and the later stages of pregnancy. The mother's body mass index is positively associated with fetal abdominal circumference throughout the pregnancy, and the baseline coefficient curves are nearly linear. Our kernel and local linear estimates for the model (3) with covariates $X_1$, $X_2$, $X_{3(s)}$ and $X_{4(s)}(t)$ has similar patterns as those shown in Figure 3, hence, are omitted from our presentation.

## 6. A SIMULATION STUDY

For the simulation design, we consider a longitudinal setting with random covariates $X_1$, $X_2$ and $X_3(t)$, where $X_1$ has a Bernoulli(0.5) distribution, $X_2$ has a $N(1, 1.5)$ distribution and, given $t$, $X_3(t)$ has a $N(2t, 2\pi)$ distribution. The coefficient functions are

$$\beta_0(t) = 5 + 0.5t, \quad \beta_1 = -1, \quad \beta_2(t) = 0.5 + 0.2t \quad \text{and} \quad \beta_3(t) = 0.5 \sin\left(\frac{1+t}{10}\right).$$

The simulation process was repeated 200 times. Within each simulation run, we randomly generated 1200 subjects, each is equally likely to have 3, 4, 5, 6 or 7 repeated measurements. For each subject $i$ and its corresponding number of repeated measurements $n_i$, we generated independent time points $t_{ij}$, $j = 1, \ldots, n_i$, from the uniform distribution on $[0, 30]$, the random covariates $X_{1i}$, $X_{2i}$ and $X_{3ij}$ from the distributions of $X_1$, $X_2$ and $X_3(t)$, respectively, and the random errors $\epsilon_{ij}$ from the mean zero Gaussian process $\epsilon(t)$ with variance $\sigma_\epsilon^2(t) = 2$ and covariance $cov[\epsilon(t_1), \epsilon(t_2)] = \exp\{-|t_1 - t_2|\}$ if $t_1 \neq t_2$. We generated the responses $Y_{ij}$ by summing up the mean terms and the random errors at time $t_{ij}$ for $i = 1, \ldots, 1200$ and $j = 1, \ldots, n_i$, and estimated $\beta_l(t)$, $l = 0, \ldots, 3$, of model (1) using the kernel and local linear estimators of Section 2.1 with the Epanechnikov kernel, the bandwidth 1.50 and the cross-validated bandwidths.

Figure 4 shows the true coefficient curves, the averages of the kernel and local linear estimates, and their corresponding lower and upper pointwise fifth percentiles computed from the 200 simulated samples. The cross-validated bandwidths were around 0.58 for the simulated samples. The large biases are visible for the kernel estimators of $\beta_0(t)$ and $\beta_3(t)$ when either the cross-validated bandwidths or the subjectively chosen bandwidth 1.50 were used. The bias further increases when the bandwidth increases from approximately 0.58 to 1.50.

When $X_2$ and $X_3(t)$ were centered or standardized, we estimated the mean and variance of $X_2$ by the sample mean and the sample variance, and estimated the local mean and the local variance of $X_3(t)$ by their corresponding kernel and local linear estimators. Following Section 2.3, we computed the kernel and local linear estimators of the coefficient curves using the Epanechnikov kernel, the cross-validated bandwidths and the subjective bandwidth choice of 1.50.

Figure 5 shows the true coefficient curves $\beta_{0(c)}(t)$, $\beta_2(t)$ and $\beta_3(t)$ for model (2), the averages of the smoothing estimates, and their lower and upper pointwise fifth percentiles computed from the 200 simulation runs. Both the kernel and the local linear estimators appear to be satisfactory in the interior of the time support. However, the local linear estimators exhibit smaller boundary biases than the kernel estimators. Results for the estimators using the standardized $X_2$ and $X_3(t)$ were similar, hence were omitted.

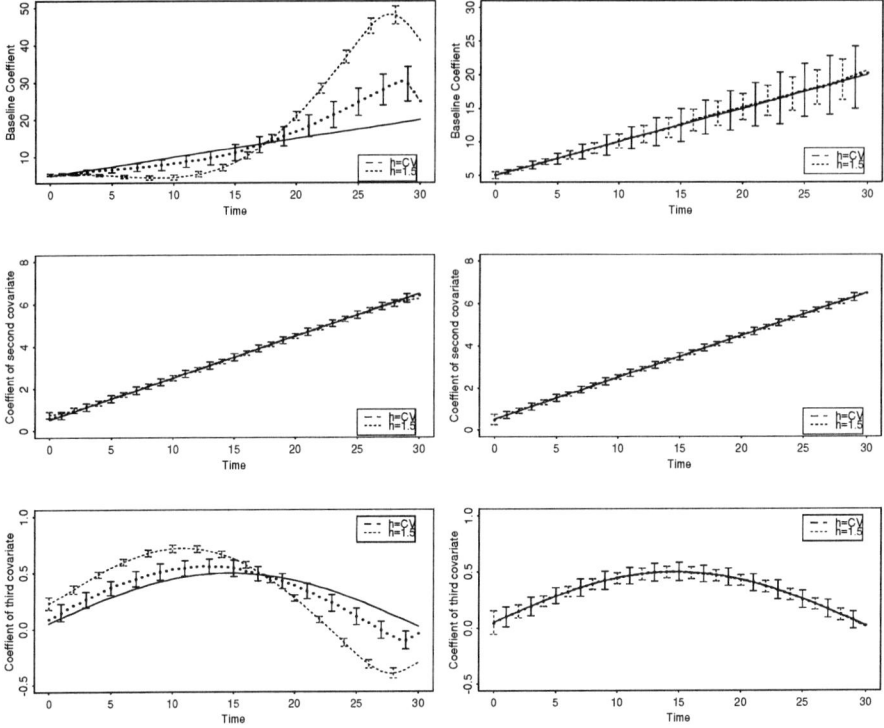

Figure 4. True coefficient curves $\beta_0(t)$, $\beta_2(t)$ and $\beta_3(t)$ (solid lines), the averages of the kernel (left panels) and local linear (right panels) estimates and their lower and upper pointwise fifth percentiles computed based on the model (1), the Epanechnikov kernel, the bandwidth 1.50 (dotted lines), the cross-validated bandwidths (dashed lines) and the 200 simulated samples.

To assess the risks of the smoothing estimators, we compared the sizes of the integrated bias squares,

$$\text{IBS} = \left\{ \int \left[ E\left(\widehat{\beta}_{l(*)}(s)\right) - \beta_{l(*)}(s) \right]^2 ds \right\},$$

the integrated variance,

$$\text{IV} = E\left\{ \int \left[ \widehat{\beta}_{l(*)}(s) - E\widehat{\beta}_{l(*)}(s) \right]^2 ds \right\},$$

and the mean integrated squared error,

$$\text{MISE} = \text{IBS} + \text{IV} = E\left\{ \int \left[ \widehat{\beta}_{l(*)}(s) - \beta_{l(*)}(s) \right]^2 ds \right\},$$

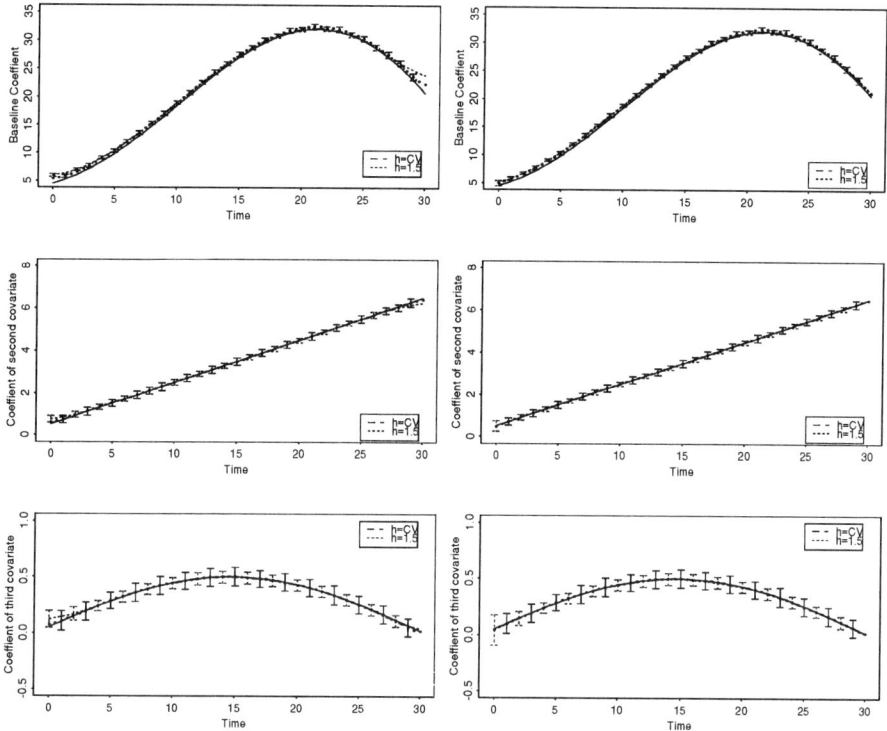

Figure 5. True coefficient curves $\beta_{0(c)}(t)$, $\beta_2(t)$ and $\beta_3(t)$ (solid lines), the averages of the kernel (left panels) and local linear (right panels) estimates and their lower and upper pointwise fifth percentiles computed based on the model (1), the Epanechnikov kernel, the bandwidth 1.50 (dotted lines), the cross-validated bandwidths (dashed lines) and the 200 simulated samples.

for each coefficient curve $\beta_{l(*)}(\cdot)$. We estimated $E[\widehat{\beta}_{l(*)}(t)]$, MISB, MIV, and MISE using the corresponding sample means over the 200 simulation runs and numerical integrations.

Table 1 shows the estimates of bias, variance and squared error terms for $\beta_2$ and $\beta_3(t)$ computed from the kernel smoothing methods with the original or the centered $X_2$ and $X_3(t)$, the subjective bandwidth 1.50 and the cross-validated bandwidths. To compare the differences between the estimators with the original covariates and the estimates with the centered covariates, the last column of Table 1 shows the estimated ratios of the mean integrated squared errors $R = \text{MISE}(\widehat{\beta}_l)/\text{MISE}(\widehat{\beta}_{l(c)})$. The results for $\beta_1$ were very similar to those for $\beta_2$, hence, were omitted from the presentation.

Table 2 shows the estimated values of IBS, IV, MISE and R for $\beta_2$ and $\beta_3(t)$ computed from the local linear method. For both the subjective bandwidth 1.50 and the cross-

Table 1
Estimates of the integrated bias squares, the integrated variances and the mean integrated squared errors for the kernel estimators of $\beta_2(t)$ and $\beta_3(t)$.

|  | Parameter | IBS | IV | MISE | R |
|---|---|---|---|---|---|
| Without centering | $\beta_2$ | 0.240 | 0.233 | 0.473 | 0.922 |
| $h = 1.50$ | $\beta_3$ | 5.391 | 0.018 | 5.409 | 96.589 |
| With centering | $\beta_2$ | 0.244 | 0.269 | 0.513 | 1.000 |
| $h = 1.50$ | $\beta_3$ | 0.014 | 0.043 | 0.056 | 1.000 |
| Without centering | $\beta_2$ | 0.020 | 0.462 | 0.482 | 0.976 |
| $h_{cv}$ | $\beta_3$ | 0.571 | 0.083 | 0.654 | 6.170 |
| With centering | $\beta_2$ | 0.021 | 0.473 | 0.494 | 1.000 |
| $h_{cv}$ | $\beta_3$ | 0.002 | 0.104 | 0.106 | 1.000 |

validated bandwidths, the kernel estimators for $\beta_3(t)$ without covariate centering have significantly larger biases, and consequently larger mean integrated squared errors, than their counterparts with centered covariates. This is consistent with the theoretical findings discussed in [12] and suggests a clear advantage for covariate centering at least for the kernel smoothing method. On the other hand, the local linear estimates shown in Table 2 have similar biases and variances either with or without covariate centering. It is also interesting to note that, when the cross-validated bandwidths are used, the biases, variances and squared errors of the kernel estimators with covariate centering are similar to theose of the local linear estimators with or without covariate centering.

Table 2
Estimates of the integrated bias squares, the integrated variances and the mean integrated squared errors for the local linear estimators of $\beta_2(t)$ and $\beta_3(t)$.

|  | Parameter | IBS | IV | MISE | R |
|---|---|---|---|---|---|
| Without centering | $\beta_2$ | $8.34 \times 10^{-4}$ | 0.228 | 0.229 | 0.991 |
| $h = 1.50$ | $\beta_3$ | $1.41 \times 10^{-3}$ | 0.046 | 0.047 | 0.904 |
| With centering | $\beta_2$ | $8.11 \times 10^{-4}$ | 0.230 | 0.231 | 1.000 |
| $h = 1.50$ | $\beta_3$ | $1.22 \times 10^{-3}$ | 0.051 | 0.052 | 1.000 |
| Without centering | $\beta_2$ | $3.80 \times 10^{-3}$ | 0.499 | 0.503 | 0.998 |
| $h_{cv}$ | $\beta_3$ | $4.13 \times 10^{-4}$ | 0.117 | 0.118 | 0.975 |
| With centering | $\beta_2$ | $3.80 \times 10^{-4}$ | 0.500 | 0.504 | 1.000 |
| $h_{cv}$ | $\beta_3$ | $4.46 \times 10^{-4}$ | 0.121 | 0.121 | 1.000 |

## REFERENCES

1. C.T. Chiang, J.A. Rice and C.O. Wu, Smoothing spline estimation for varying coefficient models with repeatedly measured dependent variables. Journal of the American Statistical Association 96 (2001), 605-19.

2. P.J. Diggle, K.Y. Liang and S.L. Zeger, Analysis of Longitudinal Data. Oxford University Press, Oxford, 1994.
3. R. L. Eubank and P. Speckman, Confidence bands in nonparametric regression. Journal of the American Statistical Association 85 (1993) 387-392.
4. J. Fan and I. Gijbels, Local Polynomial Modeling and Its Applications. Chapman & Hall, London, 1996.
5. J. Fan and J.T. Zhang, Functional linear models for longitudinal data. Journal of the Royal Statistical Society B 62 (2000) 303-22.
6. P. Hall and D.M. Titterington, On confidence bands in nonparametric density estimation and regression. Journal of Multivariate Analysis 27 (1988) 228-54.
7. W. Härdle and J.S. Marron, Bootstrap simultaneous error bars for nonparametric regression. Annals of Statistics 19 (1991) 778-796.
8. D.R. Hoover, J.A. Rice, C.O. Wu and L.P. Yang, Nonparametric smoothing estimates of time-varying coefficient models with longitudinal data. Biometrika 85 (1998) 809-22.
9. G. Knafl, J. Sacks and D. Ylvisaker, Confidence bands for regression functions. Journal of the American Statistical Association 80 (1985) 683-691.
10. J.A. Rice and C.O. Wu, Nonparametric mixed effects models for unequally sampled noisy curves. Biometrics 57 (2001) 253-59.
11. C.O. Wu, C.T. Chiang and D.R. Hoover, Asymptotic confidence regions for kernel smoothing of a varying-coefficient model with longitudinal data. Journal of the American Statistical Association 93 (1998) 1388-402.
12. C.O. Wu, K.F. Yu and C.T. Chiang, A two-step smoothing method for varying coefficient models with repeated measurements. Annals of the Institute of Statistical Mathematics 52 (2000) 519-543.
13. S.L. Zeger and P.J. Diggle, Semiparametric models for longitudinal data with application to CD4 cell numbers in HIV seroconverters. Biometrics 50 (1994) 689-99.

# Directed Peeling and Covering of Patient Rules

Michael LeBlanc[a]*, James Moon[a], and John Crowley[b]

[a]Fred Hutchinson Cancer Research Center,
1100 Fairview Ave. N., Seattle, WA, 98109

[b]Cancer Research And Biostatistics,
1730 Minor Ave, Suite 1900, Seattle, WA, 98101.

We disuss methodology for describing regions of the predictor space associated with large (or small) average values of noisy response data. The regions are expressed as a union of simple logical expressions, where each simple expression represents a box in the predictor space. Boxes are constructed by successively removing small amounts of data along coordinate axes identified by variable selection. Simulation studies are used to explore the properties of the new method and convergence of estimators based on observations falling in unions of overlapping boxes is reviewed.

## 1. INTRODUCTION

Friedman and Fisher [1] propose a powerful adaptive method called the Patient Rule Induction Method (PRIM) that finds regions where the mean of the response variable is large (or small) relative to the overall mean. PRIM identifies local extrema or "bumps" by a technique called "peeling" which repeatedly removes small fractions of the data along coordinate axes. The "peeling" technique allows calibration of the size of the identified group or the mean outcome of the group. In addition, PRIM rules tend to be interpretable because they can be represented by a union of boxes in the predictor space, where each box is a conjunctive rule. Methods utilizing peeling would be useful for describing patient outcome data. For instance, we have found that if one uses the PRIM method extended to censored survival data, the method can yield interesting prognostic groups. However, patient outcome data typically have a very low signal to noise ratio and have a moderate sample size (often limited to several hundred observations), and clinical applications demand very simple rules to be helpful in the design of new studies. We have found PRIM to give quite variable solutions in such low signal applications; therefore modifications to the PRIM algorithm may be useful for constructing improved rules.

We investigated a more structured method based on ideas from PRIM which we call "directed peeling and covering" which adaptively construct such rules including variable selection, direction of rules and thresholds or cutpoints used in each of the univariate decisions. For example, the following is a hypothetical rule of form we describe.

IF $(x_3 > 4)$ OR $[(x_5 \leq 2)$ AND $(x_8 > 3)]$ THEN $\bar{y} = 3.5$ .

---
*This work was supported by NIH through grant R01 CA90998.

The model building uses two main strategies to limit variation and help interpretations:

- *Select only a small number of variables for peeling.* This reduces variance in the adaptive removal of data along axes. For instance, best variable subsets regression or step-wise regression can be used to select a small number of potential variables for a box construction. We also include simplest box shaped rules first by using a forward stepwise model building strategy.

- *Consider monotone boxes.* Since patient outcome data (especially survival data) are usually very noisy, there is little power to detect a poor patient outcome group that is located in the middle of the range of any of the predictor variables. Therefore, each box is only defined by one sided rules as $x_1 > c$.

We view PRIM or other peeling methods as useful compliments to tree-based methods. Tree-based methods recursively partition the data into groups [2]. The resulting groups of data and partitions of the predictor space can be represented by a binary tree, where the terminal nodes represent boxes. The proposed method ("directed peeling and covering") focusses on a single poor prognostic groups with control on the size and relative prognosis of that group, while trees form multiple groups with differing prognosis [3,4,5]. In this chapter, we review the strategy for constructing less variable monotone regions which was described recently in more detail [6]. However, we provide simulation studies to better understand the performance of procedure and include convergence results for methods which construct regions based on unions of boxes.

## 2. IDENTIFYING EXTREMA

Assume there is a response variable and a vector of predictors $(y_i, x_i)$, $i = 1, ..., N$, where $y_i$ is generated as

$$y_i \sim G(f(\mathbf{x}_i))$$

and where $G(f(\mathbf{x}_i))$ is the conditional distribution of $y$, given $x = x_i$ which is assumed to have mean $f(\mathbf{x}_i)$. Obvious examples of $G$ include normal, binomial or Poisson models. Figure 1 shows a schematic of the function, $f(x)$, the region $R$ corresponding to large values of $f(x)$ and observed data.

An important quantity is the mean over that region which can be denoted by

$$\overline{y}_R = \frac{1}{N\widehat{\beta}_R} \sum_{X_i \in R} y_i,$$

and the support of the region

$$\widehat{\beta}_R = \frac{1}{N} \sum_{X_i \in R} I\{x_i \in R\}.$$

In most situations a smooth region $R$ would be the best region with "extreme" outcome. However, smooth regions in more than one dimension are difficult to understand and

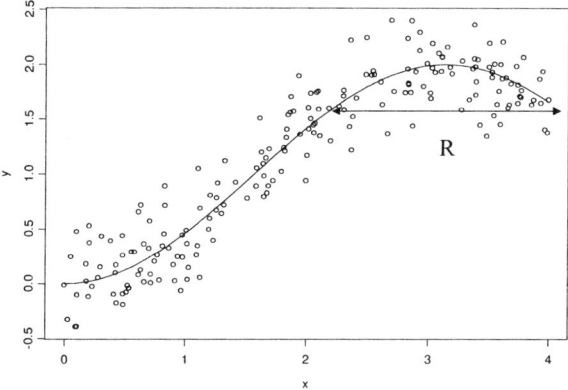

Figure 1. Regression function, region R and data.

describe. Therefore, we will construct interpretable models in terms of unions of boxes in the predictor space. Let $S_j$ represent the set of all possible values for covariate $x_j$ and let $s_{jk}$ be a subset labeled $k$ of $S_j$. Boxes can be expressed as

$$B_k = \bigcap_{j=1}^{p} \{x_j \in s_{jk}\},$$

where for ordered predictors the $s_{jk}$ are intervals of the line.

Friedman and Fisher's PRIM method builds a model to represent a region for good or poor outcome group based on the union of several boxes

$$R = B_1 \bigcup B_2 \bigcup B_3,$$

where each of the boxes is defined in terms of the predictor variables. At each step, the box is constructed by repeatedly removing ("peeling") a small fraction of the data along one of the coordinate axes. Peeling is continued until some minimum percentage of the sample remains in the box or the box has a sufficiently extreme average response value. At each step, the next box is constructed using data not contained in the previous boxes. Unlike tree-based methods, the procedure defines overlapping boxes in the predictor space. This strategy of repeated box construction is called covering.

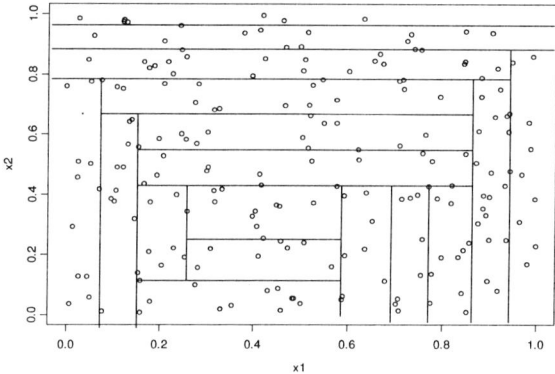

Figure 2. Schematic of unconstrained peeling

## 3. CONSTRUCTING RULES

### 3.1. Patient Rule Induction: Peeling

PRIM constructs boxes for ordered predictors by repeatedly removing small amounts of data along the coordinate axes of a box. Initially the entire data set is considered and then a fraction $\alpha$ of the data is removed along the coordinate axis (from either extreme of the variable distribution) which leaves the largest mean value for the remaining data in the box. They call this procedure "peeling". The removal of data is continued until some minimum percentage of the observations remain in the box. The peeling fraction is taken to be quite small unlike the partitioning algorithm in tree-based methods. This allows the procedure to recover from bad step-wise decisions. However, to control variance of the peeling procedure there is minimum number of observations set for the data to be removed at each step. The Figure 2 shows a hypothetical peeling trajectory for a problem with 2 predictor variables. An example of the means of the boxes induced by peeling are plotted against the support of the boxes (a trajectory plot) is given in Figure 3 for one of the simulation examples described later.

As stated above, the goal of peeling is to maximize (or minimize) the box average

$$\bar{y}_B = \frac{1}{N\widehat{\beta}_B} \sum_{X_i \in B} y_i$$

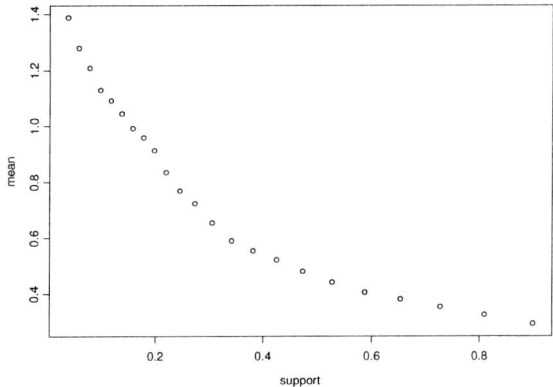

Figure 3. Box means versus the the support of the boxes for a simulated example.

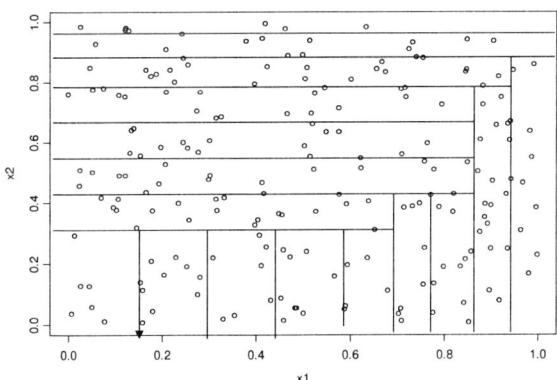

Figure 4. Schematic of monotone peeling.

for a given support for the box

$$\widehat{\beta}_B = \frac{1}{N} \sum_{X_i \in B} I\{x_i \in B\}.$$

To obtain boxes with high mean value, one can select the resulting box at each step with largest mean, or alternatively choose the box that leads to the largest rate of increase in box mean. Friedman and Fisher in [1] discuss several options for peeling in terms of patience in constructing the boxes with extreme average value. We construct boxes based on a peeling algorithm that changes the face of the box that maximizes the rate of increase in the box mean. This can also be expressed as the difference in means between the new box $B_j$ and the box to be removed $B - B_j$

$$d_j = ave_{B_j} \, y_i - ave_{B-B_j} \, y_i.$$

For categorical predictors with a limited number of levels, the definition is easily extended by removing observations equal to one of each levels of the categorical variable.

### 3.2. Directed Peeling

The above algorithm works well for large data sets. However, we have explored the addition of additional constraints to minimize variance of the procedure. In recent paper [6] we describe one strategy for minimizing variance. The full algorithm is described in that manuscript; we only detail the peeling component here and outline how multiple box rules can be constructed. While other related strategies are easy to imagine, the key ingredients are to pre-select a small number variables for box construction and to limit boxes to be monotone. We note that a function $f: \mathbf{R}^p \to \mathbf{R}$ is a monotone function if $f(y) \geq f(x)$ for all $y > x$. Here $y > x$ means $y_i \geq x_i$ and the inequality is strict for at least one of the variables. We then define a monotone region, $R$, as follows: if $x \in R$ and $y > x$ then $y \in R$. Therefore, the boxes defined by half-lines such as

$$B = \{x_1 > c_1\} \bigcap \{x_2 > c_c\} \bigcap \dots \bigcap \{x_p > c_p\}$$

satisfy the above property and hence are monotone regions.

Rather than using all $p$ predictors, we limit the potential variables to a $r$ dimensional subset of the variables, $(x_{(1)}, x_{(2)}, \dots, x_{(r)})$, selected by best subsets regression (step-wise or other variable selection methods would also work fine). Data removal is permitted along these chosen variables and in only one direction for each variable where the direction of removal of data is determined by the sign of the coefficient in the best subset regression. A schematic of monotone or "directed peeling" is given in Figure 4 . In manuscript [6] we widen the search to include peeling on several good subsets models (default 3 subsets) of $r$ variables to be involved in box construction. To choose the best box, the algorithm performs 3 peels, one peel on each one of the subsets of variables and directions. The box that has sufficiently extreme mean outcome ($\overline{y}_b > q$) with largest support is chosen to enter the model.

The constraint on the boxes to be at most $r$ dimensions allows the algorithm to focus on low dimensional boxes first before adding higher dimensional boxes to the rule. However, we have found for some problems it is useful to allow limited flexibility to consider several dimensions each step. This is important because one may not want to choose a one dimensional rule, if a 2 dimensional box with similar average outcome value includes a much larger fraction of the sample. Therefore, there is an adjustable parameter $v = 0, 1, 2$ allowing 0, 1 or 2 dimensional box "look ahead".

### 3.3. Directed Peeling Algorithm

1. Use best subset regression to select $s$ combinations of each dimension $m = r, r + 1, .., r+v$ variables for peeling where $v \geq 0$ represents the step "look ahead". Default $v = 0$.

2. Loop over the $s$ combinations for each dimension $m$ (steps 3-9).

3. Let $x_{l(1)}, x_{l(2)}, ..x_{l(m)}$ be one combination of variables to be considered for peeling, and the vector $\zeta = \{\zeta_1, \zeta_2, ..., \zeta_m\}$ be a vector of indicators of the direction of peeling. If $\zeta_j = 1$ then small values of the predictor are peeled, and if $\zeta_j = -1$ then large values of the predictor are peeled. For unordered variables $\zeta_j = 0$.

4. Start with the entire data set, $B^0$.

5. Peel a fraction, $\alpha$, of the data while the support of the box, $\beta \geq \rho$.

6. Let $k = \arg\max d_j$, the variable corresponding to the largest improvement in box mean or the value of the predictor for an unordered predictor.

7. Peel a fraction $\alpha$ of the current box along variable $k$.

8. Let the peeled box be defined by $B^{m+1} = B^m \bigcap \{x_k \geq c\}$ if $\zeta_j = 1$ or $B^{m+1} = B^m \bigcap \{x_k < c\}$ if $\zeta_j = -1$ or $B^{m+1} = B^m \bigcap \{x_k \neq c\}$ if $\zeta_j = 0$.

9. Record indexed set of boxes of dimension $i$.

10. Select the box which (at a given threshold $q$) that yields the largest support based on test data or resampling estimate of trajectory means. If $v > 0$ then pick the box with largest penalized support, $\beta + m_o \pi$, where $m_o$ is the dimension of the box and $\pi$ is the penalty per dimension. Default $\pi = .02$.

### 3.4. Cross-validation

Limiting peeling to only a small subset of the variables limits the magnitude of the selection bias of the method. However, it is still prudent to calculate less biased estimates of the box means. For this we choose to use $K$-fold cross-validation. The training data, $\mathcal{L}$, are divided up into $K$ test samples $\mathcal{L}_k$ and training samples $\mathcal{L}_{(k)} = \mathcal{L} - \mathcal{L}_k$, $k = 1, \cdots, K$ of about equal size. Boxes are constructed on each of the training samples $\mathcal{L}_{(k)}$; each test sample $\mathcal{L}_k$ is used to estimate the quantity of interest from the model grown on the training sample $\mathcal{L}_{(k)}$. We typically find that repeating 10-fold cross-validation 5 times and taking the average leads to less variable cross-validation results.

### 3.5. Ordered Covering of the Region

A single box will likely not be sufficient to describe the poor prognosis group. Therefore, the rule is refined through the addition (union) of additional boxes. We implement a covering algorithm as in PRIM where the data from the original box $B_1$ is removed before the next box is constructed. The second box is constructed on data after removing the data corresponding to $B_1$,

$$\{y_i, x_i | x_i \notin B_1\}$$

and at the $K^{th}$ iteration the box is constructed by peeling data not contained in any of the previous boxes

$$\left\{ y_i, x_i | x_i \notin \bigcup_{k=1}^{K-1} B_k \right\}.$$

At each step directed peeling (of dimension $r$) is used. The algorithm starts with one dimensional boxes, $r = 1$. If no additional boxes can be removed at that dimension the maximum dimension is increased by 1. The sequence and box yielding the largest number of observations with the mean above some threshold is chosen as the next box or logical rule in the model.

### 3.6. Overall model selection

We select the number of boxes by using the test sample (or averaged $K$-fold cross-validation) estimates of box means and support and the simple penalty function

$$\widehat{\beta}_\lambda(M) = \widehat{\beta}(M) + \lambda m$$

where $\widehat{\beta}(M)$ is the coverage corresponding to model $M$ (the sum of the supports from either the test or cross-validation estimates) and $m$ is
the number of terms in the model and $\lambda$ is the penalty per term.

### 4. SIMULATIONS

To investigate the performance of the algorithm, including covering and selection of model size, a simulation study was done. We compared tree based regression, the new directed peeling and covering (DPC) method and an unconstrained peeling and covering method (PC). We emphasize that while the PC method has strong similarities to PRIM since it uses patient unconstrained box induction and then covers the region of interest, it is not the same algorithm. We required some automatic choices to how to select a box from the peeling strategy for our simulation study. In addition, to simplify, for our simulation study we do not interactively remove variables or add data back to boxes (bottom-up pasting) as can be done in the PRIM method. Finally, we choose not to do bumping and investigate multiple trajectories since all three methods should benefit from that technique (and possibly by differing amounts).

Therefore, we think our conclusions regarding the performance of the methods should be interpreted as between trees, a less constrained box induction method and the new method which constrains the box induction process.

We generated 500 observations where the predictor variables are taken from a ten-dimensional uniform distribution $U(0,1)^{10}$ and again where the response is defined as

$$y = f(x) + \varepsilon,$$

where $\varepsilon$ is distributed $N(0, \sigma^2 k)$. We considered the following models for the mean function:

- Model A: $f(x) = 0$.
- Model B: $f(x) = 1 - [(x_1 - 1)^r + (x_2 - 1)^r + (x_3 - 1)^r]^{1/r}$ where $r = 30$.
- Model C: $f(x) = \max(\varphi(x_1; 1/4, .8) + \varphi(x_1; 1/4, .8) + \varphi(x_1; 1/4, .8), 1)$ where

$$\varphi(x; a, b) = \frac{\exp((x-a)/b)}{1 + \exp((x-a)/b)}.$$

The constant $k$ for Model A was set to 1, and chosen to make the signal to noise ratio .75 for Model B and Model C. Model A is the null model where the response is unrelated to the predictor variables. Therefore, all the procedures should fail to detect an extreme outcome group for Model A. Model B specifies a section of pyramidal shape in three dimensions with the maximum at the $(1,1,1)$ corner of the $(0,1)^3$ cube. We show a conditional contour plot at $x_3 = 1$ in Figure 5

This model should be a relatively easy test for all three procedures since regions of high response are represented by a single box of dimension 3, but it should be best suited to a method that uses undirected peeling on multiple predictor variables. Model C is a model representing a problem which has large expected values of the response when any one of the 3 predictor variables is large. Figure 6 gives a contour plot at $x_3 = 0$.

Although Model C is a more difficult problem than Model B for all three methods, it may best suited to the new method, since it could be roughly approximated by a union of three univariate rules defined on each of the predictors $x_1$, $x_2$ and $x_3$.

We generated a learning and test data set rather using resample based methods for model selection. This was done to avoid the complexity of doing comparisons of different options for model selection and allowed us to focus on the major components of the new algorithm. For both PC and DPC methods the peeling fraction was set at .1 (with a minimum of 10 observations at any step) and the minimum group size was set to .05. For the DPC method three best subsets were chosen for each dimension and no "look ahead" ($v = 0$) was chosen.

Model A specifies a problem where the response is not related to the outcome. We chose to include terms if the mean in a box was greater than .2 (equal to one standard error of a box mean for a box consisting of 5% of the sample.). Trees tended to have the smallest models grown. Note that the small tree-based models are in part due to the nature of the splitting algorithm, which does not split a node smaller than twice the minimum node size. This yields an average size of the smallest group that is larger than 5% of sample, so averages within these samples would tend to be closer to zero than obtained by the two

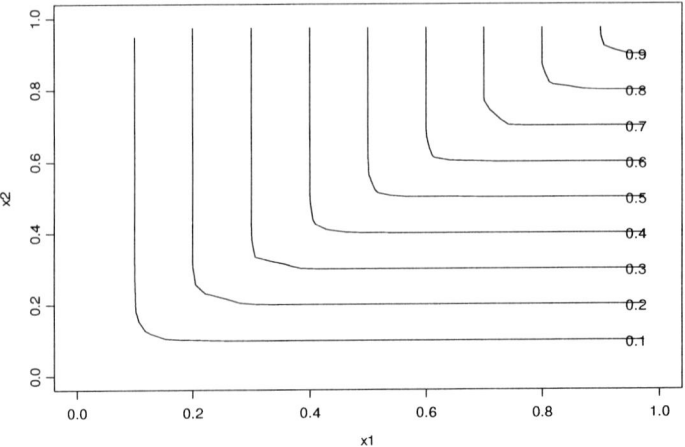

Figure 5. Contour plot of the mean function for Model B at $x_3 = 1$.

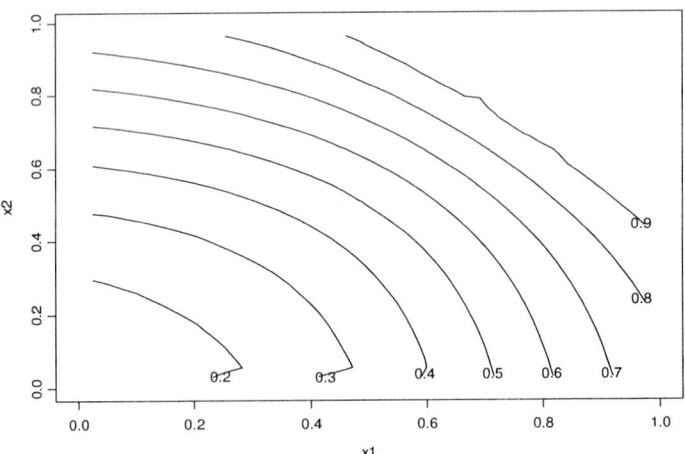

Figure 6. Contour plot of the mean function for Model C at $x_3 = 0$.

peeling methods. Both the peeling and covering (PC) and directed peeling and covering (DPC) methods lead a mean rule size of .4 boxes. We also report support and coverage (Support × Box Mean). Note that for Model A all true box means and coverage must be zero since $f(x) = 0$.

Table 1
Results for Model A

| Model A | Number of Terms | Box Support | True Box Mean | Coverage |
|---|---|---|---|---|
| Trees | 0.04 (.040) | 0.00 (.000) | 0.00 (.000) | 0.00 (.000) |
| PC | 0.40 (.100) | 0.05 (.033) | 0.00 (.000) | 0.00 (.000) |
| DPC | 0.36 (.140) | 0.08 (.022) | 0.00 (.000) | 0.00 (.000) |

Under Model B, the results for both the PC and the DPC methods are very similar. In all 25 simulated samples the two peeling methods had a single term. The trees had substantially lower average support and coverage and yet had complicated models of on average 7 nodes. Part of the poor performance for trees is because trees do not allow calibration of the average outcome in the extreme group unlike PC and DPC methods. Another component in the improved performance of the PC or DPC methods is likely due to removal of only a small fraction of data at each step rather than the strategy of splitting off large fractions of data in tree based methods.

Table 2
Results for Model B

| Model B | Number of Terms | Box Support | True Box Mean | Coverage |
|---|---|---|---|---|
| Trees | 7.04 (.492) | 0.15 (.012) | 0.55 (0.024) | 0.08 (.006) |
| PC | 1.00 (.000) | 0.27 (.010) | 0.50 (0.004) | 0.13 (.004) |
| DPC | 1.00 (.000) | 0.28 (.008) | 0.50 (0.004) | 0.14 (.003) |

The results for Model C highlight the potential for difficulty for the trees and unconstrained PC method for this structure with noisy outcome data. The support of the high response regions found by trees are worst, followed by the PC method then the new method. Trees have the additional problem of yielding very complicated models with on average 12 terminal nodes. The same model for the mean with a slightly reduced signal to noise ratio actually left unconstrained peeling (PC) with the worst performance with respect to the true box mean, support and coverage.

## 5. CONVERGENCE FOR UNIONS OF BOXES

Convergence results for methods which estimate means calculated over adaptively chosen unions of box shaped regions in the predictor space can be obtained analogous to those results for methods that are based on data dependent partitions such as regression

Table 3
Results for Model C

| Model C | Number of Terms | Box Support | True Box Mean | Coverage |
|---|---|---|---|---|
| Trees | 12.20 (.483) | 0.43 (.012) | 0.96 (.002) | 0.42 (.011) |
| PC | 1.76 (.156) | 0.49 (.015) | 0.91 (.002) | 0.44 (.013) |
| DPC | 2.32 (.095) | 0.59 (.008) | 0.90 (.002) | 0.53 (.007) |

histograms [7] or regression trees [2]. However, there some differences in the methods. Interest for peeling or directed peeling focuses on a single region, whereas for trees (or other adaptive regression models) the consistency results are often to the underlying regression function. Instead we show the mean converges to the mean to the regression values for the region defined by a union of overlapping boxes provided the number of observations falling into the region grows sufficiently fast as the sample size increases.

Let the covariates be a random vector $X \in \mathbf{R}^p$. Let $B$ be the collection of all boxes in the covariate space of the form

$$B_{Nk} = \{(x_1, ...x_p) : x_1 \in I_{1k}, ..., x_p \in I_{pk}\}$$

where $I_1, ..., I_p$ are intervals of the real line. We consider two types of sets of boxes, one using half lines $I_{jk} = [c_{jk}, \infty)$ such as those obtained from the proposed algorithm and the more general class of boxes that use intervals $I_{jk} = [c_{jk}, d_{jk})$ such as those obtained from undirected peeling algorithm. The regions of interest are the union of boxes

$$R_N = \bigcup_{k=1}^{K} B_{Nk}.$$

Let $v$ be a region in $\mathbf{R}^p$, then denote the number of observations falling in the region as

$$n_N(v) = \sum_{i=1}^{N} I\{x_i \in v\}$$

and proportion of the sample falling into that region by

$$p_N(v) = \frac{n_N(v)}{N}.$$

The mean over the region is

$$\bar{x}_N(v) = \frac{1}{n_N(v)} \sum y_i I\{x_i \in v\}$$

and the population mean for observations in the region is

$$\mu_N(v) = \frac{1}{n_N(v)} \sum E[y_i|x_i] I\{x_i \in v\}.$$

We show that the resulting mean for the region $R_N$ based on a union of overlapping boxes converges to the population mean if sufficient number of observations fall into the region as the sample size increases. For instance, if it is assumed that

$$P(p_N(v)) >= k_N \frac{\log N}{N} \text{ for } N \geqslant 1) = 1$$

for some sequence $k_N$ such that $\lim k_N = \infty$, then there is sufficient number of observations within each of the peeled boxes to converge to the population mean within the region. This result as well as many others for non-parametric techniques can use concepts from Vapnik and Chervonenkis(VC) classes. We assume response variables have bounded moments (this trivially includes random variables that are bounded).

**Theorem:** Suppose that the proportion of observations falling into $v$ are such that $p_N(v) \geq k_N N^{-1} \log N$ for $N \geqslant 1$ and the outcomes have bounded moments $E|Y_i|^m \leq M^{m-2}q/2$, for every $m \geqslant 2$ (and all $i$) for some constants $M$ and $q$ and then

$$P\left(|\bar{y}_N(v) - \mu_N(v)| > \epsilon \text{ for some } v \in B \text{ and } p_N \geq k_N N^{-1} \log N\right) \leq Q N^{s - \epsilon^2 \frac{k_N}{2q + 2M\epsilon}}$$

where $s = Kp$ for directed peeling and $s = 2Kp$ for undirected peeling.

The proof is given in the APPENDIX. The theorem is similar in spirit (Theorem 12.27) [2], except focusing on unions of overlapping boxes and instead uses Bernstein's inequality for bounding. If $\lim_N k_N = \infty$, the result above implies convergence of the box means to the mean of population for the box. We note from left side inequality, the bound is impacted in the variable $s$ by whether or not monotone or more general boxes are used in the union of boxes (the factor 1 or 2) and the number of boxes in the union, $K$.

## 6. DISCUSSION

Methods which repeatedly remove small amounts of data, such as peeling, allow one to either specify the relative outcome or the percentage of patients in a group. We think methods like PRIM and more constrained peeling methods such as the method described here and in reference [6] will be a useful complement to tree-based methods which are used to describe groups of patients with good or poor prognosis [8]. Peeling methods allow calibration of the support or relative patient outcome of the prognostic group that is not directly possible with tree-based methods.

However, these new methods still retain simple interpretable logical rules similar to those obtained from regression trees or another tool recently developed for constructing logical rules called "Logic Regression" developed by Ruzcinski, Kooperberg and LeBlanc [9].

For noisy data, we have noted our more constrained or directed peeling by selecting a small number of variables and directions to construct simple monotone boxes leads to decreased bias in box estimation and improved performance in terms of increased coverage for the region of interest. In addition, a forward step-wise model building strategy can sometimes lead to simpler descriptions. We note that the exact strategy is probably not critical but rather any good method for up-front variable selection and direction selection before peeling boxes would be helpful. We have focussed on uncensored data in this report to facilitate the discussion of the algorithm; however the extensions to survival data are easy to imagine. A simplified strategy for constrained peeling based on the proportional hazards model is the subject of current work.

## APPENDIX: Proof of Theorem

First we focus on the potential number of subsets picked out by a collection of subsets $C$. The VC index $V$ is defined as the smallest $N$ for which no set of size $N$ is shattered by $C$. It is known that a class with VC index $V$ picks out at most a polynomial number $O(N^{V-1})$ subsets. In this our case, the collection of cells (boxes) $\mathbf{R}^p$ with sides defined by $I_j = [c_j, \infty)$ have VC index $V = p+1$, for the more general class of boxes that use $I_j = [c_j, d_j)$ have VC index $V = 2p+1$. It is also well known [10] that unions of VC class sets are also VC class (Lemma 2.6.17) with $V(C \bigcup B) = V(C) + V(D)$. This leads to the more general union of $K$ boxes to pick out $O(N^{Kp})$ subsets for directed peeling and $O(N^{2Kp})$ subsets for undirected peeling.

Now consider an inequality for a mean based on a single region. It is sufficient to assume that the response variables have mean zero; however we make allowance for larger than normal tails. We use a version of Bernstein's inequality which can be expressed as follows: Let $Z_1, \ldots, Z_n$ be independent random variables with zero mean such that $E|Z_i|^m \leq m! M^{m-2} q_i/2$, for every $m \geq 2$ (and all $i$) and some constants $M$ and $q_i$. Then

$$P(|Z_1 + \ldots + Z_n| > c) \leq 2 \exp \frac{-c^2}{2(qn + Mc)}$$

where $q \geq \max(q_1, \ldots, q_p)$. Note the moment condition is trivially satisfied if the $Z_i$ are bounded. Then for a mean over $n_N(v) = p_N(v)N$ observations, $\overline{Z} = \sum_{i=1}^{n(v)} Z_i$ satisfies

$$P(|\overline{Z}| > c) \leq 2 \exp \frac{-c^2 p_N(v) N}{2(q + Mc)}$$

The mean $\overline{y}_N(v)$ and population mean $\mu_N(v)$ are defined as in Section 6.1. Since $P(\bigcup_j A_j) \leq \sum_j P(A_j)$ for sets $A_j$, one can combine the above with the multiplicity of subsets to get

$$P(|\overline{y}_N(v) - \mu_N(v)| > \epsilon \text{ for some } v \in B \text{ and } p_N \geq k_N N^{-1} \log N)$$
$$\leq Q N^{rKp - \epsilon^2 \frac{k_N}{2q + 2M\epsilon}}$$

for a constant $Q$ where $r = 1$ for directed peeling and $r = 2$ for undirected peeling.

## References

1. J. Friedman and N. Fisher, Bump Hunting in High Dimensional Data (with discussion), Computing and Statistics, 9 (1999) 123-162.

2. L. Breiman, J.H. Friedman, R.A. Olshen and C.J. Stone, *Classification and regression trees*. Wadsworth, Belmont, CA, 1984.

3. M. LeBlanc and J. Crowley, Survival trees by goodness of split. *Journal of the American Statistical Association*, 88 (1993) 457-467.

4. R. Davis and J. Anderson, Exponential survival trees. *Statistics in Medicine*, 8 (1989) 947-962.

5. M.R. Segal, Regression trees for censored data. *Biometrics*, 44 (1988) 35-48.

6. M. LeBlanc, J. Jacobson, J. Crowley, Partitioning and peeling for constructing prognostic groups. *Stat Methods Med Res,* 11 (2002) 247-274.

7. A.B. Nobel, Histogram regression estimation using data-dependent partitions. *Annals of Statistics*, 24 (1996) 1084-1105.

8. J. Crowley, M. LeBlanc, J. Jacobson and S. Salmon, Some exploratory tools for the analysis of survival data in *The First Seattle Symposium in Biostatistics: Survival Analysis.* Lecture Notes in Statistics, Springer-Verlag, New York, N.Y., 1997.

9. I. Ruczinski, C. Kooperberg and M. LeBlanc, Logic regression. *Journal of Graphical and Computational Statistics*, (2003) to appear.

10. A.W. van der Vaart and J.A. Wellner. *Weak Convergence and Empirical Processes with Applications to Statistics*, Springer-Verlag, New York, 1996.

# 10. Resampling and Subsampling

Recent Advances and Trends in Nonparametric Statistics
Michael G. Akritas and Dimitris N. Politis (Editors)
© 2003 Elsevier Science B.V. All rights reserved.

# Statistical Analysis of Survival Models with Bayesian Bootstrap

Jaeyong Lee[a] and Yongdai Kim[b]*

[a]Department of Statistics, Pennsylvania State University,
326 Thomas Building, University Park, PA 16802 U.S.A.

[b]Department of Statistics, Ewha Womans University,
11-1 Daehyun-dong, Seodaemun-gu, Seoul, 120-750, South Korea

The Bayesian bootstrap (BB) was first introduced by Rubin [1] as a Bayesian alternative to the bootstrap [2] and received much attention from both Bayesians and frequentists as tools to approximate the posterior distribution and the sampling distribution. In this paper, we study the BB procedures for various survival models. We derive the Rubin's BB in three different approaches, each of which gives a different perspective to the BB. Among the three views, we adopt the empirical likelihood view and derive the BB procedures for right censored data, proportional hazard model and doubly censored data.

## 1. Introduction

The Bayesian bootstrap (BB) was first introduced by Rubin [1] as a Bayesian alternative to the bootstrap [2]. The BB and its variants have received much attention from both Bayesians and frequentists as tools to approximate the posterior distribution and the sampling distribution of a frequentist estimator, respectively. In this paper, we study BB procedures for various survival models. In particular, we derive the Rubin's BB in three different approaches, each of which gives a different perspective to the BB, and we apply these views to right censored data, proportional hazard model and doubly censored data.

The main advantage of the BB procedure, from the Bayesian point of view, is that it can be viewed as the noninformative or default Bayesian analysis of a nonparametric model, and it is unnecessary to elicit the prior information for an infinite-dimensional parameter. This is an important advantage, for the elicitation of the prior information on nonparametric objects is not an easy task. Also, the BB procedure for a nonparametric survival model is a good approximation tool, for the full Bayesian analysis, which is easily accessible to practitioners both conceptually and computationally.

The BB procedure has advantages from the frequentist's point of view. The BB posterior is asymptotically equivalent to the sampling distribution of the maximum likelihood estimator (MLE) [3–5], and often it is computationally much simpler than standard frequentist's methods. Moreover, in some cases, the small sample frequentist property of the BB posterior is better than the standard frequentist methods. See [4] for such example

---

*The research is supported in part by KOSEF through the Statistical Research Center for Complex Systems at Seoul National University.

in the proportional hazard model.

The paper is organized as follows. In section 2, three views of BB are explained. The BB procedures for right censored data, proportional hazard models and doubly censored data are presented in sections 3,4, and 5 respectively.

## 2. Three views of Bayesian bootstrap

In this section, we will derive the Rubin's Bayesian bootstrap in three different approaches, each of which gives a different perspective to the Bayesian bootstrap. The standard version of the bootstrap proceeds as follows. Suppose $\mathcal{X} = (X_1, X_2, \cdots, X_n)$ is a random sample from an unknown distribution $F$ and a functional of $F$, $T(F)$, is of interest. For clarity of the discussion, we assume there are no ties among $\mathcal{X}$ in this section. Typically $T(F_n)$, the plug-in estimator of the nonparametric maximum likelihood estimator $F_n$ of $F$, is used as a frequentist estimator of $T(F)$. Now, to make an inference, one needs the sampling distribution of $T(F_n)$. The bootstrap is a method to get an approximate sampling distribution of $T(F_n)$. A typical bootstrap procedure consists of sampling many bootstrap samples $\mathcal{X}_1^*, \cdots, \mathcal{X}_B^*$, where each bootstrap sample is a random sample from $F_n$ and inference on $T(F)$ is based on $T(F_i^*)$s where $F_i^*$ is the empirical distribution of $\mathcal{X}_i^*$. In more detail, the distribution of $T(F_i^*)$ is used as an approximation of the sampling distribution of $T(F_n)$. The bootstrap distribution is expressed as $F_i^* \stackrel{d}{=} \sum_{j=1}^n w_j \delta_{X_j}$, where $nw = n(w_1, w_2, \cdots, w_n) \sim Multinomial(n, 1/n, \cdots, 1/n)$. Noting that the choice of multinomial distribution is not so crucial, Rubin proposed a smoother alternative, $Dirichlet(1, 1, \cdots, 1)$, for the distribution of $w$. Thus, the BB is viewed as a weighted bootstrap. It turned out that as the bootstrap distribution approximates the sampling distributions of a statistic, the Bayesian bootstrap approximates the posterior distribution. Furthermore, in the construction of the Bayesian bootstrap posterior, no prior information is necessary, hinting that the Bayesian bootstrap may be used as noninformative or default nonparametric Bayesian analysis. This is indeed the case. This can be seen easily from another construction of the Bayesian bootstrap from the nonparametric Bayesian posterior with the Dirichlet process prior. If the prior on $F$ is a Dirichlet process with parameter $\alpha G$, where $\alpha > 0$ and $G$ is a probability distribution on $\mathbb{R}$, the posterior of $F$ is again a Dirichlet process with parameter $\alpha G + \sum_{i=1}^n \delta_{X_i}$. See [6] for details. The posterior mean of $F$ is

$$E(F|\mathcal{X}) = \frac{\alpha}{\alpha + n} G + \frac{n}{\alpha + n} F_n.$$

Thus, as in the normal model, the Bayesian estimator is a weighted average of the prior guess $G$ and the empirical distribution $F_n$. Because of this fact, $\alpha$ is often called the prior sample size and is interpreted as the amount of the prior information infused into the posterior. As the prior sample size goes to 0, the posterior converges to the Dirichlet process with parameter $nF_n$. This, incidently, is the Rubin's BB posterior. As Gasparini [7] noted, this provides the basis for the use of the BB as a default nonparametric Bayesian analysis. At the same time, we can see that the BB posterior gives a good approximation to the full Bayesian posterior when $\alpha$ is small. Indeed, Weng [8] showed that the Bayesian bootstrap gives a second order correct approximation to the full Bayesian posterior of the mean functional.

The third route to construct the Bayesian bootstrap underlines the relation between the Bayesian bootstrap and the empirical likelihood [9]. Let $\mathcal{F}$ be the space of all probability distributions on $\mathbb{R}$. The likelihood function does not exist because there is no $\sigma$-finite measure dominating all members in $\mathcal{F}$. Even if we restrict our attention to only those with densities, still other delicate problems arise and usual likelihood methods can not be applied. Owen [9] gave an interesting idea to overcome these problems. First, the model $\mathcal{F}$ is reduced to a data-dependent parametric model

$$\mathcal{F}_n = \{F = \sum_{i=1}^n w_i \delta_{X_i} : w_i \geq 0 \sum_{i=1}^n w_i = 1\}.$$

Note that $w = (w_1, \ldots, w_n)$ is the only unknown parameter and we can construct the likelihood for this model

$$L_n(F) = \prod_{i=1}^n w_i.$$

Using the empirical likelihood $L_n(F)$, for a parameter of interest $T(F)$, the empirical likelihood confidence interval can be obtained by the ratio of the empirical likelihoods. See [9,10] for details. One can feed the empirical likelihood to the usual mechanism to obtain the posterior. Adopting a noninformative prior, $\prod_{i=1}^n w_i^{-1}$, the posterior is proportional to

$$L_n(F) \prod_{i=1}^n w_i^{-1} = \prod_{i=1}^n w_i^{1-1}.$$

The posterior distribution of $w$ is, then, $Dirichlet(1, 1, \ldots, 1)$. Thus, we obtain the Rubin's Bayesian bootstrap posterior

$$F|\mathcal{X} \stackrel{d}{=} \sum_{j=1}^n w_j \delta_{X_j},$$

where $w \sim Dirichlet(1, 1, \ldots, 1)$. This view is effectively summarized by

**BB posterior $\propto$ empirical likelihood $\times$ prior**. (1)

This view of the Bayesian bootstrap provides a convenient method to construct the Bayesian bootstrap, and a generalization of the usual method of the posterior construction, i.e., posterior $\propto$ likelihood $\times$ prior. This description was noted by, among others, [1,4,5,10–12]. Especially, Kim and Lee [4,5] pursued this view to construct the Bayesian bootstrap for proportional hazard model and doubly censored data.

## 3. Right Censored Data

### 3.1. Derivation of Bayesian Bootstrap

Let $X_1, \ldots, X_n$ be survival times and $C_1, \ldots, C_n$ be right censoring times. For the right censored data, we observe only $D_n = \{(T_1, \delta_1), \ldots, (T_n, \delta_n)\}$. Here, $T_i = \min(X_i, C_i)$ and $\delta_i = I(X_i \leq C_i)$, where $I(p) = 1$ if $p$ is true and 0 otherwise. To define the empirical likelihood, we need some notation introduced. Define counting processes, for $i = 1, \ldots, n$, $N_i(t) = I(T_i \leq t, \delta_i = 1)$ and $Y_i(t) = I(T_i \geq t)$. Also, let $N(t) = \sum_{i=1}^n N_i(t)$,

$\Delta N(t) = N(t) - N(t-)$, and $Y(t) = \sum_{i=1}^{n} Y_i(t)$. The cumulative hazard function (chf) $A$ of a cumulative distribution function (cdf) $F$ is defined as

$$A(t) = \int_{[0,t]} \frac{dF(s)}{1 - F(s-)}.$$

The likelihood of $F \in \mathcal{F}_n$ with censored data is best expressed in terms of cumulative hazard function (chf).

Two forms of the likelihood using the product-integration, the binomial and Poisson forms, are available for the survival model. They are denoted by $L^B(A)$ and $L^P(A)$, respectively, and given below:

$$L^B(A) = \prod_{i=1}^{n} \prod_{t \in [0,\tau]} dA(t)^{dN_i(t)} (1 - dA(t))^{Y_i(t) - dN_i(t)} \qquad (2)$$

$$L^P(A) = \prod_{i=1}^{n} \prod_{t \in [0,\tau]} (Y_i(t)dA(t))^{dN_i(t)} \exp(-\int_0^{\tau} Y_i(t)dA(t)), \qquad (3)$$

where $[0, \tau]$ is the time period of interest. Both of them are the real likelihoods when the cdf is absolutely continuous, while only the former is when the cdf is discrete. Nevertheless, both are used in the derivation of BB likelihoods since each has its own merits. For the interpretation of the two likelihoods (2) and (3), see [13].

To construct the empirical likelihood, first we need to reduce the model to a data-dependent parametric model $\mathcal{F}_n$. Define $\mathcal{T}_n = \{t_1 < \ldots < t_{q_n}\}$ be the set of all uncensored survival times and

$$\mathcal{S}_n = \begin{cases} \mathcal{T}_n, & \text{if the largest observation is uncensored,} \\ \mathcal{T}_n \bigcup \{\infty\}, & \text{otherwise.} \end{cases}$$

Let $\mathcal{F}_n$ be the set of all probability measures with support $\mathcal{S}_n$. For each $t$, define $D(t) = \{i : \Delta N_i(t) = 1\}$ and $R(t) = \{i : Y_i(t) = 1\}$. The set $D(t)$ is the set of observations which fail at time $t$ and $R(t)$ is the set of observations still at risk at time $t$. The empirical likelihoods (or BB likelihoods) can be derived from the Poisson as well as the binomial form likelihoods. Replacing $dA(t)$ with $\Delta A(t)$ at each $t \in \mathcal{T}_n$ in the binomial form likelihood (2), we obtain the binomial form BB (BFBB) likelihood

$$L_n^B(\beta, A) = \prod_{t \in \mathcal{T}_n} \Delta A(t)^{\Delta N(t)} (1 - \Delta A(t))^{Y(t) - \Delta N(t)}.$$

Similarly, we get the Poisson form BB (PFBB) likelihood

$$L_n^P(\beta, A) = \prod_{t \in \mathcal{T}_n} \Delta A(t)^{\Delta N(t)} \exp(-Y(t)\Delta A(t)).$$

It is a delicate problem to choose appropriate forms of priors. Note that the prior on $\Delta A$ used for Rubin's BB is not proper. Interestingly, no proper priors produce consistent posteriors. The following two priors for the BFBB and PFBB seem to work well. See [4] for a discussion.

$$\pi^B(A) = \prod_{t \in \mathcal{T}_n} \Delta A(t)^{-1} (1 - \Delta A(t))^{-1}, \qquad (4)$$

$$\pi^P(A) = \prod_{t \in \mathcal{T}_n} \Delta A(t)^{-1}. \tag{5}$$

A routine calculation shows

$$\pi^B(A|D_n) \propto \prod_{t \in \mathcal{T}_n} \Delta A(t)^{\Delta N(t)-1}(1 - \Delta A(t))^{Y(t)-\Delta N(t)-1}$$

$$\pi^P(A|D_n) \propto \prod_{t \in \mathcal{T}_n} \Delta A(t)^{\Delta N(t)-1} \exp(-Y(t)\Delta A(t)).$$

In summary, a posteriori, $\Delta A(t)$ for $t \in \mathcal{T}_n$ are independent and follow $Beta(\Delta N(t), Y(t) - \Delta N(t))$ for $\pi^B(A|D_n)$ and $Gamma(\Delta N(t), Y(t))$ for $\pi^P(A|D_n)$. The BFBB posterior is the same the Bayesian bootstrap posterior proposed by Lo [18].

### 3.2. Inference with Posterior Sampling

A convenient method to summarize the posterior to make an inference is to draw a random sample from the posterior, which in these cases amounts to simply sampling from beta and gamma random variables independently. For the inference on the cdf, note the relation between the cdf and chf: $F(t) = 1 - \prod_{s \leq t}(1 - dA(s))$. Depending on whether $F$ is continuous or discrete, we have the following equivalent equations

$$F(t) = \begin{cases} 1 - \prod_{s \leq t}(1 - \Delta A(s)), & \text{if } F \text{ is discrete,} \\ 1 - \exp(-A(t)), & \text{if } F \text{ is continuous.} \end{cases} \tag{6}$$

Since the empirical likelihoods of the BFBB and the PFBB are obtained by continuous and discrete approximation of the product-integration, we need to use equation (6) accordingly: use the discrete version for the BFBB and the continuous version for the PFBB. In fact, the discrete version with the PFBB will generate negative $F$, because $\Delta A(t)$ takes values in $[0, \infty)$. In summary,

$$F(t) = \begin{cases} 1 - \prod_{s \in \mathcal{T}_n, s \leq t}(1 - \Delta A(s)), & \text{for the BFBB} \\ 1 - \exp(-\sum_{s \in \mathcal{T}_n, s \leq t} \Delta A(s)), & \text{for the PFBB.} \end{cases}$$

### 3.3. Some Properties

There are close similarities between the BB posterior moments, and the frequentist chf estimator and its variance estimators. Due to the simple distributional structures of both BB posteriors, we can easily obtain the first two moments of the BB posteriors:

$$E^B(A(t)|D_n) = E^P(A(t)|D_n) = \sum_{s \in \mathcal{T}_n, s \leq t} \frac{\Delta N(s)}{Y(s)}$$

$$\text{Var}^B(A(t)|D_n) = \sum_{s \in \mathcal{T}_n, s \leq t} \frac{\Delta N(s)(Y(s) - \Delta N(s))}{Y(s)^2(Y(s)+1)}$$

$$\text{Var}^P(A(t)|D_n) = \sum_{s \in \mathcal{T}_n, s \leq t} \frac{\Delta N(s)}{Y(s)^2}.$$

Note that the posterior means of both BB posteriors are the same as the celebrated Nelson-Aalen estimator. The variances of the BB posteriors are very close to the two variance estimators of the Nelson-Aalen estimator [13]. In fact, the PFBB posterior variance is the same as (4.1.6) of [13].

## 4. Proportional Hazard Model

### 4.1. Derivation of Bayesian Bootstrap

The argument in section 2 can be generalized to construct the BB posterior for the proportional hazard model. In addition to the model assumptions in section 2, suppose that we have covariate $Z_i \in \mathbb{R}^p$, $i = 1, \ldots, n$, and that the distribution $F_i$ of $X_i$ with covariate $Z_i$ is given by $1 - F_i(t) = (1 - F(t))^{\exp(\beta^T Z_i)}$ for unknown regression parameter $\beta \in \mathbb{R}^p$ where $F$ is the baseline distribution function. Let $A$ be the chf of $F$ and $A_i$ be that of $F_i$, which is given by $dA_i(t) = 1 - (1 - dA(t))^{\exp(\beta^T Z_i)}$. If $A$ is absolutely continuous with respect to Lebesgue measure, there exists a hazard function $a$ such that $A(t) = \int_0^t a(s)ds$ and the hazard function of $A_i$ is $a(t)e^{\beta^T Z_i}$, giving the reason why the model is called the proportional hazard model. Similar argument as in section 2 leads to the BFBB and PFBB likelihoods

$$L_n^B(\beta, A) = \prod_{t \in \mathcal{T}_n} \left( \prod_{i \in D(t)} (1 - (1 - \Delta A(t))^{e^{\beta^T Z_i}})^{\Delta N_i(t)} \right) (1 - \Delta A(t))^{\sum_{i \in R(t) \setminus D(t)} e^{\beta^T Z_i}}$$

$$L_n^P(\beta, A) = \prod_{t \in \mathcal{T}_n} \exp(\sum_{i \in D(t)} \beta^T Z_i) \Delta A(t)^{\Delta N(t)} \exp(-\Delta A(t) \sum_{i \in R(t)} \exp(\beta^T Z_i)).$$

Interestingly, these two forms of the likelihood were also used by frequentists too. See [14] for the BFBB likelihood and [13] for the PFBB likelihood.

We use the same prior for $A$ as in section 2. For the prior of $\beta$, we recommend the subjective prior. But, under mild conditions one can use the improper constant prior. See [4] for a sufficient condition for the propriety of the PFBB posterior.

The BFBB posterior with prior (4) for $\Delta A(t)$ and $\pi(\beta)$ for $\beta$ is given by

$$\pi(\beta, \Delta A | \text{Data}) \propto L_n^B(\beta, A) \prod_{t \in \mathcal{T}_n} \Delta A(t)^{-1} (1 - \Delta A(t))^{-1} \pi(\beta), \tag{7}$$

and the PFBB posterior of $\beta$ and $\Delta A(t)$ with prior (5) for $\Delta A(t)$ and $\pi(\beta)$ for $\beta$, is given by

$$\pi(\beta, \Delta A | \text{Data}) \propto \pi(\beta) \prod_{t \in \mathcal{T}_n} \exp(\sum_{i \in D(t)} \beta^T Z_i) \Delta A(t)^{\Delta N(t) - 1} \exp(-\Delta A(t) \sum_{i \in R(t)} e^{\beta^T Z_i}). \tag{8}$$

There is an interesting distributional structure to the PFBB posterior (8). The jump sizes $\Delta A(t)$ for $t \in \mathcal{T}_n$, given $\beta$, are independent and follow

$$\Delta A(t) | \beta, \text{ Data} \sim Gamma(\Delta N(t), \sum_{i \in R(t)} e^{\beta^T Z_i}).$$

Integrating out $\Delta A(t)$ from the PFBB posterior, we get the marginal posterior of $\beta$,

$$\pi(\beta | \text{Data}) \propto \pi(\beta) \prod_{t \in \mathcal{T}_n} \frac{\exp(\sum_{i \in D(t)} \beta^T Z_i)}{(\sum_{i \in R(t)} \exp(\beta^T Z_i))^{\Delta N(t)}}$$

$$= \pi(\beta) \times \text{Cox's partial likelihood}.$$

When the regression coefficient $\beta$ is the only parameter of interest, the PFBB gives a justification for the Bayesian analysis with the partial likelihood and a prior on $\beta$, as was done in [15].

**Remark.** Note that when $\beta = 0$, both BB posteriors of the proportional hazard model reduce to those of the right censored data.

**Remark.** Kim and Lee [4] showed that under mild conditions, both BB posterior distributions are asymptotically equivalent to the sampling distribution of the nonparametric maximum likelihood estimator.

### 4.2. Inference with MCMC

Inference with both BB posteriors can be done by sampling posterior samples using MCMC. In the following, we present an MCMC sampling scheme.

- **Sampling $\beta$ given $A$:** In both BB posteriors, the conditional posterior distributions of $\beta$ given $A$ are not well known distributions. But, our experience shows the random walk Metropolis-Hastings sampler works reasonably well. Since the Cox's partial likelihood is well approximated by a normal distribution, we expect that acceptance-rejection method with a normal proposal with variance an integer (typically 2 or 3) multiple of the frequentist variance estimator.

- **Sampling $A$ given $\beta$:** For the PFBB, the conditional distribution of $\Delta A(t)$ given $\beta$ is $Gamma(\Delta N(t), \sum_{i \in R(t)} e^{\beta^T Z_i})$. For the BFBB, random walk Metropolis-Hastings sampler works well. An alternative sampling method can be derived by modifying the algorithm of Laud, Damien and Smith [16] for the fixed jumps of the beta process posterior.

**Remark.** Kim and Lee [4] performed a simulation, which shows that the small sample frequentist's property of the BFBB is better than the standard frequentist's methods when the survival times are grouped.

## 5. Doubly Censored Data

### 5.1. Derivation

Let $X_i$, $i = 1, \ldots, n$, be independent identically distributed positive random variables, and independent of the $X_i$'s, $Y_i \geq Z_i$ are independent pairs of right and left censoring times. Under the doubly censoring mechanism, we observe only $D_n = \{(W_i, \delta_i) : i = 1, \ldots, n\}$, where

$$(W_i, \delta_i) = \begin{cases} (X_i, 1), & if \ Z_i < X_i \leq Y_i \\ (Y_i, 2), & if \ X_i > Y_i \\ (Z_i, 3), & if \ X_i \leq Z_i. \end{cases} \quad (9)$$

The censoring indicator $\delta_i$ is 1, 2, and 3, if the survival time $X_i$ is observed, right censored, and left censored, respectively. The first step of constructing the BB likelihood is to reduce $\mathcal{F}$ to a parametric model $\mathcal{F}_n$. Unlike in sections 2 and 3, there is no intuitively obvious choice for $\mathcal{F}_n$. Kim and Lee [5] suggest $\mathcal{F}_n = \{$probability measures with support $\mathbf{U}\}$, where $\mathbf{U} = \{u_1 < \cdots < u_l\}$ are defined as follows. Let $V_1 < \ldots < V_m$ be distinct points of $\{W_1, \ldots, W_n\}$ and define, for $k = 1, \ldots, m$, $\alpha_k^* = \sum_{i=1}^n I(W_i = V_k, \delta_i = 1)$, $\beta_k^* = \sum_{i=1}^n I(W_i = V_k, \delta_i = 2)$, and $\gamma_k^* = \sum_{i=1}^n I(W_i = V_k, \delta_i = 3)$. Let $\beta_0^* = 0$.

- Case 1. If $\alpha_i^* > 0$, then $V_i \in \mathbf{U}$.
- Case 2. If $\alpha_i^* = 0, \gamma_i^* > 0$, then
  1. if $\alpha_j^* = 0$ and $\gamma_j^* = 0$ for $j < i$, then $0 \in \mathbf{U}$, or
  2. if $\beta_{i-1}^* > 0$ and $\beta_j^* = 0$ and $\alpha_j^* = 0$ for all $i \leq j \leq m$ then $(V_{i-1} + V_i)/2 \in \mathbf{U}$.
- Case 3. If $\alpha_m^* = 0$ and $\beta_m^* > 0$ then $\infty \in \mathbf{U}$.

**Remark.** If $\alpha_1^* > 0$ and $\alpha_m^* > 0$, then $\mathbf{U}$ consists of only $V_i$s with $\alpha_i^* > 0$. That is, $\mathbf{U}$ has all distinct uncensored observations. Case 2 and Case 3 deal with the cases where the smallest or largest observation is censored.

The main motivation of the above choice of $\mathbf{U}$ is that the support of the limit of the full Bayesian posteriors with Dirichlet process priors is $\mathbf{U}$ provided that $\alpha_1^* > 0$ and $\alpha_m^* > 0$. Also, the maximum likelihood estimator of $S_X$ on $\mathcal{F}_n$ is an SCE [17].

Each element of $\mathcal{F}_n$ is parameterized by $\mathbf{p} = (p_1, \ldots, p_l)$ and the likelihood of $\mathbf{p}$ is

$$L(\mathbf{p}) = \prod_{k=1}^l p_k^{\alpha_k} \left(1 - \sum_{i=1}^k p_i\right)^{\beta_k} \left(\sum_{i=1}^k p_i\right)^{\gamma_k} \qquad (10)$$

where $\alpha_k = \sum_{i=1}^n I(W_i = u_k, \delta_i = 1)$, $\beta_k = \sum_{i=1}^n I(u_{k-1} \leq W_i < u_k, \delta_i = 2)$ and $\gamma_k = \sum_{i=1}^n I(u_{k-1} < W_i \leq u_k, \delta_i = 3)$ for $k = 1, \ldots, l$.

For prior on $\mathbf{p}$, the following improper prior is used:

$$\pi(\mathbf{p}) = \prod_{k=1}^l \frac{1}{p_k}. \qquad (11)$$

This prior is the same as that used in Rubin's Bayesian bootstrap.

**Remark.** Without censored observations, the BB posterior distribution from the BB likelihood (10) and prior (11) is the same as the Rubin's BB posterior [1]. Furthermore, the proposed BB posterior is equivalent to Lo's BB [18] for right censored data.

### 5.2. Inference with MCMC

The BB posterior of $\mathbf{p}$ is given as

$$\pi(\mathbf{p}|D_n) \propto \prod_{k=1}^l p_k^{\alpha_k} \left(1 - \sum_{i=1}^k p_i\right)^{\beta_k} \left(\sum_{i=1}^k p_i\right)^{\gamma_k} \prod_{k=1}^l \frac{1}{p_k}. \qquad (12)$$

The Bayesian computation with Gibbs sampler algorithm can be done as follows.

- **Sampling p given** $X$: Given $X$, the conditional posterior distribution of $\mathbf{p}$ is the same as Rubin's BB posterior.

- **Sampling $X$ given p:** Given **p**, $X_i$ can be sampled independently. If $X_i$ is not censored, $X_i$ does not need to be generated. If $X_i$ is right censored, $X_i$ is sampled from **p** with the condition $X_i > T_i$. A left censored $X_i$ can be sampled similarly.

**Remark.** For doubly censored data, Wellner and Zhan [19] proposed a weighted bootstrap procedure. Kim and Lee [5] showed that the BB posterior in (12) is the same as the weighted bootstrap distribution of Wellner and Zhan [19], also, using this result, they proved that the large sample property of the BB posterior is the same as that of the MLE.

**Remark.** Kim and Lee [5] showed that the BB posterior in (12) can be obtained as the limit of full Bayesian posteriors with Dirichlet process priors, which justifies the proposed BB from the Bayesian point of view.

## 6. Discussion

We believe that the BB serves as a bridge connecting the two territories - Bayesians and frequentists in nonparametric problems. Using the BB, frequetists can enjoy computational advantages of Bayesian procedures and Bayesians can strengthen objectivity by avoiding choosing priors for nonparametric objects.

The choice of priors in the BB should be investigated further. Since the priors used in this paper are improper, the posterior can be improper. For example, Kim and Lee [4] presented an example where the BB posterior is improper. We may consider data-dependent priors which provide desirable properties in the resulting posterior such as consistency, propriety, etc... Further research is necessary for this matter.

There are many other possible applications of the BB in survival analysis. Examples are interval censored data and random effect models. In such problems, the estimation of the variance of the MLE is computationally demanding. In contrast, BB may be implemented easily for those models with MCMC, and the computed BB posterior variance may be used as the variance estimate of the MLE.

## REFERENCES

1. D. B. Rubin, The Bayesian bootstrap, Ann. Statist. 9, (1981) 130-134.
2. B. Efron, Bootstrap methods: another look at the jackknife, Ann. Statist. 7 (1979) 1-26.
3. A. Y. Lo, A large-sample study of the Bayesian bootstrap, Ann. Statist. 15, (1987) 360-375.
4. Y. Kim and J. Lee, Bayesian bootstrap for the proportional hazard model, to appear in Ann. Statist. (2002a) http:\\home.ewha.ac.kr/~ydkim.
5. Y. Kim and J. Lee, Bayesian bootstrap for doubly censored data, (2002b) preprint, http:\\home.ewha.ac.kr/~ydkim.
6. T. S. Ferguson, A Bayesian analysis of some nonparametric problems, Ann. Statist. 1 (1973) 209-230.
7. M. Gasparini, Exact multivariate Bayesian bootstrap distributions of moments, Ann. Statist. 23 (1995) 762-768.

8. C. Weng, On a second-order asymptotic property of the Bayesian bootstrap mean, Ann. Statist. 17 (1989) 705-710.
9. A. B. Owen, Empirical likelihood ratio confidence intervals for a single functional, Biometrika 75, (1988) 237-249.
10. A. B. Owen, Empirical likelihood ratio confidence regions, Ann. Statist. 18, (1990) 90-120.
11. N. Choudhuri, Bayesian bootstrap credible sets for multidimensional mean functional, Ann. Statist. 26 (1998) 2104-2127.
12. N.A. Lazar, Bayesian empirical likelihood, Technical report 721, Statistics department, Carnegie Mellon University. (2000).
13. P.K. Andersen, O. Borgan, R. D. Gill, and N. Keiding, Statistical models based on counting processes, Springer, New York, 1993.
14. R. L. Prentice and L. A. Gloeckler, Regression analysis of grouped survival data with application to breast cancer data, Biometrics 34, (1978) 57-67.
15. T. C. Volinsky, D. Madigan, E. A. Raftery, and A. R. Kronmal, Bayesian model averaging in proportional hazard models: assessing stroke risk, App. Statist. 46, (1997) 433-448.
16. P. W. Laud, P. Damien, and A. F. M. Smith, Bayesian nonparametric and covariate analysis of failure time data, in Practical Nonparametric and Semiparametric Bayesian Statistics, D. Dey, P. Muller, and D. Sinha (eds.) 1998.
17. P. A. Mykland and J. J. Ren, Algorithms for computing self-consistent and maximum likelihood estimators with doubly censored data, Ann. Statist. 24, (1996) 1740-1764.
18. A. Y. Lo, A Bayesian bootstrap for censored data, Ann. Statist. 21, (1993) 100-123.
19. J. A. Wellner and Y. Zhan, Bootstrapping Z-estimators, Technical Report 308, Department of Statistics, University of Washington, (1996).

Recent Advances and Trends in Nonparametric Statistics
Michael G. Akritas and Dimitris N. Politis (Editors)
© 2003 Elsevier Science B.V. All rights reserved.

# On optimal variance estimation under different spatial subsampling schemes

Daniel J. Nordman[a] and Soumendra N. Lahiri[b] *

[a]Department of Statistics, University of Dortmund,
44221, Dortmund, Germany

[b]Department of Statistics, Iowa State University
Ames, Iowa 50011, USA

We provide a comprehensive examination of subsampling methods for spatial data on a grid. The considered subsamples may be scaled-down copies of the original sampling region or have a freely chosen shape. We derive the mean square error associated with general subsampling methods for estimating the variance of a large class of estimators. This yields an expression for the optimal subsample size for a given subsample shape. However, in contrast to the time series case, we show that the optimal subsample size and performance with each spatial subsampling method depends on the geometry of the sampling and subsampling regions in a nontrivial way. Examples for a few simple cases are presented to illustrate that both subsample size and shape may be selected to optimize spatial subsampling for variance estimation.

## 1. INTRODUCTION

Here we characterize several spatial subsampling methods for estimating the variance of a real-valued statistic $\hat{\theta}_n$ on a sampling region $R_n \subset \mathbb{R}^d$. The data consist of realizations of a spatial lattice process, collected from sampling sites lying within $R_n$. For variance estimation via subsampling, the basic idea is to construct several "smaller" sampling subregions that fit inside $R_n$, evaluate the analog of $\hat{\theta}_n$ on each of these subregions, and then compute a normalized sample variance.

In the literature, subsamples usually correspond to scaled-down copies of $R_n$ (named here Version I subsamples). In this paper, we consider general subsampling schemes, allowing for subregions to differ in geometry from the original sampling region. These Verison II subsamples can have a freely selected shape. The subsampling designs may involve either maximally overlapping (OL) or non-overlapping (NOL) subregions. We show that the accuracy (e.g. variance and bias) of subsample-based estimators depends crucially on the selection of both subsample size and shape. We then give expressions for the "best" choice for subsamples for variance estimation.

---

*Research partially supported by the US National Science Foundation (grant DMS 0072571) and the Deutsche Forschungsgemeinschaft (SFB 475).

Suppose subsampling regions $_sR_n = \lambda_n^* R_0^*$ are determined by inflating a "template" set $R_0^*$ by a factor $\lambda_n^*$. This template may be based on the geometry of $R_n$ (e.g. a scaled-down copy) or on an arbitrarily chosen geometry (e.g. rectangular, elliptical). Let $\hat{\tau}_{n,\text{OL}}^2$ and $\hat{\tau}_{n,\text{NOL}}^2$ denote variance estimators based on OL and NOL subsamples, respectively. We derive expansions for the variance and bias of $\hat{\tau}_{n,\text{OL}}^2$ and $\hat{\tau}_{n,\text{NOL}}^2$ for a large class of subsample templates. In particular, we show that:

1. The bias of $\hat{\tau}_{n,\text{OL}}^2$ and $\hat{\tau}_{n,\text{NOL}}^2$ depends heavily on *subsample* geometry, not that of $R_n$.

2. The variance of $\hat{\tau}_{n,\text{NOL}}^2$ is independent of the subsample template choice.

3. The variance of $\hat{\tau}_{n,\text{OL}}^2$ depends intricately on the *subsample* geometry.

4. The subsample shape solely determines the relative efficiency of $\hat{\tau}_{n,\text{OL}}^2$ to $\hat{\tau}_{n,\text{NOL}}^2$.

Comparisons of the performance of $\hat{\tau}_{n,\text{OL}}^2$ or $\hat{\tau}_{n,\text{NOL}}^2$ are possible for different template geometries, allowing the selection of a "good" subsample template. In addition, expressions can be readily determined for optimal subsample size or scaling for general templates and sampling regions $R_n$. Our results apply to lattice data from time series, agricultural field trials, and remote sensing/image analysis (medical and satellite image processing).

Version I subsampling for variance estimation has been previously considered for some specific sampling regions. Politis and Romano [23,25] proposed subsampling variance estimators for a $\mathbb{R}^d$-rectangular sampling region $\mathbb{Z}^d \cap \prod_{i=1}^d [1, n_i]$ using integer translates of $\prod_{i=1}^d [1, m_i]$ as subsamples ($m_i \leq n_i$); see [8], [27] for similar subsampling frameworks. Sherman and Carlstein [29,30] used a more flexible Version I subsampling formulation for variance estimation in $\mathbb{R}^2$: the "inside" of a simple closed curve defines a set $D \subset [-1, 1]^2$, $\mathbb{Z}^2 \cap nD$ provides the sampling region, and translates of $mD$ within $nD$ create subsamples for some $m \leq n$. For their subsampling schemes, [23] and [29] determined the optimal *order* of subsample size (i.e., $\prod_{i=1}^d m_i = O([\prod_{i=1}^d n_i]^{d/(d+2)})$ for $\mathbb{R}^d$-rectangles and $m = O(n^{1/2})$ for inflated $\mathbb{R}^2$ curves). However, an exact determination of optimal scaling requires additional adjustments to proportion a rough order of subsample size appropriately for a given sampling region $R_n$. Considering Version I subsamples with a general $R_n \subset \mathbb{R}^d$, Nordman and Lahiri [22] derived expressions of optimal subsample sizes for a broad class of sampling regions. Here we generalize those results to include subsamples which are not necessarily scaled-down replicates of $R_n$ (i.e., Version II subsamples).

The rest of the paper is organized as follows. In Section 2, we describe the spatial subsampling methods for variance estimation. In Section 3, we state assumptions and give the variance and bias of the subsampling estimators. Theoretical optimal subsample scaling is provided in Section 4. In Section 5, subsample template choice (Version I vs Version II subsamples) is discussed and illustrated with some examples. Section 6 contains proofs of the main results.

## 2. SUBSAMPLE VARIANCE ESTIMATORS

For clarity, we give an overview of this section. In Section 2.1, we describe the sampling design and the structure of the sampling region. Section 2.2 explains the considered subsample templates. Section 2.3 frames the variance estimation problem under the smooth function model. The OL and NOL subsampling schemes with corresponding variance estimators are presented in Section 2.4.

## 2.1. Sampling design

As in [22], we assume potential sampling sites are located on a translate of the rectangular integer lattice in $\mathbb{R}^d$. For a fixed (chosen) vector $\mathbf{t} \in [-1/2, 1/2]^d$, identify the $\mathbf{t}$-translated integer lattice as $\mathbf{Z}^d \equiv \mathbf{t} + \mathbb{Z}^d$. Let $\{Z(\mathbf{s}) \mid \mathbf{s} \in \mathbf{Z}^d\}$ be a stationary weakly dependent random field (hereafter r.f.) taking values in $\mathbb{R}^p$. (Bold font will be used to denote vectors in the space of sampling $\mathbb{R}^d$ and normal font for vectors in $\mathbb{R}^p$, including $Z(\cdot)$.) We suppose that the r.f. $Z(\cdot)$ is observed at sampling sites lying within the sampling region $R_n \subset \mathbb{R}^d$. That is, the available data are $\{Z(\mathbf{s}) : \mathbf{s} \in \mathbf{Z}^d \cap R_n\}$.

We use an "increasing domain" framework to obtain the results in the paper (c.f. [5]), whereby the sampling region $R_n$ becomes unbounded as the sample size increases. Let $R_0$ be a Borel subset of $(-1/2, 1/2]^d$ containing an open neighborhood of the origin such that for any sequence of positive real numbers $a_n \to 0$, the number of cubes of the scaled lattice $a_n \mathbb{Z}^d$ which intersect the closures $\overline{R_0}$ and $\overline{R_0^c}$ is $O((a_n^{-1})^{d-1})$ as $n \to \infty$. Let $\Delta_n$ be a sequence of $d \times d$ diagonal matrices, with positive diagonal elements $\lambda_{n,1}, \ldots, \lambda_{n,d}$, such that each $\lambda_{n,i} \to \infty$ as $n \to \infty$. The sampling region $R_n$ is then obtained by "inflating" the template set $R_0$ by the directional scaling factors $\Delta_n$: $R_n = \Delta_n R_0$.

The formulation given above allows $R_n$ to have a large variety of fairly irregular shapes with the boundary condition on $R_0$ imposed to avoid pathological cases. Some common examples of such regions are convex subsets of $\mathbb{R}^d$, such as spheres, ellipsoids, polyhedrons, as well as certain non-convex subsets with irregular boundaries, such as star-shaped regions. Sherman and Carlstein [29,30] consider a similar class of regions in the plane (i.e. $d = 2$) where the boundaries of the sets $R_0$ are delineated by simple, finite-length curves. Because the origin is assumed to lie in $R_0$, the sampling region $R_n$ grows outward in all directions as $n$ increases. The location of sampling sites on $\mathbf{Z}^d$, rather than $\mathbb{Z}^d$, implies that $R_n$ does not necessarily expand around a potential sampling site at the origin.

## 2.2. Subsample verions

Before presenting variance estimators $\hat{\tau}_{n,\mathrm{OL}}^2$ and $\hat{\tau}_{n,\mathrm{NOL}}^2$, we define the subsample structure. Let $\lambda_n^*$ be a positive sequence such that $\lambda_n^* \to \infty$ and $\lambda_n^*/\lambda_{n,i} \to 0$ as $n \to \infty$, for each $i = 1, \ldots, d$. We make the "prototype" subsampling region,

$$_s R_n = \lambda_n^* R_0^*, \qquad (1)$$

based on a "template" set $R_0^* \subset (-1/2, 1/2]^d$. The set $R_0^*$ is taken to have the same interior and border properties as $R_0$ (see Section 2.1).

As in [22,29,30], a common scaling factor in all directions is used in (1) to create the subregions. (Varying amounts of scaling or inflating for $R_0^*$ in different directions can be directly incorporated in the template choice $R_0^*$.) The subsample structure in (1) also allows for lattice point counting techniques required to find bias expansions of $\hat{\tau}_{n,\mathrm{OL}}^2$ and $\hat{\tau}_{n,\mathrm{NOL}}^2$ in Section 3; see [22] for further discussion on subsample scaling issues.

We now make the distinction between two subsample versions as a function of the template choice $R_0^*$.

Version I Subsamples: scaled-down template. Most often in the literature, scaled down copies of the original observation region $R_n = \Delta_n R_0$ are used as subsamples (c.f. [22,25, 29,30]). In our subsampling formulation, this can be accommodated with the template choice $R_0^* = R_0$. (In practice, $R_0$ can be defined as the set in $\{\Delta_n^{-1} R_n \subset (-1/2, 1/2]^d :$ positive diagonal $\Delta_n\}$ with largest volume.)

Version II Subsamples: freely chosen template. We allow here for freely selected shapes to determine the subregions. For example, rectangular subsampling regions based on $R_0^* = (-1/2, 1/2]^d$ may be used regardless of the shape of $R_n \subset \mathbb{R}^d$. This variation on subsampling has received little consideration, but it can also yield consistent variance estimators, as we will show.

### 2.3. Smooth function model

Let $H : \mathbb{R}^p \to \mathbb{R}$ be a smooth function and $\bar{Z}_{N_n} = N_n^{-1} \sum_{\mathbf{s} \in \mathbb{Z}^d \cap R_n} Z(\mathbf{s})$ denote the sample mean of the $N_n$ sites within $R_n$. We suppose that the relevant statistic, whose variance we wish to estimate, can be represented as a function of the sample mean: $\hat{\theta}_n = H(\bar{Z}_{N_n})$. The statistic $\hat{\theta}_n$ estimates a population parameter of interest $\theta = H(\mu)$, involving the mean of the stationary r.f. $\mathrm{E}\{Z(\mathbf{t})\} = \mu \in \mathbb{R}^p$.

This parameter and estimator formulation is what Hall [10] calls the "smooth function" model and it has been used in other scenarios, such as with the moving block bootstrap (MBB) (see [13,18]) and empirical likelihood, for studying approximately linear functions of a sample mean (c.f. [6,14]). With suitable functions of the $Z(\mathbf{s})$'s, one can represent a wide range of estimators under the present framework. In particular, these include means, differences and ratios of means, sample moments, spatial variograms and correlograms, likelihood-based estimators of process parameters (c.f. [28], pseudo-likelihood estimators under conditional autogressive models), and test statistics for spatial autocorrelation (c.f. [4]).

The quantity which we seek to estimate nonparametrically via subsampling is the variance of the normalized statistic $\sqrt{N_n}\hat{\theta}_n$, say, $\tau_n^2 = N_n \mathrm{E}(\hat{\theta}_n - \mathrm{E}\hat{\theta}_n)^2$.

### 2.4. Subsampling variance estimators

For a given subsample template $R_0^*$, we consider variance estimators from both OL and NOL subsampling schemes.

#### Overlapping subsamples

Identify a subset of $\mathbb{Z}^d$, say $J_{\mathrm{OL}}$, corresponding to all integer translates of $_s R_n$ from (1) lying within $R_n$. That is, $J_{\mathrm{OL}} = \{\mathbf{i} \in \mathbb{Z}^d : \mathbf{i} + {_s R_n} \subset R_n\}$. The desired OL subsampling regions are precisely the translates of $_s R_n$ given by: $\mathbf{i} + {_s R_n}$, $\mathbf{i} \in J_{\mathrm{OL}}$.

Let $_s N_n = |\mathbb{Z}^d \cap {_s R_n}|$ be the number of sampling sites in $_s R_n$ and let $|J_{\mathrm{OL}}|$ denote the number of available OL subsampling regions. (The number of sampling sites within each OL subregion is the same $_s N_n = |\mathbb{Z}^d \cap (\mathbf{i} + {_s R_n})|$ for any $\mathbf{i} \in J_{\mathrm{OL}}$.) For each $\mathbf{i} \in J_{\mathrm{OL}}$, compute $\hat{\theta}_{\mathbf{i},n} = H(\bar{Z}_{\mathbf{i},n})$, using the subsample mean $\bar{Z}_{\mathbf{i},n}$ of the subregion $\mathbf{i} + {_s R_n}$. We then write the OL subsample variance estimator of $\tau_n^2$ as

$$\hat{\tau}_{n,\mathrm{OL}}^2 = |J_{\mathrm{OL}}|^{-1} \sum_{\mathbf{i} \in J_{\mathrm{OL}}} {_s N_n} \left(\hat{\theta}_{\mathbf{i},n} - \tilde{\theta}_n\right)^2, \quad \tilde{\theta}_n = |J_{\mathrm{OL}}|^{-1} \sum_{\mathbf{i} \in J_{\mathrm{OL}}} \hat{\theta}_{\mathbf{i},n}.$$

### Non-overlapping subsamples

We adopt a formulation similar to [15,22,30] to create NOL subsamples. The sampling region $R_n$ is first divided into disjoint "cubes" and the scaling factor $\lambda_n^*$ from (1) is used to determine the "window width" of partitioning cubes. Let $J_{\text{NOL}} = \{\mathbf{i} \in \mathbb{Z}^d : \lambda_n^*(\mathbf{i} + (-1/2, 1/2]^d) \subset R_n\}$ represent the set of all "inflated" subcubes that lie inside $R_n$. Denote its cardinality as $|J_{\text{NOL}}|$. For each $\mathbf{i} \in J_{\text{NOL}}$, define a subsampling region $\lambda_n^*(\mathbf{i} + R_0^*)$ by inscribing a translate of (1) such that the origin is mapped onto the midpoint of the cube $\lambda_n^*(\mathbf{i} + (-1/2, 1/2]^d)$. This provides a collection of NOL subsampling regions, which are replicates of $_sR_n$, inside $R_n$.

For each $\mathbf{i} \in J_{\text{NOL}}$, the function $H(\cdot)$ is evaluated at the subsample mean, say $\tilde{Z}_{\mathbf{i},n}$, for a corresponding NOL subsampling region $\lambda_n^*(\mathbf{i} + R_0^*)$ to obtain $\hat{\theta}_{\mathbf{i},n} = H(\tilde{Z}_{\mathbf{i},n})$. The NOL subsample estimator of $\tau_n^2$ is again an appropriately scaled sample variance:

$$\hat{\tau}_{n,\text{NOL}}^2 = |J_{\text{NOL}}|^{-1} \sum_{\mathbf{i} \in J_{\text{NOL}}} {}_sN_{\mathbf{i},n} \left(\hat{\theta}_{\mathbf{i},n} - \tilde{\theta}_n\right)^2, \quad \tilde{\theta}_n = |J_{\text{NOL}}|^{-1} \sum_{\mathbf{i} \in J_{\text{NOL}}} \hat{\theta}_{\mathbf{i},n}$$

where ${}_sN_{\mathbf{i},n} = |\mathbb{Z}^d \cap \lambda_n^*(\mathbf{i} + R_0^*)|$ denotes the number of sampling sites within a given NOL subsample.

Note that ${}_sN_{\mathbf{i},n}$ may differ between NOL subsamples, but all such subsamples will have exactly ${}_sN_{\mathbf{i},n} = {}_sN_n$ sites if $\lambda_n^*$ is integer-valued.

## 3. PERFORMANCE OF SUBSAMPLE ESTIMATORS

In this section, we discuss the effects of subsample geometry on the performance of subsample variance estimators. Namely, we give expressions for the variance and bias of $\hat{\tau}_{n,\text{OL}}^2$ and $\hat{\tau}_{n,\text{NOL}}^2$, which apply to both subsample template versions from Section 2.2.

### 3.1. Assumptions

To state the assumptions, some notation is required. For a vector $\mathbf{x} = (x_1, ..., x_d)' \in \mathbb{R}^d$, let $\|\mathbf{x}\|$ and $\|\mathbf{x}\|_1 = \sum_{i=1}^d |x_i|$ denote the usual Euclidean and $l^1$ norms of $\mathbf{x}$, respectively. Denote the $l^\infty$ norm as $\|\mathbf{x}\|_\infty = \max_{1 \leq k \leq d} |x_k|$. Define $\text{dis}(E_1, E_2) = \inf\{\|\mathbf{x} - \mathbf{y}\|_\infty : \mathbf{x} \in E_1, \mathbf{y} \in E_2\}$ for two sets $E_1, E_2 \subset \mathbb{R}^d$. We shall use $|B|$ to denote the cardinality of a countable set $B \subset \mathbb{R}^d$; for an uncountable set $A \subset \mathbb{R}^d$, $\text{vol}(A)$ will refer to the volume (i.e., the $\mathbb{R}^d$ Lebesgue measure) of $A$.

Let $\mathcal{F}_Z(T) = \sigma\langle Z(\mathbf{s}) : \mathbf{s} \in T\rangle$ be the $\sigma$-field generated by the variables $\{Z(\mathbf{s}) : \mathbf{s} \in T\}$, $T \subset \mathbb{Z}^d$. For $T_1, T_2 \subset \mathbb{Z}^d$, write

$$\tilde{\alpha}(T_1, T_2) = \sup\{|P(A \cap B) - P(A)P(B)| \ : \ A \in \mathcal{F}_Z(T_1), B \in \mathcal{F}_Z(T_2)\}.$$

Then, the strong mixing coefficient for the r.f. $Z(\cdot)$ is defined as

$$\alpha(k,l) = \sup\{\tilde{\alpha}(T_1, T_2) : T_i \subset \mathbb{Z}^d, |T_i| \leq l, \ i = 1, 2; \ \text{dis}(T_1, T_2) \geq k\} \qquad (2)$$

For proving the subsequent theorems, the assumptions below are needed along with two conditions stated as functions of a positive argument $r \in \mathbb{Z}_+ = \{0, 1, 2, \ldots\}$. In the following, $\det(\Delta)$ represents the determinant of a square matrix $\Delta$. For $\alpha = (\alpha_1, ..., \alpha_p)' \in \mathbb{Z}_+^p$, let $D^\alpha$ denote the $\alpha$th order partial differential operator $\partial^{\alpha_1 + \cdots + \alpha_p}/\partial x_1^{\alpha_1} \ldots \partial x_p^{\alpha_p}$ and

$\nabla = (\partial H(\mu)/\partial x_1, \ldots, \partial H(\mu)/\partial x_p)'$ be the vector of first order partial derivatives of $H$ at $\mu$. Limits in order symbols are taken letting $n$ tend to infinity.

**Assumptions:**

(A.1) For the scaling factors of the sampling and subsampling regions:
$$\frac{1}{\lambda_n^*} + \sum_{i=1}^{d} \frac{\lambda_n^*}{\lambda_{n,i}} + \frac{(\lambda_n^*)^{(d+1)}}{\det(\Delta_n)} = o(1), \quad \max_{1 \le i \le d} \lambda_{n,i} = O\Big(\min_{1 \le i \le d} \lambda_{n,i}\Big).$$

(A.2) There exist nonnegative functions $\alpha_1(\cdot)$ and $g(\cdot)$ such that $\lim_{k \to \infty} \alpha_1(k) = 0$, $\lim_{l \to \infty} g(l) = \infty$ and the strong-mixing coefficient $\alpha(k,l)$ from (2) satisfies the inequality
$$\alpha(k,l) \le \alpha_1(k) g(l) \quad k > 0, \ l > 0.$$

(A.3) $\sup\{\tilde{\alpha}(T_1,T_2) : T_1, T_2 \subset \mathbf{Z}^d, |T_1| = 1, \text{dis}(T_1,T_2) \ge k\} = o(k^{-d})$.

(A.4) $\tau^2 > 0$, where $\tau^2 = \sum_{\mathbf{k} \in \mathbf{Z}^d} \sigma(\mathbf{k})$, $\sigma(\mathbf{k}) = \text{Cov}(\nabla' Z(\mathbf{t}), \nabla' Z(\mathbf{t} + \mathbf{k}))$.

**Conditions:**

$D_r$: $H : \mathbb{R}^p \to \mathbb{R}$ is $r$-times continuously differentiable and, for some $a \in \mathbb{Z}_+$ and real $\mathcal{C} > 0$, $\max\{|D^\nu H(\mathbf{x})| : \|\nu\|_1 = r\} \le \mathcal{C}(1 + \|\mathbf{x}\|^a)$, $\mathbf{x} \in \mathbb{R}^p$.

$M_r$: For some $0 < \delta \le 1$, $0 < \kappa < (2r - 1 - 1/d)(2r + \delta)/\delta$, and $\mathcal{C} > 0$,
$$\mathbb{E}\|Z(\mathbf{t})\|^{2r+\delta} < \infty, \quad \sum_{x=1}^{\infty} x^{(2r-1)d-1} \alpha_1(x)^{\delta/(2r+\delta)} < \infty, \quad g(x) \le \mathcal{C} x^\kappa.$$

In Assumption A.2, we formulate a conventional bound on the mixing coefficient from (2) which is applicable to many r.f.s (c.f. [15,16]). Examples of r.f.s that meet the requirements of A.2 and $M_r$ include Gaussian fields with analytic spectral densities, some linear fields with a moving average (like $m$-dependent fields) or autoregressive representation, separable lattice processes suggested by Martin [19] for modeling in $\mathbb{R}^2$, many Gibbs and Markov fields, and important time series models (c.f. [7]). Assumption A.3 permits the CLT in Bolthausen [1] to be applied to sums of $Z(\cdot)$ on sets of increasing domain. Assumption A.4 implies a positive, finite asymptotic variance $\tau^2 = \lim_{n \to \infty} \tau_n^2$. For more details about the assumptions and the conditions, see [22].

### 3.2. Variance expansions

We now give expansions for the asymptotic variance of the subsample variance estimators $\hat{\tau}_{n,\text{OL}}^2$ and $\hat{\tau}_{n,\text{NOL}}^2$ of $\tau_n^2 = N_n \text{Var}(\hat{\theta}_n)$. These results are valid for *any* subsample shape $_sR_n = \lambda_n^* R_0^*$ determined through the template $R_0^*$. Theorem 1 essentially extends results in [22], which consider Version I subsample templates ($R_0^* = R_0$).

**Theorem 1** *Suppose that Assumptions* A.1 - A.4 *and Conditions* $D_2$ *and* $M_{5+2a}$ *hold, then*
$$\text{Var}(\hat{\tau}_{n,\text{OL}}^2) = K_* \cdot \frac{(\lambda_n^*)^d}{\text{vol}(R_n)} \left[2\tau^4\right](1 + o(1)) \quad \text{and} \quad \text{Var}(\hat{\tau}_{n,\text{NOL}}^2) = \frac{(\lambda_n^*)^d}{\text{vol}(R_n)} \left[2\tau^4\right](1 + o(1)),$$

where $K_* = \int_{\mathbb{R}^d} \left( \dfrac{\text{vol}\left[R_0^* \cap (\mathbf{x} + R_0^*)\right]}{\text{vol}(R_0^*)} \right)^2 d\mathbf{x}$ is an integral with respect to the $\mathbb{R}^d$ Lebesgue measure.

The above variance result has some important implications:

- The variance of the NOL subsample estimator $\text{Var}(\hat{\tau}_{n,\text{NOL}}^2)$ does *not* depend on the shape of the subsamples; only the subsample scaling $\lambda_n^*$ has influence.

- Under the "smooth" function model, the asymptotic variance $\text{Var}(\hat{\tau}_{n,\text{OL}}^2)$ is smaller than $\text{Var}(\hat{\tau}_{n,\text{NOL}}^2)$ by

$$K_* = \lim_{n\to\infty} \frac{\text{Var}(\hat{\tau}_{n,\text{OL}}^2)}{\text{Var}(\hat{\tau}_{n,\text{NOL}}^2)} < 1. \tag{3}$$

The geometry of subsample template $R_0^*$ solely determines the reduction $K_*$ in the variance of the OL subsample estimator.

Table 1 provides some examples of $K_*$ for various $\mathbb{R}^d$ templates.

Table 1
Examples of variance reduction factor $K_*$ for $R_0^* \subset (-1/2, 1/2]^d$

| Template shape | $R_0^*$ | $K_*$ |
|---|---|---|
| $\mathbb{R}^d$-Rectangle | $(-1/2, 1/2]^d$ | $(2/3)^d$ |
| $\mathbb{R}^3$-Sphere | radius $1/2$ about origin | $17\pi/315$ |
| $\mathbb{R}^3$-Cylinder | base radius $1/2$, height $1$ | $\pi/6 - 8/(9\pi)$ |
| $\mathbb{R}^2$-Circle | radius $1/2$ about origin | $\pi/4 - 4/(3\pi)$ |
| $\mathbb{R}^2$-Right triangle | vertices $\{(x_i, y_i)\}_{i=1}^3, x_i, y_i \in \{\pm 1/2\}$ | $1/5$ |

We remark that both $\hat{\tau}_{n,\text{OL}}^2$ and $\hat{\tau}_{n,\text{NOL}}^2$ can be shown to be (MSE) consistent for both Version I and II subsamples under Theorem 1 conditions. This extends the consistency property of subsample variance estimation to more general spatial sampling and subsampling regions than previously considered.

### 3.3. Bias expansions

We attempt to precisely describe the leading order terms in the asymptotic bias of $\hat{\tau}_{n,\text{OL}}^2$ and $\hat{\tau}_{n,\text{NOL}}^2$ for both Version I and II templates in (1). Under the smooth function model, the dominant component in the bias of either estimator can be shown to be $O(1/\lambda_n^*)$ for all dimensions of sampling and both subsample template versions (c.f. Lemma 1 of [22] for Version I subsamples). We would like to identify this $O(1/\lambda_n^*)$ bias term.

We give the required components for finding the biases of the spatial subsample variance estimators in the next theorem. Let $C_n(\mathbf{k}) \equiv |\mathbf{Z}^d \cap {}_sR_n \cap (\mathbf{k} + {}_sR_n)|$ denote the number of pairs of observations in the subsampling region ${}_sR_n$ separated by a translate $\mathbf{k} \in \mathbb{Z}^d$.

**Theorem 2** *Suppose $d \geq 2$, Assumptions A.1 - A.4, and Conditions $D_3$ and $M_{3+a}$ hold. If, in addition, $\lambda_n^* \in \mathbb{Z}_+$ for NOL subsamples and*

$$\lim_{n\to\infty} \frac{{}_sN_n - C_n(\mathbf{k})}{(\lambda_n^*)^{d-1}} = C(\mathbf{k}) \tag{4}$$

exists for all $\mathbf{k} \in \mathbb{Z}^d$, then

$$E(\hat{\tau}_n^2) - \tau_n^2 = \frac{-1}{\lambda_n^* \text{vol}(R_0^*)} \left( \sum_{\mathbf{k} \in \mathbb{Z}^d} C(\mathbf{k})\sigma(\mathbf{k}) \right) (1 + o(1))$$

where $\sigma(\mathbf{k}) = \text{Cov}\left(\nabla' Z(\mathbf{t}), \nabla' Z(\mathbf{t}+\mathbf{k})\right)$ and where $\hat{\tau}_n^2$ is either $\hat{\tau}_{n,\text{OL}}^2$ or $\hat{\tau}_{n,\text{NOL}}^2$.

The above result shows that the chosen *subsample shape*, through the template $R_0^*$, heavily influences the bias of $\hat{\tau}_{n,\text{OL}}^2$ or $\hat{\tau}_{n,\text{NOL}}^2$. These biases do not depend directly on the geometry of the original sampling region $R_n$.

To compute "closed-form" bias expansions for $\hat{\tau}_{n,\text{OL}}^2$, we approximate lattice point counts. For each $\mathbf{k} \in \mathbb{Z}^d$, we replace the numerator of (4) with the difference of corresponding Lebesgue volumes. The approximation error

$$\left(_sN_n - C_n(\mathbf{k})\right) - \left(\text{vol}(_sR_n) - \text{vol}\left[_sR_n \cap (\mathbf{k} + _sR_n)\right]\right)$$

can be shown to be small enough, for some templates $R_0^*$, to justify replacing counts with volumes. Nordman and Lahiri [22] further discuss this volume-count approximation. We use this counting technique to give bias expansions for a large class of subsampling regions in $\mathbb{R}^d$, $d \leq 3$, which are "nearly" convex. That is, the prototype subregion $_sR_n$ differs possibly from a convex set only at its boundary.

Some notation is additionally required. For $\alpha = (\alpha_1, ..., \alpha_p)' \in (\mathbb{Z}_+)^p$, $\mathbf{x} \in \mathbb{R}^p$, write $\mathbf{x}^\alpha = \prod_{i=1}^p x_i^{\alpha_i}$, $\alpha! = \prod_{i=1}^p (\alpha_i!)$, and $c_\alpha = D^\alpha H(\mu)/\alpha!$. Let $Z_\infty$ denote a random vector with a normal $\mathcal{N}(0, \Sigma_\infty)$ distribution on $\mathbb{R}^p$, where $\Sigma_\infty$ is the limiting covariance matrix of the scaled sample mean $\sqrt{N_n}(\bar{Z}_{N_n} - \mu)$. Let $B^\circ$, $\overline{B}$ denote the interior and closure of $B \in \mathbb{R}^d$, respectively. Below let $\sigma(\mathbf{k}) = \text{Cov}\left(\nabla' Z(\mathbf{t}), \nabla' Z(\mathbf{t}+\mathbf{k})\right)$, $\mathbf{k} \in \mathbb{Z}^d$.

**Theorem 3** *Suppose that $B^\circ \subset R_0^* \subset \overline{B}$ for a convex set $B$. With Assumptions A.1 - A.4, assume Conditions $D_{5-d}$ and $M_{5-d+a}$ hold for $d \in \{1, 2, 3\}$. Then,*

$$C(\mathbf{k}) = V(\mathbf{k}) \equiv \lim_{n \to \infty} \frac{\text{vol}(_sR_n) - \text{vol}\left[_sR_n \cap (\mathbf{k} + _sR_n)\right]}{(\lambda_n^*)^{d-1}}, \quad \mathbf{k} \in \mathbb{Z}^d$$

*whenever $V(\mathbf{k})$ exists and the biases $E(\hat{\tau}_{n,\text{OL}}^2) - \tau_n^2$, $E(\hat{\tau}_{n,\text{NOL}}^2) - \tau_n^2$ are equal to:*

*for $d = 1$,*
$$\frac{-1}{\lambda_n^* \text{vol}(R_0^*)} \left( \sum_{\mathbf{k} \in \mathbb{Z}} |\mathbf{k}| \sigma(\mathbf{k}) + C_\infty \right) (1 + o(1));$$

*for $d = 2$ or $3$,*
$$\left( -\sum_{\mathbf{k} \in \mathbb{Z}^d} \frac{\text{vol}(_sR_n) - \text{vol}\left[_sR_n \cap (\mathbf{k} + _sR_n)\right]}{\text{vol}(_sR_n)} \sigma(\mathbf{k}) \right) (1 + o(1))$$

*or*
$$\frac{-1}{\lambda_n^* \text{vol}(R_0^*)} \left( \sum_{\mathbf{k} \in \mathbb{Z}^d} V(\mathbf{k}) \sigma(\mathbf{k}) \right) (1 + o(1)), \text{ provided each } V(\mathbf{k}) \text{ exists};$$

where $C_\infty = \text{Var}\left(\sum_{\|\alpha\|_1=2} \frac{c_\alpha}{\alpha!} Z_\infty^\alpha\right) + 2\sum_{\substack{\|\alpha\|_1=1,\\ \|\beta\|_1=3}} \frac{c_\alpha c_\beta}{\beta!} \text{E}\left(Z_\infty^\alpha Z_\infty^\beta\right)$

$+ 2\sum_{\mathbf{k}_1,\mathbf{k}_2 \in \mathbb{Z}} \sum_{\substack{\|\alpha\|_1=1,\\ \|\beta\|_1=1, \|\gamma\|_1=1}} \frac{c_\alpha c_{(\beta+\gamma)}}{(\beta+\gamma)!} \text{E}\left([Z(\mathbf{t})-\mu]^\alpha [Z(\mathbf{t}+\mathbf{k}_1)-\mu]^\beta [Z(\mathbf{t}+\mathbf{k}_2)-\mu]^\gamma\right).$

Hence, we may write the bias of the estimators through the limiting, scaled volume differences $V(\mathbf{k})$. This simplifies the task of expanding these biases.

### 3.3.1. Further bias expansions: rectangular subsamples

The bias results of Theorem 3 also seem plausible for convex sampling regions in $\mathbb{R}^d$, $d \geq 4$, but require further study of lattice point counting techniques in higher dimensions. However, bias expansions for $\hat{\tau}_{n,\text{OL}}^2$ and $\hat{\tau}_{n,\text{NOL}}^2$ are possible for an important, widely applicable class of rectangular subsampling regions based on the template $R_0^* = (-1/2, 1/2]^d$, which can then be used with optimal subsample scaling. We give the bias expansions for $\hat{\tau}_{n,\text{OL}}^2$ and $\hat{\tau}_{n,\text{NOL}}^2$ with general rectangular subsamples, allowing also sampling sites to be potentially excluded from the borders of a subsampling region $_s R_n$.

**Theorem 4** *Let* $(-1/2, 1/2]^d \subset \Lambda_\ell^{-1} R_0^* \subset [-1/2, 1/2]^d$, $d \geq 3$ *for a* $d \times d$ *diagonal matrix* $\Lambda_\ell$ *with entries* $0 < \ell_i \leq 1$, $i = 1, ..., d$. *Suppose Assumptions* A.1 - A.4 *and Conditions* $D_2$ *and* $M_{2+a}$ *hold. If* $\hat{\tau}_n^2$ *is either* $\hat{\tau}_{n,\text{OL}}^2$ *or* $\hat{\tau}_{n,\text{NOL}}^2$, *then*

$$\text{E}(\hat{\tau}_n^2) - \tau_n^2 = -\frac{1}{\lambda_n^*}\left(\sum_{\mathbf{k} \in \mathbb{Z}^d} \sum_{i=1}^d \frac{|k_i|}{\ell_i} \sigma(\mathbf{k})\right)(1 + o(1)).$$

In particular, the main bias component of $\hat{\tau}_{n,\text{OL}}^2$ or $\hat{\tau}_{n,\text{NOL}}^2$ using the subsample template $R_0^* = (-1/2, 1/2]^d$ is simply $-(\lambda_n^*)^{-1} \sum_{\mathbf{k} \in \mathbb{Z}^d} \|\mathbf{k}\|_1 \sigma(\mathbf{k})$.

### 3.3.2. Further bias expansions: unions of convex sets

Exact bias expansions for $\hat{\tau}_{n,\text{OL}}^2, \hat{\tau}_{n,\text{NOL}}^2$ in terms of $V(\mathbf{k})$ appear possible for subsampling templates $R_0^*$ created from finite unions of convex sets whose borders do not intersect more than finitely often. More irregular subsampling regions, including polygonal and non-convex regions, could then be constructed through unions of convex sets. We do not explore this possibility at great length in this paper, but we provide a small step toward an extension in the following lemma concerning subsamples which are a union of two "nearly" convex $\mathbb{R}^2$ sets.

**Lemma 1** *Let* $R_0^* = R_1^* \cup R_2^* \subset (-1/2, 1/2]^2$. *Suppose there exists convex sets* $A_1$, $A_2$ *such that* $A_i^\circ \neq \emptyset$, $A_i^\circ \subset R_i^* \subset \overline{A_i}$ *for* $i = 1, 2$. *Assume that* $\partial A_1 \cap \partial A_2 = (\overline{A_1} \setminus A_1^\circ) \cap (\overline{A_2} \setminus A_2^\circ)$ *is empty or finite. Then, Theorem 3 still holds under the same conditions.*

*Furthermore,* $V(\mathbf{k}) = V^*(\mathbf{k})$ *whenever* $V^*(\mathbf{k}) \equiv \lim_{n\to\infty}\left(\sum_{i=1}^2 V_{i,n}^*(\mathbf{k}) - V_n^*(\mathbf{k})\right)/\lambda_n^*$ *exists for* $\mathbf{k} \in \mathbb{Z}^2$, *where:* $V_n^*(\mathbf{k}) = \text{vol}\{\lambda_n^*(R_1^* \cap R_2^*)\} - \text{vol}\{\lambda_n^*(R_1^* \cap R_2^*) \cap (\mathbf{k} + \lambda_n^*(R_1^* \cap R_2^*))\}$ *and* $V_{i,n}^*(\mathbf{k}) = \text{vol}(\lambda_n^* R_i^*) - \text{vol}\{\lambda_n^* R_i^* \cap (\mathbf{k} + \lambda_n^* R_i^*)\}$ *for* $i = 1, 2$.

The bias formulation with $V^*(\mathbf{k})$ can be determined by separately working with the areas of the (nearly) convex regions $\lambda_n^* R_1^*$, $\lambda_n^* R_2^*$ and their convex intersection $\lambda_n^*(R_1^* \cap R_2^*)$. We use this lemma to compute the main bias term of form $-B_*/\lambda_n^*$, appearing in Table 2, for the non-convex templates in Figures 1.d-1.e of Section 5.2.

## 4. OPTIMAL SUBSAMPLE SIZE

We now consider "size" or scaling selection $\lambda_n^*$ for the subsampling regions in (1) to maximize the large-sample accuracy of the subsample variance estimators.

### 4.1. Theoretical optimal subsample sizes

The best value of $\lambda_n^*$ optimizes the over-all performance of a subsample variance estimator by balancing the contributions from both the estimator's variance and bias. For a large class of subsampling regions $_sR_n$, the leading bias component of $\hat{\tau}_{n,\mathrm{OL}}^2$ or $\hat{\tau}_{n,\mathrm{NOL}}^2$ can be determined explicitly with Theorems 2-3. Theorem 1 gives the variance of these estimators. Variance and bias expansions together yield an asymptotic MSE for $\hat{\tau}_{n,\mathrm{OL}}^2$ or $\hat{\tau}_{n,\mathrm{NOL}}^2$, which can be minimized for the optimal $\lambda_n^*$ under a given OL or NOL sampling scheme, based on a subsample template $R_0^*$.

**Theorem 5** *With Assumptions A.1 - A.4, assume Conditions $D_2$ and $M_{5+2a}$ if $d \geq 2$, and Conditions $D_3$ and $D_{7+2a}$ if $d = 1$. If $B_* \equiv \mathrm{vol}(R_0^*)^{-1}\left(\sum_{\mathbf{k}\in\mathbb{Z}^d} C(\mathbf{k})\sigma(\mathbf{k}) + I_{\{d=1\}}C_\infty\right) \neq 0$, then*

$$\lambda_{n,\mathrm{OL}}^{*\mathrm{opt}} = K_*^{-1/(d+2)}\lambda_{n,\mathrm{NOL}}^{*\mathrm{opt}}(1+o(1)) \quad and \quad \lambda_{n,\mathrm{NOL}}^{*\mathrm{opt}} = \left(\frac{\mathrm{vol}(R_n)(B_*)^2}{d\tau^4}\right)^{1/(d+2)}(1+o(1)). \quad (5)$$

*For $d = 1$, $R_0^*$ is assumed above to be an interval subset of $(-1/2, 1/2]$.*

**Remark 1**: By Theorem 5, optimally scaled OL subsamples should be larger than the NOL ones by a scalar $K_*^{-1/(d+2)} > 1$, based on the limiting variance ratio $K_*$ from (3).

**Remark 2**: The *order* $\mathrm{vol}(R_n)^{1/(d+2)}$ of the optimal scaling in (5) agrees with previous findings with Version I subsampling for some specific sampling region geometries ($\mathbb{R}^d$-rectangles [23], $\mathbb{R}^2$ regions [29], and $\mathbb{R}^1$ time series [11-14,20,24]).

In the time series case, the OL subsampling scheme is known to produce an asymptotically more efficient variance estimator than its NOL counterpart. We quantify this relative efficiency of the OL/NOL subsampling procedures for arbitrary subsample shapes in $\mathbb{R}^d$. With each variance estimator respectively optimized using (1), the asymptotic relative efficiency of $\hat{\tau}_{n,\mathrm{NOL}}^2$ to $\hat{\tau}_{n,\mathrm{OL}}^2$ depends solely of the geometry of the template $R_0^*$:

$$ARE_d = \lim_{n\to\infty} \frac{\mathrm{E}(\hat{\tau}_{n,\mathrm{OL}}^2 - \tau_n^2)^2}{\mathrm{E}(\hat{\tau}_{n,\mathrm{NOL}}^2 - \tau_n^2)^2} = K_*^{2/(d+2)} < 1,$$

so that $\hat{\tau}_{n,\mathrm{OL}}^2$ is more efficient than $\hat{\tau}_{n,\mathrm{NOL}}^2$.

### 4.2. Optimal block size: bootstrap vs. subsample with time series

A length $n$ sample from a time series (i.e., $d = 1$) can be formulated in our sampling framework using $R_0 = (-1/2, 1/2]$, $\Delta_n = n$, $\mathbf{Z} = \mathbb{Z}$. For time processes, it is important to note that the optimal scaling for subsample blocks for $\hat{\tau}_{n,\mathrm{OL}}^2$ or $\hat{\tau}_{n,\mathrm{NOL}}^2$ (i.e., Version I subsamples $_sR_n$ with $R_0^* = (-1/2, 1/2]$) can differ greatly from the optimal block size for the MBB variance estimator. It follows from Theorem 3 and Table 1 that

$$\lambda_{n,\mathrm{OL}}^{*\mathrm{opt}} = (3/2)^{1/3}\lambda_{n,\mathrm{NOL}}^{*\mathrm{opt}}(1+o(1)), \quad \lambda_{n,\mathrm{NOL}}^{*\mathrm{opt}} = n^{1/3}\tau^{-4/3}\left(\sum_{\mathbf{k}\in\mathbb{Z}}|\mathbf{k}|\sigma(\mathbf{k}) + C_\infty\right)^{2/3}(1+o(1))$$

with subsample blocks, but $b^{\text{opt}} = (3n\tau^{-4}/2)^{1/3}(\sum_{\mathbf{k}\in\mathbb{Z}}|\mathbf{k}|\sigma(\mathbf{k}))^{2/3}(1+o(1))$ is the optimal block size for the MBB (c.f. [14]). In particular, the difference between $\lambda_{n,\text{OL}}^{*\text{opt}}$ and $b^{\text{opt}}$ owes to differences in the biases of subsample and MBB variance estimators. (For time series, the variances of $\hat{\tau}_{n,\text{OL}}^{2}$ and the MBB variance estimators are the same.)

For a real-valued time series sample mean $\hat{\theta}_n = \bar{Z}_n$, the well-known bias of $\hat{\tau}_{n,\text{OL}}^{2}$ (using $R_0^* = (-1/2, 1/2]$) follows from Theorem 3 as

$$\frac{-1}{\lambda_n^*}\left(\sum_{\mathbf{k}\in\mathbb{Z}}|\mathbf{k}|\text{Cov}\big(\nabla'Z(\mathbf{0}), \nabla'Z(\mathbf{k})\big)\right) \qquad (6)$$

with $\nabla = 1$ (c.f. [12,13,24]). In this special case, the term $C_\infty = 0$ and the optimal OL subsample and MBB block sizes are equal. In general though, terms in the Taylor expansion of $\hat{\theta}_{\mathbf{i},n}$ (around $\mu$) up to fourth order can contribute to the bias of $\hat{\tau}_{n,\text{OL}}^2$ (and $\hat{\tau}_{n,\text{NOL}}^2$) through $C_\infty \neq 0$ when $d=1$. In contrast, the bias of the MBB variance estimator for general "smooth" model statistics is given by (6) (c.f. [11,14]).

## 5. COMPARISON OF VERSION I, VERSION II SUBSAMPLES

As mentioned previously, Version I subsamples using scaled-down replicates of the sampling region $R_n$ appear most often in the literature. We have shown that Version II subsamples, based on templates of arbitrarily selected shape, also allow consistent subsampling variance estimation. In the following, we discuss and compare some benefits and disadvantages of both Version I and Version II subsamples.

### 5.1. MSE of subsample versions: some examples

In terms of the asymptotic MSE of variance estimation, the performance of both subsample versions depends on a combination of three factors: the shape of $R_n$, the subsample geometry $R_0^*$, and the r.f. covariances $\sigma(\mathbf{k})$, $\mathbf{k} \in \mathbb{Z}^d$. Consider two subsample templates $R_0^* =: R_0$ and $R_0^{\text{new}}$ for Version I and II subsamples, respectively, with associated bias-variance constants from (5): $B_{0*}$, $K_{0*}$, $B_{0*}^{\text{new}}$, $K_{0*}^{\text{new}}$. We can express the asymptotic relative efficiency of the subsample versions as:

$$\lim_{n\to\infty}\frac{\text{MSE}_{\text{OL}}(R_0)}{\text{MSE}_{\text{OL}}(R_0^{\text{new}})} = \left[\frac{|B_{0*}|^d K_{0*}}{|B_{0*}^{\text{new}}|^d K_{0*}^{\text{new}}}\right]^{2/(d+2)}, \qquad \lim_{n\to\infty}\frac{\text{MSE}_{\text{NOL}}(R_0)}{\text{MSE}_{\text{NOL}}(R_0^{\text{new}})} = \left[\frac{B_{0*}}{B_{0*}^{\text{new}}}\right]^{2d/(d+2)}$$

based on the optimized MSEs of $\hat{\tau}_{n,\text{OL}}^2$ and $\hat{\tau}_{n,\text{NOL}}^2$ under both subsample templates. Because template geometry solely determines $K_*$ and heavily influences $B_*$, recommendations for a subsample template (Version I vs Version II) are often possible based on geometrical considerations. We next give some examples to illustrate this.

For concreteness and simplicity, we compare Version I and II subsampling for variance estimation of the sample mean (i.e., $\nabla = 1$) from various sampling regions in $\mathbb{R}^2$ with a real-valued r.f. (i.e., $p = 1$). Table 2 provides bias $B_*$ and variance $K_*$ terms for some $\mathbb{R}^2$-templates $R_0^*$, several of which are displayed in Figure 1. These templates could correspond to the sampling region ($R_0$) or a freely chosen template ($R_0^{\text{new}}$). We often consider here rectangular Version II subsamples based on $R_0^{\text{new}} = (-1/2, 1/2]^2$.

Using only template geometry, we find that rectangular subsamples are always more efficient than their Version I counterparts for sampling regions $R_n$ based on the cross or

Figure 1. Examples of templates $R_0^* \subset (-1/2, 1/2]^2$ outlined by solid lines.

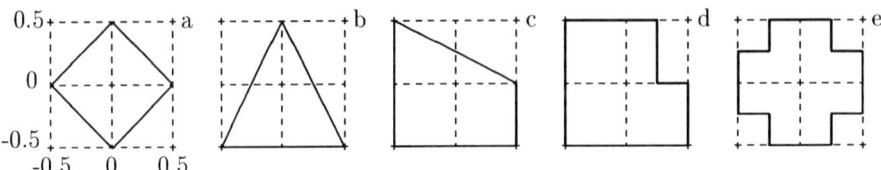

Table 2
Examples of $K_*$, $B_*$ for a subsample template $R_0^* \subset (-1/2, 1/2]^2$.

| Template $R_0^*$ | $K_*$ | $B_*$ |
|---|---|---|
| $(-1/2, 1/2]^2$ | 4/9 | $\sum_{\mathbf{k} \in \mathbb{Z}^2} \|\mathbf{k}\|_1 \sigma(\mathbf{k})$ |
| Circle, radius 1/2 | $\pi/4 - 4/(3\pi)$ | $4/\pi \cdot \sum_{\mathbf{k} \in \mathbb{Z}^2} \|\mathbf{k}\| \sigma(\mathbf{k})$ |
| Diamond (Fig 1.a) | 2/9 | $2 \sum_{\mathbf{k} \in \mathbb{Z}^2} \|\mathbf{k}\|_\infty \sigma(\mathbf{k})$ |
| Triangle (Fig 1.b) | 1/5 | $\sum_{\substack{\mathbf{k} \in \mathbb{Z}^2, \\ |k_1| \leq |k_2|/2}} 2|k_2|\sigma(\mathbf{k}) + \sum_{\substack{\mathbf{k} \in \mathbb{Z}^2, \\ |k_1| > |k_2|/2}} (|k_2| + 2|k_1|)\sigma(\mathbf{k})$ |
| Trapezoid (Fig 1.c) | $4/9 \cdot 263/360$ | $4/3 \cdot \sum_{\substack{\mathbf{k} \in \mathbb{Z}^2 \\ \mathrm{sign}k_1 = \mathrm{sign}k_2}} \|\mathbf{k}\|_1 \sigma(\mathbf{k}) + 2/3 \cdot \sum_{\substack{\mathbf{k} \in \mathbb{Z}^2 \\ \mathrm{sign}k_1 \neq \mathrm{sign}k_2}} (\|\mathbf{k}\|_\infty + \|\mathbf{k}\|_1)\sigma(\mathbf{k})$ |
| L-shape (Fig 1.d) | $4/9 \cdot 1301/1568$ | $8/7 \cdot \sum_{\mathbf{k} \in \mathbb{Z}^2} \|\mathbf{k}\|_1 \sigma(\mathbf{k})$ |
| Cross (Fig 1.e) | $4/9 \cdot 191/192$ | $4/3 \cdot \sum_{\mathbf{k} \in \mathbb{Z}^2} \|\mathbf{k}\|_1 \sigma(\mathbf{k})$ |

$\sigma(\mathbf{k}) = \mathrm{Cov}(\nabla' Z(\mathbf{t}), \nabla' Z(\mathbf{t} + \mathbf{k}))$, $\mathbf{k} \in \mathbb{Z}^2$

L-shape templates in Figure 1. Regardless of the underlying r.f., this holds true with both OL and NOL subsampling and the loss of efficiency with Version I subsamples is strongly related to the smaller volume of these templates $R_0$ compared to $(-1/2, 1/2]^2$.

If the r.f. covariances $\sigma(\mathbf{k})$ are either entirely nonnegative or nonpositive, then Table 1 shows that NOL *rectangular* subsamples are more efficient than NOL Version I subsamples associated with any template in the table (except the circle). In this case, the larger volume of the template $(-1/2, 1/2]^2$ helps to reduce the bias of $\hat{\tau}_{n,\mathrm{NOL}}^2$ relative to other subsample templates. However, for sampling regions $R_n$ based on many of these same templates, Version I subsamples may still be better than Version II rectangular ones for OL subsampling when $\sigma(\mathbf{k}) \geq 0$ (or $\leq 0$). That is, Version I subsamples can be more efficient for OL subsampling schemes while, for the same $R_0$ template, Version II can remain better for NOL subsamples.

To better appreciate this, consider a separable autoregressive (AR) process on $\mathbf{Z}^2 = \mathbb{Z}^2$:

$$\tilde{Z}(s,t) = \alpha \tilde{Z}(s-1,t) + \beta \tilde{Z}(s,t-1) - \alpha\beta \tilde{Z}(s-1,t-1) + \varepsilon(s,t), \quad |\alpha|, |\beta| < 1, \; s, t \in \mathbb{Z}$$

where $\{\varepsilon(s,t),(s,t)\in \mathbb{Z}^2\}$ is a white noise process. Martin [19] introduced this r.f. $\tilde{Z}(\cdot)$, known as an AR(1)×AR(1) lattice process with covariance function $\text{Cov}(\tilde{Z}(\mathbf{0}),\tilde{Z}(k_1,k_2)) = C\alpha^{|k_1|}\beta^{|k_2|}$, which has wide applicability for modeling in $\mathbb{R}^2$ (c.f. [9,19]). If $R_n$ is based on the diamond template in Figure 1, then

$$\lim_{n\to\infty} \frac{\text{MSE}_{\text{OL}}(R_0)}{\text{MSE}_{\text{OL}}((-1/2,1/2]^2)} = \left|\sqrt{2} - \frac{2\sqrt{2}\alpha\beta(1-\alpha)(1-\beta)}{(1-\alpha\beta)\{\alpha(1-\beta)^2+\beta(1-\alpha)^2\}}\right|$$

and this MSE ratio is less than 1 for $\alpha,\beta > 0.62$ (though the same ratio is greater than 1 for most other $\alpha,\beta \geq 0$ or $\alpha,\beta < 0$ combinations).

Examining an elliptical sampling region $R_n$, based on inflating the circle template $R_0$ in Table 1, provides an example where the the comparison of Version I and II subsamples depends more closely on the r.f. covariance structure. Consider a simple moving average process $\dot{Z}(\cdot)$ on $\mathbf{Z}^2 = \mathbb{Z}^2$ defined by:

$$\dot{Z}(s,t) = \vartheta_1\varepsilon(s+1,t) + \vartheta_2\varepsilon(s,t-3) + \varepsilon(s,t), \quad |\vartheta_1|+|\vartheta_2|<1, \ s,t\in\mathbb{Z}.$$

Compared to rectangular subsamples, the efficiency of Version I circular subsamples is

$$\lim_{n\to\infty} \frac{\text{MSE}_{\text{NOL}}(R_0)}{\text{MSE}_{\text{NOL}}((-1/2,1/2]^2)} = \frac{4|m_1|}{\pi|m_2|}, \quad \lim_{n\to\infty} \frac{\text{MSE}_{\text{OL}}(R_0)}{\text{MSE}_{\text{OL}}((-1/2,1/2]^2)} = \left(\frac{9\pi}{16}-\frac{3}{\pi}\right)^{1/2}\frac{4|m_1|}{\pi|m_2|},$$

$m_1 = (m+\vartheta_1\vartheta_2\sqrt{10})$, $m_2 = (m+4\vartheta_1\vartheta_2)$, $m = \vartheta_1 + 3\vartheta_2$. Now Version II rectangular subsamples can be better for both OL and NOL subsampling (e.g., $\vartheta_1 = \vartheta_2$), better for only NOL subsampling (e.g., $\vartheta_1 = 0.5 = -2\vartheta_2$), or worse for both OL and NOL subsampling (e.g., $\vartheta_1 = -0.5 = -3\vartheta_2/2$) compared to Version I circular subsamples.

We end this section with a small summary of MSE comparsions between Version I and II subsamples in $\mathbb{R}^2$:

1. Sampling regions $R_n$ exist where Version II subsamples (i.e., rectangles) are always better or more efficient.

2. If the r.f. covariances $\sigma(\mathbf{k}) \geq 0$ (or $\leq 0$), NOL Version II rectangular subsamples can often be better compared to NOL Version I subsamples. In such cases, this is strongly related to the smaller bias of $\hat{\tau}^2_{n,\text{NOL}}$ for subsamples using $_sR_n = \lambda_n^*(-1/2,1/2]^2$. This seems natural since NOL rectangular subsamples use all available sampling sites in a partitioning NOL cube from $J_{\text{NOL}}$ (i.e., $(-1/2,1/2]^2$ has largest possible template volume).

3. For some $R_n$ geometries, the r.f. covariance structure is more crucial for determining which subsample version performs better under either OL or NOL subsampling.

The first two statements are often verifiable based solely on comparisons of Version I ($R_0$) and Version II ($R_0^{\text{new}}$) subsample geometries.

## 5.2. Practical considerations for subsample selection

Aside from MSE comparisons, Version II subsamples generally have two major advantages over Version I subsamples:

1. simplified computations for estimators $\hat{\tau}^2_{n,\text{OL}}$ and $\hat{\tau}^2_{n,\text{NOL}}$.

2. easier empirical "plug-in" estimates of optimal subsample scaling.

Both reasons above for the use of Version II subsamples are practical in nature. For example, if $R_n$ is quite irregular in shape, Version I subsamples could become computationally undesirable and then Version II subsamples (based on, say, rectangles) may be more manageable.

The second reason above concerns optimal subsample scaling for maximizing the performance of $\hat{\tau}^2_{n,\text{OL}}$ or $\hat{\tau}^2_{n,\text{NOL}}$. A "plug-in" estimate of optimal subsample scaling for a template $R_0^*$ involves substituting estimates of unknown population quantities in (5). For time series, "plug-in" approaches have been used for estimating optimal block lengths for subsamples for $n\text{Var}(\bar{Z}_n)$ inference (c.f. [3,17,26]) as well as for the MBB (c.f. [2]).

If, again say, rectangular subsamples $R_0^* = (-1/2, 1/2]^d$ are chosen, regardless of the geometry of $R_n$, then the bias $B_s$ and variance $K_s$ expressions needed in (5) are known and quite tractable (see Table 1 and Theorem 4). This allows simple data-based estimates of optimal subsampling size or scaling $\lambda_n^{*\text{opt}}$ for OL and NOL Version II subsamples.

## 6. PROOFS

Proofs of the main results involve straightforward modifications to existing, detailed arguments found in [21,22]. We only sketch a few new aspects to save space. For two positive sequences $\{r_n\}$ and $\{t_n\}$, write $r_n \sim t_n$ if $r_n/t_n \to 1$ as $n \to \infty$.

**Proof of Theorem 1.** We consider $\text{Var}(\hat{\tau}^2_{n,\text{OL}})$. For $\mathbf{i} \in J_{\text{OL}}$, use a Taylor's expansion of $\hat{\theta}_{\mathbf{i},n} = H(\bar{Z}_{\mathbf{i},n})$ (around $\mu$) to rewrite the statistic: $\hat{\theta}_{\mathbf{i},n} = H(\mu) + Y_{\mathbf{i},n} + Q_{\mathbf{i},n}$ with $Y_{\mathbf{i},n} = \nabla'(\bar{Z}_{\mathbf{i},n} - \mu)$ and a remainder term $Q_{\mathbf{i},n}$. Theorem 1 for $\hat{\tau}^2_{n,\text{OL}}$ follows from:

(i) $\quad \text{Var}\left(\dfrac{sN_n}{|J_{\text{OL}}|} \sum_{\mathbf{i} \in J_{\text{OL}}} Y^2_{\mathbf{i},n}\right) = \dfrac{sN_n K_*}{|J_{\text{OL}}|\text{vol}(R_0^*)} \cdot [2\tau^4](1 + o(1)), \quad \begin{aligned} |J_{\text{OL}}| &\sim \text{vol}(R_n), \\ sN_n &\sim (\lambda_n^*)^d \text{vol}(R_0^*). \end{aligned}$

(ii) $\quad \left|\text{Var}(\hat{\tau}^2_{n,\text{OL}}) - \text{Var}\left(\dfrac{sN_n}{|J_{\text{OL}}|} \sum_{\mathbf{i} \in J_{\text{OL}}} Y^2_{\mathbf{i},n}\right)\right| = o\left(\dfrac{(\lambda_n^*)^d}{\text{vol}(R_n)}\right).$

Here (i)-(ii) can be shown using Theorem 8.1, Lemma 8.3, and the proof of Theorem 3.1(a) from [22]. The proof of Theorem 1 for $\text{Var}(\hat{\tau}^2_{n,\text{NOL}})$ is similar; see [22]. □

**Proofs of Theorem 2-4:** These follow from simple modifications to bias results in Section 9 of [22], which establish Theorems 2-4 for Version I subsamples. These Section 9 assertions from [22] can easily be shown to hold regardless of the shape of $R_n$. □

**Proof of Lemma 1:** Follows from simple modifications to the proof of Corollary 6.1 in [21], the result for Version I subsample templates. □

**Proof of Theorem 5:** Follows from Theorems 1-3 and Calculus involving minimization of a smooth function of a real variable. □

## REFERENCES

1. E. Bolthausen, *On the central limit theorem for stationary mixing random fields*, Ann. Probab. 10 (1982) 1047.
2. P. Bühlman and H.R. Künsch, *Block length selection in the bootstrap for time series*, Comput. Stat. Data Anal. 31 (1999) 295.
3. E. Carlstein, *The use of subseries methods for estimating the variance of a general statistic from a stationary time series*, Ann. Statist. 14 (1986) 1171.
4. A.D. Cliff and J.K. Ord, Spatial Processes Models & Applications, Pion Limited, 1981.
5. N. Cressie, Statistics For Spatial Data, John Wiley & Sons, New York, 1993.
6. T. DiCiccio, P. Hall and J.P. Romano, *Empirical likelihood is Bartlett-correctable*, Ann. Statist. 19 (1991) 1053.
7. P. Doukhan, Mixing: properties and examples, Springer-Verlag, New York, 1994.
8. P.H. Garcia-Soidan and P. Hall, *In sample reuse methods for spatial data*, Biometrics 53 (1997) 273.
9. X. Guyon, Random Fields On A Network, Springer-Verlag, New York, 1995.
10. P. Hall, The Bootstrap and Edgeworth Expansion, Springer-Verlag, New York, 1992.
11. P. Hall, J.L. Horowitz, B.-Y. Jing, *On blocking rules for the bootstrap with dependent data*, Biometrika 82 (1995) 561.
12. P. Hall and B.-Y. Jing, *On sample reuse methods for dependent data*, J. Roy. Statist. Soc. Ser. B 58 (1996) 727.
13. H.R. Künsch, *The jackknife and the bootstrap for general stationary observations*, Ann. Statist. 17 (1989) 1217.
14. S.N. Lahiri, *On empirical choice of the optimal block size for block bootstrap methods*, Preprint 14, Department of Statistics, Iowa State University, Ames, IA, 1996.
15. S.N. Lahiri, *Asymptotic distribution of the empirical spatial cumulative distribution function predictor and prediction bands based on a subsampling method*, Probab. Theory Related Fields 114 (1990) 55.
16. S.N. Lahiri, *Central limit theorems for weighted sums under some spatial sampling schemes*, (to appear) Sankhya Ser. A, 2003.
17. C. Léger, D.N. Politis and J.P. Romano, *Bootstrap technology and applications*, Technometrics 34 (1992) 378.
18. R.Y. Liu and K. Singh, *in* Exploring the Limits of Bootstrap, John Wiley & Sons, New York, 1992.
19. R.J. Martin, *The use of time-series models and methods in the analysis of agricultural field trials*, Commun. Statist. Theory Methods 19 (1990) 55.
20. M.S. Meketon and B. Schmeiser, *Overlapping batch means: something for nothing?*, Proc. Winter Simul. Conf., IEEE (1984) 227.
21. D.J. Nordman, On nonparametric methods for strongly and weakly dependent lattice data. Ph.D. Dissertation, Department of Statistics, Iowa State University, Ames, IA,

2002.
22. D.J. Nordman and S.N. Lahiri, *On optimal spatial subsample size for variance estimation*, Preprint, Department of Statistics, University of Dortmund, Germany, 2002.
23. D.N. Politis and J.P. Romano, *Nonparametric resampling for homogeneous strong mixing random fields*, J. Multivariate Anal. 47 (1993) 301.
24. D.N. Politis and J.P. Romano, *On the sample variance of linear statistics derived from mixing sequences*, Stochastic Process. Appl. 45 (1993) 155.
25. D.N. Politis and J.P. Romano, *Large sample confidence regions based on subsamples under minimal assumptions*, Ann. Statist. 22 (1994) 2031.
26. D.N. Politis, J.P. Romano and M. Wolf, Subsampling, Springer-Verlag, New York, 1999.
27. A. Possolo, *Subsampling a random field*, IMS Lecture Notes 20 (1991) 286.
28. B.D. Ripley, Spatial Statistics, John Wiley & Sons, New York, 1981.
29. M. Sherman, *Variance estimation for statistics computed from spatial lattice data*, J. Roy. Statist. Soc. Ser. B 58 (1996) 509.
30. M. Sherman and E. Carlstein, *Nonparametric estimation of the moments of a general statistic computed from spatial data*, J. Amer. Statist. Assoc. 89 (1994) 496.

# Locally Stationary Processes and the Local Block Bootstrap

Arif Dowla[a], Efstathios Paparoditis[b] and Dimitris N. Politis[c]*

[a]Department of Mathematics, University of California—San Diego,
La Jolla, CA 92093-0112, USA; email: adowla@math.ucsd.edu

[b]Department of Mathematics and Statistics, University of Cyprus, P.O. Box 20537,
Nicosia, Cyprus; email: stathisp@ucy.ac.cy

[c]Department of Mathematics, University of California—San Diego,
La Jolla, CA 92093-0112, USA; email: dpolitis@ucsd.edu

For time series that are not stationary, the block bootstrap of Künsch (1989) is not directly applicable. However, if the underlying stochastic structure is slowly changing with time, one may employ the local block bootstrap of Paparoditis and Politis (2002). We describe the local block bootstrap procedure, and show its consistency in an interesting example of a locally stationary process.

## 1. INTRODUCTION

Let $Y_1, ..., Y_n$ be an observed stretch of a real-valued time series $\{Y_t, t \in \mathbf{Z}\}$. In many applications, for instance in connection with financial or meteorological time series, the sample size $n$ may be quite large. Consequently, it may be unrealistic to assume that such a long time series is stationary; a more realistic assumption is that of a slowly-changing stochastic structure in the sense that the joint probability law of $(Y_t, Y_{t+1}, \ldots, Y_{t+k})$ changes smoothly (and slowly) with $t$ for any $k$. Such nonstationary models have been provided by the evolutionary spectra models of Priestley (1988) and the locally stationary models of Dahlhaus (1996, 1997).

Now because of the nonstationarity of $\{Y_t\}$, the block bootstrap method of Künsch (1989) is not directly applicable, and neither are its different variations such as the stationary bootstrap; see e.g. Politis (2003) for a review. Rather, a modification that takes into account the changing stochastic structure should be constructed. Such a modification has been proposed in the Local Block Bootstrap (LBB) procedure of Paparoditis and Politis (2002). The basic premise of the LBB is to only resample blocks that are close to each other, i.e., a block that starts at time $t$, can only be replaced with blocks whose starting point is close to $t$. The LBB idea is to construct a bootstrap pseudo-series by a concatenation of $q$ blocks of size $b$ (such that $qb \simeq n$), where the $j$th block of the resampled series is chosen randomly from a distribution (say, uniform) on all the size-$b$ blocks of consecutive data whose time indices are 'close' to those in the original block.

---

*Research partially supported by NSF grant DMS-01-04059.

A more detailed description of the LBB algorithm is now presented. Given the data $Y_1, ..., Y_n$, the LBB algorithm creates a bootstrap pseudo-series $Y_1^*, ..., Y_n^*$ as follows:

- Select an integer block size $b$, and a real number $B \in (0,1]$ such that $nB$ is an integer; both $b$ and $B$ are thought of as functions of $n$.

- For $h = 0, 1, \ldots, (\lceil n/b \rceil - 1)$, let $Y_{hb+j}^* := Y_{I_h+j-1}$ for $j = 1, \ldots, b$, where $I_1, I_2, \ldots$ are independent, integer-valued random variables satisfying $P(I_h = k) = W_{n,h}(k)$.

Here $\lceil k \rceil$ denotes the smallest integer that is greater or equal to $k$. Thus, in the case where $n$ is not an integer multiple of $b$, the LBB algorithm generates a bootstrap series whose length is slightly bigger than $n$, although only $Y_1^*, ..., Y_n^*$ are kept in the end.

The probability distribution $W_{n,h}(k)$ is the practitioner's choice. The easiest choice is the uniform probability over the integers in the interval $[J_{1,h}, J_{2,h}]$, where $J_{1,h} = \max\{1, hb - nB\}$ and $J_{2,h} = \min\{n - b + 1, hb + nB\}$. We will adopt this simple choice in the sequel although many other choices are possible; see Paparoditis and Politis (2002).

For an arbitrary point $k$, the range $[k - nB, k + nB]$ indicates its 'local stationarity' neighborhood; in other words, although the time series $\{X_t\}$ is not (globally) stationary, the stretch $X_{k-nB}, \ldots, X_{k+nB}$ can be thought to have been generated by an approximate stationary mechanism provided that the local window size $2nB$ is small with respect to the sample size $n$. Nevertheless, we would also need $nB \to \infty$ so that enough data accumulate in the local window. As a matter of fact, doing a block bootstrap within the local window would require a block size $b$ that is big but small with respect to the local window size $2nB$.

The required interplay between block and window size is typically given by a requirement such as eq. (2). However, the rates in eq. (2) are specific to the problem addressed in the next section, i.e., showing the consistency of the LBB in the context of data from the particular locally stationary process (1) below; all technical proofs are placed in the Appendix.

## 2. CONSISTENCY OF THE LOCAL BLOCK BOOTSTRAP

As is well-known, the applicability of bootstrap methods is usually checked on a case-by-case basis—the only exception so far seems to be the i.i.d. bootstrap with smaller resample size that is generally consistent; see e.g. Politis, Romano and Wolf (1999). Thus, we now focus on a particular interesting example of a locally stationary process in the sense of Dahlhaus (1996, 1997). Let the data be generated from the model[2]

$$Y_t = s_n(t) + \mu + v_n(t)e_t, \quad \text{for all } t \in \mathbf{Z} \tag{1}$$

where $\{e_t, t \in \mathbf{Z}\}$ is a mean-zero and variance-one, strong mixing and strictly stationary sequence satisfying:

$$E|e_t|^{6+\delta} < \infty, \quad \text{and} \quad \sum_{k=1}^{\infty} k^2 \alpha^{\delta/(6+\delta)}(k) < \infty \quad \text{for some} \quad \delta > 0,$$

---
[2] Note that under (1), the data $Y_1, ..., Y_n$ constitute the $n$th row of a triangular array; however, since no confusion arises, we will not use the usual double-index notation.

where $\alpha(k)$ indicates the strong mixing coefficient associated with $\{e_t, t \in \mathbf{Z}\}$. We also assume that $s_n(t) = m(t/n)$, $v_n(t) = \sigma(t/n)$, for some fixed functions $m, \sigma$ that are differentiable[3] with bounded derivatives on $[0,1]$. We will further assume that $\sigma(x) > 0$ for all $x$ in $[0,1]$ and that $m$ satisfies $\int_0^1 m(x)dx = 0$; due the latter, $m$ may be thought of as a 'seasonality' fluctuation about the 'grand mean' $\mu$.

Our goal is interval estimation of the unknown parameter $\mu$ based on the data $Y_1, ..., Y_n$. For this reason we require an approximation to the sampling distribution of the sample mean $\hat{\mu} = n^{-1} \sum_{t=1}^{n} Y_t$ that will serve as our estimator of $\mu$. We propose the LBB as a method for this approximation.

We first establish some of the properties of $\hat{\mu}$. The following two lemmas concern the asymptotic bias and variance of the sample mean.

**Lemma 1.** $E\left(n^{\frac{1}{2}}(\hat{\mu} - \mu)\right) = O(n^{-\frac{1}{2}})$.

**Lemma 2.** Let $c(s) = Cov(e_0, e_s)$. Then, as $n \to \infty$, $Var\left(n^{\frac{1}{2}}(\hat{\mu} - \mu)\right) \to V^2$, where $V^2 = \sum_{s=-\infty}^{\infty} c(s) \int_0^1 \sigma^2(u)du$.

In the next theorem we establish asymptotic normality of the sample mean.

**Theorem 1.** As $n \to \infty$, $\sqrt{n}(\hat{\mu} - \mu) \stackrel{\mathcal{L}}{\Longrightarrow} N(0, V^2)$.

We now turn to the LBB properties. Let $Y_1^*, ..., Y_n^*$ denote an LBB pseudo-series constructed using the algorithm of the previous section, and let $\hat{\mu}^* = n^{-1} \sum_{t=1}^{n} Y_t^*$ be the bootstrap sample mean. As usual, let $P^*, E^*, Var^*$ indicate probability, expectation, and variance under the LBB scheme (conditional on the data $Y_1, ..., Y_n$).

To investigate the consistency of the LBB we will—as previously mentioned—require some conditions on the block size, as well as the window size indicating the local neighborhood. For this reason, consider the following:

Let $n \to \infty$, $b \to \infty$ but $b = o\left(\min(n^{\frac{2}{5}}, nB)\right) \to 0$, and $nB \to \infty$ but $nB^2 \to 0$. (2)

Lemma 3 below can be compared to Lemma 1 and 2.

**Lemma 3.** Under (2), $E^*\left(n^{\frac{1}{2}}(\hat{\mu}^* - \hat{\mu})\right) = O_p(nB^2)$, and $Var^*\left(n^{\frac{1}{2}}(\hat{\mu}^* - \hat{\mu})\right) \stackrel{P}{\longrightarrow} V^2$.

Our main theorem below shows that the LBB is successful in giving a consistent approximation to the sampling distribution of the sample mean $\hat{\mu}$.

**Theorem 2.** Under (2), we have

$$\sup_x |P(\sqrt{n}(\hat{\mu} - \mu) \leq x) - P^*(\sqrt{n}(\hat{\mu}^* - E^*\hat{\mu}^*) \leq x)| \stackrel{P}{\longrightarrow} 0, \quad (3)$$

as well as

$$\sup_x |P(\sqrt{n}(\hat{\mu} - \mu) \leq x) - P^*(\sqrt{n}(\hat{\mu}^* - \hat{\mu}) \leq x)| \stackrel{P}{\longrightarrow} 0. \quad (4)$$

---

[3] Under slightly more careful arguments, the conditions on $m$ and $\sigma$ may be relaxed to just piecewise Lipschitz continuity as suggested in Paparoditis and Politis (2002).

From Theorem 2 it follows that asymptotically valid confidence intervals for $\mu$ can be based on the quantiles of either the bootstrap distribution $P^*(\sqrt{n}(\hat{\mu}^* - E^*\hat{\mu}^*) \leq x)$ or the bootstrap distribution $P^*(\sqrt{n}(\hat{\mu}^* - \hat{\mu}) \leq x)$, both of which are computable—as opposed to the quantiles of the unknown true distribution $P(\sqrt{n}(\hat{\mu} - \mu) \leq x)$.

To help delineate which of those two approximations may be preferable, recall that the distribution $P^*(\sqrt{n}(\hat{\mu}^* - \hat{\mu}) \leq x)$ has center of location $O_p(nB^2)$ by Lemma 3, while $P^*(\sqrt{n}(\hat{\mu}^* - E^*\hat{\mu}^*) \leq x)$ has center exactly zero, and $P(\sqrt{n}(\hat{\mu} - \mu) \leq x)$ has center $O(1/\sqrt{n})$ by Lemma 1. In order to satisfy (2) we may let $B = 1/n^{\epsilon+1/2}$ for some $\epsilon$ in $(0, 1/2)$. Typically, we may even have $\epsilon < 1/4$, in which case approximation (3) is preferable as its (zero) center of location is closest to that of the target distribution $P(\sqrt{n}(\hat{\mu} - \mu) \leq x)$. In addition, the validity of (3) may be proved under slightly weaker conditions, namely replacing the condition $nB^2 \to 0$ in (2) by the weaker $B \to 0$.

## 3. APPENDIX: TECHNICAL PROOFS

**Proof of Lemma 1.**

$$E\left(n^{\frac{1}{2}}(\hat{\mu} - \mu)\right) = E\left(n^{\frac{1}{2}}\left(\left(\frac{1}{n}\sum_{t=1}^n (s_n(t) + \mu + v_n(t)e_t)\right) - \mu\right)\right)$$

$$= E\left(n^{\frac{1}{2}}\left(\frac{1}{n}\sum_{t=1}^n s_n(t) + \frac{1}{n}\sum_{t=1}^n v_n(t)e_t\right)\right)$$

$$= n^{\frac{1}{2}}\left(\frac{1}{n}\sum_{t=1}^n s_n(t) + \frac{1}{n}\sum_{t=1}^n v_n(t)E(e_t)\right)$$

$$= n^{\frac{1}{2}}\left(\frac{1}{n}\sum_{t=1}^n m(t/n)\right) = n^{\frac{1}{2}}O\left(\frac{1}{n}\right) = O(n^{-\frac{1}{2}})$$

since $\frac{1}{n}\sum_{t=1}^n m(t/n)$ is the Riemann sum approximation of the integral $\int_0^1 m(u)du = 0$.

**Proof of Lemma 2.**

$$Var\left(n^{\frac{1}{2}}(\hat{\mu} - \mu)\right) = nVar\left(\frac{1}{n}\sum_{t=1}^n v_n(t)e_t\right)$$

$$= \frac{1}{n}Var\left(\sum_{t=1}^n \sigma(\frac{t}{n})e_t\right) = \frac{1}{n}\sum_{i=1}^n\sum_{j=1}^n \sigma(\frac{i}{n})\sigma(\frac{j}{n})c(i-j)$$

$$= \frac{1}{n}\sum_{s=-n+1}^{n-1}\sum_{r=1}^{n-|s|} \sigma(\frac{r}{n})\sigma(\frac{r+|s|}{n})c(s) \tag{5}$$

Using the Mean Value Theorem and letting $\xi_n^{r,s} \in [\frac{r}{n}, \frac{r+|s|}{n}]$ we can re-write (5) as

$$\frac{1}{n}\sum_{s=-n+1}^{n-1}\sum_{r=1}^{n-|s|} \sigma(\frac{r}{n})\left(\sigma(\frac{r}{n}) + \frac{|s|}{n}\sigma'(\xi_n^{r,s})\right)c(s)$$

$$= \frac{1}{n} \sum_{s=-n+1}^{n-1} \sum_{r=1}^{n-|s|} \left(\sigma(\frac{r}{n})\right)^2 c(s) + \frac{1}{n} \sum_{s=-n+1}^{n-1} \sum_{r=1}^{n-|s|} \sigma(\frac{r}{n}) \frac{|s|}{n} \sigma'(\xi_n^{r,s}) c(s)$$

$$= V_1 + V_2 \quad \text{respectively.}$$

Let us consider $|V_2|$. We have

$$\left| \frac{1}{n} \sum_{s=-n+1}^{n-1} \sum_{r=1}^{n-|s|} \sigma(\frac{r}{n}) \frac{|s|}{n} \sigma'(\xi_n^{r,s}) c(s) \right|$$

$$\leq \frac{C}{n} \sum_{s=-n+1}^{n-1} \sum_{r=1}^{n-|s|} \frac{|s|}{n} |c(s)| \leq \frac{C}{n} \sum_{s=-n+1}^{n-1} \sum_{r=1}^{n} \frac{|s|}{n} |c(s)|$$

$$= O\left( \sum_{s=-n+1}^{n-1} \frac{|s|}{n} |c(s)| \right) = o(1)$$

where $C = \max_{x,y \in [0,1]} |\sigma(x)| |\sigma'(y)|$. The limit is achieved by the mixing inequalities and Kronecker's Lemma.

We can write $V_1$ as

$$\frac{1}{n} \sum_{s=-n+1}^{n-1} \sum_{r=1}^{n} \left(\sigma(\frac{r}{n})\right)^2 c(s) - \frac{1}{n} \sum_{s=-n+1}^{n-1} \sum_{r=n-|s|+1}^{n} \left(\sigma(\frac{r}{n})\right)^2 c(s)$$

which can be written as $V_1' - V_1''$. We have that

$$|V_1''| = \left| \frac{1}{n} \sum_{s=-n+1}^{n-1} \sum_{r=n-|s|+1}^{n} \left(\sigma(\frac{r}{n})\right)^2 c(s) \right| \leq C_1 \frac{1}{n} \sum_{s=-n+1}^{n-1} |s| |c(s)| \to 0$$

where $C_1 = \max_{x \in [0,1]} (\sigma(x))^2$. Finally,

$$V_1' = \sum_{s=-n+1}^{n-1} c(s) \frac{1}{n} \sum_{r=1}^{n} \left(\sigma(\frac{r}{n})\right)^2 \to \sum_{s=-\infty}^{\infty} c(s) \int_0^1 \sigma^2(u) du$$

and the lemma is proven.

**Proof of Theorem 1.** Observe that we have a weighted sum of strong mixing random variables where the weights obey the conditions of Theorem 3.1 of Roussas, Tran and Ioannides (1992)—RTI for short—on fixed design regression. The problem is to show that $\frac{1}{\sqrt{n}} \sum_{t=1}^{n} \sigma_t e_t$ is asymptotically normally distributed under our assumptions. To do that, we need to verify the assumptions and conditions in RTI. For example, (A1) of RTI is satisfied by the definition of our model along with our assumptions. (A2) (i) of RTI is a consequence of the fact that

$$\frac{1}{n} \sum_{i=1}^{n} \sigma(\frac{i}{n}) \to C_0 \text{ as } n \to \infty.$$

for some constant $C_0 > 0$ since $\sigma$ is a positive integrable function. (A2) (ii) of RTI is a result of the fact that there exists a constant $C > 0$ such that $\sigma(u) < C$ uniformly in $u$. Furthermore,

$$\sum_{i=1}^{n} \left(\frac{\sigma(i/n)}{n}\right)^2 = \frac{1}{n}\left(\frac{1}{n}\sum_{i=1}^{n}\sigma(\frac{i}{n})^2\right) = O\left(\frac{1}{n}\right)$$

because

$$\frac{1}{n}\sum_{i=1}^{n}\sigma(\frac{i}{n})^2 \to C_1 \text{ as } n \to \infty$$

for some $C_1 > 0$. Therefore we have

$$\max_i \left|\frac{\sigma(\frac{i}{n})}{n}\right| = O\left(\sum_{i=1}^{n}\left(\frac{\sigma(\frac{i}{n})}{n}\right)^2\right).$$

Assumption (A3) of RTI is satisfied by the fact that the variance of $\hat{\mu}$ is $O(\frac{1}{n})$ by Lemma 2 which implies that

$$\sum_{i=1}^{n}\left(\frac{\sigma(\frac{i}{n})}{n}\right)^2 = O\left(Var(\hat{\mu})\right).$$

Assumption (A4) of RTI is an immediate consequence of our assumptions on the sequence $e_t$. Condition (2.21) of RTI has three parts. We first observe that the effective number of terms in our sum is $n$. We have to check that there exist $p$ and $q$ as defined in RTI where $qp^{-1} \to 0$ and where

$$nqp^{-1}\sum_{i=1}^{n}\left(\frac{\sigma(\frac{i}{n})}{n}\right)^2 \to 0,$$

$$p^2\sum_{i=1}^{n}\left(\frac{\sigma(\frac{i}{n})}{n}\right)^2 \to 0 \quad \text{and} \quad nqp^{-1}\alpha(q) \to 0 \text{ as } n \to \infty$$

since as shown earlier

$$n\sum_{i=1}^{n}\left(\frac{\sigma(\frac{i}{n})}{n}\right)^2 = O(1),$$

a possible choice is $p = n^{\frac{1}{2}-\varepsilon}$ and $q = n^{\frac{1}{4}+\varepsilon}$ for some $\varepsilon \in (0, 1/4)$. As a result we confirm that the first part of condition (2.21) of RTI holds. Finally, from our mixing assumption on $e_t$ we have that $\alpha(q) = o(1/q^2)$ and therefore $nqp^{-1}\alpha(q) = o(1)$ which satisfies the last part of condition (2.21) of RTI and the theorem is proven.

**Proof of Lemma 3.** The proof is rather long and tedious. We give just the main idea below for lack of space and refer the reader to Dowla (2002) for details. Consider the effect of the LBB on the underlying quantities in eq. (1), namely $m(\frac{t}{n})$, $\mu$, $\sigma(\frac{t}{n})$ and $e_t$. The LBB actually performs a similar re-arrangement to those unobservable quantities as the one performed on the data $Y_t$. In other words, for $h = 0, 1, \ldots, (\lceil n/b \rceil - 1)$ and $j = 1, \ldots, b$, we have $m^*(\frac{hb+j}{n}) = m(\frac{I_h+j-1}{n})$, $\sigma^*(\frac{hb+j}{n}) = \sigma(\frac{I_h+j-1}{n})$, $\mu^* = \mu$, and $e^*_{hb+j} = e_{I_h+j-1}$, where $I_1, I_2, \ldots$ are the *same* values used to generate $Y_t^*$.

Therefore, we have the LBB bootstrap version of eq. (1) given below.

$$Y_t^* = m^*(\frac{t}{n}) + \mu + \sigma^*(\frac{t}{n})e_t^*. \tag{6}$$

The key observation now is that $m^*(\frac{t}{n}) - m(\frac{t}{n}) = O(B)$ uniformly in $t$ by Taylor's theorem; thus, $n^{-1}\sum_{t=1}^{N}(m^*(\frac{t}{n}) - m(\frac{t}{n})) = O(B)$ as well. Similarly, $\sigma^*(\frac{t}{n}) - \sigma(\frac{t}{n}) = O(B)$ uniformly in $t$; recall that $B \to 0$ by eq. (2).

Thus, an expression of the type $E^*\left(n^{\frac{1}{2}}(\hat{\mu}^* - \hat{\mu})\right)$ and/or $Var^*\left(n^{\frac{1}{2}}(\hat{\mu}^* - \hat{\mu})\right)$ can be significantly simplified since $\hat{\mu}^*$ is an average of (a linear combination of) the quantities $m^*(\frac{t}{n})$, $\sigma^*(\frac{t}{n})$ and $e_t^*$ that are closely related to the quantities $m(\frac{t}{n})$, $\sigma(\frac{t}{n})$ and $e_t$ found within $\hat{\mu}$. More details can be found in Dowla (2002, Ch. 4). At the final analysis, the LBB on the errors $e_t$ is tantamount to (a version of) the regular block bootstrap for stationary sequences that is consistent for the variance of the sample mean.

**Proof of Theorem 2.** Consider the expression

$$n^{\frac{1}{2}}(\hat{\mu}^* - E^*\hat{\mu}^*) = n^{\frac{1}{2}}\left(\sum_{i=1}^{n}\frac{1}{n}(Y_i^* - E^*Y_i^*)\right).$$

We can rewrite the above in terms of a sum of blocks of size $b$, i.e.,

$$\sum_{i=0}^{\lfloor\frac{n}{b}\rfloor-1}\left(n^{-\frac{1}{2}}\sum_{j=1}^{b}(Y_{ib+j}^* - E^*Y_{ib+j}^*)\right) = \sum_{i=0}^{\lfloor\frac{n}{b}\rfloor-1}\xi_{i,n}^*.$$

Note that the random variable $\xi_{i,n}^*$ is a function of the $i^{th}$ independent block of data of size $b$. We need to show that the $\xi_{i,n}^*$'s satisfy the Central Limit Theorem for a triangular array sum of independent, non-identically distributed random variables. Since we have finite moments of order $6+\delta$ we can use the Lyapunov condition. From Lemma 3, and the independence of the $\xi_{i,n}^*$'s, it follows that

$$\sum_{i=0}^{\lfloor\frac{n}{b}\rfloor-1}Var^*(\xi_{i,n}^*) \xrightarrow{P} \sum_{s=-\infty}^{\infty}c(s)\int_0^1\sigma^2(u)du.$$

We also have

$$E^*(\xi_{i,n}^{*(6)}) = E^*\left(n^{-\frac{1}{2}}\sum_{j=1}^{b}(Y_{ib+j}^* - E^*Y_{ib+j}^*)\right)^6 = \frac{1}{n^3}O_p(b^6)$$

Therefore we obtain

$$\frac{\sum_{i=0}^{\lfloor\frac{n}{b}\rfloor-1}E^*(\xi_{i,n}^{*6})}{\left(\sum_{i=0}^{\lfloor\frac{n}{b}\rfloor-1}Var^*(\xi_{i,n}^*)\right)^2} = \frac{\frac{n}{b}\left(\frac{1}{n^3}O_p(b^6)\right)}{\left(\sum_{s=-\infty}^{\infty}c(s)\int_0^1\sigma^2(u)du\right)^2} = O_p\left(\frac{b^5}{n^2}\right)$$

Using (2), we can invoke Lyapunov's condition to establish that—over an event whose probability tends to one—the random variable $n^{\frac{1}{2}}(\hat{\mu}^* - E^*\hat{\mu}^*)$ is asymptotically normal with mean zero. Furthermore, from Lemma 3 we know that $\hat{\mu}^*$ has the correct asymptotic variance. Thus, the theorem is proven.

## REFERENCES

1. Dahlhaus, R. (1996). On the Kullback-Leibler information divergence of locally stationary processes, *Stoch. Proc. Appl.*, 62, 139-168.
2. Dahlhaus, R. (1997). Fitting time series models to nonstationary processes, *Annals of Statistics*, 25, 1-37.
3. Dowla, A. (2002). *Local Block Bootstrap Based Inference for Nonstationary Time Series*, Ph.D. Thesis, Department of Mathematics, UCSD.
4. Künsch, H. R. (1989). The Jackknife and the Bootstrap for General Stationary Observations, *Annals of Statistics*, 17, 1217-1241.
5. Paparoditis, E. and Politis, D. N. (2002). Local block bootstrap, *C. R. Acad. Sci. Paris, Ser. I*, 335, 959-962.
6. Politis, D. N. (2003). The impact of bootstrap methods on time series analysis, to appear in *Statistical Science*.
7. Politis, D.N., Romano, J.P., and Wolf, M. (1999). *Subsampling*. Springer, New York.
8. Priestley, M.B. (1988). *Non-linear and non-stationary time series analysis*, Academic Press, London.
9. Roussas, G.G., Tran, L.T. and Ioannides, D.A. (1992). Fixed design regression for time series: asymptotic normality, *J. Multivariate Analysis*, 40, 262-291.

# 11. Time Series and Stochastic Processes

# Spectral Analysis and a Class of Nonstationary Processes

Murray Rosenblatt[a]

[a]Department of Mathematics, University of California at San Diego,
La Jolla, CA 92093-0112, USA.

A class of nonstationary processes having a Fourier representation is considered, and some results are given towards consistent estimation.

Let us consider a class of nonstationary processes having a Fourier representation. We try to determine when their structure can be estimated consistently from a sequence $x_{-n/2}, \ldots, x_{n/2}$ as $n \to \infty$ ($n$ even). Comments are made on a few recent results in this direction. A plausible choice for such a process is given by a harmonizable sequence (see [2])

$$x_t = \int_{-\pi}^{\pi} e^{it\lambda} dz(\lambda), \; Ex_t \equiv 0,$$

$$Edz(\lambda)\overline{dz(\mu)} = dH(\lambda, \mu)$$

with the complex spectral measure $H(\lambda, \mu)$ of bounded variation

$$\int |dH(u, v)| < \infty.$$

The covariances $r_{j,k}$ are given by

$$r_{j,k} = \iint_{-\pi}^{\pi} e^{ij\lambda - ik\mu} dH(\lambda, \mu).$$

At this point we can already see that if $x_t$ is a Gaussian process with $r_{0,0} > 0$,

$$dH(\lambda, \mu) = h(\lambda, \mu) d\lambda \, d\mu$$

and $h(\lambda, \mu)$ a finite nonzero trigonometric polynomial

$$h(\lambda, \mu) = \sum_{|j|, |k| \leq m} c_{j,k} e^{ij\lambda - ik\mu},$$

such consistent estimation is not possible since $x_k = 0$ for $|k| > m$. For this would mean that the $r_{j,k}$'s can be precisely determined using only a finite number of observations. The weakly stationary processes correspond to the spectral measure concentrated on the main diagonal

$$dH(\lambda, \mu) = \delta_{\lambda - \mu} dF(\lambda).$$

There has also been interest in processes with almost periodic covariance function (see [1], [3], [5]) that is, processes whose covariance function $r_{j,k}$ is for each value $j$ an almost periodic function (in the sense of Bohr) in $k$. If the process has an almost periodic covariance function and is harmonizable, the spectral mass $dH(\lambda, \mu)$ is concentrated on a countable number of lines $\lambda = \mu + b_i$ parallel to the main diagonal $\lambda = \mu$. If the covariance function is not periodic but almost periodic, we have an example of processes that are not stationary or even locally stationary in the usual sense (see [4], [6]). An example of a process with almost periodic covariance function is given by

$$x_j = \sum_{k=1}^{m} a_k \cos(\lambda_k j + \varphi_k) y_j^{(k)}$$

where the $y_i^{(k)}, k = 1, \ldots, m$ are jointly stationary processes with finite second moments.

We consider real harmonizable processes with spectral mass concentrated on lines $\lambda = a\mu + b$ having slope $a$ and intercept $b$. For convenience *assume only a finite number of lines of support $u = av + b$* all with positive slope. Let $\mathcal{L}$ be the lines of spectral support $(a, b)$ with slope $a$ and intercept $b$. The spectrum on the line $\lambda = a\mu + b$ is assumed to be given by a *continuously differentiable density* $f_a(\mu)$. Such processes we call processes of class A. An example of such a process with continuous time parameter is $x_t = \sum_{s=1}^{k} \beta_s y_{\alpha_s t}$ with $y_t$ stationary and the constants $\beta_s, \alpha_s$ positive. This could represent a signal propagating along different paths with different speeds ($\alpha_s$ Doppler stretches).

There are spectral symmetries in the case of a real-valued, harmonizable process of the type we are considering. If $u = av + b$ is a line of spectral support, the lines $u = a^{-1}v - a^{-1}b, u = av - b$ and $u = a^{-1}v + a^{-1}b$ are also lines of spectral support with

$$f_{a,b}(v) = a\overline{f}_{a^{-1},-a^{-1}b}(av + b) = \overline{f}_{a,-b}(-v) = af_{a^{-1},a^{-1}b}(-av - b).$$

One should note that if $u = av + b$ is a line of spectral support with $f_{a,b}(v) \neq 0$, then $f_{1,0}(v)$ and $f_{1,0}(av + b)$ are positive. Thus, if there is a line of nontrivial support the diagonal will be a line of spectral support.

Given observations $x_{-n/2}, \ldots, x_{n/2}$ (assume $n$ even) let $F_n(\lambda)$ be the finite Fourier transform

$$F_n(\lambda) = \sum_{t=-n/2}^{n/2} x_t e^{it\lambda}.$$

A periodogram is given by

$$I_n(\lambda, \mu) = \frac{1}{2\pi n} F_n(\lambda) \overline{F_n(\mu)}.$$

If $u = \alpha v + \omega$ is a line of spectral support, a possible estimate of $f_{\alpha,\omega}(\eta)$ is given by

$$\hat{f}_{\alpha,\omega}(\eta) = \int_{-\pi+}^{\pi+} I_n(\alpha\mu + \omega, \mu) K_n(\mu - \eta) d\mu$$

where it is understood that $K(\eta)$ is a continuous nonnegative symmetric weight function of finite support with $\int K(x) dx = 1$ and

$$K_n(\eta) = b_n^{-1} K(b_n^{-1}\eta).$$

Here $b_n$ is a bandwidth such that $b_n \to 0$ and $nb_n \to \infty$ as $n \to \infty$. Further $\pi+ = \pi + \delta (\delta > 0)$.

It can be shown that as $n \to \infty$

$$E\hat{f}_{\alpha,\omega}(\eta) = o(b_n) + O\left(\frac{\log n}{nb_n}\right)$$

if $\alpha$ is not the slope of a spectral support line. If $\alpha$ is the slope of a spectral support line

$$E\hat{f}_{\alpha,\omega}(\eta) = \ell(\alpha) \sum_{(a,b) \in \mathcal{L}} f_{a,b}(\eta) \frac{\sin((n+1)/2)\ell(\alpha))}{((n+1)/2)y'\ell(a)} + o(b_n) + O\left(\frac{\log n}{n}\right)$$

with

$$\ell(\alpha) = \frac{\min(\alpha, 1)}{\alpha}$$

and

$$y' = ((b - \omega) \bmod 2\pi).$$

Also if $\xi = \alpha\eta + \omega, \xi' = \alpha'\eta' + \omega'(\alpha, \alpha' > 0)$ and $\log n/(nb_n) \to 0$ as $n \to \infty$

$$\text{cov}\left(\hat{f}_{\alpha,\omega}(\eta), \hat{f}_{\alpha',\omega'}(\eta')\right) = O\left(\frac{1}{nb_n}\right)$$

as $n \to \infty$ if the processes are Gaussian and the lines of $\mathcal{L}$ are known. Detailed estimates for the covariance are given in [4]. If $0 < \alpha \leq 1$ the estimates are consistent and since $f_{\alpha,\omega}(\eta) = \alpha \bar{f}_{\alpha^{-1},-\alpha^{-1}\omega}(\alpha\eta + \omega)$ if $\alpha \geq 1$ consistent estimates can also be determined for $\alpha \geq 1$. Estimation procedures for the values $b$ in $\mathcal{L}$ for processes with almost periodic covariance function are given in [4]. It is plausible that similar procedures will be affective in the general context.

In time frequency analysis a definition of a time varying spectrum is sometimes introduced (see [7]). This is thought of as a formal Fourier transform of

$$r(t, t - \tau)$$

in $\tau$. For the class of processes $A$ that we deal with, this is well defined and is given by

$$q(t, \eta) = \sum_{(a,b) \in \mathcal{L}} e^{itb} e^{it(\alpha-1)\eta} f_{a,b}(\eta).$$

It is clear that the estimates we have discussed can be used to obtain estimates of $q(t, \eta)$. However, estimates of the $(a, b) \in \mathcal{L}$ of sufficient precision are required. In the paper dealing with the almost periodic case, as already noted such estimates are obtained for $b \in \mathcal{L}$. As remarked earlier, the expectation is that similar methods can be used for estimation of $(a, b) \in \mathcal{L}$ in the more general case.

# REFERENCES

[1] Gardner, W.A., "Cyclostationarity in Communications and Signal Processing", IEEE Press, (1994).
[2] Gerr, N. and Allen, J., "The generalized spectrum and spectral coherence of a harmonizable time series", Digital Signal Processing **4**(1994), 222-238.
[3] Hurd, H., "Nonparametric time series analysis for periodically correlated processes", IEEE Trans. Inform. Theory **35**(1989), 350-359.
[4] Lii, K.S. and Rosenblatt, M., "Spectral analysis for harmonizable processes", Ann. Statist. **30**(2002), 258-297.
[5] Lii, K.S. and Rosenblatt, M., "Estimation for almost periodic processes", unpublished manuscript.
[6] Lii, K.S. and Rosenblatt, M., "Line spectral analysis for harmonizable processes", Proc. Nat. Anal. Sci. USA **95**(1998), 4800-4803.
[7] Scharf, L. and Friedlander, B., "Toeplitz and Hankel kernels for estimating time - varying spectra for discrete - time random processes", IEEE Trans. Signal Proc., **49**(2001), 179-189.

# Curve estimation for locally stationary time series models

Rainer Dahlhaus

Institut für Angewandte Mathematik, Universität Heidelberg,
Im Neuenheimer Feld 294, 69120 Heidelberg, Germany

Locally stationary processes are models for nonstationary time series whose behavior can locally be approximated by a stationary process. In this situation the classical characteristics of the process such as the covariance function at some lag k, the spectral density at some frequency $\lambda$, or for example the parameters of an AR(p)-process are curves which change slowly over time. We discuss different methods of nonparametric curve estimation for locally stationary processes, including stationary methods on segments, local likelihood methods and nonparametric maximum likelihood estimates. Furthermore we point out the close relation of many estimation problems to nonparametric regression.

## 1. Introduction

Stationarity has always played a major role in time series analysis. There exist many linear and nonlinear stationary models and many powerful techniques such as bootstrap methods or nonparametric methods based on the spectral density or on higher order spectra. Furthermore, the properties of methods for stationary models are usually well investigated. This has mainly been done in asymptotic considerations by using such powerful results as the ergodic theorem or the central limit theorem for stationary mixing sequences or martingale differences. The existence of such tools for asymptotic considerations has lead over the last decades to an overemphasis of stationary models while in practice many time series show a nonstationary behavior.

In contrast the situation for nonstationary time series is much more difficult: First, there exists no natural generalization from stationary to nonstationary time series and second, it is often not clear how to set down a reasonable asymptotics for nonstationary processes. An exception are nonstationary models which are generated by a time invariant generation mechanism - for examples integrated or cointegrated models. In general asymptotic considerations are often contradictory to the idea of nonstationarity since future observations of a general nonstationary process may not contain any information at all on the probabilistic structure of the process at present.

This is in particular the case for a process which is locally close to a stationary one but whose characteristics (parameters, covariances, spectral density, etc) gradually change in an unspecific way as time evolves. A simple example is the time varying AR(1)-process

$$X_t = \alpha_t X_{t-1} + \sigma_t \varepsilon_t \quad \text{with } \varepsilon_t \text{ iid } \mathcal{N}(0,1) \tag{1}$$

which we assume to be observed for t=1,...,n. Suppose we have some estimator for $\alpha_t$ and $\sigma_t$ and we want to approximate its finite sample distribution by using some asymptotic distribution. It is obvious that any asymptotics where just $n \to \infty$ will not lead to a meaningful approximation of this distribution. In order to overcome this problem we rescale the function $\alpha_t$ to the unit interval and assume that the triangular array $X_{t,n}$ $(t = 1, \ldots, n)$ is observed where

$$X_{t,n} = \alpha(\frac{t}{n}) X_{t-1,n} + \sigma(\frac{t}{n}) \varepsilon_t \quad \text{with } \varepsilon_t \text{ iid } \mathcal{N}(0,1). \tag{2}$$

This rescaling is a standard approach in nonparametric statistics. However, in other areas like engineering or econometrics such a model is rather unusual since it does no longer describe the generating mechanism of the data in a physical sense as $n \to \infty$. Instead it is a framework for an asymptotic statistical theory leading to meaningful results such as approximations for the distribution of estimates etc. In the present situation it allows for investigation of different estimates for the parameter curves $\alpha(\cdot)$ and $\sigma(\cdot)$ by using well established statistical tools.

Instead of deriving results for the special model (2) it is our goal to derive results for a broader class of nonstationary processes parameterized by one or several curves in time (even infinitely many). Hence the class of processes should be as large as possible leading to the definition of a locally stationary process given in Section 2. Furthermore, many other issues can be studied for such processes such as spectral estimation, discriminant analysis, bootstrap techniques, etc.

In Section 3 we study local inference by stationary methods for such processes. In particular we discuss local Whittle estimation. In Section 4 we discuss inference by local likelihood methods, both with the conditional likelihood and with a frequency domain local likelihood.

There are several aspects of locally stationary processes which are not treated in this paper, for example results on large deviations for quadratic forms of locally stationary processes (cf. Zani [25]), on bootstrap methods (cf. Paparoditis and Politis [19]), on multivariate time varying linear systems (cf. Chang and Morettin [2] and Ombao et. al. [18]) and on prediction (cf. Fryzlewicz et.al. [14]).

## 2. Locally stationary processes

Below we refer to various results on locally stationary processes. All theses results have been proved under slightly different assumptions - mainly under the assumption of a time varying spectral representation as in (22) and (23). We now give a definition of local stationarity in terms of a time varying MA($\infty$) (tvMA($\infty$)) - representation as it is used in Dahlhaus and Polonik [12]. This reflects our present view of a weak and simple assumption. Let

$$V(g) = \sup\Big\{\sum_{k=1}^{m}|g(x_k) - g(x_{k-1})| : 0 \leq x_o < \ldots < x_m \leq 1,\, m \in \mathbf{N}\Big\}$$

be the total variation of $g$. We call a sequence of stochastic processes $X_{t,n}(t=1,\ldots,n)$ *locally stationary* if it fulfills the following assumption.

**Assumption 2.1** *The sequence of stochastic processes $X_{t,n}$ has a representation*

$$X_{t,n} = \mu(\frac{t}{n}) + \sum_{j=-\infty}^{\infty} a_{t,n}(j)\,\varepsilon_{t-j} \tag{3}$$

*where $\mu$ is of bounded variation and the $\varepsilon_t$ are iid with $E\varepsilon_t = 0$, $E\varepsilon_s\varepsilon_t = 0$ for $s \neq t$, $E\varepsilon_t^2 = 1$. Let*

$$\ell(j) := \begin{cases} 1, & |j| \leq 1 \\ |j|\log^{1+\kappa}|j|, & |j| > 1 \end{cases}$$

*for some $\kappa > 0$ and*

$$\sup_{t}|a_{t,n}(j)| \leq \frac{K}{\ell(j)} \quad \text{(with $K$ indep. of $n$)}. \tag{4}$$

*Furthermore we assume that there exist functions $a(\cdot,j) : (0,1] \to \mathbf{R}$ with*

$$\sup_{u}|a(u,j)| \leq \frac{K}{\ell(j)}, \tag{5}$$

$$\sup_{j} \sum_{t=1}^{n} |a_{t,n}(j) - a(\frac{t}{n}, j)| \leq K, \tag{6}$$

$$V(a(\cdot, j)) \leq \frac{K}{\ell(j)}. \tag{7}$$

The above assumptions are weak in the sense that only bounded variation is required for the coefficient functions. This is sufficient for certain global results. For local results stronger smoothness assumptions have to be imposed - for example in addition for some $i$

$$\sup_{u} |\frac{\partial^i \mu(u)}{\partial u^i}| \leq K, \tag{8}$$

$$\sup_{u} |\frac{\partial^i a(u, j)}{\partial u^i}| \leq \frac{K}{\ell(j)} \quad \text{for } j = 0, 1, \ldots \tag{9}$$

and instead of (6) the stronger assumption

$$\sup_{t,n} |a_{t,n}(j) - a(\frac{t}{n}, j)| \leq \frac{K}{n\,\ell(j)}. \tag{10}$$

The construction with $a_{t,n}(j)$ and $a(\frac{t}{n}, j)$ looks complicated at first glance. The function $a(\cdot, j)$ is needed for rescaling and to impose necessary smoothness conditions while the additional use of $a_{t,n}(j)$ makes the class rich enough to cover interesting cases such as tvAR - models (which otherwise were not included). Furthermore, additional moment conditions on $\varepsilon_t$ are usually required while the independence assumption on $\varepsilon_t$ can be relaxed.

The name *locally stationary process* can easily be justified: Suppose Assumption 2.1 and (8)-(10) hold for i=1. Let u ∈ (0,1] be fixed and define

$$\tilde{X}_t(u) := \mu(u) + \sum_{j=-\infty}^{\infty} a(u, j)\, \varepsilon_{t-j}. \tag{11}$$

Then it follows directly that

$$|X_{t,n} - \tilde{X}_t(u)| \leq K\{|\frac{t}{n} - u| + \frac{1}{n}\} U_t \tag{12}$$

with the stationary process

$$U_t := 1 + \sum_{j=-\infty}^{\infty} \frac{1}{\ell(j)} |\varepsilon_{t-j}|. \tag{13}$$

This means that the process $\tilde{X}_t(u)$ is a stationary approximation of $X_{t,n}$ in a local neighborhood about u, for example on the segment $\{t : |t/n - u| \leq b/2\}$ where $b \to 0$ and $bn \to \infty$ as $n \to \infty$. $\tilde{X}_t(u)$ can for example be used for proving results on the asymptotic behavior of estimates. This was done in Dahlhaus and Subba Rao [11].

Before we give some examples of locally stationary processes we define the time varying spectral density by

$$f(u, \lambda) := \frac{1}{2\pi} |A(u, \lambda)|^2 \tag{14}$$

where

$$A(u, \lambda) := \sum_{j=-\infty}^{\infty} a(u, j) \exp(-i\lambda j), \tag{15}$$

and the time varying spectral covariance of lag $k$ at rescaled time $u$ by

$$c(u, k) := \int_{-\pi}^{\pi} f(u, \lambda) \exp(i\lambda k) d\lambda = \sum_{j=-\infty}^{\infty} a(u, k+j) a(u, j). \tag{16}$$

$f(u, \lambda)$ and $c(u, k)$ are the spectral density and the covariance function of the stationary approximation $\tilde{X}_t(u)$. Under Assumption 2.1 and (10) it can be shown that

$$\text{cov}(X_{[un],n}, X_{[un]+k,n}) = c(u, k) + O(n^{-1}) \tag{17}$$

uniformly in $u$ and $k$.

**Example 2.2** *(i) A simple example of a process $X_{t,n}$ which fulfills the above assumptions is $X_{t,n} = \mu(\frac{t}{n}) + \phi(\frac{t}{n}) Y_t$ where $Y_t = \Sigma_j a(j) \varepsilon_{t-j}$ is stationary with $|a(j)| \leq K/\ell(j)$ and $\mu$ and $\phi$ are of bounded variation. If $Y_t$ is an AR(2)-process with (complex) roots close to the unit circle then $Y_t$ shows a periodic behavior and $\phi$ may be regarded as a time varying amplitude modulating function of the process $X_{t,n}$. If $Y_t$ is an i.i.d. sequence then this is the classical situation of nonparametric regression.*

*(ii) The tvARMA(p,q) process*

$$\sum_{j=0}^{p} \alpha_j(\frac{t}{n}) X_{t-j,n} = \sum_{k=0}^{q} \beta_k(\frac{t}{n}) \sigma(\frac{t-k}{n}) \varepsilon_{t-k} \tag{18}$$

*where $\varepsilon_t$ are iid with $E\varepsilon_t = 0$ and $E\varepsilon_t^2 < \infty$ and all $\alpha_j(\cdot), \beta_k(\cdot)$ and $\sigma^2(\cdot)$ are of bounded variation and $\sum_{j=0}^{p} \alpha_j(u) z^j \neq 0$ for all $u$ and all $|z| \leq 1 + \delta$ for some $\delta > 0$,* is locally stationary, ie it fulfills Assumption 2.1. If the parameters are differentiable with bounded derivatives then also (8)-(10) are fulfilled (for i=1). The time varying spectral density is

$$f(u,\lambda) = \frac{\sigma^2(u)}{2\pi} \frac{|\sum_{k=0}^{q} \beta_k(u)\exp(i\lambda k)|^2}{|\sum_{j=0}^{p} \alpha_j(u)\exp(i\lambda j)|^2}. \tag{19}$$

This is proved in Dahlhaus and Polonik [12] and under stronger assumptions in Dahlhaus [3].

(iii) It is obvious from Assumption 2.1 that nonlinear processes with time varying parameters are not covered by our definition. In Dahlhaus and Subba Rao [11] tvARCH-models have been studied by using tvVolterra expansions. It it obvious that tvVolterra expansions can be used for other nonlinear models as well - however this has not been investigated yet.

We conclude this section by discussing the relation of Assumption 2.1 to time varying spectral representations. The time varying MA($\infty$)-representation (3) can easily be transformed into a time varying spectral representation as used e.g. in Dahlhaus [6], [3] and [7]. If the $\varepsilon_t$ are assumed to be stationary then there exists a Cramér representation

$$\varepsilon_t = \frac{1}{\sqrt{2\pi}} \int_{-\pi}^{\pi} \exp(i\lambda t) d\xi(\lambda) \tag{20}$$

where $\xi(\lambda)$ is a process with mean 0 and orthonormal increments (cf. [1]). Let

$$A_{t,n}(\lambda) := \sum_{j=-\infty}^{\infty} a_{t,n}(j) \exp(-i\lambda j). \tag{21}$$

Then

$$X_{t,n} = \frac{1}{\sqrt{2\pi}} \int_{-\pi}^{\pi} \exp(i\lambda t) A_{t,n}(\lambda) d\xi(\lambda). \tag{22}$$

(10) now implies

$$\sup_{t,\lambda} |A_{t,n}(\lambda) - A(\frac{t}{n},\lambda)| \leq Kn^{-1} \tag{23}$$

which was assumed in the above cited papers of the author. Conversely, if we start with (22) and (23) then we can conclude from adequate smoothness conditions on $A(u, \lambda)$ to the conditions of Assumption 2.1.

A time varying spectral representation was first used by Priestley [20] in his theory of processes with an evolutionary spectral representation. However, in contrast to the present work Priestley did not rescale the parameter curves. We also mention work on locally stationary wavelet processes which is based on a time varying wavelet representation of the process instead of (22) (cf. [16], [18]).

Tests for locally stationary processes have been derived in von Sachs and Neumann [23] and Sakiyama and Taniguchi [22].

In the next two sections we mainly restrict ourselves to the case $\mu(\cdot) \equiv 0$. Results with nonzero mean curve can for example be found in [6], [3] and [4].

## 3. Local inference by stationary methods

Suppose we have observations $X_{1,n}, \ldots, X_{n,n}$ and want to estimate the parameter curves of a locally stationary model such as (2). There are striking similarities of this problem to nonparametric regression with equidistant design and many methods from traditional curve estimation can be transferred to inference for locally processes. These similarities will be explored in more detail in the next section. In this section we describe a standard technique which often is equivalent to kernel estimation.

The simplest way to do inference for locally stationary processes is via stationary methods on segments. The idea is that the process is close to the stationary process $\tilde{X}_t(u)$ on a reasonably small segment $\{t : |t/n - u| \leq b/2\}$. The parameter of interest (or the correlation, spectral density, etc) is estimated by some classical method and the resulting estimate is assigned to the midpoint u of the segment. By shifting the segment this finally leads to a curve estimate of the unknown parameter curve (time varying correlation, time varying spectral density, etc). An obvious and useful modification of this method is obtained when more weight is put on data in the center of the interval than on data at the edges. This can either be achieved by using a data taper on the segment or by using a kernel type estimate.

Since we use observations from the process $X_{t,n}$ (instead of $\tilde{X}_t(u)$) the procedure causes a bias which depends on the degree of instationarity of the process on the segment. It is possible to evaluate this bias and to use the resulting expression for an adaptive choice of the segment length (although the latter has not been investigated yet). To demonstrate this and to show the similarities to nonparametric regression we now briefly discuss the estimation of the time varying covariance function.

Consider the estimation of the covariance of lag $k$ at time $u$ as defined in (16). For simplicity we assume $\mu(u) \equiv 0$. It is natural to consider as an estimate all products of lag $k$ in a neighbourhood of $u$, i.e.

$$\hat{c}_n(u,k) := \frac{\sum_t K\left(\frac{u-(t+k/2)/n}{b}\right) X_{t,n} X_{t+k,n}}{\sum_t K\left(\frac{u-(t+k/2)/n}{b}\right)} \qquad (24)$$

where $K : \mathbf{R} \to [0,\infty)$ is a kernel with $K(x) = K(-x)$, $\int K(x)dx = 1$ and $K(x) = 0$ for $x \notin [-1/2, 1/2]$. $b$ is the bandwidth in rescaled time and $bn$ is the segment length. For a rectangular kernel this is a stationary segment method. As usual we assume $b \to 0$ and $bn \to \infty$ as $n \to \infty$.

Suppose $X_{t,n}$ is a locally stationary process with mean 0 which fulfills Assumption 2.1 and (8)-(10) for i=3. Then it is not very difficult to prove the following bias expansion for $\hat{c}_n$

$$E\hat{c}_n(u,k) = c(u,k) + \frac{1}{2}b^2 \int x^2 K(x)\,dx \left[\frac{\partial^2}{\partial u^2} c(u,k)\right] + o(b^2) + O\left(\frac{1}{bn}\right). \qquad (25)$$

Note the similarity to nonparametric regression. The bias of order $b^2$ is solely due to nonstationarity of $X_{t,n}$ on the segment. This nonstationarity is measured by $\frac{\partial^2}{\partial u^2} c(u,k)$. If the process is stationary this second derivative gets zero and the bias disappears. The bias of order $(bn)^{-1}$ is due to the procedure and can be calculated as well (it varies if the mean is unknown). Note, that $bn$ is the segment length i.e. the 'effective sample length' used for the calculation of $\hat{c}_n(u,k)$. Without proof we remark that

$$\mathrm{var}(\hat{c}_n(u,k)) = \frac{1}{bn} \int_{-1/2}^{1/2} K(x)^2 dx \sum_{\ell=-\infty}^{\infty} c(u,\ell)[c(u,\ell) + c(u,\ell+2k)]. \qquad (26)$$

It is now straightforward to calculate the mean squared error and to minimize the leading terms with respect to $b$. This leads to the (theoretically) optimal segment length $bn$. It depends on the unknown covariance function and its second derivative. An data adaptive segment choice may now be derived by a plug-in strategy.

It is heuristically clear that an improvement may be obtained by a local polynomial fit on the segment $v \in [u - b/2, u + b/2]$, that is by the estimate $\bar{c}_n(u,k) := \hat{c}_o$ where

$$\hat{c} = \arg\min_c \frac{1}{bn} \sum_t K\left(\frac{u-(t+\frac{k}{2})/n}{b}\right) \left(X_{t,n} X_{t+k,n} - \sum_{j=0}^{q} c_j \left(\frac{t+\frac{k}{2}}{n} - u\right)^j\right)^2. \qquad (27)$$

An alternative is the ordinary tapered covariance on a segment of length $m$ ($\approx bn$)

$$\tilde{c}_n(u,k) = H_m^{-1} \sum_{1 \leq t, t+k \leq m} h\left(\frac{t}{m}\right) h\left(\frac{t+k}{m}\right) X_{[un]-m/2+t} X_{[un]-m/2+t+k} \tag{28}$$

where $h : [0,1] \to \mathbf{R}$ is a data taper with $h(x) = h(1-x)$ and $H_m = \sum_{t=1}^m h(t/m)^2$. This estimate has the same bias expansion and the same asymptotic variance as $\hat{c}_n(u,k)$ with

$$K(x) := \left\{ \int_0^1 h(x)^2 dx \right\}^{-1} h(x+1/2)^2 \qquad x \in [-1/2, 1/2] \tag{29}$$

and $b = m/n$.

The use of data tapers has priority in spectral estimation or frequency based likelihood estimation since there exists no elegant choice of a spectral estimate with a time domain kernel (note however the suggestion below (52) based on the preperiodogram). We use the tapered periodogram

$$I_m(u, \lambda) = \frac{1}{2\pi H_m} \left| \sum_{s=1}^m h\left(\frac{s}{m}\right) X_{[un]-m/2+s,n} \exp(-i\lambda s) \right|^2 \tag{30}$$

as a row estimate and, as for stationary time series, the spectral domain kernel estimate

$$\hat{f}_n(u, \lambda) = \frac{1}{b_f} \int_{-\pi}^{\pi} K_f\left(\frac{\lambda - \mu}{b_f}\right) I_m(u, \mu) \, d\mu \tag{31}$$

in order to obtain a consistent estimate of $f(u, \lambda)$. Note that implicitly two kernels are involved: The kernel $K_f$ with bandwidth $b_f$ in frequency direction and the kernel $K$ from (29) with bandwidth $b = m/n$ in time direction.

This is nicely reflected in the expressions for the bias and the variance of the estimate which have been calculated in Dahlhaus [5], Theorem 2.2. The adaptation problem now consists in selecting h,m,$K_f$ and $b_f$. This has been solved (only theoretically!) in Dahlhaus [5], Theorem 2.3. We briefly mention at this point that estimation of $f(u, \lambda)$ with wavelet methods has been considered in Neumann and von Sachs [17].

We now discuss the estimation of parameter curves $\boldsymbol{\theta}(\cdot) = (\theta_1(\cdot), \ldots, \theta_d(\cdot))'$ of a locally stationary model with spectral density $f_{\boldsymbol{\theta}(u)}(\lambda)$. We assume that $\boldsymbol{\theta}(\cdot)$ is identifiable from $f_{\boldsymbol{\theta}(u)}(\lambda)$ (which in particular is true for Gaussian processes). As a stationary method on segments we may use the local Whittle-estimate [24]

$$\widehat{\boldsymbol{\theta}}(u) = \underset{\boldsymbol{\theta}}{\operatorname{argmin}} \, \hat{\mathcal{L}}_m(\boldsymbol{\theta}, u) \tag{32}$$

with

$$\hat{\mathcal{L}}_m(\boldsymbol{\theta}, u) := \frac{1}{4\pi} \int_{-\pi}^{\pi} \left\{ \log 4\pi^2 f_{\boldsymbol{\theta}}(\lambda) + \frac{I_m(u, \lambda)}{f_{\boldsymbol{\theta}}(\lambda)} \right\} d\lambda. \tag{33}$$

For stationary processes the (nontapered) Whittle estimate is an approximative Gaussian maximum likelihood estimate. $\hat{\mathcal{L}}_m(\boldsymbol{\theta}, u)$ is a minimum distance functional between the nonparametric spectral estimate $I_m(u, \lambda)$ and the parametric spectral density $f_{\boldsymbol{\theta}}(\lambda)$. For autoregressive processes $\widehat{\boldsymbol{\theta}}(u)$ is equal to the local tapered Yule-Walker estimate.

The properties of $\widehat{\boldsymbol{\theta}}(u)$ have been investigated in Dahlhaus and Giraitis [8], Section 3. In particular the asymptotic variance has been calculated and a bias expansion has been given. Furthermore, the asymptotic mean squared error has been minimized with respect to the segment length m and the data taper $h(\cdot)$. In Section 2 of the same paper the case of local Yule Walker estimates for tvAR processes has been investigated in detail.

Suppose now we want to make a parametric fit of the curves $\boldsymbol{\theta}(\cdot)$ that is we assume $\boldsymbol{\theta}(\cdot) = \boldsymbol{\theta}_\eta(\cdot)$ with $\eta \in \mathbf{R}^q$. An example is the situation where the curves $\theta_j(\cdot)$ are polynomials in time and the $\eta_i$ are the coefficients. One way of estimation is to extend the above local Whittle likelihood to a global one by summing over (possibly overlapping) data segments.

The shift from segment to segment is denoted by $S$, i.e. we calculate the tapered periodogram $I_m(u, \lambda)$ as in (30) over segments with midpoints $t_j := S(j-1) + m/2$ $(j=1,\ldots,J)$ where $n = S(J-1) + m$, or written in rescaled time, at time points $u_j = t_j/n$. We now set

$$\tilde{\mathcal{L}}_n(\eta) = \frac{1}{J} \sum_{j=1}^{J} \hat{\mathcal{L}}_m(\boldsymbol{\theta}_\eta(u_j), u_j) \tag{34}$$

$$= \frac{1}{4\pi} \frac{1}{J} \sum_{j=1}^{J} \int_{-\pi}^{\pi} \left\{ \log 4\pi^2 f_{\boldsymbol{\theta}_\eta(u_j)}(\lambda) + \frac{I_N(u_j, \lambda)}{f_{\boldsymbol{\theta}_\eta(u_j)}(\lambda)} \right\} d\lambda \tag{35}$$

and

$$\hat{\eta}_n = \underset{\eta}{\operatorname{argmin}} \, \tilde{\mathcal{L}}_n(\eta). \tag{36}$$

$\tilde{\mathcal{L}}_n(\eta)$ converges in probability to

$$\mathcal{L}(\eta) = \frac{1}{4\pi} \int_0^1 \int_{-\pi}^{\pi} \left\{ \log 4\pi^2 f_{\boldsymbol{\theta}_\eta(u)}(\lambda) + \frac{f(u,\lambda)}{f_{\boldsymbol{\theta}_\eta(u)}(\lambda)} \right\} d\lambda\, du \qquad (37)$$

where $f(u,\lambda)$ is the true time varying spectral density of the process. In [6] we prove asymptotic normality and efficiency of $\hat{\eta}_n$. It is interesting that the data taper does not lead to an increase of the asymptotic variance in the case of overlapping segments. If the model is misspecified then $\hat{\eta}_n$ converges in probability to

$$\eta_0 = \underset{\eta}{\operatorname{argmin}}\, \mathcal{L}(\eta). \qquad (38)$$

Furthermore, the asymptotic distribution under misspecification is derived. These results include the case where a stationary model is fitted to the data but the true process is only locally stationary.

The related expression

$$\frac{1}{4\pi} \int_0^1 \int_{-\pi}^{\pi} \left\{ \log \frac{\tilde{f}(u,\lambda)}{f(u,\lambda)} + \frac{f(u,\lambda)}{\tilde{f}(u,\lambda)} - 1 \right\} d\lambda\, du \qquad (39)$$

is the Kullback-Leibler information divergence between two locally stationary Gaussian processes with time varying spectral densities $f(u,\lambda)$ and $\tilde{f}(u,\lambda)$ respectively. This has been proved in [3]. Sakiyama and Taniguchi [21] use the Kullback-Leibler divergence for discriminant analysis of locally stationary processes.

## 4. Inference by local likelihood methods

In this section we discuss the estimation of parameter curves $\boldsymbol{\theta}(\cdot) = (\theta_1(\cdot), \ldots, \theta_d(\cdot))'$ by local likelihood methods which are not based on segments. Suppose as an example that we want to estimate the curves $\boldsymbol{\theta}(\cdot) = (\alpha(\cdot), \sigma(\cdot))'$ of the model (2) by a wavelet method or a local polynomial fit. A naive idea would be to start with a preliminary raw estimate (eg with a Yule Walker estimate on a small segment) and to smooth the resulting estimate by using wavelets or a local polynomial. However, with such a procedure some information on the smoothness of the coefficient curves were already lost after the first step since any preliminary estimate already involves some smoothing.

In order to avoid such a two step procedure one could use a least squares criterion for the estimation of $\alpha(\cdot)$ which, for a local polynomial fit, takes the form

$$\hat{c} = \underset{c}{\operatorname{argmin}}\, \frac{1}{bn} \sum_{t=1}^{n} K\!\left(\frac{u - t/n}{b}\right) \left( X_{t,n} - \sum_{j=0}^{d} c_j \left(\frac{t}{n} - u\right)^j X_{t-1,n} \right)^2 \qquad (40)$$

with $\hat{a}(u) = \hat{c}_0$, or more generally a local (conditional) likelihood approach

$$(\hat{c}_0,\ldots,\hat{c}_d) = \underset{c_0,\ldots,c_d}{\operatorname{argmin}} \frac{1}{bn} \sum_{t=1}^{n} K\left(\frac{u - t/n}{b}\right) \ell_{t,n}\left(\sum_{j=0}^{d} c_j (\frac{t}{n} - u)^j\right) \qquad (41)$$

with $\hat{\boldsymbol{\theta}}(u) = \hat{c}_0$ where

$$\ell_{t,n}(\boldsymbol{\theta}(\frac{t}{n})) = -\log f_{\boldsymbol{\theta}(\frac{t}{n})}(X_{t,n}|X_{t-1,n},\ldots,X_{1,n}) \qquad (42)$$

is the conditional log-likelihood. For the present tvAR(1) example we have

$$\ell_{t,n}(\boldsymbol{\theta}(\frac{t}{n})) = \frac{1}{2}\log\left(2\pi\sigma^2(\frac{t}{n})\right) + \frac{1}{2\sigma^2(\frac{t}{n})}\left(X_{t,n} - \alpha(\frac{t}{n})X_{t-1,n}\right)^2. \qquad (43)$$

Besides the local polynomial estimate (38) many other estimates can be constructed by using the local likelihood:

1. A kernel estimate by

$$\hat{\boldsymbol{\theta}}(u) = \underset{\boldsymbol{\theta}}{\operatorname{argmin}} \frac{1}{bn} \sum_{t=1}^{n} K\left(\frac{u - t/n}{b}\right) \ell_{t,n}(\boldsymbol{\theta}); \qquad (44)$$

2. an orthogonal series estimator (e.g. wavelets) by

$$\bar{\alpha} = \underset{\alpha}{\operatorname{argmin}} \frac{1}{n} \sum_{t=1}^{n} \ell_{t,n}\left(\sum_{j=1}^{J} \alpha_j \psi_j(\frac{t}{n})\right) \qquad (45)$$

together with some shrinkage of $\bar{\alpha}$;

3. a nonparametric maximum likelihood estimator by

$$\hat{\boldsymbol{\theta}}(\cdot) = \underset{\boldsymbol{\theta}(\cdot) \in \Theta}{\operatorname{argmin}} \frac{1}{n} \sum_{t=1}^{n} \ell_{t,n}(\boldsymbol{\theta}(\frac{t}{n})) \qquad (46)$$

where $\Theta$ is an adequate function space, for example a space of curves under shape restrictions;

4. a parametric fit for the curves $\boldsymbol{\theta}(\cdot) = \boldsymbol{\theta}_\eta(\cdot)$ with $\eta \in \mathbf{R}^q$ by

$$\hat{\eta} = \operatorname*{argmin}_{\eta} \frac{1}{n} \sum_{t=1}^{n} \ell_{t,n}(\boldsymbol{\theta}_\eta(\frac{t}{n})). \tag{47}$$

Apart from the parametric fit (see [4]) the above estimates have hardly been investigated for general models so far, nor have the computational problems been addressed. There exists only some work for tvAR models. Local polynomial fits for tvAR models have been considered by Kim [15], the orthogonal series estimator for a truncated wavelet series expansion for tvAR models by Dahlhaus, Neumann and von Sachs [9], and nonparametric maximum likelihood estimates under shape restrictions by Dahlhaus and Polonik [13].

There exists an interesting modification of the above estimates based on a different local likelihood function $\ell_{t,n}(\cdot)$. This new likelihood seems to have computational advantages over the exact conditional Gaussian likelihood. It is a frequency domain likelihood function related to the Whittle likelihood for stationary processes as defined in (33). However, instead of the periodogram we use a localized version of the periodogram, namely the so called preperiodogram introduced by Neumann and von Sachs [17] as a starting point for a wavelet estimate of the time-varying spectral density. In the case of a process with mean zero it takes the form

$$\tilde{I}_n(u,\lambda) := \frac{1}{2\pi} \sum_{\substack{k \\ 1 \leq [un+0.5+k/2], [un+0.5-k/2] \leq n}} X_{[un+0.5+k/2],n} X_{[un+0.5-k/2],n} \exp(-i\lambda k) \tag{48}$$

where $u \in [0,1]$ is the rescaled time. There exists a nice relation between the preperiodogram and the ordinary nontapered periodogram:

$$I_n(\lambda) = \frac{1}{2\pi n} |\sum_{s=1}^{n} X_s \exp(-i\lambda s)|^2 \tag{49}$$

$$= \frac{1}{2\pi} \sum_{k=-(n-1)}^{n-1} \left( \frac{1}{n} \sum_{t=1}^{n-|k|} X_t X_{t+|k|} \right) \exp(-i\lambda k) \tag{50}$$

$$= \frac{1}{n} \sum_{t=1}^{n} \frac{1}{2\pi} \sum_{\substack{k \\ 1 \leq [t+0.5+k/2], [t+0.5-k/2] \leq n}} X_{[t+0.5+k/2],n} X_{[t+0.5-k/2],n} \exp(-i\lambda k) \tag{51}$$

$$= \frac{1}{n} \sum_{t=1}^{n} \tilde{I}_n(\frac{t}{n}, \lambda), \tag{52}$$

i.e. the periodogram is the average of the preperiodogram over time. (50) means that the periodogram $I_n(\lambda)$ is the Fourier transform of the covariance estimator of lag $k$ over the whole segment while the preperiodogram $\tilde{I}_n(\frac{t}{n}, \lambda)$ just uses the pair $X_{[t+0.5+k/2],n} X_{[t+0.5-k/2],n}$ as a kind of "local estimator" of the covariance of lag $k$ at time $t$ (note that $[t+0.5+k/2] - [t+0.5-k/2] = k$).

A classical kernel estimator of the spectral density of a stationary process at some frequency $\lambda_0$ therefore can be regarded as an average of the preperiodogram over all time points and over the frequencies in the neighbourhood of $\lambda_0$. It is therefore plausible that averaging the preperiodogram about some frequency $\lambda_0$ and about some time-point $u_0$ gives an estimate of the time-varying spectrum $f(u_0, \lambda_0)$.

We now define the approximate local likelihood at time t by

$$\tilde{\ell}_{t,n}(\eta) = \frac{1}{4\pi} \int_{-\pi}^{\pi} \left\{ \log 4\pi^2 f_\eta(\frac{t}{n}, \lambda) + \frac{\tilde{I}_n(\frac{t}{n}, \lambda)}{f_\eta(\frac{t}{n}, \lambda)} \right\} d\lambda, \tag{53}$$

where $f_\eta(\frac{t}{n}, \lambda)$ is the time varying spectral density of the locally stationary model. For a parametric model we now use the global likelihood

$$\frac{1}{n} \sum_{t=1}^{n} \tilde{\ell}_{t,n}(\eta) = \frac{1}{4\pi} \frac{1}{n} \sum_{t=1}^{n} \int_{-\pi}^{\pi} \left\{ \log 4\pi^2 f_\eta(\frac{t}{n}, \lambda) + \frac{\tilde{I}_n(\frac{t}{n}, \lambda)}{f_\eta(\frac{t}{n}, \lambda)} \right\} d\lambda. \tag{54}$$

For stationary models, where $f_\eta(\frac{t}{n}, \lambda) = f_\eta(\lambda)$, we obtain with (52) that this is exactly equal to the Whittle likelihood

$$\frac{1}{4\pi} \int_{-\pi}^{\pi} \left\{ \log 4\pi^2 f_\eta(\lambda) + \frac{I_n(\lambda)}{f_\eta(\lambda)} \right\} d\lambda. \tag{55}$$

This means that the global likelihood from (54) is a true generalization of the Whittle likelihood to locally stationary processes. In Dahlhaus [7], Theorem 2.1 we have shown that it is indeed an approximation of the true Gaussian likelihood for a locally stationary process.

We now can also consider the nonparametric estimates (41) and (44) -(47) from above with $\tilde{\ell}_{t,n}(\boldsymbol{\theta}(\frac{t}{n}))$ instead of $\ell_{t,n}(\boldsymbol{\theta}(\frac{t}{n}))$ where (with a slight abuse of notation)

$$\tilde{\ell}_{t,n}(\boldsymbol{\theta}(\frac{t}{n})) = \frac{1}{4\pi} \int_{-\pi}^{\pi} \left\{ \log 4\pi^2 f_{\boldsymbol{\theta}(\frac{t}{n})}(\lambda) + \frac{\tilde{I}_n(\frac{t}{n}, \lambda)}{f_{\boldsymbol{\theta}(\frac{t}{n})}(\lambda)} \right\} d\lambda. \tag{56}$$

The properties of the parametric estimator (47) with this local likelihood have been studied in Dahlhaus [7] where asymptotic normality and efficiency has been proved and the misspecified case has been studied (for multivariate processes). The orthogonal series estimator (45) has been investigated for a truncated wavelet series expansion together with nonlinear threshholding in Dahlhaus and Neumann [10]. The method is fully automatic and adapts to different smoothness classes. It is shown that the usual rates of convergence in Besov classes are attained up to a logarithmic factor. The nonparametric estimator (46) is studied in Dahlhaus and Polonik [12]. Rates of convergence, depending on the metric entropy of the function space, are derived. This includes in particular maximum likelihood estimates derived under shape restriction. The main tool for deriving these results is the so called empirical spectral processes. The kernel estimator (44) and the local polynomial fit (41) have not been investigated in combination with this likelihood.

## REFERENCES

1. D.R. Brillinger. Time Series: Data Analysis and Theory. Holden Day, San Francisco, 1981.
2. C. Chang and P. Morettin. Estimation of time varying linear systems. Statist. Inference Stoch. Proc. 2 (1999) 253-285.
3. R. Dahlhaus. On the Kullback-Leibler information divergence of locally stationary processes. Stoch. Proc. Appl. 62 (1996) 139-168.
4. R. Dahlhaus. Maximum likelihood estimation and model selection for locally stationary processes. J. Nonparam. Statist. 6 (1996) 171-191.
5. R. Dahlhaus. Asymptotic statistical inference for nonstationary processes with evolutionary spectra. In: Athens Conference on Applied Probability and Time Series Analysis, Volume II (Eds. P.M. Robinson und M. Rosenblatt), Springer Verlag, New York, 1996.
6. R. Dahlhaus. Fitting time series models to nonstationary processes. Ann. Statist. 25 (1997) 1-37.
7. R. Dahlhaus. A likelihood approximation for locally stationary processes. Ann. Statist. 28 (2000) 1762-1794.
8. R. Dahlhaus and L. Giraitis. On the optimal segment length for parameter estimates for locally stationary time series. J. Time Series Anal. 19 (1998) 629-655.
9. R. Dahlhaus, M.H. Neumann and R. von Sachs. Nonlinear wavelet estimation of time-varying autoregressive processes. Bernoulli 5 (1999) 873-906.
10. R. Dahlhaus and M.H. Neumann. Locally adaptive fitting of semiparametric models to nonstationary time series. Stoch. Proc. and Appl. 91 (2001) 277-308.
11. R. Dahlhaus and S. Subba Rao. Statistical inference for time-varying ARCH pro-

cesses. Preprint Universität Heidelberg, 2003.
12. R. Dahlhaus and W. Polonik. Nonparametric maximum likelihood estimation and empirical spectral processes for locally stationary processes. In preparation (2003).
13. R. Dahlhaus and W. Polonik. Inference under shape restrictions for time varying autoregressive models. In preparation (2003).
14. P. Fryzlewicz and S. Van Bellegem and R. von Sachs. Forecasting non-stationary time series by wavelet process modelling. Discussion Paper 02/08 de l'Institut de Statistique, UCL, Louvain-la-Neuve, 2002.
15. W. Kim. Nonparametric kernel estimation of evolutionary autoregressive processes. SFB 373 Discussion Paper 103, Berlin, 2001.
16. G. Nason, R. von Sachs and G. Kroisandt. Wavelet processes and adaptive estimation of the evolutionary wavelet spectrum. J. Royal Statistical Society Series B 62 (2000) 271-292
17. M.H. Neumann and R. von Sachs. Wavelet thresholding in anisotropic function classes nand applications to adaptive estimation of evolutionary spectra. Ann. Statist. 25 (1997) 38 - 76.
18. H. Ombao, J. Raz, R. von Sachs and W. Guo. The SLEX Model of a non-stationary random process. Ann. Inst. Stat. Math. 54 (2002) 171-200.
19. E. Paparoditis, D. N. Politis, Local block bootstrap. C. R. Acad. Sci. Paris, Ser. I 335 (2002) 959-962.
20. M.B. Priestley. Evolutionary spectra and non-stationary processes. J. Roy. Statist. Soc. Ser. B 27 (1965) 204-237.
21. K. Sakiyama and M. Taniguchi. Discriminant analysis of locally stationary processes. Preprint, Osaka University (2001).
22. K. Sakiyama and M. Taniguchi. Testing composite hypotheses for locally stationary processes. J.Time Ser.Anal. (2003) to appear.
23. R. von Sachs and M. Neumann. A wavelet-based test for stationarity. J. Time Series Analysis 21 (2000) 597-613
24. P. Whittle. Estimation and information in stationary time series. Ark. Mat. 2 (1953) 423-434.
25. M. Zani. Large deviations for quadratic forms of locally stationary processes. J. Multivar. Anal. 81 (2002) 205-228.

*Recent Advances and Trends in Nonparametric Statistics*
Michael G. Akritas and Dimitris N. Politis (Editors)
© 2003 Elsevier Science B.V. All rights reserved.

# Assessing Spatial Isotropy

Michael Sherman, Yongtao Guan and James A. Calvin[a]*

[a]Dept. of Statistics, Texas A&M University
College Station, TX, USA

In a spatial data analysis the analyst typically requires knowledge of the underlying correlation structure in order to effectively model data. A common assumption for this structure is one of isotropy, i.e., direction independent correlation.

While graphical techniques are useful to check for isotropy, they are often difficult to assess and cannot be interpreted objectively. We propose an objective test of the isotropy assumption that applies to a large class of spatial structures. Specifically, no explicit knowledge of marginal or joint distributions of the process is necessary, and the shape of the random field can be quite irregular. We consider observations on a subset of $Z^2$. Our method can also be extended to irregularly-spaced data.

## 1. Introduction

Data that are spatially close are often correlated. For example, crop yields on adjacent plots in an agricultural study tend to be positively correlated due to similar environmental conditions. Similarly in an epidemiological study, disease rates may be similar for geographic regions which are close. In order to perform proper spatial analyses precise knowledge of this correlation is necessary. For example, optimal prediction (Kriging), depends entirely on knowledge of the correlation function.

A common assumption on correlation is that it is isotropic. Correlation is isotropic if correlation between observations at any two sites depends only on the Euclidean distance between those sites and not on their relative orientation. The assumption of isotropy is often made due to the simpler interpretations this allows and to the more efficient estimation of the dependence structure possible under isotropy. Thus, it is advantageous to make this assumption if it holds.

When the correlation structure is not isotropic, it is said to be anisotropic. In many applications isotropy does not seem to be a reasonable assumption. For example, in an agricultural study, wind plays an important role in transporting pollens and thus in determining crop yields. Correlation of crop yields might be stronger in the major wind direction than perpendicular to that direction. In an epidemiological study contaminants may be transported by a river and this likely leads to anisotropic covariances.

---

*Michael Sherman's research supported by National Cancer Institute grant R29CA72015-01 and by the Texas A&M Center for Rural and Environmental Health via National Institute of Environmental Health Sciences ES09106. The authors thank Hsiao-Chuan Lu and Dale L. Zimmerman for their kindness in sharing a preprint of their paper.

To see concretely how an assumption of isotropy impacts an analysis consider the following covariance model $C(x,y) = exp(-x^2-4y^2)$ where $C(x,y)$ denotes the covariance between two observations separated by spatial lag $(x,y)$. Data coming from this spatial process yields an anisotropic correlation whose spatial dependence in the $x$ direction is much stronger than that in the $y$ direction. Suppose four points have been observed on a stationary random field following this covariance structure at locations $(0,.65)$, $(0,-.65)$, $(.65,0)$ and $(-.65,0)$ and assume that the goal is to predict the value at the point $(0,0)$ which is equally distant from the observed four points.

Under isotropy, the optimal weights are equal to 0.25 for each of the four observations; under the true model, the optimal weights for the two observations on the $x$ axis are 0.4997 and on the $y$ axis 0.0003, reflecting the fact that correlation is much stronger in the $x$ direction than in the $y$ direction. A researcher who assumed isotropy a priori would not detect this dependence of correlation on direction and thus miss the underlying spatial structure. A comparison of the kriging variances shows that the prediction error under an assumed isotropic model is almost twice as large as that under the true model. This shows a second aspect of misspecifying an anisotropic covariance model as isotropic, a dramatic increase in prediction error. In addition, in a spatial regression model with errors given by $C(x,y)$ the appropriateness of inferences drawn to regression parameters depends on correctly specifying the correlation structure.

In this article we present a new method of testing for spatial isotropy that removes restrictions of earlier methods. Our methodology applies quite generally: no model is necessary for either marginal or joint distributions, the shape of the random field can be quite irregular, and correlation is allowed for observations separated by arbitrarily long distances.

After giving our setup in Section 2, we discuss previous methods of testing for isotropy in Section 3. Semivariogram estimation is discussed in Section 4, while Section 5 gives our main asymptotic distribution result for the semivariogram estimator. Section 6 gives our test statistic based on the result in Section 5, while Section 7 discusses some extensions of the methodology.

## 2. Notation, Setup and Background

Let $\{Z(s) : s \in D\}$ be a strictly stationary random field defined over a domain of interest $D$ in $\Re^2$. Define the variogram function as

$$2\gamma(t) := \text{Var}\{Z(s) - Z(s+t)\},$$

where $t$ is an arbitrary spatial lag in $\Re^2$.

The main features of the semivariogram, $\gamma(t)$, are the sill, range, and nugget. The sill in direction $t$ is defined as $\lim_{b\to\infty} \gamma(bt)$, given that the limit exists. If the sill is attained by $\gamma(t)$ at a finite distance, $d_t$, then the semivariogram is said to have range $d_t$ in the direction $t$. If the sill is attained only asymptotically, then the distance at which some (large) fixed percentage of the sill is reached is often called the "effective range". The nugget is defined to be $\lim_{b\to 0} \gamma(bt)$.

A random field that is second order stationary with covariance function $C(t)$ satisfies $\gamma(t) = C(0) - C(t)$. However, due to the fact that the semivariogram is more generally defined and easier to estimate under a variety of settings, we consider the semivariogram in what follows (see, e.g., Cressie, 1993). If there is a nonconstant mean function, we assume that this function has been estimated and that the residuals can be assumed to be approximately stationary.

Our goal is to test the hypothesis:

$$H_0 : \gamma(t) = \gamma_0(||t||) \text{ for all } t \in H,$$

where $H := \{t = u - v : u, v \in D\}$ is the set of all possible lags and $\gamma_0(\cdot)$ is a function of one variable.

## 3. Prior Methods

A popular method of assessing isotropy is to assess plots of direction-specific estimated sample semivariograms. For example, Isaaks and Srivastava (1989) discuss the use of a rose diagram in detecting anisotropy. In another example Diggle (1981) draws contour plots of the empirical correlation function for a binary mosaic and concludes isotropy. These graphical diagnostics are often useful in practice. However, they are subjective and open to interpretation.

A more formal testing procedure for investigating isotropy was proposed by Baczkowski & Mardia (1990). They assume that observations are generated by a "doubly geometric" model and demonstrate the efficacy of their approach under this model. However, their method is not appropriate for other covariance models.

Lu & Zimmerman (2001) have proposed a more general procedure that works well for a large class of covariance models. They noted that the test for isotropy can be written as a general linear hypothesis. Their approach is more general than that of Baczkowski and Mardia but their method applies only to equally-spaced Gaussian processes.

In this article, we consider an extension of Lu & Zimmerman's testing approach to a much wider range of applications. For example, no explicit knowledge of the marginal distribution of the underlying process is needed in order to use our approach, the shape of the random field can be quite irregular, and correlation is permitted at arbitrarily large spatial lags. We also discuss how to extend our method to irregularly-spaced data locations.

## 4. Semivariogram Estimation

### 4.1. Equally Spaced Observations

Assume that we have observed the variable of interest at $n$ locations in the domain $D$. Denote the observed data by $[\{s_1, Z(s_1)\}, ..., \{s_n, Z(s_n)\}]$.

Due to the assumption of a constant mean $2\gamma(t) = E\{(Z(s) - Z(s+t))^2\}$. Thus, for equally spaced observations at locations in a subset of $Z^2$, a natural moment estimator of the semivariogram at spatial lag $t$ is given by:

$$\hat{\gamma}(t) := \frac{1}{2|D(t)|} \times \sum \{Z(s_i) - Z(s_j)\}^2,$$

where the sum is over $D(t) := \{(i,j) : s_i, s_j \in D, s_i - s_j = t\}$ and $|D(t)|$ is the cardinality of $D(t)$. The collection of estimates $\{\hat{\gamma}(t_i) : i = 1, ..., k\}$ is the sample semivariogram. It is seen that each component of the sample semivariogram is an unbiased estimator of the corresponding component in the true semivariogram under the assumption of a stationary mean function.

### 4.2. Irregularly-spaced Observations

We consider the points at which $Z(\cdot)$ is observed to be *random* in number and location; specifically they are generated from a homogeneous two-dimensional point process with intensity parameter $\nu$. Denote the random point process by $N$. For a given Borel set $B$, $N(B)$ is the random number points of $N$ contained in $B$. We further assume $N$ to be independent of $Z(\cdot)$. For unequally spaced data we consider kernel smoothing the observed squared differences to obtain an estimator of the semivariogram.

We adopt the same notations as in the equally spaced situation but now $|D|$ denotes the volume of $D$, $\partial D$ its boundary, and $|\partial D|$ the length of $\partial D$. Let $h$ be a positive constant and $w(\cdot)$ be a nonnegative, bounded, isotropic density function which has positive values only on a finite support, say $C$ (e.g., a circular field centered at the origin with finite radius $r$). Let $dx$ denote an infinitesimally small disc centered at $x$. Define $N^{(2)}(dx_1, dx_2) \equiv N(dx_1)N(dx_2)1(x_1 \neq x_2)$, where $1(x_1 \neq x_2) = 1$ if $x_1 \neq x_2$ and 0 otherwise. Then a kernel based semivariogram estimator is given by

$$\hat{\gamma}(t) = \frac{1}{2\nu^2|D|} \int_{x_1 \in D} \int_{x_2 \in D} h^{-2} w\left(\frac{t - x_1 + x_2}{h}\right) \times \left[Z(x_1) - Z(x_2)\right]^2 N^{(2)}(dx_1, dx_2) \quad (\star)$$

In practice, $\nu$ is usually replaced by $N(D)/|D|$, which is a consistent estimator for $\nu$ (Stoyan and Stoyan, 1994).

## 5. Asymptotic Normality of the Semivariogram Estimator

Let $\{Z(s) : s \in D\}$ be a strictly stationary random field defined over a domain of interest $D$ in $\Re^2$. In what follows, we assume that locations where we have observations are perfectly aligned on a grid with unit spacing as discussed in Section 4.1. Observe that if two lags have the same length, i.e., they have the same Euclidean distance from the origin, their corresponding semivariograms will be the same under isotropy. Thus a test for isotropy may be obtained by comparing semivariograms at lags with the same length but in different directions. Because the semivariograms are typically unknown, we form a test based on estimators of the semivariograms. We use the classical estimator of the semivariogram given in Section 4.1.

Let $\Lambda$ be a set of spatial lags for which we want to compare the sample semivariograms. Define $\mathbf{G} := \{\gamma(t) : t \in \Lambda\}$ to be the vector of semivariograms at lags contained in $\Lambda$. Consider a sequence of random fields $D_n$ and corresponding observations $\{Z(s) : s \in D_n\}$ and let $\hat{\gamma}_n(t)$ and $\hat{\mathbf{G}}_n := \{\hat{\gamma}_n(t) : t \in \Lambda\}$ denote the estimators of $\gamma(t)$ and $\mathbf{G}$ obtained over data in $D_n$, respectively. In order to formally state asymptotic results we need to specify the form of the sets $D_n$ and quantify the dependence in the underlying random field.

## 5.1. Domains of Observation

Define the boundary of a set $D$ to be the set $\partial D := \{s \in D : \exists s' \notin D \text{ with } d(s,s') = 1\}$ where we assume the maximal distance $d[(s_x, s_y), (s'_x, s'_y)] := max(|s_x - s'_x|, |s_y - s'_y|)$. Letting $|\partial D|$ denote the cardinality of the boundary, we assume

$$|D_n| = O(n^2) \text{ and } |\partial D_n| = O(n). \tag{1}$$

Equation (1) is satisfied by a large class of shapes. For example, let $A \subset (0,1] \times (0,1]$ denote the interior of a simple closed curve in the unit square of finite length. Now multiple the set $A$ by $n$, to obtain the set $A_n \subset (0,n] \times (0,n]$, i.e., the shape $A$ inflated by a factor of $n$. If we define $D_n := \{s : s \in A_n \cap Z^2\}$, we see that (1) is satisfied. This formulation allows for a wide variety of shapes on which the data can be observed. This includes rectangles, circles, and any starshape.

## 5.2. Quantifying Dependence

In the nonparametric spirit, we quantify the strength of dependence in the random field by a model free mixing condition. Following Rosenblatt (1956), we use a particular type of strong mixing coefficients defined by

$$\alpha_p(k) \equiv \sup\{|P(A_1 \cap A_2) - P(A_1)P(A_2)| : A_i \in F(E_i), |E_i| \leq p, i = 1, 2, d(E_1, E_2) \geq k\}$$

where $|E|$ is the cardinality of the index set $E$, $F(E)$ is the $\sigma$-algebra generated by the random variables $\{Z(s) : s \in E\}$ and $d(E_1, E_2)$ is the minimal city block distance between $E_1$ and $E_2$.

If the observations are independent, then $\alpha_p(k) = 0$ for all $k \geq 1$. Here we will need $\alpha_p(k)$ to approach 0 for large $k$, at some rate depending on the cardinality $p$. Following Sherman & Carlstein (1994), we assume the following mixing condition

$$\sup_p \frac{\alpha_p(k)}{p} = O(k^{-\epsilon}) \text{ for some } \epsilon > 2. \tag{2}$$

Condition (2) says that at a fixed distance $k$, as the cardinality increases, we allow dependence to increase at a rate controlled by $p$. As the distance increases, the dependence must decrease at a polynomial rate in $k$. This condition is satisfied by any $m$-dependent random field (i.e., observations separated by distance $m$ or more are independent). From Kunsch's (1982) proposition 3.1 and Remark 3.5 iv, it can be deduced that it holds for a natural class of Gibbs (i.e., Markov) random fields which are useful in statistical mechanics and image processing. Bradley (1993) has also justified the use of mixing condition (2).

## 5.3. Moment Conditions

We require the following mild moment condition

$$\sup_n E\left\{\left|\sqrt{|D_n|} \times [\hat{\gamma}_n(t) - \gamma(t)]\right|^{2+\delta}\right\} \leq C_\delta \text{ for some } \delta > 0, C_\delta < \infty \tag{3}$$

To appreciate the mildness of this condition note that (3) with $\delta = 0$ is necessary for the existence of the (standardized) asymptotic variance of $\hat{\gamma}_n(t)$. If the random process is $m$-dependent, then the existence of its marginal $(4+2\delta)$th moment will be sufficient for condition (3) to hold.

## 5.4. Theorem 1

Let $\{Z(s) : s \in D_n\}$ be a strictly stationary random field which is observed at lattice point in $D_n$ satisfying (1). Assume

$$\sum_{s \in Z^2} |Cov\{[Z(0) - Z(s_1)]^2, [Z(s) - Z(s+s_2)]^2\}| < \infty \text{ for all } s_1, s_2. \qquad (4)$$

Then $\Sigma \equiv \lim_{n \to \infty} |D_n| \times \text{Var}(\hat{\mathbf{G}}_n)$ exists, the $(i, j)$th element of which is:

$$\sum_{s \in Z^2} Cov\{[Z(0) - Z(t_i)]^2, [Z(s) - Z(s+t_j)]^2\}.$$

Further, if $\Sigma$ is positive definite and (2) and (3) hold, then the limiting distribution of $\sqrt{|D_n|}(\hat{\mathbf{G}}_n - \mathbf{G})$ is multivariate normal with mean $\mathbf{0}$ and covariance matrix $\Sigma$.

For a proof of this result see Guan, Sherman and Calvin (2002).

## 6. Construction of the test statistic

### 6.1. Choice of the lag set

To assess the hypothesis of isotropy, the lag set, $\Lambda$ needs to be specified. The choice of $\Lambda$ depends on a number of factors including the configuration of the spatial locations, the goal of the analysis, and the underlying physical phenomenon of interest. Thus, there can be no unique rule for choosing $\Lambda$. In general, empirical studies show that small lags are preferable to large ones. This is due to two facts. The sample semivariograms at large lags are based on fewer observations than estimates at small lags and thus are more variable. Secondly, the observations at small lags are typically more correlated and identification of the correlation among these lags is more important for spatial prediction.

For regularly spaced observations, note that the two components of a lag $t$ should both be integers so that $\hat{\gamma}_n(t)$ can be calculated. For irregularly spaced observations, sample semivariograms can in principle be calculated at any lag.

### 6.2. Asymptotic Distribution

We can only detect departures from isotropy where we estimate the semivariogram and thus we actually test the hypothesis

$$H_0 : \gamma(t_1) = \gamma(t_2) : t_1, t_2 \in \Lambda, ||t_1|| = ||t_2||.$$

Under this hypothesis there exists a full row rank matrix $\mathbf{A}$ such that $\mathbf{AG} = \mathbf{0}$ (Lu and Zimmerman, 2001). For example, if $\Lambda = \{(1,0), (0,1), (1,1), (-1,1)\}$, then

$$\mathbf{A} = \begin{pmatrix} 1 & -1 & 0 & 0 \\ 0 & 0 & 1 & -1 \end{pmatrix}$$

Let $d$ denote the row rank of $\mathbf{A}$. From Theorem 1 we have under the null hypothesis that

$$|D_n| \times (\mathbf{A}\hat{\mathbf{G}}_\mathbf{n})'(\mathbf{A}\Sigma\mathbf{A}')^{-1}(\mathbf{A}\hat{\mathbf{G}}_\mathbf{n}) \xrightarrow{D} \chi_d^2 \text{ as } n \to \infty.$$

We see that in order to use this result, we need an estimator of $\Sigma$.

## 6.3. Covariance Matrix Estimation

In order to obtain a test statistic we require an estimate of the asymptotic covariance matrix of the sample semivariograms. We apply the subsampling technique. Subsampling has been widely employed, for example, to estimate for the variance of a general spatial statistic (e.g., Hall, 1985, Possolo 1991, Sherman and Carlstein, 1994), to obtain general confidence intervals (e.g., Politis and Romano (1994), and to estimate the variance of estimators derived from estimating functions (e.g., Heagerty and Lumley, 2000). Here we apply subsampling to estimate the covariance matrix of $\hat{\mathbf{G}}_n$.

Towards this end, we divide the original field $D_n$ into overlapping subblocks. These subblocks are obtained by moving an $l(n) \times l(n)$ square window across $D_n$, where $l(n) = cn^\alpha$ for some $c > 0$ and $\alpha \in (0, 1)$. If the window is fully contained in $D_n$, then a "replicate" is obtained. The replicates are chosen to be congruent to $D_n$ both in configuration and orientation so as to retain the same dependence structure as the original data. In what follows, we denote the total number of subblocks by $k_n$, the $i$th subblock by $D^i_{l(n)}$. Let $S(D_n)$ denote a vector statistic, with replicates $S(D^i_{l(n)})$. Then, an estimator of $\Sigma$ is given by:

$$\hat{\Sigma}_n \equiv \frac{1}{k_n} \times \sum_{i=1}^{k_n} \left\{ |D^i_{l(n)}|(S(D^i_{l(n)}) - \overline{S_n})(S(D^i_{l(n)}) - \overline{S_n})' \right\} \text{ with } \overline{S_n} \equiv \sum_{i=1}^{k_n} S(D^i_{l(n)})/k_n$$

We apply this estimator with $S(D_n) = \hat{\mathbf{G}}_n$.

The following theorem gives conditions under which, $\hat{\Sigma}_n$ is an $L_2$ consistent estimator of $\Sigma$, in the sense that every element of $\hat{\Sigma}_n$ is an $L_2$ consistent estimator for the counterpart in $\Sigma$.

**Theorem 2**:

Assume that condition (3) holds for some $\delta > 2$ and all the rest conditions in *Theorem 1* hold. Then

$$\hat{\Sigma}_n \xrightarrow{L_2} \Sigma,$$

in the sense that every element of $\hat{\Sigma}_n$ is an $L_2$ consistent estimator of its counterpart in $\Sigma$.

For a proof of this result see Guan, Sherman, and Calvin (2002).

An important practical issue with the subsampling method is determination of an appropriate subshape size, i.e., choice of $l(n)$. A common definition of optimal is the value of $l(n)$ that minimizes the mean squared error. In general, the optimal subblock size depends on the dependence structure of the underlying process and the shape of the domain of observation (Nordman and Lahiri, 2002). Politis and Romano (1994) gave $cn^{d/(d+2)}$ for some $c > 0$ as the optimal rate for estimating the variance of a statistic on a $d$-dimensional stationary mixing random field and Sherman (1996) showed that the rate of $cn^{1/2}$ is correct for spatial lattice data. Garcia-Soidan and Hall (1997) also discussed the optimal rates for estimating the one- and two-sided distribution functions of a studentized statistic in the spatial context.

## 6.4. The Test Statistic

Given the results in the previous section, we have that

$$TS := |D_n| \times (\mathbf{A\hat{G}_n})'(\mathbf{A\hat{\Sigma}}_n\mathbf{A}')^{-1}(\mathbf{A\hat{G}_n}) \xrightarrow{D} \chi_d^2 \text{ as } n \to \infty.$$

Thus for a large enough field, an approximate size-$\alpha$ test for isotropy is to reject the hypothesis of isotropy if $TS$ is bigger than $\chi_{d,\alpha}^2$, where $\chi_{d,\alpha}^2$ is the upper $\alpha$ percentage point of a $\chi^2$ distribution with $d$ degrees of freedom.

## 7. Extensions

### 7.1. Functions of the Semivariogram

Often a smooth function of the semivariogram will yield a quicker approach to normality and thus make the limiting $\chi^2$ distribution more accurate. Specifically

$$\sqrt{|D_n|} \times (h(\mathbf{\hat{G}}_n) - h(\mathbf{G}))$$

should be asymptotically normal under smoothness conditions on $h(\cdot)$ and then $TS := |D_n| \times (\mathbf{A}h(\mathbf{\hat{G}}_n))'(\mathbf{A\hat{\Sigma}}_h\mathbf{A}')^{-1}(\mathbf{A}h(\mathbf{\hat{G}}_n)) \xrightarrow{D} \chi_d^2$ as $n \to \infty$, where $\mathbf{\hat{\Sigma}}_h$ denotes an estimator of the asymptotic variance of $\sqrt{|D_n|} \times \hat{h}(\mathbf{G_n})$.

For example, Baczkowski and Mardia (1987) show that when $h(\cdot) = log(\cdot)$ the distribution of the transformed sample semivariogram approaches normality faster than sample semivariogram.

### 7.2. More General Locations

It will be worthwhile to extend Theorem 1 (and thus the distribution theory given in Section 6.4) to formally cover irregularly spaced locations generated by general stationary processes. For example, a Poisson process is the natural starting point. Preliminary results show that asymptotic normality of the semivariogram estimator holds in this situation.

Also, useful will be locations generated by Poisson cluster processes or Inhibition processes. These are interesting situations as, unlike in the Poisson case, we no longer have $E(N^2(ds_1, ds_2)) = E(N(ds_1))E(N(ds_2))$. Now the former expectation depends on the second order intensity function of the point process, and this function needs to be estimated to construct an estimator of the semivariogram.

## REFERENCES

1. Baczkowski, A. J. and Mardia, K. V. (1987), Approximate Lognormality of the sample semivariogram under a Gaussian process, *Communications in Statistics-Simulation and Computation*, 19, 571-585.
2. Baczkowski, A. J. and Mardia, K. V. (1990), A test of spatial symmetry with general application, *Communications in Statistics-Theory and Methods*, 19, 555–572.
3. Bernstein, S. (1927), Sur l'extension du théorèm limite du calcul des probabilités aux sommes de quantités dépendantes, *Mathematische Annalen*, 97, 1–59.

4. Bradley, R. C. (1993), Some examples of mixing random fields, *Rocky Mountain Journal of Mathematics*, 23, 495–519.
5. Cressie, N. (1993), *Statistics for Spatial Data*, Wiley, New York.
6. Diggle, P. J. (1981), Binary mosaics and the spatial pattern of heather, *Biometrics*, 37, 531–539.
7. Garcia-Soidan, P. H. and Hall, P. (1997), On sample reuse methods for spatial data, *Biometrics*, 53, 273–281.
8. Guan, Y., Sherman, M. and Calvin, J.A. (2002), A nonparametric test for spatial isotropy usning subsampling, *preprint*.
9. Hall, P. (1985), Resampling a coverage pattern, *Stochastic Processes and their Applications*, 20, 231–246.
10. Ibragimov I.A. and Linnik, Y.V. (1971), *Independent and stationary sequences of random variables*, Wolters-Noordhoff Publishing, Grongigen.
11. Isaaks, E. H. and Srivastava, R. M. (1989), *An introduction to Applied Geostatistics*, Oxford University Press, Oxford.
12. Karr, A. F. (1986), Inference for stationary random fields given Poisson samples, *Advanced Applied Probability*, 18, 406–422.
13. Kunsch, H. (1982), Decay of correlations under Dobrushin's uniqueness condition and its applications, *Communications in Mathematical Physics*, 84, 207–222
14. Lu, H. and Zimmerman D. L. (2001), Testing for isotropy and other directional symmetry properties of spatial correlation, *preprint*.
15. Nordman, D. and Lahiri, S.N. (2002), On optimal spatial subsample size for variance estimation, *preprint*.
16. Politis, D. N. and Romano, J. P. (1994), Large sample confidence regions based on subsamples under minimal assumptions, *Annals of Statistics*, 22, 2031–2050
17. Possolo, A. (1991), Subsampling a random field, in *Spatial Statistics and Imaging*, I.M.S. Lecture Notes-Monograph Series, Vol. 20 (A.Possolo, editor) 286–294.
18. Rosenblatt, M. (1956), A central limit theorem and a strong mixing condition, *Biometrics*, 37, 531–539.
19. Sherman, M. and Carlstein, E. (1994), Nonparametric estimation of the moment of a general statistic computed from spatial data, *Journal of the American Statistical Association*, 89, 496–500.
20. Sherman, M. (1996), Variance estimation for statistic computed from spatial lattice data, *Journal of the Royal Statistical Society*, ser. B, 58, 509–523.
21. Stoyan, D. and Stoyan, H. (1994), *Fractals, Random Shapes and Point Fields*, Wiley, N.Y.

# 12. Wavelet and Multiresolution Methods

# Automatic landmark registration of 1D curves

Jérémie Bigot[a]

[a]Laboratoire IMAG-LMC, Université Joseph Fourier, BP 53, 38041 Grenoble Cedex 9, France

This paper is concerned with the problem of landmark-based matching. A nonparametric approach is proposed to estimate the landmarks of a signal via the zero-crossings lines of its continuous wavelet transform. This approach yields to a new technique to automatically decide which pair of landmarks should be associated when comparing two functions that have to be aligned. The usefulness of this method is illustrated on a real example, and a small simulation study to compare its performances with a dense matching approach is proposed.

## 1. Introduction

When studying some biological or physical process in different subjects, we usually see that the observed curves or images have a common structural pattern. An important matter consists in determining the typical shape of the observed process or in testing whether there is any statistically significant difference among two subsets of subjects. However, to compare similar objects, it is necessary to find a common referential to represent them. A possible approach consists in finding, for each observed curve, a warping function in order to synchronize all the curves before performing the average or applying any other statistical inferential procedure. This work is then closely related to the analysis of deformations and warping for 1D structures. Matching two functions can be done by aligning individual locations of corresponding structural points (or landmarks) from one curve to another. Landmark-based registration of smooth curves has been studied from a statistical point of view by Kneip and Gasser [12] using kernel estimators to retrieve the locations of the structural points of a smooth function. A survey of recent developments in the analysis of deformations and warping can be found in a tutorial by Younes [24], while extensive references on curves alignment for functional data analysis can be found in Ramsay and Silverman [20]. In this paper, we propose to use the scale space approach developed in Bigot [2] to detect the structural points of a smooth function observed with noise. In [2], a nonparametric approach is proposed to estimate the zero-crossings lines of the continuous wavelet transform a signal corrupted by noise, and a new tool, the "structural intensity" is introduced to identify the limits of these lines which correspond to the landmarks of the unknown signal. The main contribution of this paper is then to show that the the structural intensity can be used to define an automatic landmark-based registration method.

The paper is organized as follows: in section 2, we present the problem of matching two curves and review some methods for computing a warping function. Then, in section 3, we define the problem of estimating a warping function to align two curves that are observed with noise. In section 4, we briefly recall the methodology proposed in [2] to identify the landmarks of an unknown signal. A method to automatically decide which pair of landmarks should be matched, when comparing the structural points of two functions, is presented in section 5. The usefulness of this approach is illustrated on a real example, and a small simulation study to compare the performances of our method with those of the dynamic time warping algorithm suggested in Wang and Gasser ([22],[23]) is proposed.

## 2. A matching problem in one dimension

Suppose we are given two functions $f_1, f_2 : [0,1] \to \mathbb{R}$, and that we want to synchronise them. The issue is then to find for each point $x \in [0,1]$ a point $x' \in [0,1]$ such that $f_1(x) \approx f_2(x')$. If every point of one curve is uniquely matched to some point of the other curve, the problem is then to find a one-to-one function $u : [0,1] \to [0,1]$ such that:

- $u$ is not "too far" from the identity function.

- $f_1 \approx f_2 \circ u$.

Two approaches can be investigated to solve this problem. One consists in aligning individual locations of corresponding structural characteristics from one curve to another, while the second approach is based on the minimization of an appropriate functional. To have a sufficiently smooth warping function, we will require that $u$ is an increasing diffeomorphism of $[0,1]$.

### 2.1. Landmark-based matching

A first approach to determine a warping function $u$ is to find the landmarks of each curve $f_j, j = 1, 2$. Landmarks are ordered series of locations (in $[0,1]$) that are associated to typical features of a curve. Typical examples of landmarks are the locations of extrema or inflection points of a function. The first issue encountered by landmark-based matching methods is the correspondence problem between two sets of features extracted from two functions to be aligned. For the 1D case, it is generally assumed that one is given two sets of labelled landmarks which consist of pair of structural points to be matched. The identification of the landmarks that should correspond may be done manually (generally by an expert or from *a priori* informations, see e.g. Munoz Maldonado *et. al.* [16]) or by determining the typical features that occurs consistently in a set of curves (see Gasser and Kneip [9]). When two unknown signals are observed with noise, the problem of labelling two sets of estimated landmarks is further complicated by the presence of outliers and the fact that there could be many features in either set that have no counterparts in the other. In section 5, we therefore proposed a method to automatically solve this correspondence problem. However, before presenting this approach, we will first explain how to compute a warping function when one is given two sets of labelled landmarks $(\tau_{1,1}, \ldots, \tau_{1,N})$ and $(\tau_{2,1}, \ldots, \tau_{2,N})$ extracted from two curves $f_1$ and $f_2$. The purpose of landmark-based registration is then to find an increasing diffeomorphism $u$ such that for all $i = 1, \ldots, N$: $\tau_{2,i} = u(\tau_{1,i})$ (*exact matching*) or $\tau_{2,i} \approx u(\tau_{1,i})$ (*inexact matching*). Generally, it will be also required that $u(0) = 0$ and $u(1) = 1$. The problem of determining an increasing warping function can be formulated as a constrained smoothing problem (see e.g [14]) or be solved by the method of Ramsay [18] (details for the practical implementation of this approach can be found in Ramsay and Li [19], while Matlab programs for creating such warping functions are available from the ftp site: *http://www.psych.mcgill.ca/faculty/ramsay/ramsay.html*).

### 2.2. Dense matching

Let $F$ be a function defined on $]0, \infty[ \times \mathbb{R} \times \mathbb{R}$. Another approach to find a warping function to align two curves $f_1$ and $f_2$, is to search the minimum, among all increasing diffeomorphisms $u$ of $[0, 1]$, of the functional:

$$L_{f_1, f_2}(u) = \int_0^1 F(u'(x), f_1(x), f_2 \circ u(x)) dx.$$

The function $F$ and therefore the functional $L_{f_1, f_2}$ have to satisfy some specific properties to guarantee that a solution of this minimization problem corresponds to a solution of our matching problem. For further references on matching 1D structures by variational methods see the paper by Trouve and Younes [21]. The problem of minimizing the functional $L_{f_1, f_2}(u)$ by dynamic time warping has been studied from a statistical point of view by Wang and Gasser ([22],[23]) using kernel estimators. In section 5.2, we briefly explained the method proposed in [22], and a small simulation study is carried out to compare our automatic landmark-based registration technique with the algorithm designed in [23].

## 3. A nonparametric regression setting

Suppose we are given two unknown signals $f_1$ and $f_2$ (with $f_j : [0, 1] \to \mathbb{R}$, $j = 1, 2$) observed with noise at the same discrete time positions $t_i = i/n$:

$$y_{j,i} = f_j(t_i) + \sigma_j \epsilon_{j,i}, \ j = 1, 2, \ i = 1, \ldots, n,$$

where $\epsilon_{j,i}$ ($j = 1, 2$) are independent identically distributed (i.i.d.) normal variables with zero mean and variance 1, and $\sigma_j$ are unknown noise level parameters. The problem that we consider in this paper is the alignment of the functions $f_1$ and $f_2$, i.e. estimating a warping function to match the curves $f_1$ and $f_2$ in the sense of section 2.1. Suppose that it is possible to find a warping function $u$ via landmark-based matching when the curves $f_1$ and $f_2$ are noise free. A natural idea to estimate the function $u$ is to first estimate the landmarks of the curves $f_1$ and $f_2$ when they are observed with noise, and then to use the resulting estimators to perform one of the landmark-based matching procedures proposed in section 2. This issue has already been studied from a statistical point of view by Kneip and Gasser [12] using kernel estimators to retrieve the locations of the structural points of a function. Let $H_N$ denote the set of all $\tau = (\tau_1, \ldots, \tau_N)' \in [0, 1]^N$ with $\tau_i < \tau_j$ for all $i < j$, and $C^1([0, 1])$ be the set of all continuously differentiable functions on $[0, 1]$. Following the notations in [12], we will say that an operator $G : H_N^2 \to C^1([0, 1])$ is a *warping operator* if the following conditions are satisfied:

- (i) For all $(\tau_1, \tau_2) \in H_N^2$, $G_{(\tau_1, \tau_2)}(.)$ is a strictly monotonically increasing real function. Furthermore, for all $t \in [0, 1]$, $G_{(.,.)}(t)$ is continuous.

- (ii) For all $\tau_1 = (\tau_{1,1}, \ldots, \tau_{1,N})' \in H_N$ and all $\tau_2 = (\tau_{2,1}, \ldots, \tau_{2,N})' \in H_N$, either:

$$G_{(\tau_1,\tau_2)}(\tau_{1,i}) = \tau_{2,i}, \text{ for all } i = 1, \ldots, N,$$

and in this case $G$ is called an *exact warping operator*, or:

$$G_{(\tau_1,\tau_2)}(\tau_{1,i}) \approx \tau_{2,i}, \text{ for all } i = 1, \ldots, N,$$

while in this case $G$ is called an *inexact warping operator*.

Let $\tau_1 = (\tau_{1,1}, \ldots, \tau_{1,N})$ and $\tau_2 = (\tau_{2,1}, \ldots, \tau_{2,N})$ be two sets of $N$ landmarks for the two curves $f_1$ and $f_2$. For some warping operator $G$, $u(.) = G_{(\tau_1,\tau_2)}(.)$ is called a warping function to synchronize the curves $f_1$ and $f_2$. In practice, landmarks and warping functions have to be estimated. Assume that one has a method to estimate the landmarks of two unknown functions (see section 4) and to decide which pair of features should correspond (see section 5), then, based on a warping operator $G$, an estimate $\hat{u}$ of the warping function $u$ is obtained by $\hat{u}(.) = G_{(\hat{\tau}_1,\hat{\tau}_2)}(.)$, where $\hat{\tau}_1 = (\hat{\tau}_{1,1}, \ldots, \hat{\tau}_{1,\hat{N}})$ and $(\hat{\tau}_2 = \hat{\tau}_{2,1}, \ldots, \hat{\tau}_{2,\hat{N}})$ denote two sets of estimated and labelled landmarks of the curves $f_1$ and $f_2$ respectively.

## 4. Estimating the landmarks of a function

### 4.1. Wavelets and structure of a signal

In many practical experiments, typical examples of landmarks are extrema and inflections points of a smooth function (see e.g. [20], [12], [16]). More generally, the landmarks considered in this paper are the points $y_0$ where a function $f$ is $r$-times ($r \geq 1$) continuously differentiable in a neighborhood of $y_0$ and such that the $r$th derivative of $f$ has a zero at $y_0$ To characterize such landmarks, we propose to use the framework considered in [2] which consists in following the propagation at fine scales of the zero-crossings and the wavelet maxima lines of the continuous wavelet transform of a signal. In [2], it is shown how to detect the singularities of a function via its wavelet maxima lines, but in this paper, we restrict our study to the estimation of the zeros of the $r$th derivative of a smooth signal.

#### 4.1.1. The continuous wavelet transform

We assume that we are working with an admissible real-valued wavelet $\psi$ with $r$ vanishing moments ($r \in \mathbb{N}$). We will suppose that the wavelet $\psi$ has a fast decay and has no more than $r$ vanishing moments which implies (see Theorem 6.2 of Mallat [15]) that there exists $\theta$ with a fast decay such that: $\psi(u) = (-1)^r \frac{d^r \theta(u)}{dt^r}$, and $\int_{-\infty}^{+\infty} \theta(u) du \neq 0$. Moreover, we will assume that the wavelet $\psi$ is normalized to one i.e.: $\int_{-\infty}^{+\infty} (\psi(u))^2 du = 1$. Then, by definition, the continuous wavelet transform of a function $f \in L^2(\mathbb{R})$ at a given scale $s$ is: $W_s(f)(x) = \int_{-\infty}^{+\infty} f(u) \psi_s(u-x) du$, where $\psi_s(u) = \frac{1}{\sqrt{s}} \psi(\frac{u}{s})$.

#### 4.1.2. Zero-crossings of the wavelet transform

Suppose that a function $f \in L^2(\mathbb{R})$ is $r$-times continuously differentiable in an interval $[a, b]$. By commuting the convolution and differentiation operators, and given that $\theta$ has a fast decay, we have that for all $x \in ]a, b[$,

$$\lim_{s \to 0} \frac{W_s(f)(x)}{s^{r+1/2}} = \lim_{s \to 0} f^{(r)} \star \frac{1}{s} \theta\left(\frac{-x}{s}\right) = K f^{(r)}(x), \text{ where } K = \int_{-\infty}^{+\infty} \theta(u) du \neq 0 \quad (1)$$

The term *zero-crossings* will be used to describe any point $(z_0, s_0)$ in the time-scale space such that $z \mapsto |W_{s_0}(f)(z)|$ has exactly one zero at $z = z_0$ in a neighborhood of $z_0$. Equation (1) proves that the zero-crossings of the wavelet transform converge to the zeros of the $r$th derivative of $f$ that are located in $[a, b]$ when the scale $s \to 0$. Hence, one can find the location of the extrema (resp. the points of inflexion) of a function by following the propagation of the zero-crossings of its wavelet transform when the scale $s$ decreases and when the mother wavelet has $r = 1$ (resp. $r = 2$) vanishing moments.

### 4.1.3. Some properties of the zero-crossings lines

We will call *zero-crossings line* any connected curve $z(s)$ in the time-scale plane $(x, s)$ along which all points are zero-crossings. From equation (1), one might think that it is sufficient to follow any zero-crossings line to detect the landmarks of a function. However, we are not guaranteed that, for any wavelet $\psi = (-1)^r \theta^{(r)}, r \geq 0$, a zero-crossings located at $(z_1, s_1)$ belongs to a zero-crossings line that propagates towards finer scales. A zero-crossings line might be interrupted when the scale decreases. However, if $\theta$ is a Gaussian (i.e $\theta(x) = K \frac{1}{\sqrt{2\pi}\beta} e^{-\frac{x^2}{2\beta^2}}$, for some non-negative reals $K$ and $\beta$), then for any $f \in L^2(\mathbb{R})$, the zero-crossings of $W_s(f)(x)$ belong to connected curves that are never interrupted when the scale decreases (see e.g. Proposition 6.1 of Mallat [15]). In Figure 1, we plotted the zero-crossings of the Time Shifted Sine $f_T = 0.3 \sin 3\pi[g(g(g(g(x)))) + x] + 0.5$, $x \in [0, 1]$, where $g(x) = (1 - \cos(\pi x))/2$, for a Gaussian wavelet with $r = 1$ and $\beta = 0.5$. It can be seen in Figure 1b that all the zero-crossings lines propagate to fine scales. In [2], the properties of the zero-crossings lines are investigated in detail by using the notion of causality arising from the scale-space literature (see Lindeberg [13]). However, in this paper, we will not investigate the properties of these lines for any wavelet but will restrict our study to wavelets that are derivatives of a Gaussian. Then, in this case, one obtain that the limits of the zero-crossings lines correspond the landmarks of a signal (see Proposition 3.3 in [2]).

### 4.1.4. Structural intensity of the zero-crossings lines

As pointed out in [2], we only have a visual representation of the zero-crossings lines in the time-scale plane that indicates "where" the landmarks are located, and the analytical expression of these lines is generally unknown. To overcome this drawback, a new tool called "structural intensity" has been introduced in [2] (see Proposition 3.4) to identify the limits of the zero-crossings lines computed for a scale-space representation derived from B-Spline wavelets. The following proposition extends the concept of structural intensity to wavelets that are not compactly supported:

**Proposition 1** *Let $f \in L^2(\mathbb{R})$ and $\psi = (-1)^r \theta^{(r)}, r \geq 1$ where $\theta$ is a Gaussian. Suppose that there exists $p$ zero-crossings lines $z_i(s)$ that respectively converge to $y_i \in \mathbb{R}$ as $s$ tends to zero and that are $C^1$ in a neighborhood of $y_i, i = 1, \ldots, p$. For $x \in \mathbb{R}$, define the structural intensity of the zero-crossings $G_z(x)$ as:*

$$G_z(x) = \sum_{i=1}^{p} \int_0^{s_{z_i}} \frac{1}{s} \theta\left(\frac{x - z_i(s)}{s}\right) ds,$$

*where $[0, s_{z_i}]$ is the support of the zero-crossings line $z_i(.)$. Then, $G_z$ is continuously differentiable on $\mathbb{R} \setminus \{y_1, \ldots, y_p\}$ and such that $G_z(x) \to +\infty$ as $x \to y_i, i = 1, \ldots, p$.*

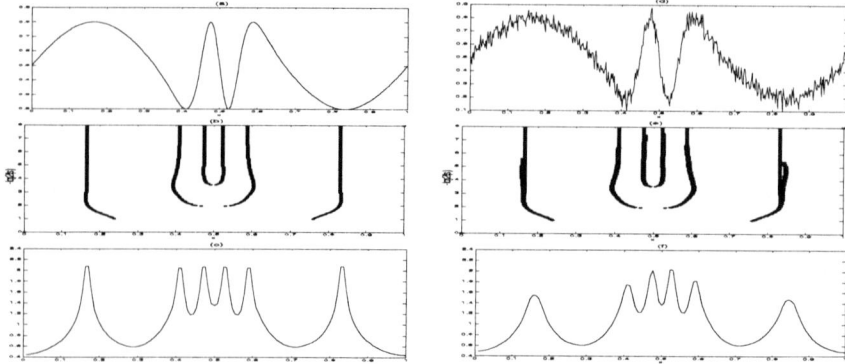

Figure 1. (a) Time function $f_T$. (b) Zero-crossings lines of $f_T$ computed with a Gaussian wavelet ($\beta = 0.5$) with $r = 1$ vanishing moments. (c) Structural intensity of the zero-crossings. (d) Noisy signal $f_T$ with $n = 512$. (e) Estimated zero-crossings (thick lines) and true zero-crossings (thin lines) of $f_T$ for $r = 1$. (f) Structural intensity of the estimated zero-crossing of $f_T$.

**Proof.** Let $f \in L^2(\mathbb{R})$ satisfying the above assumptions. Define:

$$g_{z_i}(x) = \int_0^{s_{z_i}} \frac{1}{s} \theta(\frac{x - z_i(s)}{s}) ds \text{ for } x \in \mathbb{R} \text{ and } 1 \leq i \leq p.$$

Given that $\theta$ has a fast decay, there exists a constant $C$ such that for all $x \in \mathbb{R}$ and all $s \in [0, s_{z_i}]$, $\theta'(\frac{x - z_i(s)}{s}) \leq \frac{C}{1 + (\frac{x - z_i(s)}{s})^2}$. Let $a < y_i$ and $x \in ]-\infty, a[$. Given that $z_i(s) \to y_i$ as $s \to 0$, there exists a constant $M$ and a scale $s_0 > 0$ such that for all $x \in ]-\infty, a[$ and all $s \leq s_0$, $|x - z_i(s)| \geq M$ which implies that:

$$\theta'(\frac{x - z_i(s)}{s}) \leq \frac{Cs^2}{M^2}, \qquad (2)$$

for all $x \in ]-\infty, a[$ and all $s \leq s_0$. Now, note that for all $s \in ]0, s_{m_i}]$, $x \mapsto \frac{1}{s}\theta(\frac{x - z_i(s)}{s})$ is continuously differentiable on $]-\infty, a[$. From equation (2), we have that $s \mapsto \frac{1}{s^2}\theta'(\frac{x - z_i(s)}{s})$ is bounded on $[0, s_{m_i}]$ and the differentiation theorem under the integral sign finally implies that $g_{z_i}$ is continuously differentiable on $]-\infty, a[$. We can similarly show that $g_{z_i}$ is continuously differentiable on $]a, +\infty[$ for $a > y_i$.

Now, since $z_i(.)$ is $C^1$ in a neighborhood of $y_i$, we can define $A = \max_{s \in [0, s_{m_i}]}\{\frac{|y_i - z_i(s)|}{s}\}$. Given that $\theta$ is a Gaussian, we obtain that for all $s \in ]0, s_{m_i}]$, $\theta(\frac{y_i - z_i(s)}{s}) \geq \theta(A)$ which implies that $s \mapsto \frac{1}{s}\theta(\frac{y_i - z_i(s)}{s})$ is not integrable on $[0, s_{m_i}]$ and finally shows that $g_{z_i}(x) \to +\infty$ as $x \to y_i$. Since $G_z(x) = \sum_{i=1}^p g_{z_i}(x)$, the result immediately follows. □

Roughly speaking, calculating the structural intensity consists in computing the "density" of the zero-crossings along various scales. Then, Proposition 1 shows that in practice, the modes of the resulting "density" are located at the landmarks of the signal $f$ as it can be seen in Figures 1c where the local maxima of $G_z(.)$ correspond to the limits of the zero-crossings lines of the signal $f_T$. Note that in practice, the structural intensities are normalized to be probability density functions.

## 4.2. The nonparametric regression and white noise models

In [2], the estimation of the zero-crossings lines has been considered in the white noise model, i.e. when $f$ is observed according to:

$$Y(dx) = f(x)dx + \tau B(dx), \ x \in [0,1], \quad (3)$$

where $\tau$ is a noise level parameter, $f$ an unknown function which may have various landmarks, and $B$ is a standard Brownian motion. When $\tau = \frac{\sigma}{\sqrt{n}}$, the white noise model (3) is closely related to the following nonparametric regression problem (see Brown and Low [3], Donoho and Johnstone ([8], [5]) for further details):

$$y_i = f(x_i) + \sigma \epsilon_i, \ i = 1, \ldots, n, \quad (4)$$

where $x_i = \frac{i}{n}$, $f$ is an unknown function, $\sigma$ is the level of noise and $\epsilon_i$ are i.i.d. normal variables with zero mean and variance 1. Assume that $f \in L^2([0,1])$ is observed according to the white noise model (3). If we modify a function that belongs to $L^2(\mathbb{R})$ by multiplying it by the indicator function of $[0,1]$, we do not modify its regularity and its landmarks on $]0,1[$. The wavelet transform of $f$ at a scale $s > 0$ is then equal to: $W_s(f)(x) = \int_0^1 f(u)\psi_s(u-x)du$, and the wavelet transform of the white noise $B(du)$ is defined to be: $W_s(B)(x) = \int_{-\infty}^{+\infty} \psi_s(u-x)B(du)$, for $x \in [0,1]$. Then, the wavelet transform of $Y$ is:

$$W_{s,n}(Y)(x) = \int_{-\infty}^{+\infty} \psi_s(u-x)Y(du) = W_s(f)(x) + \frac{\sigma}{\sqrt{n}}W_s(B)(x) \quad (5)$$

In the following sections, we will assume that the wavelet $\psi$ is the $r$-th derivative of a Gaussian and will show how to estimate the landmarks of $f$ from the zero-crossings of $W_{s,n}(Y)(x)$.

## 4.3. Estimation of the zero-crossings

In [2] (see section 6.1), the following statistical test is considered to estimate the zero-crossings of $W_s(f)(x)$ at a given scale $s$: let $q_{1-\alpha}^n$ be such that $P(\max_{x \in [0,1]} |W_{n^{-1}}(B)(x)| \leq q_{1-\alpha}^n) \to 1 - \alpha$ as $n \to \infty$ for some $0 < \alpha < 1$, and define the statistical test that accepts the null hypothesis $H_0^{s,x} : W_s(f)(x) = 0$ if $|W_{s,n}(Y)(x)| \leq \frac{\sigma}{\sqrt{n}}q_{1-\alpha}^n$ or concludes significant evidence for $W_s(f)(x)$ being positive or negative if $W_{s,n}(Y)(x) > \frac{\sigma}{\sqrt{n}}q_{1-\alpha}^n$ or $W_{s,n}(Y)(x) < -\frac{\sigma}{\sqrt{n}}q_{1-\alpha}^n$ respectively. Then, if $H_0^{s,x}$ is rejected, we have statistically significant evidence for $W_s(f)(x)$ being positive or negative depending of the sign of $W_{s,n}(Y)(x)$. Each point of significant zero-crossings of $W_s(f)(x)$ at a given scale $s$ is then located between a pair of points $x_1, x_2 \in [0,1]$ such that $H_0^{s,x_1}$ and $H_0^{s,x_2}$ are rejected, and $W_s(f)(x_1)$ and $W_s(f)(x_2)$ have opposite signs. Let $s_{min}$ be the largest scale at which one wants to estimate the zero-crossings of $W_s(f)(x)$. The following empirical estimation of the zero-crossings is then proposed is [2]: for $s \leq s_{min}$, let $\{(\hat{x}_{1,s,i}, \hat{x}_{2,s,i}); 1 \leq i \leq \hat{p}_s\}$ be the sequence of points of $]0,1[\times]0,1[$ such that:

- $H_0^{s,\hat{x}_{1,s,i}}$ and $H_0^{s,\hat{x}_{2,s,i}}$ are rejected.

- $W_{s,n}(Y)(\hat{x}_{1,s,i})$ and $W_{s,n}(Y)(\hat{x}_{2,s,i})$ have opposite signs.

- all the hypothesis $H_0^{s,x}$ for $\hat{x}_{1,s,i} < x < \hat{x}_{2,s,i}$ are accepted.

For $1 \leq i \leq \hat{p}_s$, if $W_{s,n}(Y)$ has **only one zero-crossings** $z_{1,s,i}$ in $[\hat{x}_{1,s,i}; \hat{x}_{2,s,i}]$ define $\hat{z}_i(s) = z_{1,s,i}$ else discard the zero-crossings located in $[\hat{x}_{1,s,i}; \hat{x}_{2,s,i}]$ since most of them are likely to be due to the fluctuations of the noise. Once various zero-crossings have been estimated at different scales, the landmarks of the unknown signal $f$ are defined as the local maxima of the structural intensity of the estimated zero-crossings lines.

To complete this brief review of the methodology developed in [2], we give an example of the estimation of the local extrema of the function $f_T$ when it is observed from the model (4) with $n = 512$ (see in Figure 1d). Throughout this paper, the continuous wavelet transform of the discrete signal $y_i, i = 1, \ldots, n$ is computed at dyadic scales $s = 2^{-j}$ with 20 voices per octave ($1 \leq j \leq (\log_2(n) - 1)$) for a Gaussian wavelet with $r = 1$ and $\beta = 0.5$. To estimate the noise level $\sigma$ we used the robust estimate suggested by Donoho and Johnstone [7] based on the *median absolute deviation* of the empirical wavelet coefficients associated with an orthonormal wavelet basis of $L^2([0,1])$ (we took the Symmlet 8 wavelet basis (as described on page 198 of Daubechies [6])).

The result is very satisfactory, since it can been in Figure 1e that all the lines are correctly estimated. The local maxima of the structural intensity correspond to the local extrema of the Time signal, and the shape of the peaks in Figure 1f is related to the length and the location along the scale axis of the zero-crossings lines. A sharp peak corresponds to a line that propagates to fine scales, while a flat peak is related to a line that has been estimated only at some coarse scales.

## 5. Automatic landmark registration

We will now explain how the structural intensities of the zero-crossings lines of two curves $f_1$ and $f_2$ can be used to construct a method to automatically decide if a landmark of $f_1$ (resp. $f_2$) corresponds to a landmark of $f_2$ (resp. $f_1$). To present this method, we propose to illustrate, on a real example, the problem of identifying the pair of landmarks that should correspond. The two curves plotted in Figure 2ad are Medulla density profiles for two old rats extracted from the data set analyzed in [16]. The samples curves in Figure 2ad have been rescaled on the unit interval $[0, 1]$ and interpolated with cubic Splines to form sequences of observations of $n = 128$ points. The number of points is chosen to avoid smoothing, so that the fitted curves $\tilde{g}_1$ and $\tilde{g}_2$ agreed with the shape of the original data (see [16]). Suppose that for aligning the two functions $\tilde{g}_1$ and $\tilde{g}_2$, one wants to match the significant local extrema of each curves. To retrieve the location of these landmarks, we used the procedure described in section 4 to estimate the zero-crossings that correspond to the local maxima of $\tilde{g}_1$ and $\tilde{g}_2$ (computed for a Gaussian wavelet with $r = 1$ vanishing moment and $\beta = 0.5$). The estimated zero-crossings lines are given in Figure 2 with their corresponding structural intensities. In this example, the structural intensity $\hat{G}_{z1}^+$ in Figure 2c has seven modes while the structural intensity $\hat{G}_{z2}^+$ in Figure 2f has six modes. Obviously, one cannot easily decide if the $i$th modes of $\hat{G}_{z1}^+$

should correspond to the $j$th modes of $\hat{G}^+_{z2}$ for $i = 1, \ldots, 7$ and $j = 1, \ldots, 6$. In [16], five main peaks are identified in each density profile to register the fitted curve. Clearly, one can say that $\tilde{g}_1$ and $\tilde{g}_2$ have five main maxima located at $x = 0.04, 0.28, 0.42, 0.63, 0.86$ and $x = 0.04, 0.35, 0.51, 0.72, 0.89$, and the shape of these peaks suggests that these five landmarks should be matched. When looking at the zero-crossings lines converging to these five main peaks, it can be seen that the shape of the lines corresponding to the $i$th peaks in $\tilde{g}_1$ and $\tilde{g}_2$ are similar for $i = 1, \ldots, 5$. In some sense, to identify the pair of landmarks that should correspond, it would be natural to try to associate the zero-crossings lines whose shape and length in the time-scale plane are similar. Since the width and the height of the modes of the structural intensities are related to the shape of the zero-crossings lines, we propose to align $\hat{G}^+_{z1}$ and $\hat{G}^+_{z2}$ by dynamic time warping (DTW) to measure similarity between these lines in an automatic way. Since DTW is a technique that automatically aligns structural points of two functions, it is expected that the common modes of $\hat{G}^+_{z1}$ and $\hat{G}^+_{z2}$ would be aligned which would therefore indicate which pair of landmark should be matched.

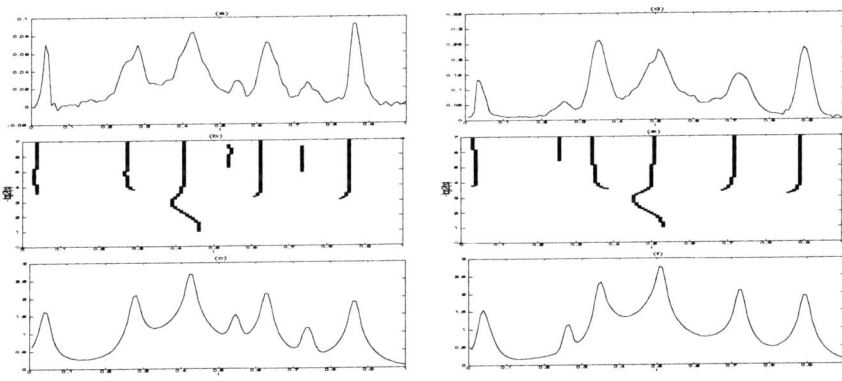

Figure 2. (a) Fitted curve $\tilde{g}_1$. (b) Estimated zero-crossings of $\tilde{g}_1$ that correspond to local maxima with their associated structural intensity (c). (d) Fitted curve $\tilde{g}_2$. (e) Estimated zero-crossings of $\tilde{g}_2$ that correspond to local maxima with their associated structural intensity (f).

### 5.1. Alignment of the structural intensities by dynamic time warping

In a continuous setting, the alignment of two functions $\theta_1$ and $\theta_2$ by dynamic time warping can be formulated as a dense matching problem. For the classical quadratic cost function, a possible formulation of this problem is the following: let $\theta_1, \theta_2 : [0, 1] \to \mathbb{R}$, the problem of aligning the two curves $\theta_1$ and $\theta_2$, is to search the minimum, among all increasing diffeomorphisms $h$ of $[0, 1]$, of the functional: $L_{\theta_1, \theta_2}(h) = \int_0^1 (\theta_1(x) - \theta_2 \circ h(x))^2 dx$. One of the main difficulty of this approach is to guarantee that the solution of this variational problem is an increasing function. Piccioni et. al. [17] have studied in detail the variational problem associated with this sort of functional. To study the existence of a minimum point, they suggested to reformulate this variational problem on the set of all probability measures on $[0,1]$. We proposed to follow their methodology to study the problem of aligning two structural intensities by dynamic time warping where the objective is to match common modes. To this aim, let $M_1$ denote the set of probability measures on the Borel sets of $[0,1]$. This is a compact space when endowed with the topology of weak convergence. For an element $\nu \in M_1$, we shall write its distribution function as: $F(\nu)(x) = \int_0^x d\nu$, $x \in [0,1]$. Let $\theta_1$ and $\theta_2$ be bounded functions that are in $L^2([0,1], \mathbb{R}, dx)$, where $dx$ denotes the Lebesgue measure on $[0,1]$. Then for $\nu \in M_1$ set:

$$L_{\theta_1,\theta_2}(F(\nu)) = \int_0^1 (\frac{\theta_1(x)}{\|\theta_1\|_\infty} - \frac{\theta_2 \circ F(\nu)(x)}{\|\theta_2\|_\infty})^2 dx. \tag{6}$$

If $\theta_2$ is continuous on $[0,1]$, one can show that $L_{\theta_1,\theta_2}(.)$ is continuous and has therefore a minimum since $M_1$ is a compact (see the proof of Proposition 4.2 in [17]). The normalization with the sup-norm $\|.\|$ in (6) is intended to reduce the differences in amplitudes of the curves when calculating the function that minimizes $L_{\theta_1,\theta_2}(.)$. Since this warping function tends to align the local maxima of $\theta_1$ and $\theta_2$ (see [22],[23] where a discretized finite-dimensional approximation of the variational problem (6) is presented), we propose to align the structural intensities of the zero-crossings lines of two functions $f_1$ and $f_2$ with various landmarks to decide which pair of structural points should be matched. This idea is formulated in the following proposition where a slightly modified version of the structural intensities is considered in order to deal with continuous and bounded functions on $[0,1]$:

**Proposition 2** *Let $f_1, f_2 \in L^2(\mathbb{R})$ and $\psi = (-1)^r \theta^{(r)}, r \geq 1$ where $\theta$ is a Gaussian. Suppose that there exists $p_1$ zero-crossings lines $z_{1,i}(s), i = 1, \ldots, p_1$ and $p_2$ zero-crossings lines $z_{2,i}(s), i = 1, \ldots, p_2$ associated to $f_1$ and $f_2$ respectively. For $x \in \mathbb{R}$, define the structural intensities of the zero-crossings $G_{z_1}(x)$ and $G_{z_2}(x)$ as:*

$$G_{z_1}(x) = \sum_{i=1}^{p_1} \int_0^{s_{z_{1,i}}} |log(s)|^3 \theta(\frac{x - z_{1,i}(s)}{s}) ds$$

$$G_{z_2}(x) = \sum_{i=1}^{p_2} \int_0^{s_{z_{2,i}}} |log(s)|^3 \theta(\frac{x - z_{2,i}(s)}{s}) ds,$$

*where $[0, s_{z_{j,i}}]$ is the support of the zero-crossings line $z_{j,i}(.)$. Then, $G_{z_1}(.)$ and $G_{z_2}(.)$ are continuous functions on $[0,1]$.*

Now, let $\tau_1 = (\tau_{1,1}, \ldots, \tau_{1,p_1^*})$ be the set of local maxima of $G_{z_1}$ and $\tau_2 = (\tau_{2,1}, \ldots, \tau_{2,p_2^*})$ be the set of local maxima of $G_{z_1}$. Define $\mu = \text{argmin}_{\nu \in M_1} L_{G_{z_1}, G_{z_2}}(F(\nu))$. Let $\epsilon > 0$. Then, an $\epsilon$-correspondence between the two sets of landmarks $\tau_1$ and $\tau_2$ is defined by the following scheme:

- for $i = 1, \ldots, p_1^*$ let $\tau_{2,i}^* = (\tau_{2,j_1}, \ldots, \tau_{2,j_i})$ be the set of landmarks that are such that $F(\mu)(\tau_{2,k}) \in [\tau_{1,i} - \epsilon, \tau_{1,i} + \epsilon], \tau_{2,k} \in \tau_{2,i}^*$ and that have not already been associated to another landmark $\tau_{1,k}, k = 1, \ldots, i - 1$.

- if $\tau_{2,i}^*$ is not empty then the landmark $\tau_{2,j_1}$ is associated to the landmark $\tau_{1,i}$.

**Proof:** We only need to prove that $G_{z_1}(.)$ and $G_{z_2}(.)$ are continuous functions on $[0,1]$ which implies that $L_{G_{z_1},G_{z_2}}(.)$ has a minimum. Let:

$$g_{z_1,i}(x) = \int_0^{s_{z_1,i}} |log(s)|^3 \theta(\frac{x - z_{1,i}(s)}{s}) ds, \ x \in [0,1].$$

For almost every $s \in [0, s_{z_1,i}]$, $x \mapsto |log(s)|^3 \theta(\frac{x-z_{1,i}(s)}{s})$ is continuous on $[0,1]$ and for all $x \in [0,1]$, $|log(s)|^3 \theta(\frac{x-z_{1,i}(s)}{s}) \leq \|\theta\|_\infty |log(s)|^3$. Since $s \mapsto |log(s)|^3$ is integrable on $[0, s_{z_1,i}]$, the continuity theorem under the integral sign implies that $g_{z_1,i}$ is continuous on $[0,1]$. Since $G_{z_1}(x) = \sum_{i=1}^{p_1} g_{z_1,i}(x)$, it is therefore continuous on $[0,1]$. The proof to show that $G_{z_2}(.)$ is continuous on $[0,1]$ is identical. □

**Remark:** In Proposition 2, the definition of the structural intensity is slightly different from the one given in Proposition 1. To make $G_{z_1}(.)$ continuous on $[0,1]$, we replaced the weight function $s \mapsto \frac{1}{s}$ by $s \mapsto |log(s)|^3$ in Proposition 2. Since the curves of these weight functions are close to each other on $[0,1]$, the shape of the new structural intensity introduced in this section is similar to the one defined previously. In all the following figures, the various structural intensities are computed with $\frac{1}{s}$ replaced by $|log(s)|^3$.

For reasons of space and because proving limit theorems for "argmin" estimators is a difficult task, we will not discuss the consistency of $\epsilon$-correspondence when the curves $f_1$ and $f_2$ are unknown and observed with noise. We rather propose to apply our automatic landmark-based matching method to a real example and to run some simulations to illustrate the very nice properties of this approach in practice.

### 5.2. Practical implementation and simulations
#### 5.2.1. A real example

We now return to the problem of aligning the Medulla density profiles for two old rats (see Figure 2). The alignment by dynamic time warping of the structural intensities $\hat{G}_{z_1}^+$ and $\hat{G}_{z_2}^+$ of the estimated zero-crossings that correspond to the significant local maxima of $\tilde{g}_1$ and $\tilde{g}_2$ is given in Figure 3a. In [16], the five main peaks of $\tilde{g}_1$ located at $x = 0.04, 0.28, 0.42, 0.63, 0.86$ are matched with the five main peaks of $\tilde{g}_2$ which are located $x = 0.04, 0.35, 0.51, 0.72, 0.89$. From Figure 3a, it can be seen that if one performs $\epsilon$-correspondence with $\epsilon = 0.01$ between the modes of $\hat{G}_{z_1}^+$ and $\hat{G}_{z_2}^+$, then the vectors of landmarks that should be matched are $\tau_1^+ = (0.04, 0.28, 0.42, 0.63, 0.86)$ and $\tau_2^+ = (0.04, 0.35, 0.51, 0.72, 0.89)$. Hence, in this example, $\epsilon$-correspondence correctly aligns the five main peaks of $\hat{G}_{z_1}^+$ and $\hat{G}_{z_2}^+$ and therefore agrees with the matching that one would have made "by the eyes".

For the purpose of curves alignment, it is often required to match both common local maxima and minima. In Figure 3b, we plotted the alignment by dynamic time warping of the the structural intensities $\hat{G}_{z_1}^-$ and $\hat{G}_{z_2}^-$ of the estimated zero-crossings that correspond to significant minima of $\tilde{g}_1$ and $\tilde{g}_2$. With $\epsilon = 0.01$, the vectors of local minima that should be matched are $\tau_1^- = (0.12, 0.34, 0.52, 0.79)$ and $\tau_2^- = (0.13, 0.41, 0.63, 0.83)$.

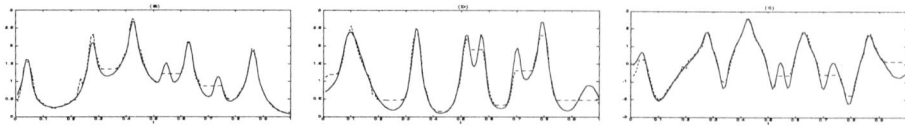

Figure 3. (a) Alignment by DTW of the structural intensities: (a) $\hat{G}^+_{z1}$ (solid line) and $\hat{G}^+_{z2}$ (dotted line), (b) $\hat{G}^-_{z1}$ (solid line) and $\hat{G}^-_{z2}$ (dotted line), (c) $\hat{G}^*_{z1}$ (solid line) and $\hat{G}^*_{z2}$ (dotted line)

Since a significant maxima should be followed by a significant minima, one could define $\tau_1 = \tau_1^+ \cup \tau_1^-$ and $\tau_2 = \tau_2^+ \cup \tau_2^-$ as the vectors of landmarks to be matched, in such a way that the elements of $\tau_1$ and $\tau_2$ are sorted in increasing order, i.e $\tau_1 = (0.04, 0.12, 0.28, 0.34, 0.42, 0.52, 0.63, 0.79, 0.86)$ and $\tau_2 = (0.04, 0.13, 0.35, 0.41, 0.51, 0.63, 0.72, 0.83, 0.90)$. In our example, this method works since it preserves the correspondence between local maxima and local minima. However, if for some $i > j$ one has $\tau_1^+(i) < \tau_1^-(j)$ and $\tau_2^+(i) > \tau_2^-(j)$ this procedure would associate a maxima with a minima. Hereby, to match both local minima and local minima of two curves that one wishes to align, the following approach is proposed:

- let $\hat{G}^*_{z1} = \hat{G}^+_{z1} - \hat{G}^-_{z1}$ and $\hat{G}^*_{z2} = \hat{G}^+_{z2} - \hat{G}^-_{z2}$.

- define $\mu^* = argmin_{\nu \in M_1} L_{\hat{G}^*_{z1}, \hat{G}^*_{z2}}(F(\nu))$ and $w^*(x) = F(\mu^*)(x)$.

- let $\tau_1^{*,+}$ and $\tau_2^{*,+}$ be the vectors obtained by $\epsilon$-correspondence between the *local maxima* of $\hat{G}^*_{z1}$ and $\hat{G}^*_{z2}$, and $\tau_1^{*,-}$ and $\tau_2^{*,-}$ be the vectors obtained by $\epsilon$-correspondence between the *local minima* of $\hat{G}^*_{z1}$ and $\hat{G}^*_{z2}$, with the warping path $w^*$.

- define $\tau_1^* = \tau_1^{*,+} \cup \tau_1^{*,-}$ and $\tau_2^* = \tau_2^{*,+} \cup \tau_2^{*,-}$ in such a way that the elements of $\tau_1^*$ and $\tau_2^*$ are sorted in increasing order. Then, take $\tau_1^*$ and $\tau_2^*$ as the vectors of landmarks that should be matched.

Figure 3c gives an illustration of this procedure for the structural intensities of $\tilde{g}_1$ and $\tilde{g}_2$. With $\epsilon = 0.01$, the correspondence between the local extrema of $\hat{G}^*_{z1}$ and $\hat{G}^*_{z2}$ yields to $\tau_1^* = (0.04, 0.10, 0.27, 0.34, 0.42, 0.52, 0.63, 0.79, 0.87)$ and $\tau_2^* = (0.04, 0.12, 0.35, 0.41, 0.51, 0.63, 0.72, 0.83, 0.90)$. Hence, this method preserves the correspondence between local maxima and local minima, and it can be seen that the vectors $\tau_j^*$ and $\tau_j$ ($j = 1, 2$) are very closed to each other.

### 5.2.2. Simulations

We propose to run a small simulation study to evaluate the performances of our automatic landmark-based matching method and compare it with the dynamic programming algorithm proposed by Wang and Gasser ([22],[23]). Before describing the model used in the simulations, we briefly present the approach investigated in [22] and [23]. Let $f_1$ and $f_2$ be two regression functions observed from noisy data. The kernel method of Gasser and

Muller [11] is used to estimate $f_1$ and $f_2$ and their first derivative, while the optimal bandwidth is chosen by the plug-in method described in [10]. In our simulations we used the subroutine glkern.f, which can be obtained from the Web site www.unizh.ch/biostat/, to compute the kernel estimators. The estimation of a warping function $\hat{u}$ is then given by the minimum, among all increasing diffeomorphisms $u$ of $[0,1]$, of the functional $L_{\hat{f}_1,\hat{f}_2,\alpha}(u) = \int_0^1 \alpha^2 \left( \frac{\hat{f}_1(x)}{\|\hat{f}_1\|_\infty} - \frac{\hat{f}_2(u(x))}{\|\hat{f}_2\|_\infty} \right)^2 + (1-\alpha)^2 \left( \frac{\hat{f}'_1(x)}{\|\hat{f}'_1\|_\infty} - \frac{\hat{f}'_2(u(x))}{\|\hat{f}'_2\|_\infty} \right)^2 + 2\phi(u(x))dx$, where $\alpha \in [0,1]$ and $\phi(u(x)) = 0.001 \frac{(u(x)-x)^2}{1+(u(x)-x)^2}$ penalizes too irregular warping functions. Following the computational considerations discussed in [23], we solve the above variational problem by dynamic programming for three values $\alpha = 0.3, 0.5, 0.7$ and then choose the best alignment which yields smallest cost $L_{\hat{f}_1,\hat{f}_2,\alpha}(u)$. We also impose the following slope constraint: $|u(x) - x| \leq 1/2$ and the derivatives of the sample curves are estimated with oversmoothing.

The evaluation of the two registration methods is performed for the basic problem of warping one function to a second one. The shape function $s(t)$ shown in Figure 4a is the one used by Wang and Gasser [22] in their simulations. We consider the shape invariant model investigated in [22]: $f_{1,i}(t) = s(t)$ and $f_{2,i}(t) = a_i s(h_i(t)) + d_i, i = 1,\ldots,100$, where $h_i(t) = t + \frac{\alpha_i}{2\pi} \sin(2\pi(\beta_i t + \gamma_i))$, $a_i = \max(1 + N(0,1), 0.5)$, $d_i = 4N(0,1)$ ($N(0,1)$ stands for a standard normal variable), and the parameters $\alpha_i, \beta_i, \gamma_i$ are randomly generated as in [22] to ensure that $h_i$ is strictly increasing. The functions $f_{1,i}$ and $f_{2,i}$ have ten extrema denoted by $\tau_{1,i} = (\tau_{1,j,i}, j = 1,\ldots, 10)$ and $\tau_{2,i} = (\tau_{2,j,i} = h_i(\tau_{1,j,i}), j = 1,\ldots, 10)$ respectively for $i = 1,\ldots, 100$. At each run, two sequences of noisy data of length $n$ are then formed as: $y_{j,i,k} = f_{j,i}(t_k) + \sigma_{j,i}\epsilon_{j,i,k}$, $k = 1,\ldots, n$, where $t_k = \frac{k}{n}$, $\epsilon_{j,i,k}$ are i.i.d. normal variables with zero mean and variance 1, and $\sigma_{j,i}$ are noise level parameters. The factors are the samples sizes $n$ and the values of $\sigma_{j,i}$. In this paper, we report the results for two samples sizes, $n = 128, 512$ and a root signal-to-noise ratio RSNR equal to 7, 5 and 3 where: $RSNR(f_{j,i}, \sigma_{j,i}) = \frac{\sqrt{\int_0^1 (f_{j,i}(x) - \bar{f}_{j,i})^2 dx}}{\sigma_{j,i}}$, with $\bar{f}_{j,i} = \int_0^1 f_{j,i}(x)dx$.

For $i = 1,\ldots, 100$, let $\hat{h}_i$ be the warping function obtained by the method of Wang and Gasser [22]. For the $i$th run and $j = 1, 2$, define $G^+_{z_{j,i}}$ and $\hat{G}^+_{z_{j,i}}$ respectively as the structural intensities of the true and estimated zero-crossings that correspond to local maxima, and define $G^-_{z_{j,i}}$, $\hat{G}^-_{z_{j,i}}$ respectively as the structural intensities of the true and estimated zero-crossings that correspond to local minima. Then, we calculate $G^*_{z_{j,i}} = G^+_{z_{j,i}} - G^-_{z_{j,i}}$ and we define $\tau^*_{1,i}$ and $\tau^*_{2,i}$ as the vectors of true landmarks obtained by $\epsilon$-correspondence (with $\epsilon = 4/n$) between the local extrema of $G^*_{z_{1,i}}$ and $G^*_{z_{2,i}}$. Similarly, let $\hat{G}^*_{z_{j,i}} = \hat{G}^+_{z_{j,i}} - \hat{G}^-_{z_{j,i}}$ and define $\hat{\tau}^*_{1,i}$ and $\hat{\tau}^*_{2,i}$ as the vectors of estimated landmarks obtained by $\epsilon$-correspondence (with $\epsilon = 4/n$) between the local extrema of $\hat{G}^*_{z_{1,i}}$ and $\hat{G}^*_{z_{2,i}}$. Then, we define $u_i(t) = G_{(\tau^*_{1,i},\tau^*_{2,i})}(t)$ and $\hat{u}_i(t) = G_{(\hat{\tau}^*_{1,i},\hat{\tau}^*_{2,i})}(t), t \in [0,1]$, where $G$ is a warping operator (we took the method of Ramsay [18] to compute the warping functions). A typical run, for which the quality of the estimation of $h_i$ and $u_i$ is very good, is shown in Figure 4 (with $n = 512$ and $RSNR = 5$).

To compare our method with the one of Wang and Gasser, we used two error cri-

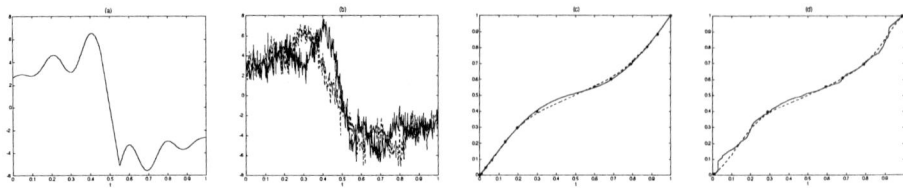

Figure 4. (a) Shape function (b) Functions $f_{1,i}$ (solid line) and $f_{2,i}$ (dotted line) observed with noise ($n = 512$, $RSNR = 5$) (c) True warping function $h_i$ (solid line) and landmark-based warping function $u_i$ (dotted line) with true landmarks (*) associated by $\epsilon$-correspondence (d) Estimated warping function $\hat{h}_i$ (solid line) by DTW and landmark-based warping function $\hat{u}_i$ (dotted line) with estimated landmarks (*) associated by $\epsilon$-correspondence.

terions for the $i$th run ($i = 1, \ldots, 100$): $WMSE_i = \int_0^1 \left( f_{1,i}(t) - \frac{\hat{f}_{2,i}(\hat{w}_i(t)) - d_i}{a_i} \right)^2 dt$ and $\rho_i = \frac{1}{10} \sum_{j=1}^{10} \left( \hat{w}_i(\tau_{1,j}) - h_i(\tau_{1,j}) \right)^2$, where $\hat{w}_i$ is either equal to $\hat{h}_i$ or $\hat{u}_i$ and $\hat{f}_{2,i}$ is an estimation of $f_{2,i}$ either obtained by kernel smoothing or by the translation invariant wavelet estimator with hard thresholding (TI-H) of Coifman and Donoho [4]. We chose the TI-H procedure because the simulative comparison study between wavelet estimators carried out in Antoniadis et. al. [1], shows that TI-H is markedly better than other methods for smooth functions. $WMSE_i$ is a global measurement of the quality of the alignment while $\rho_i$ is an error measurement for aligning extremes. In Table 1, the columns "WG", "Auto TI-H" and "Auto Kernel" give the sample means and sample standard deviation of $WMSE_i$ respectively for the method of Wang and Gasser with $\hat{f}_{2,i}$ obtained by kernel smoothing, for our method with $\hat{f}_{2,i}$ obtained by the TI-H procedure and for our method with $\hat{f}_{2,i}$ obtained by kernel smoothing. The columns "WG $\rho$" and 'Auto Land $\rho$" gives the sample means and sample standard deviation of $\rho_i$ respectively for the method of Wang and Gasser and for our method. Obviously, the performance of our method depends upon the number of estimated landmarks that are put into correspondence. Hence, we plotted in Figure 5, the locations (in $[0, 1]$) of the estimated landmarks of $f_1$ that are put into correspondence with estimated landmarks of $f_2$ at each run and for different combinations of the factor levels.

The sample mean of $WMSE_i$ obtained with "WG" is always smaller than the one obtained with "Auto TI-H". For $n = 128, RSNR = 7, 5$ the mean and standard deviation of $WMSE_i$ for "Auto Kernel" and "WG" have the same amplitude. For $n = 512, RSNR = 7$, "Auto Kernel" performs better than "WG" for the $WMSE$ criterion. This case correspond to the best approximation of the landmarks of $f_1$ put into correspondance with estimated landmarks of $f_{2,i}, i = 1, \ldots, 100$ (see Figure 5d: at each run, six landmarks (in $]0, 1[$) are labelled to perform the registration). For $RSNR = 3$, the mean of $WMSE_i$ for "WG" is much smaller than the one obtained either with 'Auto Kernel" or "Auto TI-H", which can be explained by Figure 5cf (at each run, only two landmarks in $]0, 1[$ (around 0.4 and 0.7) are used to calculate $\hat{u}_i$).

The error of aligning extreme ($\rho_i, i = 1, 100$) is smaller for our method with $n = 128$, $RSNR = 7, 5, 3$ and $n = 512, RSNR = 7$. The mean and standard deviation of $\rho_i$ obtained with both methods are of the same order for $n = 512, RSNR = 5$, while the method of Wang and Gasser is better for $n = 512, RSNR = 3$. Hence, the results given in columns "WG $\rho$" and 'Auto Land $\rho$" shows that the warping function obtained with our method gives a better alignement of the landmarks of two noisy functions.

When a sufficient number of landmarks can be estimated and labelled, both methods performs similarly for the global measurement error when $f_1$ and $f_2$ are estimated with kernel smoothing. One drawback of our method is the choice of a denoising procedure which will generally give rise to estimated functions whose landmarks are not located at the modes of the structural intensity of the estimated zero-crossings. This could explain why "Auto Kernel" is better than "Auto TI-H". The estimation of the landmarks might be more accurate with kernel smoothing. Note that generally, wavelet thresholding performs poorly at the boundaries of $[0, 1]$, which could explain the difference in mean squared error between "Auto TI-H" and "Auto Kernel". Since our method seems to be an effective tool to automatically associate the landmarks of two functions, it would be interesting to define a functional $L_{f_1,f_2}(u)$ that would combine a global measurement error between $f_1$ and $f_2 \circ u$ with an error measurement for aligning landmarks.

Table 1
Sample mean (M) and standard deviation (STD) of $WMSE_i$ and $\rho_i, i = 1, \ldots, 100$ for different values of $n$ and $RSNR$.

|  |  | WG | Auto TI-H | Auto Kernel | WG $\rho$ | Auto Land $\rho$ |
|---|---|---|---|---|---|---|
| $n = 128, RSNR = 7$ | M | 0.36 | 0.50 | 0.36 | $6.48e^{-4}$ | $5.49e^{-4}$ |
|  | STD | 0.62 | 0.42 | 0.38 | $1.40e^{-3}$ | $9.40e^{-4}$ |
| $n = 128, RSNR = 5$ | M | 0.52 | 0.74 | 0.52 | $1.07e^{-3}$ | $6.28e^{-4}$ |
|  | STD | 1.35 | 0.64 | 0.62 | $3.12e^{-3}$ | $1.11e^{-3}$ |
| $n = 128, RSNR = 3$ | M | 0.84 | 2.37 | 2.08 | $2.24e^{-3}$ | $1.89e^{-3}$ |
|  | STD | 1.37 | 2.33 | 2.35 | $4.81e^{-3}$ | $2.57e^{-3}$ |
| $n = 512, RSNR = 7$ | M | 0.23 | 0.24 | 0.20 | $2.14e^{-4}$ | $1.59e^{-4}$ |
|  | STD | 0.30 | 0.30 | 0.29 | $9.22e^{-4}$ | $2.07e^{-4}$ |
| $n = 512, RSNR = 5$ | M | 0.310 | 0.59 | 0.53 | $3.48e^{-4}$ | $3.90e^{-4}$ |
|  | STD | 0.90 | 1.08 | 1.06 | $1.32e^{-3}$ | $8.45e^{-4}$ |
| $n = 512, RSNR = 3$ | M | 0.38 | 0.97 | 0.87 | $4.86e^{-4}$ | $7.04e^{-4}$ |
|  | STD | 0.97 | 1.24 | 1.21 | $1.17e^{-3}$ | $1.02e^{-3}$ |

## 6. Conclusion

An important problem in landmark-based registration consists in deciding which pair of structural points should be associated when two functions, corrupted by noise, have to aligned. A scale-space approach with wavelets has been proposed to estimate the landmarks of a function which yields a new method, named $\epsilon$-correspondence, to auto-

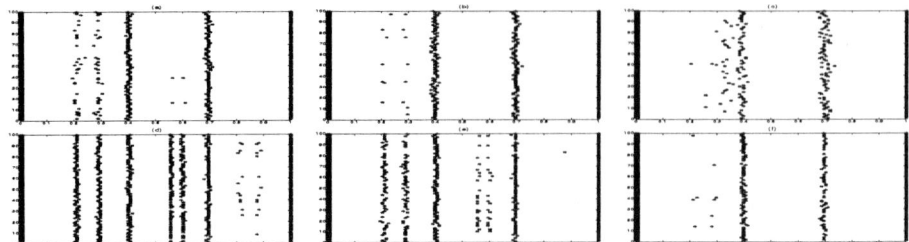

Figure 5. Location of the estimated landmarks (in $[0,1]$) of $f_1$ that are put into correspondence with estimated landmarks of $f_{2,i}$ ($i = 1, \ldots, 100$). In each figure the vertical axis gives the indice of the $i$th run, while the horizontal axis is the interval $[0,1]$: (a) $n = 128, RSNR = 7$, (b) $n = 128, RSNR = 5$, (c) $n = 128, RSNR = 3$, (d) $n = 512, RSNR = 7$, (e) $n = 512, RSNR = 5$, (f) $n = 512, RSNR = 3$.

matically put into correspondence pair of landmarks from two functions. The usefulness of this approach has been illustrated on a real example, and a small simulation study to compare this landmark-based registration technique with a dense matching approach to curve alignment has also been proposed. The results are very satisfactory and show that our method is an effective technique for aligning the structural points of two functions. A possible extension of this work, that could lead to better results, would be to combine our method with a dense matching approach.

### Acknowledgements

I gratefully acknowledge Yolanda Munoz Maldonado for having accepted to provide the Medulla density profiles. I particularly thank Anestis Antoniadis for helpful comments, for the various references he gave me and for having corrected a previous version of this paper.

### REFERENCES

1. Antoniadis, A. Bigot, J. and Sapatinas, T. (2001). Wavelet estimators in nonparametric regression: a comparative simulation study. *Journal of Statistical Software*, Vol. 6, Issue 6, 1–83.
2. Bigot, J. (2002). A scale-space approach to landmark detection. Technical Report TR2046, PAI (Interuniversity Attraction Pole network), http://www.stat.ucl.ac.be/IAP/.
3. Brown, L.D. and Low, M.G. (1996). Asymptotic equivalence of nonparametric regression and white noise, *Ann. Statist.* **3** No.6, 2384–2398.
4. Coifman, R.R. and Donoho, D.L. (1995). Translation-invariant de-noising. In *Wavelets and Statistics*, Antoniadis, A. and Oppenheim, G. (Eds.), Lect. Notes Statist., **103**, pp. 125150, New York: Springer-Verlag.
5. Donoho, D.L. and Johnstone, I.M. (1994). Ideal spatial adaptation by wavelet shrink-

age, *Biometrika* **81**, 425–455.
6. Daubechies, I. (1992). Ten Lectures on Wavelets, *Philadelphia : SIAM*.
7. Donoho, D.L. and Johnstone, I.M. (1998). Minimax estimation via wavelet shrinkage, *Ann. Statist.* **26**, 879–921.
8. Donoho, D.L. and Johnstone, I.M. (1999). Asymptotic minimality of wavelet estimators with sampled data, *Stat. Sinica* **9**, 1–32.
9. Gasser, T. and Kneip, A. (1995). Searching for Structure in Curve Samples, *J. Am. Statist. Ass.* **90** No.432, 1179–1188.
10. Gasser, T. Kneip, A. and Kohler, W. (1991). A flexible and fast method for automatic smoothing, *J. Am. Statist. Ass.* **86** , 643–652.
11. Gasser, T. and Muller, H. (1984). Estimating regression functions and their derivatives by the kernel method, *Scand. J. Statist.* **11** 171–185.
12. Kneip, A. and Gasser, T. (1992). Statistical tools to analyze data representing a sample of curves, *Ann. Statist.* **20** No.3 , 1266–1305.
13. Lindeberg, T. (1994). Scale Space Theory in Computer Vision, *Kluwer, Boston*.
14. Mammen, E. Marron, J.S. Turlach, B.Aànd Wand, M.P. (2001). A general projection framework for constrained smoothing, *Statistical Science* 16(3): 232-248.
15. Mallat, S. (1998). A Wavelet Tour of Signal Processing, *Academic Press*.
16. Munoz Maldonado, Y. Staniswalis, J.G. Irwin, L.N. and Byers, D. (2002). A similarity analysis of curves, *The Canadian Journal of Statistics* **30** No.3, ???–???.
17. Piccioni M. , Scarlatti S. and Trouvé, A. (1998) A variational problem arising from speech recognition, *SIAM Journ. Appl. Math.*, V.58. No.3, pp. 753771.
18. Ramsay, J.O. (1998) Estimating smooth monotone functions. *Journal of the Royal Statistical Society*, Series B, 60, 365-375.
19. Ramsay, J.O. and Li, X. (1998) Curve registration. *Journal of the Royal Statistical Society*, Series B, 60, 351-363.
20. Ramsay, J.O. and Silverman, B.W. (1997). Functional data analysis, *New York: Springer Verlag*.
21. Trouve, A. and Younes, L. (2000). Diffeomorphic matching problems in one dimension: designing and minimizing matching functionals, *Proceedings ECCV 2000*.
22. Wang, K. and Gasser, T. (1997). Alignment of curves by dynamic time warping, *Ann. Statist.* **25** No.3, 1251–1276.
23. Wang, K. and Gasser, T. (1999). Synchronizing sample curves nonparametrically, *Ann. Statist.* **27** No.2, 439–460.
24. Younes, L. (2000). Deformations, Warping and Object Comparison, *Tutorial* http://www.cmla.ens-cachan.fr/ỹounes.

# Stochastic Multiresolution Models for Turbulence

B. Whitcher[a,*], J. B. Weiss[b], D. W. Nychka[a] and T. J. Hoar[a]

[a]Geophysical Statistics Project, National Center for Atmospheric Research,
P.O. Box 3000, Boulder, Colorado 80307-3000, United States

[b]Program in Atmospheric and Oceanic Sciences,
Department of Astrophysical, Planetary and Atmospheric Sciences,
University of Colorado at Boulder, Boulder, Colorado 80309, United States

The efficient and accurate representation of two-dimensional turbulent fields is of interest in the geosciences because the fundamental equations that describe turbulence are difficult to handle directly. Rather than extract the coherent portion of the image from the background variation, as in the classical signal plus noise model, we present a statistical model for individual vortices using the non-decimated discrete wavelet transform. A template image, supplied by the user, provides the features we want to extract from the observed field. By transforming the vortex template into the wavelet domain specific characteristics present in the template, such as size and symmetry, are broken down into components associated with spatial frequencies. Multivariate multiple linear regression is used to fit the vortex template to the observed vorticity field in the wavelet domain.

## 1. Introduction

The efficient and accurate representation of two-dimensional turbulent fields is of interest in the geosciences because the fundamental equations that describe turbulence are difficult to handle directly. Recent advances in computing allow scientists to produce realistic simulations of turbulent flows. The new emphasis of research in turbulence is on the individual coherent structures in the flow instead of the signal versus background noise. Our approach to this problem is to develop a flexible model for a single coherent structure (vortex) using a multiresolution analysis and then efficiently represent the entire vorticity field through a collection of these structures. Defining the vortex utilizes specific information from the scientist via a template function, but is flexible enough to capture a broad range of features associated with observed coherent structures. Modeling the vorticity field is done using classic regression methodology, allowing the information in the observed data to dictate what is and what is not a vortex. This paper concentrates on the computational algorithms used in our methodology and is a companion to [6].

---

*The author is now in the Research Statistics Unit, GlaxoSmithKline, New Frontiers Science Park (South), Third Avenue, Harlow CM19 5AW, United Kingdom (*bjw34032@mh.uk.sbphrd.com*). Support for this research was provided by the National Center for Atmospheric Research (NCAR) Geophysical Statistics Project, sponsored by the National Science Foundation (NSF) under Grants DMS98-15344 and DMS93-12686.

Wavelet-based analysis of two-dimensional turbulence has been performed previously; see, for example, [1], [7], [8] and the summary article [2]. The focus of the methodology used in these papers has a signal processing orientation; that is, the wavelet transform discriminates between "signal" and "noise" using compression rate as a classifier. The idea being that the coherent structures (vortices) observed in the turbulent flow will be well represented by a few wavelet coefficients and therefore most coefficients may be discarded. From one vorticity field, two fields are produced from such a procedure: one based on the largest wavelet coefficients that is meant to capture the coherent structures in the original field, and another based on the remaining wavelet coefficients which captures the filament and background structure. Our methodology differs significantly from the signal-plus-noise model, in that we formulate a statistical model for an individual coherent structure and extract a fixed number of isolated vortices from the original field. From these estimates a completely different set of summary statistics may be calculated; for example, instead of global summary statistics like the enstrophy spectrum we are able to look at local statistics for each structure such as the average size, amplitude, circulation and enstrophy.

Previous work in identifying vortices in a turbulent flow has produced a wide variety of methods. We summarize two of them here. Let $\zeta(\mathbf{x})$ define the vorticity at pixel $\mathbf{x} \in \mathbb{R}^2$. A simple approach to isolating individual structures within the vorticity field is to define a vorticity threshold $\delta$. Any observed values such that $\zeta(\mathbf{x}) > |\delta|$ would describe a region of "anomalously high vorticity" and, hence, a coherent structure.

There are several problems with a thresholding approach. Coherent structures are stripped of any vorticity beneath the threshold thus introducing bias and reduced variability into vortex statistics, the threshold must be adjusted *a posteriori* from time step to time step whereas we would prefer a method that adapts to the set of coherent structures in each individual image, and lastly there is no clear way to generalize the technique to more complex problems (e.g., multipole or noisy structures).

An interesting method developed in the late 1980s is the *objective observer* approach [4]. The method is as follows: determine the pixel with the largest (in magnitude) vorticity, descend/ascend the peak in vorticity in the positive $x$-direction until a vortex boundary is encountered, record this point and search for additional vortex boundary points at the base of this coherent structure until a closed curve is produced. The set of pixels found in this way define a boundary for the vortex. Repeat this procedure for all local minima/maxima in the observed vorticity field. The limitations of this procedure is that it depends too heavily on properly defining what is and what is not a coherent structure.

A wavelet-packet census algorithm was developed where the discrete wavelet packet transform is applied to simulated turbulent flow [5]. The first step of this procedure is an iterative denoising method that partitions the wavelet packet coefficients into two groups – large and small coefficients. The large wavelet packet coefficients correspond to the coherent portion of the observed vorticity field and the small wavelet coefficients correspond to the disorganized, background information. On the coherent portion, wavelet coefficients are grouped together through their ideal support in the spatial domain to form individual coherent structures. The wavelet-based method presented here is an attempt to replace the *ad hoc* and iterative schemes in [5] with sound statistical procedures.

Section 2 introduces the two-dimensional multiresolution analysis of an image and its

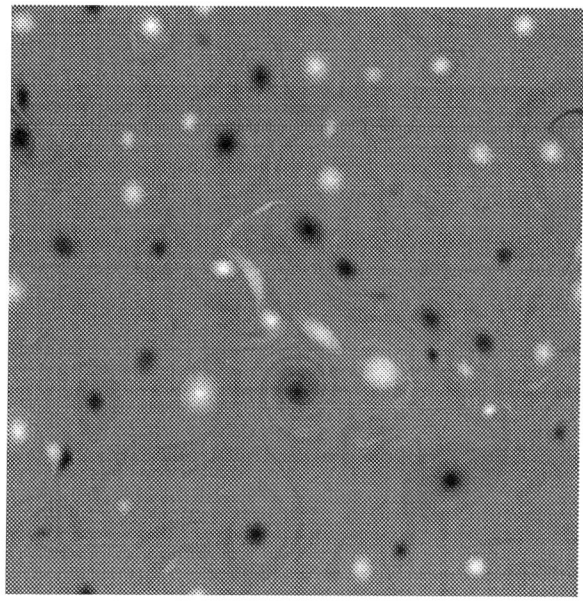

Figure 1. Simulated vortex field (512 × 512) at time $t = 15$. The gray scales are artificial and distinguish positive and negative vorticity from the background. Negative vortices are seen as white and positive vortices are seen as black. Zero vorticity is a uniform gray.

decomposition into unique spatial scales and directions. Section 3 outlines our modeling strategy by discussing vortex template functions and models for both single and multiple coherent structures. Section 4 provides the methodology we use to estimate a single coherent structure and also multiple structures in the vorticity field. Section 5 examines how well the technique performs at estimating multiple coherent structures in a sample vorticity field.

## 2. The Two-Dimensional Discrete Wavelet Transform

### 2.1. Multiresolution Analysis

To simplify the vortex dynamics within a given realization of a vorticity field, we rely on the two-dimensional discrete wavelet transform (2D DWT) and assume the vorticity field $\zeta(\mathbf{x})$ has finite energy; i.e., $\int \zeta^2(\mathbf{x})d\mathbf{x} < \infty$. Let $\phi$ be a scaling function and $\psi$ be the corresponding wavelet generating an orthonormal basis on $L^2(\mathbb{R})$ and define the three separable wavelets

$$\psi^{\mathrm{h}}(x_1, x_2) = \phi(x_1)\psi(x_2), \quad \psi^{\mathrm{v}}(x_1, x_2) = \psi(x_1)\phi(x_2), \quad \psi^{\mathrm{d}}(x_1, x_2) = \psi(x_1)\psi(x_2), \tag{1}$$

corresponding to horizontal, vertical and diagonal directions, respectively. More information concerning the interpretation of "spatial directions" captured by the separable wavelet transform may be found in Section 2.2. The separable scaling function $\phi(x_1, x_2) = \phi(x_1)\phi(x_2)$ is associated with the approximation space. For $m \in \{\mathrm{h, v, d}\}$, let

$$\psi^m_{j,k,l}(x_1, x_2) = \frac{1}{2^j}\psi^m\left(\frac{x_1 - 2^j k}{2^j}, \frac{x_2 - 2^j l}{2^j}\right),$$

such that

$$\{\psi^{\mathrm{h}}_{j,k,l}(x_1, x_2), \psi^{\mathrm{v}}_{j,k,l}(x_1, x_2), \psi^{\mathrm{d}}_{j,k,l}(x_1, x_2)\}_{(j,k,l)\in\mathbb{Z}^3}$$

is an orthonormal basis of $L^2(\mathbb{R}^2)$. The triplet $(j, k, l)$ denotes the $j$th wavelet scale at the spatial position $(k, l)$. In practice the depth of the 2D DWT is determined by the finite dimension $M \times N$ of $\zeta(\mathbf{x})$, such that $1 \leq j \leq J = \lfloor \log_2 \min\{M, N\} \rfloor$. Efficient implementation of the separable 2D DWT is performed via a series of convolutions and subsampling [3, Sec. 7.7].

Whereas it would be most efficient to represent the vorticity field using a decimated transform such as the 2D DWT, we find it advantageous to utilize the 2D maximal overlap DWT (2D MODWT). Unlike the orthonormal 2D DWT, the 2D MODWT produces a redundant non-orthogonal transform. The reason for this discrepancy is that the 2D MODWT does not subsample in either dimension, it only filters the original image. The advantages are that the transform is translation invariant to integer shifts in space and it reduces artifacts caused by the choice of a specific wavelet filter.

Let $\bar{\phi} = \phi/\sqrt{2}$ and $\bar{\psi} = \psi/\sqrt{2}$ be rescaled versions of the scaling and wavelet functions. The three separable wavelet functions associated with spatial directions $m \in \{\mathrm{h, v, d}\}$ are now

$$\bar{\psi}^m_{j,k,l}(x_1, x_2) = \frac{1}{2^{2j}}\psi^m\left(\frac{x_1 - k}{2^j}, \frac{x_2 - l}{2^j}\right).$$

Hence, each level in the transform will have the same spatial dimension as the original field $(M \times N)$ and represent a redundant set of wavelet coefficients. The 2D MODWT begins with the original vorticity field $\zeta(\mathbf{x})$, and at all scales we denote $\bar{v}_j(k, l) = \langle \zeta, \bar{\phi}_{j,k,l} \rangle$ and $\bar{w}_j(k, l) = \langle \zeta, \bar{\psi}^m_{j,k,l} \rangle$ for $m \in \{\mathrm{h, v, d}\}$, where $\langle x, y \rangle$ is the two-dimensional inner product of $x(\cdot, \cdot)$ and $y(\cdot, \cdot)$. The vorticity field $\zeta(\mathbf{x})$ may now be decomposed into $3J + 1$ sub-fields,

$$\left[\{\bar{w}^{\mathrm{h}}_j, \bar{w}^{\mathrm{v}}_j, \bar{w}^{\mathrm{d}}_j\}_{1 \leq j \leq J}, \bar{v}_J\right]; \tag{2}$$

three fields of wavelet coefficients corresponding to distinct spatial directions and one field containing the scaling coefficients at the final level. The scaling (approximation) field for level $j$ may be obtained from the four fields at level $j + 1$ via

$$\bar{v}_j(k, l) = \langle \bar{v}_{j+1}, \bar{\phi}^*_{j,k,l} \rangle + \sum_m \langle \bar{w}^m_{j+1}, \bar{\psi}^{m*}_{j,k,l} \rangle. \tag{3}$$

The 2D MODWT *multiresolution analysis* (MRA) of the vorticity field is an additive decomposition given by recursively applying (3) over all $j$; i.e.,

$$\zeta(x_1, x_2) = \langle \bar{v}_J, \phi^*_{J,x_1,x_2}\rangle + \sum_{j=1}^{J}\sum_m \langle \bar{w}_j^m, \psi^{m*}_{j,x_1,x_2}\rangle = \bar{a}_J(x_1, x_2) + \sum_{j=1}^{J}\sum_m \bar{d}_j^m(x_1, x_2),$$

where $\bar{a}_J(x_1, x_2)$ is the wavelet approximation field and $\bar{d}_j^m(x_1, x_2)$ is the wavelet detail field associated with the spatial direction $m \in \{\text{h}, \text{v}, \text{d}\}$. The MRA of $\zeta(\mathbf{x})$ provides a convenient way of isolating features at different scales and directions with coefficients in the spatial domain versus the wavelet domain. This is advantageous since reconstruction is now reduced from the full inverse 2D MODWT to simple addition.

## 2.2. Spatial Directions of the 2D MRA

Estimating the coefficient associated with a basis function is equivalent to a filtering operation on the original image. The qualitative features of these filters help to interpret the wavelet coefficients. Two-dimensional wavelets (denoted by $\psi^\text{h}$, $\psi^\text{v}$, and $\psi^\text{d}$ in (1)) are associated with horizontal, vertical and diagonal features in the vorticity field. This follows from the fact that the two-dimensional filters are the outer product of two one-dimensional filters, where the scaling filter averages (smooths) across its spatial direction and the wavelet filter differences across its spatial direction. The two-dimensional wavelet $\psi^\text{h}(x_1, x_2) = \phi(x_1)\psi(x_2)$ will therefore smooth across the first dimension ($x_1$-axis) and difference across the second dimension ($x_2$-axis), thus favoring horizontal features. The two-dimensional filter $\psi^\text{v}(x_1, x_2) = \psi(x_1)\phi(x_2)$ differences across the $x_1$-axis and smooths across the $x_2$-axis, thus favoring vertical features. Finally, the $\psi^\text{d}(x_1, x_2) = \psi(x_1)\psi(x_2)$ differences across both directions and favors non-vertical/non-horizontal (i.e., diagonal) features.

Figures 2 and 3 display the six scales from a 2D MRA of the sample vorticity field in Figure 1, only a portion is shown centered at ($x_1 = 399, x_2 = 101$). Each row displays the wavelet detail fields associated with the three spatial directions: horizontal, vertical and diagonal. It is clear that each of the two-dimensional wavelet filters captures distinct spatial directions at a fixed spatial scale. Given the filaments from this particular vortex are elliptical in shape, it is not surprising to see the detail coefficients of the filament structures strongest in the northeast-southwest directions.

It is interesting to note that the coherent structure is not seen until the third scale (third row in Figure 2) and then only in the horizontal and vertical directions. At higher scales, corresponding to larger spatial areas and lower spatial frequencies, the coherent structure is apparent in all directions. These figures indicate the usefulness of a multiscale representation in separating stylized features of a coherent structure.

## 3. Models for Vorticity Fields

Let $\zeta_t(\mathbf{x})$ denote an observed vorticity field at time $t$. The spatial locations $\{\mathbf{x}\}$ are assumed to fall on a regular grid for both the $x_1$ and $x_2$ dimensions, denoted by $M$ rows and $N$ columns, being dyadic in length ($M$ and $N$ are not required to be equal).

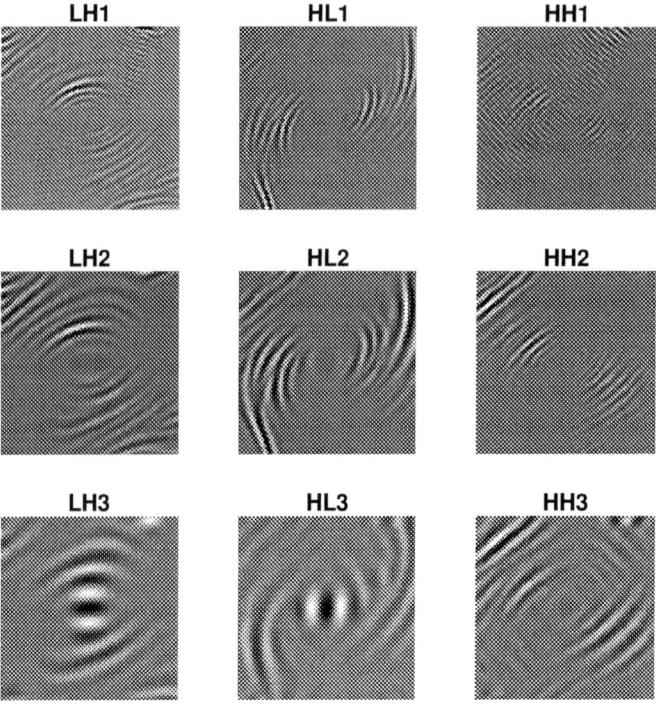

Figure 2. 2D multiresolution analysis ($j = 1, 2, 3$) of vorticity field centered at ($x_1 = 399, x_2 = 101$). The Daubechies least asymmetric wavelet filter ($L = 8$) was used in the computation.

### 3.1. Selecting the Vortex Template

Our goal in this work is to identify and accurately model individual structures in an observed vorticity field. Arguably there must be some user-defined information provided in order to attain this goal, specifically the definition of what a coherent structure should look like from the observed vorticity field. This provides focus for the regression model and yields scientifically interesting results. By framing our modeling technique in terms of a MRA, the influence of this user-defined information is important but not too restrictive.

One choice of a template function $\tau(\mathbf{x})$ is the outer product of two identical Gaussian curves. This appeared to capture the relevant features of an ideal vortex even if the observed vortices decay back to zero at a faster rate than a Gaussian curve. Although this template is too simple or naive for modeling an observed vortex directly its MRA is strikingly similar to the MRA of individual structures in the observed vorticity field [6].

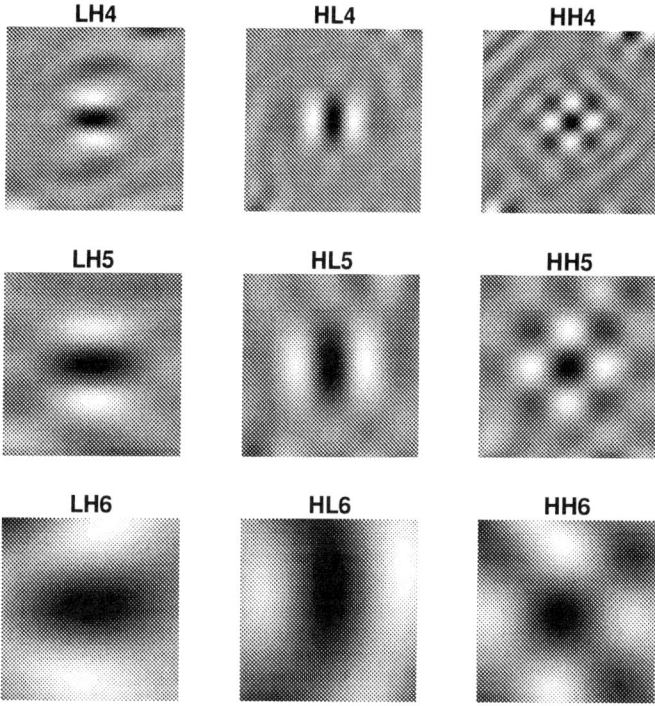

Figure 3. 2D multiresolution analysis ($j = 4, 5, 6$) of vorticity field centered at ($x_1 = 399, x_2 = 101$). The Daubechies least asymmetric wavelet filter ($L = 8$) was used in the computation.

There are obvious differences between the template function and coherent structures in the observed vorticity field, some of them are handled automatically by the MRA while others require further discussion. Limitations of the template function that are taken care of by framing our linear regression model in terms of a MRA include: fixed spatial size, amplitude, and radial symmetry. The fixed spatial size is taken care of by levels of the MRA being associated with different spatial ranges. The multivariate multiple linear regression is allowed to pick any spatial size of the template function and associate it with any spatial size in the observed data. The amplitude of $\tau(\mathbf{x})$ is similarly handled through the magnitude of the regression coefficients. The radial symmetry of $\tau(\mathbf{x})$ is another obvious drawback. This is dealt with through the MRA since within each level of the decomposition there are three spatial directions (Section 2.2).

## 3.2. Single Vortex Model

Our statistical model for a single coherent structure, or *vortex*, in the vorticity field $\zeta(\mathbf{x})$ (we drop the time index to simplify notation) is based on a multivariate multiple linear regression model. The observational model for a single coherent structure is given by

$$\mathbf{Y} = \mathbf{Z}\boldsymbol{\beta} + \boldsymbol{\epsilon}, \tag{4}$$

where $\boldsymbol{\beta}$ is the $(3J+1)\times(3J+1)$ matrix of regression coefficients and $\boldsymbol{\epsilon}$ is the length $3J+1$ vector of errors. The data matrix $\mathbf{Y}$, of dimension $(MN)\times(3J+1)$, is based on a multiresolution analysis (MRA) of $\zeta(\mathbf{x})$; see Section 2.1. The $3J+1$ columns of $\mathbf{Y}$ are generated by column-scanning the 2D wavelet detail fields and concatenating each column. The $(MN)\times(3J+1)$ matrix of covariates (design matrix) $\mathbf{Z}$ is generated by column-scanning and concatenating the user-defined template function $\tau(\mathbf{x})$. The template function is the same dimension as the observed vorticity field.

The implied regression model for each response vector $Y_i$ (column of $\mathbf{Y}$) is

$$Y_i = \beta_{1,i} Z_1 + \beta_{2,i} Z_2 + \beta_{3,i} Z_3 + \cdots + \beta_{3J+1,i} Z_{3J+1} + \epsilon_i,$$

for $i = 1, \ldots, 3J+1$. By transforming the template function via a MRA, a function that is idealized and simplistic in nature provides the foundation of our semi-parametric fitting procedure.

For example, $\tau(\mathbf{x})$ may be a two-dimensional Gaussian bump of fixed size. This is reasonable as an intuitive description of the coherent structures seen in the simulations of interest, but does not allow for the full range of vortices observed; e.g., different sizes, distorted shapes, etc. The MRA of $\tau(\mathbf{x})$ separates features of the idealized vortex by direction (horizontal, vertical and diagonal) and size (ranges of spatial frequency) through the use of a collection of basis functions (wavelets).

## 3.3. Vortex Field Model

The previous section proposed modeling an isolated coherent structure in the vorticity field $\zeta(\mathbf{x})$. To more accurately model multiple coherent structures that may be spatially correlated, we propose a vortex field model where the single vortex design matrix is replicated $k$ times ($k$ is assumed to be known)

$$\mathbf{Y} = \mathbf{Z}_1 \boldsymbol{\beta}_1 + \mathbf{Z}_2 \boldsymbol{\beta}_2 + \cdots + \mathbf{Z}_k \boldsymbol{\beta}_k + \boldsymbol{\epsilon}, \tag{5}$$

where each $\mathbf{Z}_i$ is of dimension $(MN)\times(3J+1)$, each regression coefficient matrix $\boldsymbol{\beta}_i$ is $(3J+1)\times(3J+1)$, and $\boldsymbol{\epsilon}$ is defined as before. The significance of having $k$ distinct design matrices is that each one is centered at the spatial location $(x_{1,k}, x_{2,k})$. Thus, the model in (5) allows for the simultaneous estimation of multiple coherent structures that interact in space.

## 4. Model Estimation

### 4.1. Multiresolution-Based Census Algorithm

The following algorithm is our implementation of a vortex extraction procedure for an observed vorticity field of arbitrary finite dimension. The procedure requires at input: a wavelet filter, the maximum depth of the 2D MODWT, and the template function.

1. Compute single vortex models at each spatial location $(x_1, x_2)$ using the efficient implementation in Section 4.2.

2. From the single vortex fits in step 1, we compute the distance measure $\Lambda$ for each spatial location $(x_1, x_2)$ and obtain the local maxima (Section 4.3). The pixels associated with these local maxima constitute our collection of *candidate points* $\mathcal{M}$.

3. Compute the vortex field model, via backfitting, for spatial locations $k \in \mathcal{M}$. We order these candidate points via their distance measure $\Lambda_k$ from step 2, and use forward stepwise selection to include coherent structures into the model (Section 4.4).

4. Statistics may be computed from the individual models $\zeta_t(\mathbf{x})$ of the vortex field model, such as:

   - # vortices
   - Circulation: $\Gamma_t = \int \zeta_t(\mathbf{x})\,d\mathbf{x}$
   - Enstrophy: $\Omega_t = \int \zeta_t^2(\mathbf{x})\,d\mathbf{x}$
   - Peak amplitude: $\alpha_t = \max_{\mathbf{x}} \zeta_t(\mathbf{x})$
   - Size: the area associated with a 2D Gaussian curve-fitting procedure.

   These calculations are not provided in this paper, see [6] for more details.

### 4.2. Fitting the Single-Vortex Model

From the definition of $\mathbf{Z}$ in (4), the template does not shift in the horizontal or vertical and a MRA does not change this fact. Parameters of interest include the spatial position $(x_1, x_2)$ of each coherent structure, along with its matrix $\boldsymbol{\beta}$ of regression coefficients. In order to capture all possible spatial locations and prepare for the eventual fitting of multiple coherent structures, the estimated regression matrix

$$\hat{\boldsymbol{\beta}}^{(x_1, x_2)} = (\mathbf{Z}'\mathbf{Z})^{-1}\mathbf{Z}'\mathbf{Y}, \quad 1 \leq x_1 \leq M, \ 1 \leq x_2 \leq N, \tag{6}$$

must be performed $M \cdot N$ times; that is, centered at every spatial location. If a sample vorticity field is 512×512, that means (6) must be evaluated approximately 2.6 million times. Even using popular methods for least-squares fitting, such as the QR or singular value decomposition, this is a daunting computational task.

To fit a multivariate multiple linear regression at each pixel in an observed vorticity field, we make use of the fact that regression may be viewed as smoothing, and utilize the Fourier transform. First, let $\mathbf{UDV'} = \mathbf{Z}$ be the singular value decomposition of the $(MN) \times (3J+1)$ covariate matrix in (4) and replace $\mathbf{Z}$ with the orthogonal matrix $\mathbf{U}$ in (6) to obtain the matrix of regression coefficients $\hat{\boldsymbol{\beta}}_{\text{SVD}}^{(x_1,x_2)} = \mathbf{U}'\mathbf{Y}$. The regression matrix of interest in (6) is recovered by undoing the SVD via $\hat{\boldsymbol{\beta}}^{(x_1,x_2)} = \mathbf{V}\mathbf{D}^{-1}\hat{\boldsymbol{\beta}}_{\text{SVD}}^{(x_1,x_2)}$. This eliminates the explicit matrix inversion, but does not replace iterating over $MN$ pixels in the observed vorticity field. If we regard regressing $Y_i$ on $U_k$ at each pixel as just "smoothing," then the discrete Fourier transform $\mathcal{F}$ may be utilized to perform this convolution; i.e.,

$$\widehat{\mathcal{B}}_{k,i} = \mathcal{F}^{-1}\left\{\mathcal{F}[\text{image}(U_k)] \cdot \mathcal{F}^*[\text{image}(Y_i)]\right\}, \quad 1 \leq i, k \leq 3J+1, \tag{7}$$

where image($\cdot$) reconstructs a matrix from a vector, $\widehat{\mathcal{B}}_{k,i}$ is of dimension $M \times N$ and $z^*$ denotes the complex conjugate of $z$. Since both dimensions were defined to be dyadic, the two-dimensional fast Fourier transform may be used to greatly increase computational efficiency. Perform (7) $(3J+1) \times (3J+1)$ times to obtain the alternative regression matrix $\hat{\boldsymbol{\beta}}_{\text{SVD}}^{(x_1,x_2)}$ for all spatial positions $(x_1, x_2)$; e.g., letting $J = 6$ implies iterating (7) $19^2 = 361$ times versus 2.6 million by direct computation.

### 4.3. Candidate Point Selection

For an observed vorticity field with $M = N = 512$ and $J = 6$, the above procedure will produce approximately $95,000,000$ regression coefficients. We would like to compute a regression model where multiple coherent structures are involved. This is not possible when the number of possible coherent structures is $M \cdot N$, so we introduce a simple procedure for discarding a large number of spatial locations. From the single vortex fits we obtained in (7), we compute

$$\Lambda_k = \min\left\{\text{tr}[(\hat{\boldsymbol{\beta}}_k - \mathbf{I})^2], \text{tr}[(\hat{\boldsymbol{\beta}}_k + \mathbf{I})^2]\right\} \tag{8}$$

for each spatial location $(x_1, x_2)$. The criterion $\Lambda_k$ is small when the regression matrix $\hat{\boldsymbol{\beta}}_k$ is similar to the identity matrix, and thus, the single-vortex model is similar to the template function. Two comparisons are made, one to the positive identity matrix and one to the negative identity matrix, so that positive and negative spinning vortices are favored equally. For specific vorticity fields, the set of candidate points was well partitioned into coherent and non-coherent structures using $\Lambda_k$.

We denote the eight-member neighborhood around the pixel $(x_1, x_2)$ by $\mathcal{N}$ and look for local maxima in the $\Lambda_k$ field. The collection of *candidate points* is defined via $\mathcal{M} = \{(a, b) : \|\hat{\boldsymbol{\beta}}^{(a,b)}\|^2 > \|\hat{\boldsymbol{\beta}}^{(x_1,x_2)}\|^2$, for all $(x_1, x_2) \in \mathcal{N}\}$. When applying this criterion on observed vorticity fields, the collection of candidate points $\mathcal{M}$ not only includes almost all coherent structures in the field but also additional points. The point being that the size of $\mathcal{M}$ is much less, by at least two orders of magnitude, than the number of pixels in the original field.

### 4.4. Fitting the Vortex Field Model

Instead of directly calculating $\hat{\boldsymbol{\beta}}$ via linear algebra for (5), we utilize backfitting to solve the $k$ simultaneous linear equations

$$\mathbf{Y}_1 - (\cdot + \widehat{\mathbf{Y}}_2 + \widehat{\mathbf{Y}}_3 + \cdots + \widehat{\mathbf{Y}}_{k-1} + \widehat{\mathbf{Y}}_k) = \mathbf{Z}_1 \boldsymbol{\beta}_1$$
$$\mathbf{Y}_2 - (\widehat{\mathbf{Y}}_1 + \cdot + \widehat{\mathbf{Y}}_3 + \cdots + \widehat{\mathbf{Y}}_{k-1} + \widehat{\mathbf{Y}}_k) = \mathbf{Z}_2 \boldsymbol{\beta}_2$$
$$\vdots$$
$$\mathbf{Y}_k - (\widehat{\mathbf{Y}}_1 + \widehat{\mathbf{Y}}_2 + \cdots + \widehat{\mathbf{Y}}_{k-1} + \cdot) = \mathbf{Z}_k \boldsymbol{\beta}_k.$$

The $k$th vortex model is associated with the spatial location $(x_{1,k}, x_{2,k})$ in $\zeta(\mathbf{x})$, where $k \in \mathcal{M}$. Convergence for this algorithm is achieved by looking at the absolute difference between each regression coefficient matrix $\hat{\boldsymbol{\beta}}_k$ from iteration $i-1$ to $i$. For any $k$, if $|\sum_{m,n} \hat{\beta}_{m,n,k}^{(i)} - \hat{\beta}_{m,n,k}^{(i-1)}| > \delta$, for some $\delta > 0$, then the backfitting method was applied again. The number of iterations was found to be small across a wide range of $k$, usually two or four iterations sufficed. We attribute the small number of iterations to the fact that the coherent structures in an observed vorticity field are relatively isolated (Figure 1) and therefore estimation is not seriously affected by fitting them simultaneously.

To determine the number $K$ of coherent structures in the field we utilize a penalized likelihood criterion, specifically the generalized cross validation (GCV) function. We would like to search the entire model space for the best combination of coherent structures (all subset regression), but this would be too computationally intensive for several hundred candidate points. Instead, an ordering is imposed on the candidate points and a model for the vorticity field is found by only searching along the ordering. After single-vortex models are obtained for all candidate points, they are ranked by their similarity to the identity matrix of dimension $(3J+1) \times (3J+1)$ via $\Lambda_k$ (8). After the candidate points are ordered according to $\Lambda_k$, additional coherent structures are added to the vortex-field model one by one when the GCV is reduced. The number of free parameters associated with a single-vortex model is simply $(3J+1) \times (3J+1)$. If the added vortex model does not reduce the GCV it is discarded and the next model is added. This procedure is iterated until no more candidate models exist.

## 5. Application to Two-dimensional Turbulent Flows

We return to the observed vorticity field in Figure 1 and compute the vortex statistics using our multiresolution census algorithm outlined in Section 4.1. This vorticity field provides a reasonable example of multiple coherent structures embedded within relatively quiescent background structure. A total number of 53 coherent structures were identified. To illustrate the ability of our procedure to find coherent structures of varying sizes and shapes, Figure 4 shows the fitted vortex field model $\widehat{\mathbf{Y}}$ and the residuals for the vorticity field. Although the template function is of fixed spatial size and amplitude, the fitted coherent structures are quite different from the template function and from each other. Comparing the residual field to the original data (Figure 1), it appears that all visually obvious structures have been identified. In addition, one spurious structure was included in the model.

## 6. Discussion

Using a multiresolution regression model, we have been able to accurately model and extract coherent structures from a simulated two-dimensional fluid flow. This technique allows for a fast, objective analysis of observed turbulent fields. Although these results have been achieved by previous algorithms, our method provides flexibility to the researcher through the choice of a template function and modular implementation of the algorithm while stability and precision of the results are guaranteed by using sound statistical techniques. We plan on applying this method to a variety of data sets including more realistic simulations, climate model output, and satellite imagery. The method may also be applied to simulations of three-dimensional quasi-geostrophic turbulence.

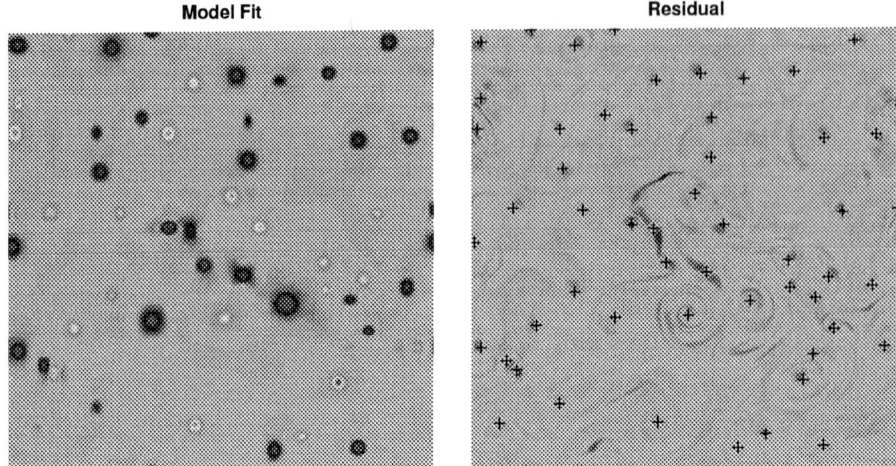

Figure 4. Fitted vortex field model $\widehat{\mathbf{Y}}$ and residual field for the vorticity field at time $t = 15$, plotted on the same color map. The spatial locations of the estimated coherent structures are plotted on the residual field for ease of comparison.

## REFERENCES

[1] Farge, M., Goirand, E., Meyer, Y., Pascal, F., and Wickerhauser, M. V. (1992). Improved predictability of two-dimensional turbulent flows using wavelet packet compression. *Fluid Dynamics Research*, 10, 229–250.

[2] Farge, M., Kevlahan, N., Perrier, V., and Goirand, E. (1996). Wavelets and turbulence. *Proceedings of the IEEE*, 84 (4), 639–669.

[3] Mallat, S. (1998). *A Wavelet Tour of Signal Processing*. Academic Press, San Diego, 1st edition.

[4] McWilliams, J. C. (1990). The vortices of two-dimensional turbulence. *Journal of Fluid Mechanics*, 219, 361–385.

[5] Siegel, A. and Weiss, J. B. (1997). A wavelet-packet census algorithm for calculating vortex statistics. *Physics of Fluids*, 9 (7), 1988–1999.

[6] Whitcher, B., Weiss, J. B., Nychka, D. W., and Hoar, T. J. (2002). A multiresolution census algorithm for calculating vortex statistics. Geophysical Statistics Project, National Center for Atmospheric Research.

[7] Wickerhauser, M. V., Farge, M., Goirand, E., Wesfreid, E., and Cubillo, E. (1994). Efficiency comparison of wavelet packet and adapted local cosine bases for compression of a two-dimensional turbulent flow. In C. K. Chui, L. Montefusco, and L. Puccio,

editors, *Wavelets: Theory, Algorithms, and Applications*, pages 509–531. Academic Press, San Diego, California.

[8] Wickerhauser, M. V., Farge, M., and Goirand, E. (1997). Theoretical dimension and the complexity of simulated turbulence. In W. Dahmen, P. Oswald, and A. J. Kurdila, editors, *Multiscale Wavlet Methods for Partial Differential Equations*, volume 6 of *Wavelet Analysis and Applications*, pages 473–492. Academic Press, Boston.

# Author Index

| | | | |
|---|---|---|---|
| Akritas, M.G. | 79 | Khumbah, N.-A. | 35 |
| Asgharian, M. | 367 | Kim, Y. | 411 |
| Bathke, A. | 109 | Kreiss, J.-P. | 303 |
| Bigot, J. | 481 | Lahiri, S.N. | 421 |
| Brunner, E. | 79, 109 | LeBlanc, M. | 393 |
| Bühlmann, P. | 19 | Lee, J. | 411 |
| Cai, Z. | 217, 283 | Linton, O. | 203 |
| Calvin, J.A. | 469 | Liu, R.Y. | 155 |
| Cao, R. | 139 | Mammen, E. | 203 |
| Cheng, M.-Y. | 315 | Meerschaert, M.M. | 265 |
| Crowley, J. | 393 | Michailidis, G. | 169 |
| Cuevas, A. | 251 | Moon, J. | 393 |
| Dahlhaus, R. | 451 | Munk, A. | 123 |
| Dowla, A. | 437 | Nordman, D.J. | 421 |
| Fan, J. | 315 | Nychka, D.W. | 499 |
| Ferraty, F. | 61 | Paparoditis, E. | 437 |
| Franke, J. | 303 | Politis, D.N. | 335, 437 |
| Freitag, G. | 123 | Rodríguez-Casal, A. | 251 |
| Gijbels, I. | 183 | Rosenblatt, M. | 447 |
| Gluhovsky, I. | 93 | Scheffler, H.-P. | 265 |
| González Manteiga, W. | 139 | Schölkopf, B. | 3 |
| Guan, Y. | 469 | Sen, P.K. | 351 |
| Härdle, W. | 303 | Sherman, M. | 469 |
| Heckman, N.E. | 49 | Spokoiny, V. | 315 |
| Hoar, T.J. | 499 | van de Geer, S.A. | 235 |
| Hong, Y. | 283 | Vieu, P. | 61 |
| Iglesias Pérez, C. | 139 | Vogt, M. | 123 |

| | | | |
|---|---|---|---|
| Wegman, W.J. | 35 | Wu, C.O. | 377 |
| Weiss, J.B. | 499 | | |
| Whitcher, B. | 499 | Yu, K.F. | 377 |
| Wolfson, D.B. | 367 | Yuan, V.W.S. | 377 |